Advances in the Biomechanics of the Hand and Wrist

NATO ASI Series

Advanced Science Institutes Series

A series presenting the results of activities sponsored by the NATO Science Committee, which aims at the dissemination of advanced scientific and technological knowledge, with a view to strengthening links between scientific communities.

The series is published by an international board of publishers in conjunction with the NATO Scientific Affairs Division

A	Life Sciences	Plenum Publishing Corporation
B	Physics	New York and London
C	Mathematical and Physical Sciences	Kluwer Academic Publishers
D	Behavioral and Social Sciences	Dordrecht, Boston, and London
E	Applied Sciences	
F	Computer and Systems Sciences	Springer-Verlag
G	Ecological Sciences	Berlin, Heidelberg, New York, London,
H	Cell Biology	Paris, Tokyo, Hong Kong, and Barcelona
I	Global Environmental Change	

Series A: Life Sciences

Advances in the Biomechanics of the Hand and Wrist

Edited by
F. Schuind
Erasme University Hospital
Brussels, Belgium

K. N. An and W. P. Cooney III
Mayo Clinic and Mayo Foundation
Rochester, Minnesota

and
M. Garcia-Elias
Hospital General de Catalunya
Sant Cugat, Barcelona, Spain

Plenum Press
New York and London
Published in cooperation with NATO Scientific Affairs Division

Proceedings of a NATO Advanced Research Workshop on
Advances in the Biomechanics of the Hand and Wrist,
held May 21–23, 1992,
in Brussels, Belgium

NATO-PCO-DATA BASE

The electronic index to the NATO ASI Series provides full bibliographical references (with keywords and/or abstracts) to more than 30,000 contributions from international scientists published in all sections of the NATO ASI Series. Access to the NATO-PCO-DATA BASE is possible in two ways:

—via online FILE 128 (NATO-PCO-DATA BASE) hosted by ESRIN, Via Galileo Galilei, I-00044 Frascati, Italy

—via CD-ROM "NATO Science and Technology Disk" with user-friendly retrieval software in English, French, and German (©WTV GmbH and DATAWARE Technologies, Inc. 1989). The CD-ROM also contains the AGARD Aerospace Database.

The CD-ROM can be ordered through any member of the Board of Publishers or through NATO-PCO, Overijse, Belgium.

Library of Congress Cataloging-in-Publication Data

Advances in the biomechanics of the hand and wrist / edited by F.
  Schuind ... [et al.].
       p.   cm. -- (NATO ASI series. Series A, Life sciences ; vol.
  256)
     "Proceedings of a NATO Advanced Research Workshop on Advances in
  the Biomechanics of the Hand and Wrist, held May 21-23, 1992, in
  Brussels, Belgium"--T.p. verso.
     "Published in cooperation with NATO Scientific Affairs Division."
     Includes bibliographical references and index.
     ISBN 0-306-44580-8
     1. Human mechanics--Congresses. 2. Hand--Congresses. 3. Wrist-
  -Congresses.   I. Schuind, Frédéric. II. North Atlantic Treaty
  Organization. Scientific Affairs Division. III. NATO Advanced
  Research Workshop on Advances in the Biomechanics of the Hand and
  Wrist (1992 : Brussels, Belgium) IV. Series: NATO ASI series.
  Series Am, Life sciences ; v. 256.
  QP303.A328 1993
  612'.97--dc20   93-43655
                                                                 CIP

ISBN 0-306-44580-8

©1994 Plenum Press, New York
A Division of Plenum Publishing Corporation
233 Spring Street, New York, N.Y. 10013

Printed in the United States of America

FOREWORD

William P. Cooney III, R.A. Berger, and K.N. An

Orthopedic Biomechanics Laboratory
Department of Orthopedic Surgery
Mayo Clinic and Mayo Foundation
Rochester, MN 55905, U.S.A.

As surgeons struggle to find new insights into the complex diseases and deformities that involve the wrist and hand, new insights are being provided by applied anatomy, physiology and biomechanics to these important areas. Indeed, a fresh new interaction of disciplines has immersed in which anatomists, bioengineers and surgeons examine together basic functions and principles that can provide a strong foundation for future growth. Clinical interest in the hand and wrist are now at a peak on an international level. Economic implications of disability affecting the hand and wrist are recognized that have international scope crossing oceans, cultures, languages and political philosophies.

As with any struggle, a common ground for understanding is essential. NATO conferences such as this symposium on Biomechanics of the Hand and Wrist provides such a basis upon which to build discernment of fundamental postulates. As a start, basic research directed at studies of anatomy, pathology and pathophysiology and mechanical modeling is essential. To take these important steps further forward, funding from government and industry are needed to consider fundamental principles within the material sciences, biomechanical disciplines, applied anatomy and physiology and concepts of engineering modeling that have been applied to other areas of the musculoskeletal system. Until recently, the hand and wrist were neglected and not recognized, at least in the United States, as essential areas of research funding. By education of the Federal and Private funding agencies, the importance of our first tactile organs, the hand and wrist, has been recognized as equal to sports injuries, osteoporosis, and cartilage chemistry. However, funding levels for both clinical and basic research must be stimulated to greater levels within regional as well as international forums in order to achieve a global commitment to study, apply and interchange new methodologies for an improved understanding of disease and injury that affect the hand and wrist.

Basic research, for example, in trauma to the upper extremity, is lacking with only a few studies addressing the forces associated with fracture and dislocations of finger and thumb joints or mechanism of wrist instabilities. The precise role of internal and external fixation of metacarpal and phalangeal fractures, the importance of soft tissue injuries and their repair, as well as the need to maintain dynamic support provided by tendons are not understood as part of the total management of complex bone and soft tissue injuries to the hand and wrist. Both motion and force analysis of normal and pathologic conditions are needed to provide the building blocks for clinical studies involving reconstructive surgery. Not simple mechanical equivalents of hinge or ball-and-socket joints as in other parts of the body, the wrist, thumb and finger joints are complex anatomical structures that are not easily simulated or computer modeled. The wrist to some resembles a rubic cube wherein the rotation of one carpal bone effects the alignment and rotation of all other carpal elements - particularly those within the same carpal row. New sophisticated techniques of three dimensional imaging and motion analysis now available make understanding of these complex structures more realistic. Space age technology combined with anatomy, histomorphometry, and biomechanics provides the potential to grasp difficult three dimensional concepts of spatial rotations and force displacements that can be applied to clinical problems by the integration of different disciplines which are addressing seemingly different but actually inter-related solutions to the same clinical problem. The engineer with mathematical modeling, the anatomist with descriptions of detailed ligament origins and insertions and the physiologist with concepts of joint cartilage surface interactions can meet with the surgeon to develop careful methods of analysis, understanding, and hopefully applications to improve clinical treatment of diseased hand and wrists.

By means of international workshops sponsored by NATO organizations, scientists from sixteen countries have gathered together to report past efforts, learn new techniques, and to explore future avenues of interactive research in anatomy, biomechanics, physiology, and radiology as applied to improved understanding of the workings of the hand and wrist. That this forum for international exchange is not lost or scientifically "wasted' has inspired this monograph in which the major presentations can be re-assessed and reviewed for future discussion and clarification. The combination of distinct disciplines from different countries provides unique and exciting avenues of intellectual exchange. Where else, for example, could established concepts of tendon repair be challenged by studies on the anatomy and mechanical properties of tendons from Japan, Great Britain, Sweden and the United States with the investigators agreeing that new ideas on tendon repair, tendon excursion and effects of tendon adhesions must be pursued if advancements in this field are to be expected. Studies regarding the tendon pulley system, tendon morphology, and *in vivo* forces required for normal hand function are reported, combining both basic research with clinical research to provide practical value to hand surgeons. Within the hand, the 3-D anatomy of finger and thumb joints was also explored. Joint surface anatomy and arthritic deformity are assessed by sophisticated techniques such as stereo-photogrammetry, finite analysis, and computer simulation with three dimensional reconstruction in efforts to predict joint failure mechanisms and potential methods for prevention. For the wrist, detailed analysis of pressure and force distributions were described using specialized Fuji film, rigid body spring models, pressure sensitive film in pathologic conditions, and dynamic modeling are compared and contrasted with resolution as to ways that each can contribute more to our understanding of basic precepts of wrist function. New techniques such as cine-computed tomography, dynamic magnetic resonant imaging, and 3-D reconstruction are applied to examine normal and pathologic kinematic analyses of wrist function. Within the wrist, the role of intrinsic and extrinsic ligaments are modeled and the mechanical and biologic role of the triangular fibrocartilage is appraised. Wrist simulators are reported that can nearly duplicate normal joint motion and forces to study reconstructive ligament procedures or to analyze current total wrist joint prostheses. Finally, a number of studies are reported regarding the clinical applications of anatomy, physiology and mechanics as they apply to new joint replacements in the hand and wrist; better methods of fracture fixation; causes of carpal instability after Colles fractures and the rationale for persistent distal radioulnar joint laxity despite reconstruction.

Within the work place, several analyses were presented that should help us clarify problems of the injured worker. Ergonomic aspects of wrist mechanics were addressed. Repetitive trauma in carpal tunnel syndrome as a function of crowding of soft tissues within a closed compartment was considered. More physiologically beneficial hand and wrist positions during repetitive activities such as key boarding and more functional "out of plane" wrist motions combining radial deviation and wrist extension ("dart throwers wrist motion") were presented and justified with respect to their clinical significance. The effects of joint cartilage damage and instability after finger and wrist ligament injuries were assessed with respect to repetitive strain and endurance. All in all, a very broad and open forum for discussion of internationally similar problems was available which appeared to provide the catalyst for advances in our understanding of disease mechanisms. More importantly, the symposium served as source of stimulus for basic investigators of all nations to join together in their efforts to advance basic and clinical research and to share an understanding with others who have a common bond and interest in basic research involving mechanics of the hand and wrist. This ability to come together to determine future directions and areas of commitment was most exciting and stimulating. It provided an arena to broaden cultural, intellectual and personal interactions that potentially can serve all of mankind in making work, leisure activities, sports and life in general more enjoyable and rewarding. All who participated in this NATO advance research workshop, Brussels, Belgium, May 1992 wish to particularly thank the sponsorship of the NATO organization and the leadership and planning conjointly provided by Drs. F. Schuind (Belgium), M. Garcia-Elias (Spain) and K. An (USA) who tirelessly worked together to bring this first conference to fruition. Representatives from Belgium, Netherlands, Great Britain, Germany, Spain, Italy, France, Canada, United States, Croatia, Sweden, Switzerland, Japan, South-Africa, Turkey and Poland are to be thanked for taking the time and interest to participate in this inaugural meeting.

CONTENTS

II. FORCE ANALYSIS

III. MOTION ANALYSIS

IV. CLINICAL APPLICATIONS

THE NATO ADVANCED RESEARCH WORKSHOP:

ADVANCES IN THE BIOMECHANICS OF THE HAND AND WRIST,

GENVAL (BELGIUM), MAY 1992

Frédéric A. Schuind

Université libre de Bruxelles
Cliniques Universitaires de Bruxelles
808 route de Lennik
1070 Brussels, Belgium

INTRODUCTION

Through the support of the NATO Science Committee, Drs An and Cooney, from the Mayo Clinic (Rochester, MN, USA), Garcia-Elias, from Barcelona (Spain) and the author were able to organize in May 1992 the first international workshop related to the Biomechanics of the Hand and Wrist. One hundred five scientists from 16 different countries attended this 2 and 1/2 day meeting (Table 1). The participants were accommodated in a lovely quiet location, the 'Château du Lac', 15 km from Brussels (Figure 1). Fifty-five original papers were presented. We have selected for this book the most significant contributions to the Biomechanics of the Hand and Wrist.

Table 1. Country of origin of the participants.

Belgium	28	Canada	5
Croatia	1	France	2
Germany	8	Italy	5
Japan	14	Poland	1
Netherlands	9	South-Africa	1
Spain	1	Sweden	3
Switzerland	1	Turkey	1
United Kingdom	1	United States of America	24

NATO AND SCIENCE

The North Atlantic Treaty was signed in 1949. The signatory countries stated their desire to live in peace with all peoples. The concept of alliance and mutual security includes a broad range of security concerns of a global nature, including the protection of the physical environment, the management of natural resources and the welfare of the peoples. This philosophy led to the institution in 1957 of the NATO Science Committee, the only cooperative international institution embodying multilateral government support for advancing the frontiers of modern science. Three main programs have been established since the beginning: the Science Fellowships, the Advanced Study Institutes, and the Collaborative Research Grants Programmes. The Science Committee had in 1989 a budget of about US $ 20 million, with 6 % allocated to the Advanced Research Workshops like the one organized in Genval.[1]

Figure 1. Participants (May 23th, 1992).

THE BIOMECHANICS LABORATORY OF THE MAYO CLINIC

The author would like to pay a special tribute to the Biomechanics Laboratory of the Department of Orthopedics of the Mayo Clinic where he was himself a research fellow in 1989-1990 (Figure 2). Edmund Y.S. Chao, Kai-Nan An, William P. Cooney III and others were among the first to be interested in the Biomechanics of the Hand and Wrist. Besides contributing a tremendous amount of knowledge in this field, Edmund Chao and coworkers either developed or perfected many research methods, for example the Rigid Body Spring Modeling technique, the technique of in vivo measurement of tendon forces, etc ... Many former fellows of the Laboratory have subsequently worked on exciting research at Mayo and have become well known internationally. Many of them were in Genval to present original contributions.

Figure 2. Biomechanics Laboratory, Department of Orthopedics, Mayo Clinic, February 1990 (courtesy of E.Y.S. Chao, PhD).

The Laboratory was first instituted in October 1972, initially as a one-room facility. The staff consisted of 2 research fellows, J.D. Opgrande and E. Weber, who are now both renowned hand surgeons, and Edmund Y.S. Chao. It is noteworthy that since the very beginning, Edmund Y.S. Chao and coworkers were involved in Hand and Wrist research, and Chao was the Director of the Laboratory until 1992.[8] The present Director is Kai-Nan An, himself a former research fellow in the Laboratory from 1975 to 1977, whose first research project was in finger joint force analysis.

During the past twenty years, this Laboratory has trained more than 165 orthopedic residents, medical students, research fellows, engineering students, and visiting scientists. There have been 34 residents and research fellows who have worked on hand and wrist projects. Since 1975, the Hand research was continuously (except for a 1-year lapse in 1975) supported by an NIH grant on which Edmund Chao and then William Cooney were the principal investigators. In 1991, the Laboratory received another NIH grant to study the biomechanics of carpal instability, on which Ronald Linscheid was the principal investigator. In addition to these grants, the Hand and Wrist research program was also supported by grants from the Arthritis Foundation, the Orthopedic Research and Education Foundation and others.[3]

BELGIUM AND ORTHOPEDICS BIOMECHANICAL RESEARCH

The NATO Advanced Research Workshop 'Advances in the Biomechanics of the Hand and Wrist' was organized in Belgium. We recalled during the opening ceremony of the workshop that Belgium has always been a country of pioneers in Orthopedics and Biomechanics, starting with Vesalius. The author will limit himself to the 20th century.

Albin Lambotte (1866-1955), as chief of surgery at Antwerp, was from the beginning interested in orthopedic surgery. In 1907 Lambotte wrote: "my aim is primarily to study bony suture, or to be more precise, **osteosynthesis**", and by this sentence he created the word so widely used.[5,6,9] Moreover, Lambotte created the techniques and the instruments necessary for all common procedures that we still use nowadays: external fixation, also used by Lambotte in the hand (Figure 3), cerclage, screws and plates. Lambotte himself modeled his instruments in wood before asking manufactures to make them in metal. He also constructed beautiful violins; one is displayed in our Department, a gift of the late Edouard Van der Elst (Figure 4). Lambotte also had some preliminary ideas about what is now sophisticated Bimechanics. In Antwerp, Lambote was visited on many occasions by his friends, the Mayo brothers from Rochester, and he was himself one of the founders of the Belgian Society of Orthopedics and Traumatology (SOBCOT).[9]

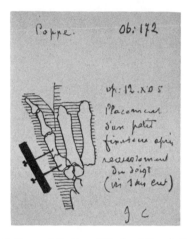

Figure 3. External fixation of the hand (Lambotte, 1912).

Figure 4. Violin constructed by Lambotte around 1930, on display in the Department of Orthopedics of Erasme University Hospital (courtesy of F. Burny, MD, PhD).

One of the pupils of Lambotte, Jean Verbrugge (1896-1964) from Brussels, who held a Mayo Clinic fellowship in 1922, remained famous because of the instruments he developed.[6]

Robert Danis (1880-1962) was the famous professor of surgery in Brussels in the Twenties. Dissatisfied with the available techniques, Danis invented his own procedures of internal fixation, insisting on the importance of achieving axial compression and primary healing of fractures.[4] One can say that the initiation of the AO movement in

Figure 5. Painting by Robert Danis, on display in the Department of Orthopedics of the Erasme University Hospital (courtesy of F. Burny, MD, PhD).

Switzerland was largely inspired by the ideas of Danis. Danis developed many instruments of internal fixation, the most famous being his coaptor. Danis was also an artist (Figure 5).

The history of Biomechanics in Belgium is also very rich. The work of Paul Maquet from Liège, related to the Biomechanics of the lower extremity, particularly of the knee, is well known.[7] In Brussels, there was also an early interest in Biomechanics. A strong group, the 'Centre interdisciplinaire de Biomécanique osseuse' (CIBO) was established in the early Seventies. The group is now led by Franz Burny, head of the Department of Orthopedics and Traumatology of the Erasme University Hospital (Université libre de Bruxelles). At the CIBO, original techniques of research were developed, particularly using strain gauges. On the basis of such experimental work, Burny was able to propose an original concept of fracture healing and elastic fixation.[2] Burny was among the founders and the first Secretary General of the European Society of Biomechanics (Figures 6,7), and he is also the Chairman of the Biomaterials Group of the Council of Europe.

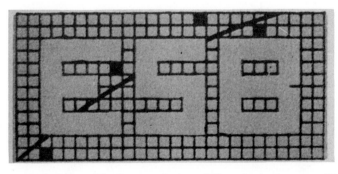

Figure 6. Original logo of the European Society of Biomechanics (courtesy of F. Burny).

Figure 7. First council of the European Society of Biomechanics (Brussels, 1960).
Upper row: F. Burny (left), J. Wagner (†), P.M. Calderale (right).
Lower row: Quenada (left), J.T. Scales, R. Huiskes (right).

REFERENCES

1. "The North Atlantic Treaty Organisation. Facts and Figures", NATO Information Service, Brussels (1989).
2. F. Burny, "Biomécanique de la Consolidation des Fractures. Mesure de la Rigidité du Cal In Vivo. Etude Théorique, Expérimentale et Clinique. Application à la Théorie de l'Ostéosynthèse", Thèse d'Agrégation, Université libre de Bruxelles (1976).
3. E.Y.S. Chao, Personal communication, (1992).
4. R. Danis, "Théorie et Pratique de l'Ostéosynthèse", Masson, Paris, (1949).
5. A. Lambotte, "L'Intervention Opératoire dans les Fractures Récentes et Anciennes envisagée plus particulièrement au point-de-vue de l'Ostéosynthèse avec la Description de plusieurs Techniques Nouvelles", Lamertin, Bruxelles (1907).
6. D. Le Vay, "The History of Orthopedics", Roche, Basel (1989).
7. P.G.J. Maquet, "Biomechanics of the Knee, with application to the Pathogenesis and the Surgical Treatment of Osteoarthritis", Springer-Verlag, Berlin (1976).
8. B.F. Morrey, "Orthopedic Surgery at Mayo (1910-1990)", Rochester (1991).
9. E. Vander Elst, "Les débuts de l'ostéosynthèse en Belgique", Imprimerie des Sciences, Bruxelles (1971).

Part I

MATERIAL PROPERTIES

THE ANATOMIC, CONSTRAINT AND MATERIAL PROPERTIES OF THE

SCAPHOLUNATE INTEROSSEOUS LIGAMENT: A PRELIMINARY STUDY

Richard A. Berger, Toshihiko Imaeda, Lawrence Berglund, Kai-Nan An, William P. Cooney, Peter C. Amadio, and Ronald L. Linscheid

Section of Hand Surgery
Orthopedic Biomechanics Laboratory
Department of Orthopedic Surgery
Mayo Clinic and Mayo Foundation
Rochester, MN 55905, U.S.A.

INTRODUCTION

The scapholunate interosseous ligament normally connects the scaphoid and lunate carpal bones by spanning the proximal, dorsal and palmar margins of the scapholunate joint. It has been felt to be a critical structure for maintaining structural integrity within the proximal carpal row, and thus important in maintaining normal carpal mechanics.[1,2,3,4,5,6,7,8,9,10,11,12,13,14,15,16,17,18,19,20,21] Complete disruption of the scapholunate ligament has been associated with a clinical condition of instability termed dorsal intercalated segmental instability (DISI), which is felt to be a precursor of advancing degenerative disease,[11] but there is no clear concensus regarding the effects of partial scapholunate ligament disruption. In the laboratory, efforts have been made to study the kinematic influences of this ligament, however conflicting results have been published.[2,18] Previous anatomic investigations have revealed the complexity of the interface between the scapholunate and radioscapholunate ligaments, but specific investigations of the scapholunate ligament anatomy have not been published.[3] Additionally, material property studies have been carried out on the scapholunate ligament in two published studies.[13,16] In both studies, distraction to failure was applied to the entire ligament, which may not represent physiologic failure. In order to obtain a better understanding of the role of the scapholunate ligament on the mechanics of the wrist and the relative contributions of the anatomic subregions of the ligament, we designed an investigation to evaluate the detailed gross and histologic anatomy and mechanical properties of the subregions of the scapholunate interosseous ligament.

Advances in the Biomechanics of the Hand and Wrist
Edited by F. Schuind *et al.*, Plenum Press, New York, 1994

MATERIALS AND METHODS

Anatomic Study

Forty wrists from twenty formation-fixed adult cadaver specimens and eight wrists from four fresh adult cadaver specimens were dissected from a dorsal approach through the radiocarpal joint capsule. Observations were made on the gross anatomy of the scapholunate ligament. In eight additional fixed specimens, the scapholunate bone-ligament-bone complex was carefully excised en-bloc through a dorsal approach and all bone other than the subchondral cortex of the scapholunate joint cleft was carefully excised. The specimens were then decalcified, dehydrated and mounted in paraffin. Eight micron thick serial sections, stained with hematoxylin and eosin, were obtained in either transverse, coronal or sagittal planes.

Constraint and Material Property Study

Twenty-four fresh cadaver specimens were dissected to harvest intact scapholunate bone-ligament-bone complexes. These specimens were mounted in polymethyl-methacrylate plugs designed to fit in the clamps of the testing machine. The specimens were tested on an MST servohydrolic testing machine under load control. Twelve specimens were oriented to allow testing of dorsal and palmar translation of the scaphoid relative to a fixed lunate with load limits set at +/- 4 lbs. Twelve specimens were oriented for testing of dorsiflexion and palmar flexion of the scaphoid relative to a fixed lunate with load limits set at +/- 4 in.lbs. In both experiments, output recorded included loads and either angular or linear displacements. The specimens were tested with all regions of the ligament intact, and then subjected to repeat testing with either the dorsal, palmar or membranous region divided with a scalpel. Testing was repeated with a second region divided. Finally, all specimens were oriented for failure testing of the remaining intact region, which was carried out at a rate of 5 mm/sec. The end points of displacement were digitized and compared with similar points from each ligament division. These differences in displacement were then analyzed using paired T testing with Bonferroni adjustments for multiple comparisons.

RESULTS

Anatomy

The scapholunate ligament normally forms a complete barrier between the midcarpal and radiocarpal joints by connecting the dorsal, proximal and palmar aspects of the mutually articulating surfaces of the scaphoid and lunate. The palmar region is just deep to, but distinctly separate from the long radiolunate ligament. The radioscapholunate ligament, actually a synovial covered neurovascular termination of the radial artery and anterior interosseous artery and nerve, enters the radiocarpal joint space between the long and short radiolunate ligaments. It merges with the scapholunate ligament complex in the interval between the palmar and proximal regions. Dorsally, a stout deep capsular ligament connecting the scaphoid, lunate and triquetrum forms the distal edge of the dorsal region of the scapholunate ligament.

The dorsal region of the scapholunate ligament is short and thick, measuring approximately 5 mm at its widest point. Normally, its fibers are oriented transversely. Transverse sections show this region to be somewhat trapezoidal in shape, averaging 3 mm in depth. The proximal region is not defined in terms of fiber orientation, as it has an amorphous appearance. Fifteen of the forty-eight specimens dissected in this experiment had vertical defects in the membranous region, all from specimens greater than fifty years of age. Of these, three had what appeared to be mineral deposits associated with these defects. A variably sized meniscus extended into the scapholunate joint space, parallel with the joint surfaces. The cross-sectional geometry of the meniscus is an apex-distal triangle. The palmar region is quite thin, averaging one millimeter of thickness and five millimeters of width. It is obliquely oriented between the palmar surface of the articular surfaces, coursing from proximal and ulnar at the lunate to distal and radial at the scaphoid. It attaches to the lunate just deep to the attachment of the long radiolunate ligament.

Histology

The dorsal and palmar regions of the scapholunate ligament are histologically classified as true ligaments. As such, they are composed of collagen fascicles, largely oriented in a parallel fashion, surrounded by perifascicular loose connective tissue containing nerves and blood vessels. The palmar region is covered on both its deep and superficial surfaces by a layer of synoviocytes, whereas the dorsal ligament has synoviocytes on its deep surface only.

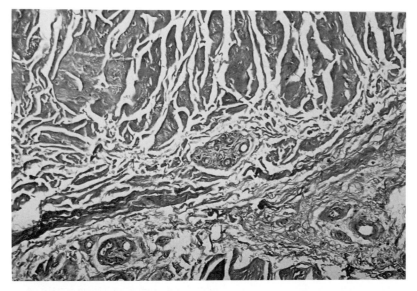

Figure 1. Light micrograph of a section of the dorsal region of the scapholunate ligament, illustrating typical ligament morphology with dense collagen fascicles surrounded by loose connective tissue, nerves and blood vessels. Hematoxylin and Eosin 45X.

The proximal region of the scapholunate ligament is composed of fibrocartilage. There are no nerves or blood vessels in this region. Additionally, there is no synovial lining on either the deep or superficial surfaces of this region. In the sagittal sections, the radioscapholunate ligament overlaps the superficial surface of the palmar extent of this region, and interdigitates with the palmar region fibers.

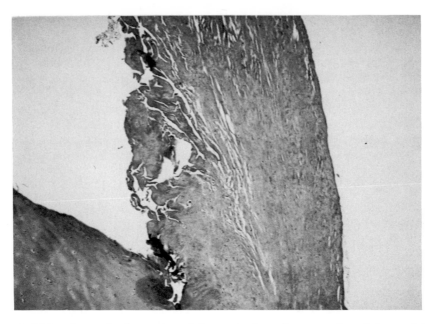

Figure 2. Light micrograph of a section of the proximal membranous region of the scapholunate ligament, illustrating its atypical composition of fibrocartilage. There are no blood vessels or nerves and no fascicular organization. Hematoxylin and Eosin 45X.

CONSTRAINT AND MATERIAL PROPERTIES

The results of the constraint and material property testing experiments are described as follows. Data presented represents the difference in translation or rotation from the intact state caused by the division of a single region or combination of two regions. In order to qualify as a significant difference, a value of $p < 0.05$ must be found.

Translation

The only statistically significant difference effected during dorsal translation of the scaphoid relative to a fixed lunate is caused by division of the dorsal region (2.2 ± 0.6 mm). No other region division or combination of region divisions caused a statistically significant change in translation.

During palmar translation of the scaphoid relative to a fixed lunate, significant changes in translation excursion were seen with division of the dorsal region (5.7 ± 1.5 mm) and with the combination of the proximal and palmar regions (5.3 ± 0.6 mm).

Rotation

Statistically significant differences during dorsal rotation of the scaphoid relative to the fixed lunate were found between the intact state and sectioning of the dorsal region (2.8 ± 0.7 degrees). Additionally, sectioning both the palmar and proximal region produced significant rotational changes (5.1 ± 1.5 degrees). During palmar rotation of the scaphoid relative to the fixed lunate, differences were noted only with sectioning of the dorsal and palmar regions (17.6 ± 2.5 degrees).

Failure Testing

Statistically significant differences between the failure strengths of the dorsal, palmar, and proximal regions were noted at the $p < 0.05$ level. The strongest region was the dorsal region (260.3 ± 118.1 N), followed by the palmar region (117.9 ± 21.3 N) and the proximal region (62.7 ± 32.2 N).

DISCUSSION

Scapholunate dissociation has been recognized as a distinct entity for over twenty years, and has been linked to the development of degenerative disease in the wrist. The most common pattern of instability arising from scapholunate dissociation is the dorsal intercalated instability (DISI) pattern.[11] This pattern produces a relatively palmarflexed scaphoid with concurrent lunotriquetral dorsiflexion. This posture redistributes axial force transmission such that the radioscaphoid load shifts dorsally, as does the capitolunate load.[19] It has been demonstrated that chronic instability universally leads to degenerative changes, now referred to as a scapholunate advanced collapse (SLAC) wrist.[20]

The scapholunate ligament has been identified theoretically as the principal stabilizer of the scapholunate joint. Disruption of this structure has been identified as the first step in producing a progressive perilunate dislocation in the laboratory setting, and disruption of this ligament is often the most striking finding during reconstructive surgery for patients with scapholunate dissociation.[8,10,12,17] Treatment options for scapholunate dissociation have been confusing in terms of indications, technique, and rehabilitation.[1,5,6,7,9,10,17,20,21] Additionally little information exits regarding objective long-term patient follow-ups with the various procedures.

Scapholunate dissociation is largely a clinical and radiographic diagnosis.[1,2,3,5,6,7,8,9,10,11,12,13,14,16,17,18,19,20,21] Tenderness in the scapholunate region and pain with provocative maneuvers, as well as a history of pain, weakness, tenderness, and trauma related to the scapholunate region propel scapholunate dissociation to a prominent position in the differential diagnosis. Radiographically, advanced scapholunate dissociation may show abnormal intercarpal angles and carpal height ratios, but acute scapholunate dissociation may appear deceptively normal.[11] Even clenched-fist views often fail to demonstrate any abnormalities in the carpal relationships.[14]

Radiocarpal and midcarpal arthrography has been used to demonstrate tears in the scapholunate ligament. When competent, the scapholunate ligament should prevent any communication between the radiocarpal and midcarpal joints. Thus, any communication of contrast material through this region has been held as evidence of injury to the ligament, and typically labelled as scapholunate dissociation. It has been demonstrated, however, that advancing age produces degenerative tears in the central region of the scapholunate ligament, not associated with instability patterns.[15] Additionally, contralateral communications have been demonstrated in patients with asymptomatic wrists.

This investigation has defined three distinct anatomic regions of the scapholunate ligament, each with distinctly different material properties. The least strong region in distraction, translation and rotation, is the proximal or membranous region. It is this region which is most often demonstrated to have a defect with arthrography or magnetic resonance studies. Any other statement about the integrity of the palmar or dorsal region of a scapholunate ligament with an arthrographically proven membranous defect cannot be reliably made. The data from this study do not suggest that the membranous proximal region of the scapholunate ligament contributes to the stability of the scapholunate complex. Rather, the dorsal and palmar regions demonstrate greater resistance to rotation and translation, and possess substantially greater failure strengths than does the proximal region. Although strain was not measured during these experiments, it was evident that substantially larger displacements were allowed when the dorsal and palmar regions were cut, leaving the membranous region intact, compared with the condition in which the membranous region was cut with adjacent regions intact. It is conceivable, therefore, that a defect in the membranous region is indeed an indicator that substantial damage has been imparted to the dorsal or palmar region, thus indicating a potentially unstable situation. In other words, the amount of displacement, not load, necessary to disrupt the proximal region of the scapholunate ligament probably will exceed the amount of displacement necessary to disrupt either the dorsal or palmar regions. Thus, an arthrogram or magnetic resonance image demonstrating disruption of the proximal region of the scapholunate ligament would be most valuable as an indicator of potential damage to the contiguous regions.

Just as arthrography and magnetic resonance imaging evaluate the proximal region of the scapholunate ligament, arthroscopy is also limited in its ability to visualize the entire ligament. First, the palmar region is in a deep recess distal to the plane of the radiocarpal joint. This makes access difficult if not impossible using a standard arthroscopic lens. Second, the dorsal region is partially a capsular ligament, thus in part beyond the limits of the radiocarpal joint space. This leaves the proximal region most accessible for arthroscopic evaluation, which may have negligible benefits in assessing instability.

Treatment modalities for scapholunate dissociation have ranged from cast immobilization to partial intercarpal arthrodesis.[1,5,6,7,9,10,11,17,20,21] A technique which has slowly been gaining favor is direct repair of the torn scapholunate ligament.[10] This technique employs an open dorsal approach where the scapholunate ligament is re-approximated to the scaphoid through intraosseous drill holes through the scaphoid. As described, a substantial portion of this repair utilized the tissue of the proximal region of the scapholunate ligament. This investigation casts theoretical doubt on the rationale of including this region in the repair zone for two reasons. First, the proximal region is composed of fibrocartilage, and is therefore without blood supply. Thus, it is unclear how completely this tissue will heal, even if surgically reapproximated. Second, the material property testing demonstrated the significantly weaker behavior of the proximal region relative to the dorsal and palmar regions. Perhaps effort should be placed at repairing or reconstructing the dorsal region. Further basic and clinical evaluation will be required before these questions are answered.

The results of this study correlate well with previously published reports of the material properties of the scapholunate ligament. Mayfield et al. reported an ultimate strength of approximately 400 Newtons, and Nowak reported an ultimate strength of approximately 250 Newtons.[13,16] It is clear from the current investigation that the majority of this strength is found in the dorsal region of the ligament.

The methods used to evaluate the anatomy of the scapholunate ligament have also proven useful in other ligament systems of the wrist.[3,13] It has been shown that what appears to be a ligament grossly may not have any characteristics of a ligament histologically. Additional information regarding the cellular composition of a ligament, or region of that ligament, may influence our understanding of the mechanical and biologic behavior of that ligament. The material property tests were designed to evaluate the ligament regions in modes other than pure distraction. This was intended to improve our understanding of the behavior of the ligament in more "physiologic" terms. The authors acknowledge the assumptions necessary and limitations imposed by such a protocol. However, this represents the next step in understanding the sophisticated behavior of small bone-ligament-bone systems. Further evaluations will be necessary to evaluate the influence on the kinematic behavior of the carpal bones that each region of the scapholunate ligament exerts. Additionally, it will be informative to evaluate the various repair and reconstruction techniques on carpal bone kinematics.

ACKNOWLEDGEMENTS

This work was supported in part by grants from the American Association for Hand Surgery and the Orthopaedic Research and Education Foundation and Grant Number AR40242 from the National Institutes of Health.

REFERENCES

1. E.E. Almquist, A.N. Bach, J.T. Sack, S.E. Fuhs, and D.M. Newman, Four-bone ligament reconstruction for treatment of chronic complete scapholunate separation, *J. Hand Surg.* 16A:322 (1991).

2. R.A. Berger, W.F. Blair, R.D. Crowninshield, and A.E. Flatt, The scapholunate ligament, *J. Hand Surg.* 7:87 (1982).

3. R.A. Berger, J.M.G. Kauer, and J.M.F. Landsmeer, The radioscapholunate ligament: a gross anatomic and histologic study of fetal and adult wrists, *J. Hand Surg.* 16A:350 (1991).

4. R.A. Berger, and J.M.F. Landsmeer, The palmar radiocarpal ligaments: a study of adult and fetal human wrist joints, *J. Hand Surg.* 15A: 847 (1990).

5. G. Blatt, Capsulodesis in reconstructive hand surgery: dorsal capsulodesis for the unstable scaphoid and volar capsulodesis following excision of the distal ulna, *Hand Clin.* 3:81 (1987).

6. D.J. Conyers, Scapholunate interosseous reconstruction and imbrication of palmar ligaments, *J. Hand Surg.* 15A: 690 (1990).

7. S. Hom, and L.K. Ruby, Attempted scapholunate arthodesis for chronic scapholunate dissociation, *J. Hand Surg.* 16A:334 (1991) .

8. J.M.G. Kauer, The interdependence of carpal articular chains, *Acta Anat. Scand.* 88:481 (1974).

9. W.B. Kleinman, and C. Caroll IV, Scapho-trapezio-trapezoid arthrodesis for treatment of chronic static and dynamic scapholunate instability: a 10-year perspective on pitfalls and complications, *J. Hand Surg.* 15A:408 (1990).

10. C.L. Lavernia, M.S. Cohen, and J. Taleisnik, Treatment of scapholunate dissociation by ligamentous repair and capsulodesis, *J. Hand Surg.* 17A:354 (1992).

11. R.L. Linscheid, J.H. Dobyns, J.W. Beabout, and R.S. Bryan, Traumatic instability of the wrist, *J. Bone Joint Surg.* 54A:1612 (1972).

12. J.K. Mayfield, R.P. Johnson, and R.F. Kilcoyne, The ligaments of the human wrist and their functional significance, *Anat. Rec.* 186:417 (1976).

13. J.K. Mayfield, W.J. Williams, A.G. Erdman, W.J. Dahlof, M.A. Wallrich, W.A. Kleinhenz, and N.R. Moody, Biomechanical properties of human carpal ligaments, *Orthop. Trans.* 3:143 (1979).

14. T.D. Meade, L.H. Schneider, and K. Cherry, Radiographic analysis of selective ligament sectioning at the carpal scaphoid: a cadaver study, *J. Hand Surg.* 15A:855 (1990).

15. Z.D. Micik, Age changes in the triangular fibrocartilage of the wrist joint, *J. Anat.* 126:367 (1978).

16. M.D. Nowak, Material properties of ligaments, in "Biomechanics of the wrist joint" Edited by K.N. An, R.A. Berger, and W.P. Cooney, Springer-Verlag, New York, 139 (1991).

17. A.K. Palmer, J.A. Dobyns, and R.L. Linscheid, Management of post-traumatic instability of the wrist secondary to ligament rupture, *J. Hand Surg.* 3:507 (1978).

18. L.K. Ruby, K.N. An, R.L. Linscheid, W.P. Cooney, and E.Y.S. Chao, The effect of scapholunate ligament section on scapholunate motion, *J. Hand Surg.* 12A:767 (1987).

19. S.F. Viegas, A.F. Tencer, J. Cantrell, M. Chang, P. Clegg, C. Hicks, C. O'Meara, and J.B. Williamson, Load transfer characteristics of the wrist: Part II, perilunate instability, *J. Hand Surg.* 12A:978 (1987).

20. H.K. Watson, and F.L. Ballet, The SLAC wrist: scapholunate advanced collapse pattern of degenerative arthritis, *J. Hand Surg.* 9A:358 (1984).
21. H.K. Watson, and R.F. Hempton, Limited wrist arthrodesis, I. the triscaphoid joint, *J. Hand Surg.* 5:320 (1980).

EFFECT OF REPETITIVE LOADING ON THE WRIST OF YOUNG RABBITS

Hideo Kawai,[1] Tsuyoshi Murase,[1] Takashi Masatomi,[2] Masakazu Murai,[2] Ryoichi Shibuya,[1] Yukio Terada,[2] Keiro Ono,[2] and Kozo Shimada[3]

[1]Department of Orthopaedic Surgery
Hoshigaoka Koseinenkin Hospital
4-8-1 Hoshigaoka, Hirakata-shi, Osaka 573, Japan
[2]Department of Orthopaedic Surgery
Osaka University Medical School
1-1-50 Fukushima, Fukushima-ku, Osaka 553, Japan
[3]Department of Orthopaedic Surgery
Osaka Koseinenkin Hospital, 4-2-78 Fukushima
Fukushima-ku, Osaka 553, Japan

INTRODUCTION

Weight-bearing joints are subjected to shear stress and axial loading. Clinical conditions under which either of these stresses is abnormally increased may thus induce degenerative changes. Repetitive movement of the wrist is thought to be responsible for wrist disorders such as Kienbock's disease, triangular fibrocartilage complex damage and osteoarthritis of the wrist. The effect of physical exercise and overuse alone on osteoarticular cartilage damage and growth plate behavior remains to be elucidated. This study was done to clarify how repetitive movement of the wrist of electrically stimulated young rabbits has an effect on the articular cartilage and growth plate.

MATERIALS AND METHODS

Thirteen white Japanese rabbits, aged 8 to 9 weeks and weighing 2.0 kg to 2.5 kg, were anesthetized with pentobarbital (30 mg/kg). Two wire electrodes,0.5 mm in diameter, which were made from seven strands of stainless steel wire with a 17-µm diameter were inserted into the right forearm flexor muscles at a distance of 1cm. All wire leads were exteriorized in the neck and connected to an electrical stimulator (Figure 1). The rabbits were returned to their cages and cyclic wrist flexion movement was applied through forearm flexor muscle stimulation with the electrical stimulator (SEN-2201, Nihon Kohden, Japan). The wrist movement was limited to a physiologic range by adjusting the voltage of the signal with rectangular impulse duration within a

Advances in the Biomechanics of the Hand and Wrist
Edited by F. Schuind *et al.*, Plenum Press, New York, 1994

range of 0.5 ms to 1.0 ms. The frequency of the stimulation was 30 cycles/min. and the average time of loading 11 hours/day. The total volume of cyclic loading was approximately 20,000 cycles/day. Three rabbits were loaded with approximately 800,000 cycles over 60 days, four rabbits with 400,000 cycles over 20 days, four rabbits with 200,000 cycles over 10 days and two rabbits with 100,000 cycles over five days. An antibiotic (cefotiam dihydrochloride; 200 mg/day) was administered intramuscularly for five days after the surgery. The rabbits developed and their body weight increased normally without any difference with controls. After the load had been completed, the rabbits were killed with pentobarbital anesthesia intravenously, and their forelimbs were removed together with the opposite, unloaded forelimbs for radiologic and histologic examination. Radiologic examination was done to evaluate the wrist joint and the distal radial and ulnar growth plates. The wrist was stripped of the soft tissues, decalcified, and stained with haematoxylin and eosin, masson-trichrome, toluidine blue, and safranin-O to examine the articular cartilage of the wrist and the radial and ulnar growth plate.

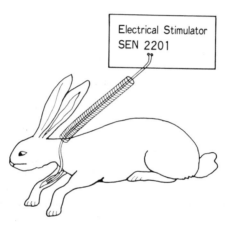

Figure 1. Schematic drawing of electrical stimulation of the right forelimb flexor musculature of rabbit.

RESULTS

The results are summarized in table 1.

Radiologic findings

In the experimented groups, cystic changes of the ulnar head were found in three out of seven cases loaded with 400,000 cycles or more (Figure 2). There were no significant changes in the groups loaded with 100,000 cycles and 200,000 cycles. Earlier closing of the radial and narrowing of the ulnar growth plate were observed in the unloaded wrist opposite which was loaded with 400,000 cycles, whereas the loaded side showed open radial and ulnar growth plates (Figure 3).

Table 1. Repetitive wrist flexion in rabbits.

Loaded cycles	Number of cases	Radiologic findings	Histologic findings
100000	2	Normal	No changes
200000	4	Normal	No changes
400000	4	Cystic changes of ulnar epiphysis in the loaded wrist (1 case) Growth plate closure in the unloaded wrist (1case)	Orderly arrangement of primary spongiosa in the loaded wrist (4 cases) Growth plate closure and narrowing in the unloaded wrist (1 case)
800000	3	Cystic changes of ulnar epiphysis in the loaded wrist (2 cases)	Orderly arrangement of primary spongiosa in the loaded wrist (3 cases) Erosive changes of ulnar head cartilage in the loaded wrist (1 case)

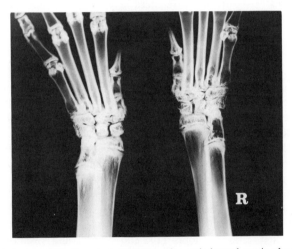

Figure 2. Right wrist was volarly flexed at 400,000 cycles and showed cystic changes in the epiphysis of the ulnar head.

Histologic findings

In the group loaded with 800,000 cycles, the superficial layer of the ulnar head cartilage was abraded (Figure 4). There were no remarkable articular changes in either wrist in the groups loaded with 100,000, 200,000 and 400,000 cycles. The calcified cartilaginous bars (primary spongiosum) were arranged orderly in the case of groups loaded with 400,000 cycles or more but were in disorder in the case of the unloaded ones (Figures 5 and 6). However, the growth plate width of the radius and ulna

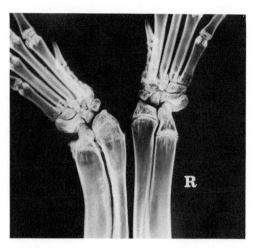

Figure 3. A group loaded with 400,000 cycles on the right side. The unloaded side showed earlier closure of the radial growth plate and narrowing of the ulnar growth plate. On the other hand, the loaded side showed an open growth plate.

disclosed no statistically significant differences between the loaded and unloaded wrists except one case with closure of the growth plate in the unloaded wrist.

DISCUSSION

Few experimental in vivo loadings under physiological conditions have been reported in the field of biomechanics.[1,5] This study represents an attempt at in vivo loading of the wrist of young rabbits under physiological conditions. Frictional overuse alone is reported not to be responsible for the production of the osteoarthritis which can be produced by a fatigue mechanism.[6] Radin and Paul[3] described an *in vitro* system of cartilage disruption in which bovine metacarpophalangeal joints were stressed at 40 cycles/min with constant and pulsed loading. With constant loading alone, no significant joint damage occurred even when specimens were loaded with 1000 pounds for up to 500 hours (1,200,000 cycles). The addition of a single pulsed load of 500 pounds at the midpoint of each constant-load oscillating cycle of 500 pounds developed partial cartilage thickness loss after 72 hours (172,800 cycles) and resulted in exposure of subchondral bone after 192 hours (460,800 cycles). With greater loads associated with pulsed loading, the mechanical disruption of cartilage occurs with fewer cycles. It is estimated that the physiologic load of the bovine metatarsophalangeal joint is in the order of 250 pounds. Therefore, it is difficult to produce articular cartilage erosion in a bovine metatarsophalangeal joint under physiologic stress. In rabbits, however, the wrist has nine carpal bones including os centrale, while the ulnotriquetral joint has a ball and socket joint with direct cartilage contact and without a triangular fibrocartilage complex; it makes up a major portion of the wrist. In our *in vivo* study, repetitive wrist flexion caused minimal cartilage changes in the ulnar head only in the groups loaded with 800,000 cycles. The repetition of joint movement within a physiologic range therefore does not seem to cause a deleterious effect on the articular cartilage.

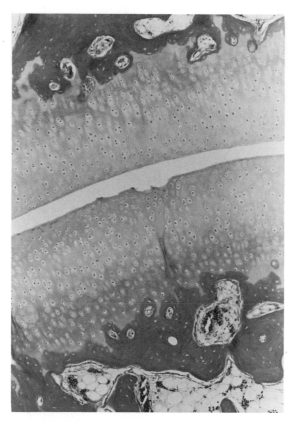

Figure 4. A group loaded with 800,000 cycles. The superficial layer of the ulnar head cartilage was abraded (haematoxylin and eosin).

It has been suggested that subchondral bone plays an important role in the initiation of the osteoarthritic process.[2] The cystic changes in ulnar head are supposed to be an epiphyseal response to repetitive stresses although osteoarthritis did not occur in our series. The effect of repetitive loading on the wrist, especially on the growth plate, was observed in young rabbits. This loading had a growth-stimulating effect, as seen in an open growth plate and orderly arranged primary spongiosa on the loaded side.[4] Physical exercise within the physiological range can therefore have a stimulating effect on growth potential in the loaded limb, possibly due to an increase in the blood flow.

CONCLUSIONS

The effect of repetitive loading on the wrist of young rabbits was observed. Our observation focused on the growth plate where the primary spongiosa was orderly arranged on the loaded side and the thickness was well preserved. Repetitive wrist flexion within a physiologic range did not have any deleterious effect on the articular cartilage and disclosed only a minimal change even in the groups loaded with 800,000 cycles.

Figure 5. A group loaded with 800,000 cycles. The primary spongiosa of the radial growth plate are orderly arranged on the loaded side (toluidine blue).

Figure 6. The growth plate of unloaded radius opposite to one loaded with 800,000 cycles. The primary spongiosa are in disorder (toluidine blue).

REFERENCES

1. S. Deckel, and S.C. Weissman, Joint changes after overuse and peak overloading of rabbit knees in vivo, *Acta Orthop. Scand.*, 49:519 (1978).
2. E.L. Radin, M.G. Ehrlich, R. Chernack, P. Abernethy, and I.L. Paul, Effect of repetitive impulsive loading on the knee joints of rabbits, *Clin. Orthop.*, 131:288 (1978).
3. E.L. Radin, and I.L. Paul, Responses of joints to impact loading, I. In vitro wear, *Arthritis Rheum.*, 14: 356 (1971).
4. F. Seinsheimer III, and C.B. Sledge, Parameters of longitudinal growth rate in rabbit epiphyseal growth plates, *J. Bone Joint Surg.*, 63-A:627 (1981).
5. E. Wada, S. Ebara, S. Saito, and K. Ono, Experimental spondylosis in rabbit spine, *Spine*, 17, (In press).
6. B.O. Weightman, M.A.R. Freeman, and S.A.V. Swanson, Fatigue of articular cartilage, *Nature*, 244:303 (1973).

CEMENTLESS TOTAL TRAPEZIO-METACARPAL PROSTHESIS

PRINCIPLE OF ANCHORAGE

Pascal Ledoux

Centre S.O.S. Main Bruxelles
Clinique du Parc Léopold
Bruxelles, Belgium

INTRODUCTION

Strength, stability, and mobility are the most important requirements for good thumb function. However, conventional treatment of osteoarthritis of the trapezio-metacarpal joint does not provide total satisfaction.

The long-term tolerance of silastic implants seems to be unsatisfactory. Trapeziectomy does not correct loss of strength because of increased collapse and metacarpal shortening. Arthrodesis suppresses the main function of the thumb: the mobility which allows good opposition of the thumb with regard to the fingers. Cemented thumb total prosthesis gives the best results regarding strength and mobility, but with a risk of loosening, probably due to the heat of the bone cement polymerization. To achieve good results for a total thumb joint implant it is important to: [1] abolish pain, [2] conserve the height of the trapezium, and to [3] conserve physiologic mobility.

In order to address these problems a study was initiated to design a cementless total trapezio-metacarpal prosthesis. The design that resulted is made of titanium and polyethylene. It is a ball-and-socket joint with the center of rotation in the trapezium.

Because it is cementless, the design of the components is very important for immediate and long-lasting anchorage.

STUDY OF LOADS

During a key pinch, there is a dorsal shear stress of the base of the thumb metacarpal. In the conditions of an articulated prosthesis, this stress is counterbalanced by a reaction force of the cup on the head of the stem (Figure 1).

The different forces produce moments in function of the fulcra of the prosthesis. The position of the fulcrum is a very important point. In the case of a classical prosthesis (Figure 2), the fulcrum of the stem is distal. Under these conditions, the lever arm of the force acting on the head of the stem is important in order to prevent loosening.

Advances in the Biomechanics of the Hand and Wrist
Edited by F. Schuind *et al.*, Plenum Press, New York, 1994

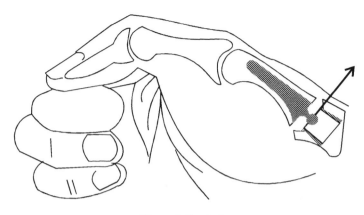

Figure 1. See text.

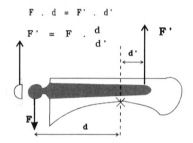

$$F \cdot d = F' \cdot d'$$

$$F' = F \cdot \frac{d}{d'}$$

Figure 2. See text.

Under such conditions, the stress between the distal extremity of the stem and the dorsal cortical bone of the metacarpal is concentrated and may be the first area to show loosening. The dorsal migation of the distal extremity observed with the cemented prosthesis is the result of this excessive stess due to a long lever arm and too small a contact area between bone and prosthesis.

For this new generation prosthesis, the fulcrum of the stem on the bone is near to the fulcrum of the cup on the head, and under these conditions, the lever arm is shorter and the distal stress is greatly decreased (Figure 3).

It is important to note that the moments on the bone and on the prosthesis are in opposite directions, consequently the stress between bone and prosthesis is important. In order to solve the problem of bone-prosthesis interface stresses, the contact surface in the zone of stress must be as great as possible to decrease the bone-prosthesis surface load.

PRINCIPLE OF ANCHORAGE

The first metacarpal can roughly be compared to a cone. A stem with an anatomical shape which fits into two concentric cones produces a mechanical locking press fit (Figure 4).

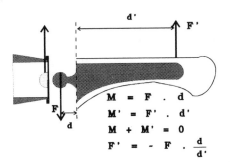

$$M = F \cdot d$$
$$M' = F' \cdot d'$$
$$M + M' = 0$$
$$F' = -F \cdot \frac{d}{d'}$$

Figure 3. See text.

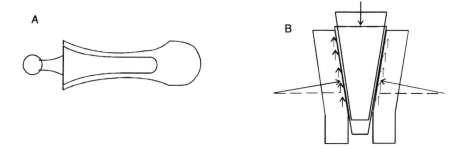

A

B

Figure 4A and B. See text.

The trapezial component is made of titanium and polyethylene. The titanium cup is conical inside, cylindrical outside and has six small wings arranged longitudinally. When the cylindrical polyethylene nucleus is introduced, the wings expand, producing a "plug" effect. This mechanism allows immediate anchorage of the trapezium implant (Figure 6).

When reduced in an articulated relationship, the joint allows for 66° of angulatory motion in any direction and free unrestricted rotation. This range of motion is a little above the physiological range of motion (Figure 6).

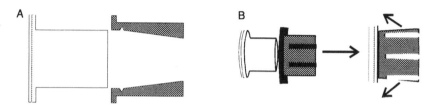

Figure 5A and B. See text.

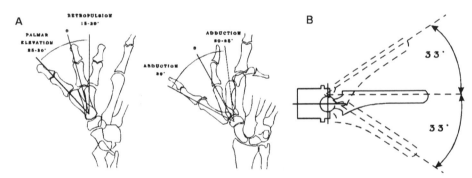

Figure 6A and B. See text.

The principle of the self-locking stem and immediate anchorage of the cup is the basis for immediate motion and use of the cementless implant. Surgical technique is of vital importance. It is therefore necessary to have different sizes of implants and to have a precisely gauged trial implantation. It is of prime importance that the trapezial cup is well oriented. The trapezial cavity is, of course, gauged on the cup before the wings are expanded.

There are five sizes of metacarpal stem and five sizes of rasp calibrated to the size of the metacarpal stem. It is important to use the largest possible size.

RESULTS

Seventy-two patients have been operated, and fifty-one have a follow-up of more than six months. With respect to mobility, the results are as good as for the cemented prosthesis. Strength is a good test for the stability and the tolerance of the implant. Pinch strength was measured pre-and postoperatively. The preoperative average was 3.5 kg, and we observed an increase to 7 kg after 12 months (the value in a normal population is 8 kg - Figure 7).

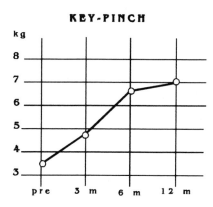

Figure 7. See text.

X-RAY ANALYSIS

With the exception of one case of prosthesis loosening into the trapezium and into the metacarpal owing to incorrect choice of implant size, no loosening occurred in this series. At the start of our experiment, four patients showed ossifications in the area of the "new joint". This is now avoided by thorough washing of the fragments of bone during the operation. No sclerotic bone reaction was observed in the other cases and no lucent lines have appeared between bone and prosthesis. This absence of bone reaction is due to a good position of the fulcrum and to a wide distribution of the loads with increased contact area between bone and prosthesis.

COMPLICATIONS

There was unfortunately one subluxation of the implant as a result of bad orientation of the cup in the trapezium. We had one prosthesis loosen into the trapezium and into the metacarpal, owing to the smallness of the prosthesis.

Two patients had titanium synovitis due to wear of the titanium head on the polyethylene. The prosthesis is now treated by nitrogen ionization of the prosthetic head which is titanium. This procedure reduces the coefficient of friction and prevents titanium wear on the polyethylene.

CONCLUSION

The results in the short and medium term are as good as with the cemented prosthesis. The stability of the anchorage seems to be established. Surgical technique is very important: it is vital that the cup be well oriented and the prosthesis size must be properly selected. Consequently we believe that this technique is a good choice for the treatment of trapezio-metacarpal arthritis, and, with the absence of bone cement, we anticipate less problems with prosthetic loosening.

REFERENCES

1. Y. Allieu, B. Lussiez, and B. Martin, Résultats à long terme de l'implant de Swanson dans le traitement de la rhizarthrose, *Rev. Chir. Orthop.* 76:437 (1990).
2. R.M. Braun, Total joint replacement at the base of the thumb. Preliminary report, *J. Hand Surg.* 7A:245 (1982).
3. J.J. Creighton, J.B. Steichen, and J.W. Strickland, Long-term tolerance of silastic trapezial arthroplasty in patients with osteoarthritis, *J. Hand Surg.* 16A:510 (1991).
4. P. Saffar, "La Rhizarthrose." Monographies du G.E.M., Expansion Scientifique Française (1990).
5. C. Van Mow, and W.C. Hayes, "Basic Orthopedic Biomechanics", Raven Press (1991).

PASSIVE FORCE-LENGTH PROPERTIES OF CADAVERIC HUMAN

FOREARM MUSCULATURE

R. Wells, D. Ranney, and P. Keir

Department of Kinesiology
University of Waterloo
Waterloo, Ontario, Canada, N2L 3G1

INTRODUCTION

The passive force/length properties of the forearm musculature, notably the extrinsic flexors and extensors of the fingers, have been found to be important in unloaded prehensile activities of the hand.[11] A knowledge of these passive properties is of interest in surgery, in understanding hand function and in mathematical modelling of hand function.

In surgery, knowledge of these properties is potentially useful during tendon transfer procedures since tension in a transferred muscle/tendon unit is used to gauge the best length to affix the tendon to its new insertion. The passive properties of the extrinsic musculature exert a strong influence on unloaded finger control and positioning.[11] Mathematical models of the hand, especially simulation models, require quantitative description of the mechanical and geometrical properties of the hand and forearm tissues.

This paper reports the passive force/length properties of the major muscles in the forearm: each head of the flexor digitorum profundus, superficialis and extensor digitorum, extensor indicis, extensor digiti minimi, flexor pollicis longus, abductor pollicis longus, extensor pollicis longus, palmaris longus, extensors carpi radialis (longus and brevis) and ulnaris, as well as flexores carpi ulnaris and radialis. Using the dimensions of the wrist and finger joints, the movement of the tendons to achieve a number of hand and wrist postures were predicted from published studies. These data were then used to estimate the range of passive tensions incurred in achieving these positions.

METHODS

Two fresh/frozen cadavers were used to develop the passive properties of the forearm muscles. The hand and forearm were strapped to a splint in a standard posture after rigor mortis had passed. The wrist and fingers were straight, the thumb adjacent to the lateral aspect of the second metacarpal and the elbow at 90 degrees flexion. The arm was removed at the mid humerus level wrapped and frozen. The specimens were frozen at -25° for approximately two months. On the day of testing the frozen specimen was cut through at the mid metacarpal level. This process allows the length of each tendon to be related to a known posture when the end of the tendon approximates the cut end of the metacarpal concerned. The anthropometric characteristics of the two cadavers are shown in table 1.

Table 1. Characteristics of Specimens.

Subject	Age	Gender	Height (cm)	Left wrist depth (mm)	Left wrist width (mm)	Cause of death
91047	80	F	< 150	28.3	41.7	CVA
92051	77	M	179	37.0	60.0	respiratory arrest.

The elbow was maintained at 90 degrees by a screw through the olecranon into the humerus. The forearm was maintained in the mid-prone position during testing by a screw from the testing apparatus into the head of the radius. The wrist was stabilised by a 14 gauge wire through the third metacarpal, the carpal bones and into the distal end of the radius.

The following muscles of the forearm were tested: the flexor digitorum profundus (FDP2-5), superficialis (FDS2-5) and extensor digitorum (ED2-5) for each digit, extensor indicis (EI), extensor digiti minimi (EDM), flexor pollicis longus (FPL), abductor pollicis longus (APL), extensors pollicis longus and brevis (EPL, EPB), palmaris longus (PL), extensors carpi radialis longus and brevis (ECRL, ECRB) and ulnaris (ECU), as well as flexores carpi ulnaris (FCU) and radialis (FCR). Thus 24 muscles were investigated on each cadaver arm.

Upon thawing of the wrist and forearm, the tendons of the major muscles were identified, tagged and connected to a wire bridle by either wire suture or small clamps. The clamps were constructed of thin walled aluminium tubing which was crimped and glued using cyanoacrylic adhesive.[5] Sufficient tendon was cleared to allow application of the clamps. Tendons of the thumb musculature, ie. FPL, APL, EPL, and EPB, were cleared sufficiently to allow an axial pull but all tendons were retained within their respective flexor or extensor retinacula. Junctura tendineae of the extensor tendons were divided. Temperature of the deep tissue was monitored and ranged from 15 to 24.2°C during testing.

When all tendons were identified and tagged the arm was fixed in the testing frame by screws through the epicondyles and into the head of the radius (Figure 1). Force was measured by 4 strain gauge cantilever transducers and their common displacement by a precision potentiometer. The strain gauge transducer signals were conditioned using a standard strain gauge bridge amplifier and analog to digital converted at 10Hz. The signals were calibrated using shunt calibration and validated using test weights and displacements. The software also allowed real time display of force/displacement profiles during testing. Tendons were connected in functional groups with up to 4 tendons connected simultaneously; for example the 4 tendons of the flexor digitorum profundus were tested together. The tendons were connected to the testing fixture and the wire link adjusted to bring the tendons to the same level as the cut ends of the metacarpal. This was designated as the reference length and was considered equivalent to the anatomical position.

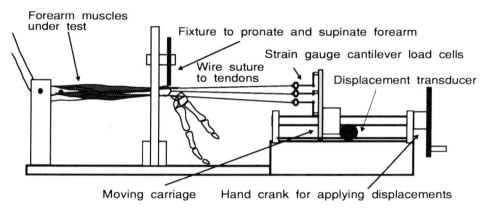

Figure 1. Testing apparatus and experimental setup.

The passive force/length properties were determined by applying slow displacement of approximately 2 mm/sec. This displacement was applied until force levels reached approximately 20 N or lower if it was felt that damage might occur if displacement continued. The load was then removed for one minute and then reapplied until failure of the muscle/tendon unit or the clamp occurred.

The dimensions of the wrist, metacarpophalangeal, proximal and distal interphalangeal joints were measured using calipers. Tendon excursions to achieve a number of wrist and finger postures (Table 2) were determined from the regression equations of Armstrong and Chaffin[2] and the angular excursion method of Brand.[4] The estimated excursions for the postures are summarised for specimen 92051L in Table 3.

RESULTS

An interesting observation was made on the starting position of the finger flexors and extensors (Figures 2-5). The flexors tended to retract upon cutting them while the extensors stayed in place or slid forward, beyond the length in the straight position. This would lead us to expect that the reference position used does not result in muscle lengths close to L_0 for the muscles tested.

Table 2. Postures used for tendon displacements, joint angles in degrees.

Posture	Wrist	MP	PIP	DIP
A) Straight	0	0	0	0
B) Extended wrist, straight fingers	-45	0	0	0
C) Flexed wrist, straight fingers	45	0	0	0
D) Extended wrist, flexed fingers	-35	90	110	70
E) Typical rest posture	-20	30	50	10

Table 3. Estimated tendon/muscle excursions required for the postures in Table 2. Positive excursions indicate lengthening of the muscle and negative excursions indicate shortening of the muscle.

	EXCURSION (cm)				
MUSCLE	A	B	C	D	E
ED 2	0	-1.45	1.45	2.21	0.65
3	0	-1.45	1.45	2.40	0.73
4	0	-1.45	1.45	1.81	0.50
5	0	-1.45	1.45	1.43	0.35
EI	0	-1.45	1.45	2.21	0.65
EDM	0	-1.45	1.45	1.43	0.35
FDP 2	0	0.79	-1.27	-3.30	-1.08
3	0	0.79	-1.27	-3.37	-1.11
4	0	0.79	-1.27	-3.11	-1.00
5	0	0.79	-1.27	-2.89	-0.92
FDS 2	0	0.95	-1.43	-3.69	-1.03
3	0	0.95	-1.43	-3.76	-1.07
4	0	0.95	-1.43	-3.49	-0.96
5	0	0.95	-1.43	-3.27	-0.88
FCR,FCU,PL	0	1.45	-1.45	1.13	0.65
ECRB,ECRL,ECU	0	-1.45	1.45	-1.13	-0.65

In some instances, the tendons had slippage between the crimping sleeve and the tendon. This resulted in somewhat lower force levels but they may still be considered valid in the low force region which we were investigating.

Greater tensions were obtained from the larger cadaver (92051) in the deep flexor tendons with similar excursions or strains (Figure 3). This was not found with the superficial flexors (Figure 4) where the smaller cadaver arms (91047R) achieved higher tension.

The finger extensor force-length curves are found in figures 4 and 5 for the common extensors and the extensor arising from the interosseous membrane (EI), respectively. The force-length curves for the extrinsic thumb musculature is shown in figure 6.

With the wrist flexors (Figure 7), the FCR and FCU show very similar trends with respect to the passive force-length curves for each of the cadaver arms. Palmaris longus tendons show nearly identical curves with the exception of 91047L. An aberration can be seen in the right PL tendon of specimen 92051 at approximately 0.5 cm of stretch, this is attributed to slippage of the connectors. The wrist extensor passive force-length curves are plotted in figure 8.

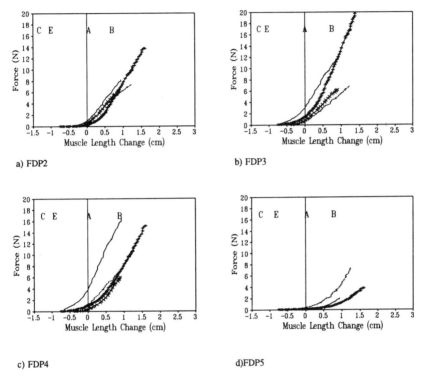

a) FDP2

b) FDP3

c) FDP4

d)FDP5

Figure 2. Force-displacement curves for the flexor digitorum profundus. Specimens 92051L (---), 92051R (-+-), 91047L (- -), and 91047R (-X-).

After testing the lower range of passive tension, the intact muscles were displaced until failure of the clips or muscle. Many clips slipped when the displacement was continued until failure. This was not surprising as the fasteners were intended only for the range of 0-50 N. Many muscles ruptured, most of these occurred at the muscle-tendon junction. Tendon failure was found on three muscles, all in one trial, possibly due to overtight crimping with the fastener, resulting in possible damage to the tendon. All other muscle failures occurred at the muscle-tendon junction. All results reported are for the initial loading phase.

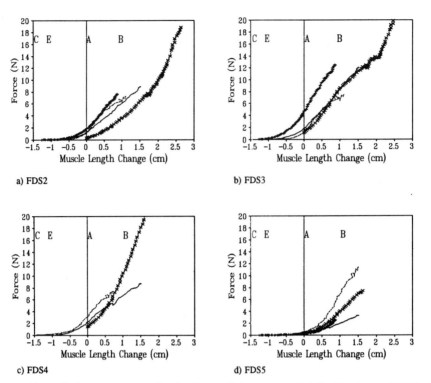

Figure 3. Force-displacement curves for the flexor digitorum superficialis. Specimens 92051L (---), 92051R (-+-), 91047L (- -), and 91047R (-X-).

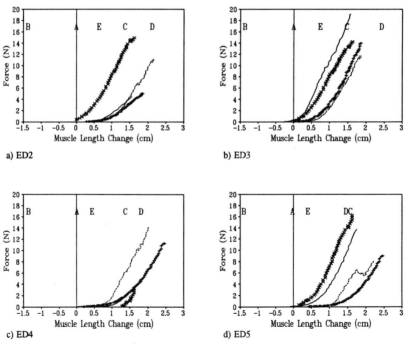

Figure 4. Force-displacement curves for the extensor digitorum. Specimens 92051L (---), 92051R (-+-), 91047L (- -), and 91047R (-X-).

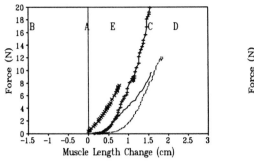

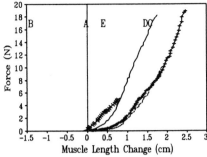

a) Extensor indicis.　　　　　　　　　　b) Extensor digiti minimi.

Figure 5. Force-displacement curves for extensor indicis and extensor digiti minimi. Specimens 92051L (---), 92051R (-+-), 91047L (- -), and 91047R (-X-).

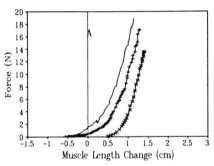

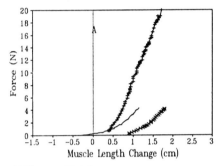

a) Abductor pollicis longus.　　　　　　　b) Flexor pollicis longus.

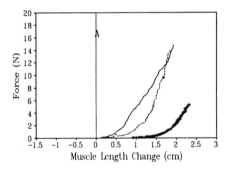

c) Extensor pollicis longus.

Figure 6. Force-displacement curves for the thumb musculature. Specimens 92051L (---), 92051R (-+-), 91047L (- -), and 91047R (-X-).

DISCUSSION

There has been a sizable amount of research directed at the stress-strain relationships in connective tissues such as ligaments and tendons but there is a scarcity of similar research on the passive properties of muscle. Knowledge of the passive properties of muscle are extremely important in hand and finger function: the resting position of the fingers is determined by the passive muscle properties themselves.

37

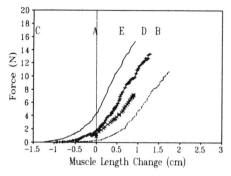
a) Flexor carpi radialis.

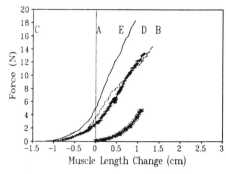
b) Flexor carpi ulnaris.

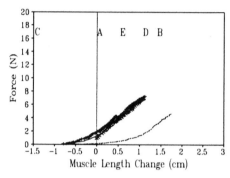

Figure 7. Force-displacement curves for the wrist flexors. Specimens 92051L (---), 92051R (-+-), 91047L (- -), and 91047R (-X-).

Several studies have determined the effects of freezing biological tissues, unfortunately it seems that striated muscle tissue has not received much attention in this area. The specimens were thawed at room temperature and with the aid of radiant heat. Due to the required dissection and labelling of muscles, this thawing procedure necessitated some two to three hours. Research in the area of freezing of tissues is important for those tissues often used surgically. Several groups have found no significant differences due to freezing in ligaments.[9,12,13] Similar to ligaments, tendons have been found to have similar stress-strain cycles in fresh and previously frozen specimens, however, the elastic moduli of previously frozen tendons were significantly lower than the fresh tendons.[8] With this evidence we have concerns with respect to the aponeuroses and muscle tissue proper, and the effects of freezing on them.

Figure 9 has been included to compare the data of the present study to that of pre-rigor mortis cadaver data.[11] Visual inspection indicates that a possible definition difference in muscle resting length may be present. In spite of procedural differences, the curves are similar for FDP2, FDS2, and ED2. The curves of Ranney et al.[11] were performed individually which may have allowed greater consistency for a given muscle. The similarity with the present study, although requiring further investigation, indicates that the freezing procedure may have a relatively small effect on the passive properties of muscle.

A finding of functional interest was that APL on the left side (subject 92051) was stiffer than its counterpart (Figure 6), possibly due to an apparent injury to the EPB on

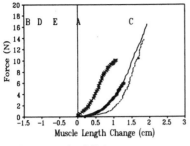

a) Extensor carpi radialis longus.

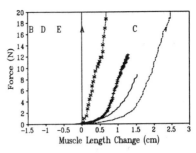

b) Extensor carpi radialis brevis.

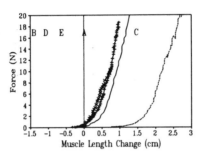

c) Extensor carpi ulnaris.

Figure 8. Force-displacement curves for the wrist extensors. Specimens 92051L (---), 92051R (-+-), 91047L (- -), and 91047R (-X-).

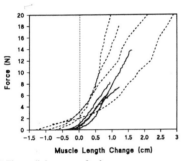

a) Flexor digitorum profundus.

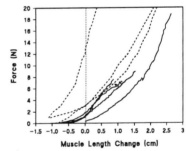

b) Flexor digitorum superficialis.

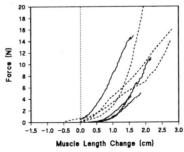

c) Extensor digitorum.

Figure 9. Force-displacement curves for the extrinsic musculature of the index finger comparing data from the present study (---) and from Ranney et al.[11] with pre-rigor mortis muscle (- -).

the left side. This apparent injury left the EPB bound to the metacarpal and non-functional. The APL tendon was noted to be thicker than expected and the results show it to be stiffer, apparently a response due to the loss of the other muscle.

As can be seen from the plots of flexor muscles (Figures 2,3), the flexor muscles are not subjected to the same strain levels as the extensor muscles in functional terms. The finger flexor musculature are required to stretch under one centimetre (from the anatomical position) for a wrist in 45° of extension. For most of their functional range, the extrinsic digit flexors have little or no passive tension. Some of the extensor muscles, in contrast, are required to undergo stretches of almost 3.5 cm in a fully flexed finger position with extended wrist for the larger hand (Table 3).

There appears to be some difference of opinion on the source of passive tension in muscle. Evidence has been shown supporting the endomysium and perimysium to be responsible for the resting tension in muscle.[1,3,10] Purslow[10] concluded that the perimysial collagen network is present to prevent over-stretching of the muscle fibre bundles. However, it has also been shown that the myofibrils themselves bear most of the resting (passive) tension in frog striated muscle.[7] Due to our purpose, hence protocol, we do not distinguish between these structures. We have an "overall" passive tension which relates to the muscle as a functional unit rather than its constituent subunits.

REFERENCES

1. M.A. Alnaqeeb, N.S. Al Zaid, and G. Goldspink, Connective tissue changes and physical properties of developing and ageing skeletal muscle., *J. Anat.* 139:677 (1984).

2. T.J. Armstrong, and D.B. Chaffin, An investigation of the relationship between displacements of the finger and wrist joints and the extrinsic finger flexor tendons, *J. Biomech.* 11:119 (1978).

3. T.K. Borg, and J.B. Caulfield, Morphology of connective tissue in skeletal muscle, *Tissue & Cell* 12:197 (1980).

4. P.W. Brand, Clinical Mechanics of the Hand, C.V. Mosby (1985).

5. S.A. Goldstein, T.J. Armstrong, D.B. Chaffin, and L.S. Matthews, Analysis of cumulative strain in tendons and tendon sheaths, *J. Biomech.* 20:1 (1987).

6. M. Maes, V.J. Vanhuyse, W.F. Decraemer, and E.R. Raman, A thermodynamically consistent constitutive equation for the elastic force-length relation of soft biological materials, *J. Biomech.* 22:1203 (1989).

7. A. Magid, and D.J. Law, Myofibrils bear most of the resting tension in frog skeletal muscle, *Science* 230:1280 (1985).

8. L.S. Matthews, and D. Ellis, Viscoelastic properties of cat tendon: effects of time after death and preservation by freezing, *J. Biomech.* 1:65 (1968).

9. F.R. Noyes, and E.S. Grood, The strength of the anterior cruciate ligament in humans and Rhesus monkeys. Age-related and species-related changes, *J. Bone Joint Surg.* 58A:1074 (1976).

10. P.P. Purslow, Strain-induced reorientation of an intramuscular connective tissue network. Implications for passive muscle elasticity, *J. Biomech.* 22:21 (1976).

11. D.A. Ranney, R.P. Wells, and J. Dowling, Lumbrical function: interaction of lumbrical contraction with the elasticity of the extrinsic finger muscles and its effect on metacarpophalangeal equilibrium, *J. Hand Surg.* 12A:566 (1987).

12. A. Viidik, and T. Lewin, Changes in tensile strength characteristics and histology of rabbit ligaments induced by different modes of postmortal storage. *Acta Orthop. Scand.* 37:141 (1966).

13. S.L.-Y. Woo, C.A. Orlando, J.F. Camp, and W.H. Akeson, Effects of postmortem storage by freezing on ligament tensile behaviour, *J. Biomech.* 19:399 (1986).

THE MECHANICAL PROPERTIES OF FINGER FLEXOR TENDONS AND

DEVELOPMENT OF STRONGER TENDON SUTURING TECHNIQUES

Andrew A. Amis

Biomechanics Section
Mechanical Engineering Department
Imperial College
London SW7 2BX, England

INTRODUCTION

As the finger flexor tendons pass distally from the flexor muscles of the forearm through the wrist and fingers to their insertions in the phalanges they are progressively constricted into tighter surrounding tissues. This passage through the carpal tunnel and into the fibro-osseous canals of the fingers is accompanied by progressively increasing difficulty in repairing the tendons and obtaining good functional results afterwards, so much so that the area between the distal crease of the palm of the hand and the distal interphalangeal joint was known as "no man's land" by surgeons for many years.[50,51] The paucity of tendon vascularity in much of this zone[23] causes a tendency for the surrounding tissues to become linked to the repair site by scar tissue adhesions, which can bring nutrition by vascular perfusion but prevent the tendons from moving adequately.[37] Anything less than perfection in surgical technique encourages this tendency, and factors such as rough handling of the surface of the tendon and the presence of suture material on the surface of the tendon have been implicated in causing poor results.[33] A further factor is that the tightness of the pulleys, which hold the tendons close to the bones when the finger is flexed,[14] means that the surgeon must not make a bulky repair. Knots in the suture material or protruding edges of the cut tendon ends will catch on the edges of the pulleys, preventing finger flexion.[3] The requirements for lack of repair bulk and little handling of the tendon mean that surgeons cannot use many strands of thick sutures to try to make their repairs strong, and this leads to repairs failing after the operation. Because of this, surgeons have to be cautious about mobilising their patients' fingers post operatively, waiting for the repair to gain some strength by healing and avoiding activities which will load the tendons and disrupt the repairs. If the surgeon is too cautious, the hand will be stiff and the outcome will not be a success.[18] It follows from the above observations that it would be a great help to surgeons if finger flexor tendon repairs were stronger, yet still respecting the physiology of the tissues, so that the patients can have their hand mobilised rapidly after operation without the fear of the repair being pulled apart. It was the objective of the work in this chapter to develop a stronger flexor tendon suturing technique, which would lead to better results after suturing of finger flexor tendons.

MECHANICAL PROPERTIES OF HUMAN FLEXOR TENDONS

Background

The frequent disruption of flexor tendon sutures shows that the repairs have only a small part of the strength of the intact tendon, which itself must have a safety margin above the loads imposed in normal use. Flexor tendons can be transected by penetrating wounds[9] or be ruptured by sudden loads,[17,34] although excessive force often leads to tendon avulsion from the bone insertion.[8,28] Tendon failure can also follow fraying at angulations or sharp edges caused by bone fractures or rheumatoid deformities.[19,46,47]

The strength of tendons in the lower limb and in other species has been studied,[1,5,16,21] but the human finger flexor tendons have only had the properties of embalmed specimens reported.[13,20]

Materials and Methods

The flexor digitorum superficialis (FDS) and flexor digitorum profundus (FDP) tendons were removed from seventeen hands of cadavers with a mean age of 72 years, within 24 hours of death. They were immersed in Hartmann's lactate solution and either tested within six hours or frozen at -20C until used. Short term storage at room temperature[52] and freezing[48] do not affect properties of collagenous tissues significantly. In ten hands, the middle finger tendons were removed complete with their respective phalanges, so that insertion strength could be found, otherwise tendons were cut from the bone. The entire tendon sheet was removed from the muscle proximally, then split into strips corresponding to individual fingers.

To test the insertion strength, the bone was potted in pmma bone cement in a metal container. The bone was orientated so that a load on the tendon would pull in a physiological direction. The edge of the cement was removed from around the tendon, so that it would not cause cutting under load. The specimen was mounted in an Instron tensile test machine as in figure 1, with the bone pot clamped to the moving crosshead and the proximal tendon sheet clamped in a freezing jaw.[41] The use of liquid CO_2 gave a rapid freezing effect within the metal jaw; this did not spread along the tendon because of low conductivity. The tendon did not slip under loads to failure, which were applied using a crosshead speed of 100 mm/minute. All specimens with bone attached failed distally. They were retested by fixing both tendon ends in freezing clamps, as were the remaining 116 specimens without bones.

Results

The bone insertion strengths were 558 ± 69N and 657 ± 135N for FDP and FDS, respectively (mean ± 1 standard deviation). The FDP always avulsed the palmar surface of the distal phalanx; the FDS ruptured close to the bone, at the two insertion slips.

The mean tensile behaviour of the FDP and FDS tendons is shown in figure 2. With a mean failure strain of 13% and mean lengths around 190 mm, the tendons elongated approximately 25 mm to failure. The middle finger tendons are strongest, with the index and ring finger tendons all very similar. The FDS tendon of the little finger was significantly less stiff and strong than the FDP tendon. Greater detail of these results has been published previously.[40]

The tendon strength was greater than the bone insertion strength by 2.12 times for the FDP and 2.04 times the insertion slip strength in the FDS.

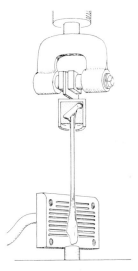

Figure 1. Tendon tensile test, with one jaw removed and pot sectioned to show specimen mounting.

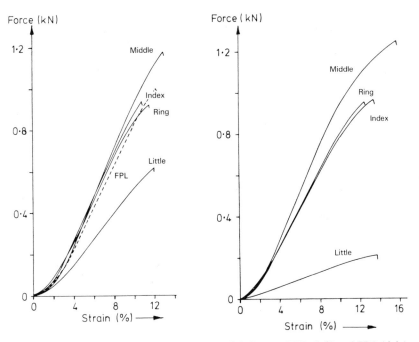

Figure 2. Mean tensile behaviour of FDP and flexor pollicis longus (FPL) (left) and FDS (right) tendons.

Discussion

The relative weakness of the tendon insertions was expected from clinical experience, although the strength could have been influenced by the high age of the specimens. The relative weakness of the profundus insertion is reflected by profundus avulsion being by far the more common failure.[8] This is the 'rugby finger' lesion, caused by sudden hyperextension.[28] A previous study of this insertion found a mean strength of only 387N.[22] At first sight, it does not seem a good design to have the tendon so much stronger than the insertion. Possible reasons for this include: the need to reduce tendon elongation under load, which would require greater muscle length changes for a given gripping action; or the need for a large cross-section to avoid excessive localised pressures (a cutting action) on the tissues of the pulleys when the finger joints are flexed.

The relative weakness of the little finger FDS tendon suggests that it is less heavily loaded than the FDP tendon, and it is often left unrepaired.

In general, the tendons are stronger than has been thought previously. This is reflected by the low strengths of the different artificial tendons and their anastomoses that have been produced in the past,[35,42] and their mechanical inadequacy may have been part of the reason for their lack of success.

TENDON LOADS IN-VIVO

Estimates of tendon loads in-vivo will show what sutured repairs ought to be able to withstand if patients are to exercise their hands and return rapidly to normal function. In order to make this estimate, data is needed for the external force actions, the internal geometry of the load-bearing structures, and on which muscles act during the activity being analysed.

The complex structure of the fingers includes many redundancies, meaning that the loads on internal structures are not statically determinate.[12] This has led to different schemes for trying to find an optimum or most likely solution.[56] Most work has been on the finger tip pinch action, which only involves one external force action.[6,11,45] Although the fingers are, of course, three-dimensional objects, some authors have analysed their actions in a two dimensional sagittal plane. This has been estimated to cause only a small error in the magnitudes of the major tendon tensions that must resist the extension moments imposed by external forces in pinching or grasping, and will certainly be adequate for obtaining a goal for suture repair strengths to aim at.[56] A review of prior work on pinching strength[55,56] and further analysis using the geometry of An et al.[4] has suggested that tendon tensions of 220N act during maximal pinching actions.

Since pinch actions require the muscles to stabilise all the joints of the cantilevered finger simultaneously, estimates of tendon tensions during grasping actions, when the fingers wrap around the object, may well show higher tensions. This is because any lack of equilibrium about a joint axis in grasping can be accommodated by redistribution of the loads acting on the three finger pads, whereas lack of equilibrium in pinch means collapse of the finger if the intrinsic muscles cannot modify the moment balance of the extrinsic muscles sufficiently. Data for grasping forces of the fingers onto cylinders has been published,[2] showing that the grasp forces increase as cylinder diameter decreases, down to 25-30 mm. The force applied by the distal phalanx was 80N, by the middle phalanx 75N, and 65N by the proximal phalanx. Use of an existing two-dimensional model[56] on this data led to predictions of tendon tensions in maximal grasping actions of 225N for the FDP and 240N for the FDS.

INVESTIGATION OF EXISTING SUTURING METHODS

Background

Many different suturing methods have been described, and some have been analysed mechanically. Unfortunately, it is difficult to compare the results of different authors, because they have usually used different tendon specimens or suture materials.

For many years the most popular suture method was the Bunnell suture[9] (Figure 3), which used a criss-cross arrangement to grasp the tendon away from the cut ends. However, Caroll and Match[10] noted that early motion caused the figure of eight suture to convert into a single loop by longitudinal separation of tendon fibres, leading to failure. More recently, the Kessler core stitch has become popular,[24] following publication of tests that showed greater strength. This uses 'locking loops' at the corners of the otherwise square arrangement of the suture, to grasp the tendon fibres. The original Kessler arrangement used a double loop at each corner and two sutures, leading to two knots on the surface of the tendon. This was simplified by Kleinert[26] to a single loop at each corner, with only one knot buried in the gap between the tendon ends, thus reducing sites for adhesion attachment or catching on the pulleys when the finger was flexed. Kleinert added a circumferential running stitch, in order to tidy-up the repair, avoiding the cut edges of the tendon from catching on the pulleys. This is currently the most popular technique, often referred to as the 'modified Kessler' method (Figure 3).

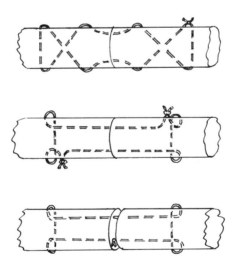

Figure 3. The Bunnell, Kessler and Kleinert 'modified Kessler' core stitches.

Until recently, the peripheral stitch was perceived to be only for making a tidy repair, with the core stitch assumed to be the major load carrier. Thus, the peripheral stitch was inserted in the epitenon layer, to avoid eversion of the cut edges, which might also cause snagging.[27] However, even though the peripheral stitch is of finer suture material than the core, there are usually more strands crossing the repair site, so the strength conferred by the peripheral stitch is multiplied. A study by Wade[53] showed that the peripheral stitch was important for delaying the onset of gap formation at the repair site. This is a more important failure criterion than the ultimate strength of the repair, because the tendon cannot heal well if there is a large gap between the tendon ends, even if the suture has not ruptured. Ketchum[25] said that the critical gap was 1 mm, while Seradge[44] and Ejeskar[15] stated that a 3 mm gap impaired results. Previous investigators had only reported the ultimate strength of their techniques.[25,49]

It was decided to investigate two aspects that might lead to better results: firstly, to examine the strength of the core stitch with different suture materials, to find which material gave the strongest result; secondly, to compare all the existing peripheral stitch techniques in the literature, to see if they could add to the force at which gaps appeared at the repair site when compared to the standard over and over stitch of the Kleinert method.

Materials and Methods

All experiments used human cadaver tendons as in the previous section, with 15 specimens in each group. All repairs were done using loupes giving x4 magnification, with the tendons kept moist with Ringer's solution to prevent changes due to dehydration.

The Kleinert version of the Kessler core stitch was inserted into six groups of tendons using 4/0 gauge sutures. The suture materials were Maxon (polytrimethylene carbonate), PDS (polydioxanone), stainless steel, Prolene (polypropylene), Ethibond (braided polyester) and Novafil (polybutester).

For the second experiment, the peripheral stitches all used 5/0 gauge Ethibond, with four stitches applied to each repair. Seven techniques were examined, as seen in figure 4.

The repairs were extended to failure at 50 mm/minute, with gap formation assessed by eye, using a ruler alongside the tendon. This was judged to be accurate enough for comparative purposes.

Results

The effect of different materials for the core stitch is shown in figure 5, which shows that 2 mm gap formation occured at low forces for all materials in the absence of a peripheral stitch. The ultimate strength was similar for Maxon, PDS and steel, with a lower strength for Prolene, Ethibond and Novafil, between which there was also no significant difference.

Figure 6 shows the force needed to cause a 2 mm gap and the maximum failure strength of the peripheral repairs. The continuous Halsted technique was significantly stronger than the over and over method ($p < 0.05$, Student's unpaired t test) for both criteria.

Discussion

Of the three core stitch materials to show higher strength, only stainless steel is accepted for tendon surgery at present. There is a fear that the absorbable materials will lose strength before the tendon has healed. The literature suggests that Maxon retains it's strength longer than PDS,[7,31] but experimental use in animal tendons has shown a tissue reaction that blocked tendon movement.[31] Therefore, stainless steel appears to be

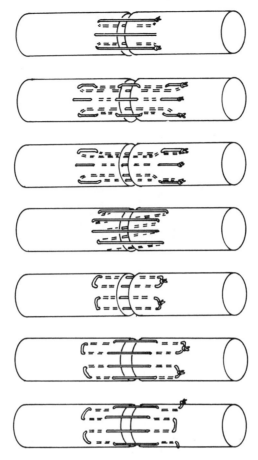

Figure 4. Peripheral stitches, arranged in order of strength: interrupted, interrupted Lembert, interrupted vertical mattress, over and over, interrupted horizontal mattress, interrupted Halsted, continuous Halsted.

the best core stitch material, even though it is difficult to handle. The other materials are all easy to use, and gave similar strengths, so all could be chosen for peripheral stitches.

Although the force to cause rupture is of interest, the gap created before reaching failure is too large for tendon healing (often more than 10 mm), so it is not relevant to clinical use. Different gap sizes have been associated with failure of normal healing, in the range 1 to 3 mm, so the use of a 2 mm gap as a criterion to examine differences between suture methods appears valid.

Continuous sutures were marginally stronger than interrupted sutures, which was expected. Continuous sutures are more evenly loaded than interrupted, which fail sequentially, at 2N increments of load: it is difficult to tie individual stitches to the same tension. Further, interrupted peripheral stitches leave multiple suture knots on the tendon surface, forming sites for adhesions.

It is clear that the stitches which pull parallel to the collagen fibres allow gap formation at low forces. It was concluded from this survey of existing stitches that the peripheral sutures should be continuous and pass perpendicular to the tendon fibres. However, the Halsted stitch was originally designed for intestinal anastomoses,[36] and leaves a lot of suture material on the tendon surface, which may lead to adhesions.

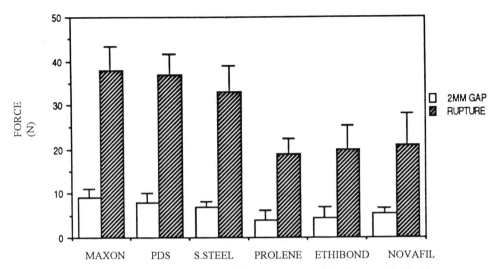

Figure 5. Core stitch strength with different suture materials. Mean + s.d., N= 15.

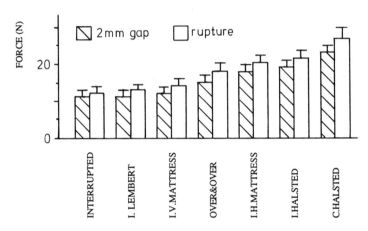

Figure 6. Forces to cause 2 mm gap and rupture of repair with a range of peripheral repairs. Mean + s.d., N=15.

An earlier study of peripheral stitches in human cadaver flexor tendons[54] used a 5/0 stainless steel core stitch, either alone, with a continuous over and over peripheral, or with a continuous Halsted peripheral. The peripheral stitches used 5/0 braided polyester (Ethibond), monofilament polypropylene (Prolene) or monofilament polydioxanone (PDS). The results of that study were in line with those reported above: the load for a 2mm gap increased significantly with the addition of a continuous peripheral stitch, and was increased further by changing to the Halsted peripheral stitch: 9N to 37N to 55N. The maximum strengths increased from 22N to 42N to 75N. There were no significant differences between the strengths of Prolene, PDS, or Ethibond repairs.

The findings so far suggest that stainless steel gives the strongest core stitch in an acceptable material, although the configuration of that stitch must be simple in view of the difficulty of handling the material. For the peripheral stitch, it appears that several suture materials are suitable, and that a continuous suture that has transverse grasping features will be best.

EVALUATION OF CORE STITCH FAILURE MECHANISMS

The previous study had shown that the overall strength of the repair could be increased significantly by changing the peripheral stitch configuration when using the 'modified Kessler' core stitch. There have, however, been a number of different core stitches described in the literature.[9,32,43] Also, other workers have added to the number of locking loops on the Kessler core stitch, to try to get a better grip on the tendon. Thus, Verdan[51] used two pairs of locking loops in each tendon end, and Ketchum et al.[25] used three pairs of loops in each tendon end.

Since it would be difficult to derive a mathematical model for predicting suture failure in tendons, in view of the non-homogeneous fibre material, it was decided to study suture failure mechanisms by means of multiple radiographs taken as the sutures were loaded. This would allow the failure mechanisms under load to be visualised, allowing identification of weak and strong features, and hence determine which particular features, such as locking loops or transverse passes, should be incorporated if a stronger suture method were to be designed.

Materials and Methods

This experiment used 40 mm lengths of fresh human cadaveric digital flexor tendons, with 15 specimens with each suture technique and suture material. Multifilament stainless steel and monofilament polybutester (Novafil) were used in 4/0 and 5/0 gauges. The suture techniques used are shown in figures 3 and 7. Locking loops were inserted with the cross-over configuration described by Pennington,[38] ensuring that they grasped the tendon fibres and thus resisted collapse.

In order to visualise failure mechanisms, the repairs using stainless steel sutures were x-rayed at 10 second intervals during tensile tests at 50mm/min. on an Instron 1122 materials test machine.

Small, high-contrast dental films were fixed into colour slide mounts and loaded into the magazine of a slide projector which had had it's lenses removed. The x-ray beam was fired through the tendon and into the lens aperture. A remote control was used to change films between each x-ray exposure, and precise timing of exposures allowed the load and gap formation values to be found for each image.[29]

Results

The forces required to cause a 2 mm gap at the repair and ultimate rupture were significantly higher with stainless steel than with Novafil in each of the 4/0 and 5/0 gauges. In each suture material, the thicker 4/0 gauge gave significantly greater strength than 5/0 repairs and there was a change in failure mechanisms: most of the 5/0 sutures broke, while most of the 4/0 sutures pulled through the tendon ends.

The failure mechanisms were shown clearly by the sequential radiographs (Figure 8). In particular, the locking loops untwisted and collapsed at low forces, and the addition of extra pairs of locking loops in the Verdan and Ketchum techniques did not alter the load at which this happened (Figure 9): figure 8 shows clearly that the pairs of loops were loaded and collapsed sequentially. Thus, the addition of more locking loops had the deleterious effect of liberating greater amounts of suture material from the tendon end under a given load, thus increasing the gap between the tendon ends at each load.

The addition of locking loops did not give a significant increase in gap formation or rupture forces over those recorded for a simple transverse pass of suture material, apart from the rupture force with 4/0 stainless steel, which was significantly increased. It was also found that slanting or 'criss-cross' features, exemplified by the Bunnell stitch, did not add to the repair strength beyond that contributed by the transverse component alone; the tendency was for the slanting components to split the tendon fibres apart and to swing until parallel to the tendon. This, of course, liberated suture material from the tendon ends, causing gap formation at the repair site.

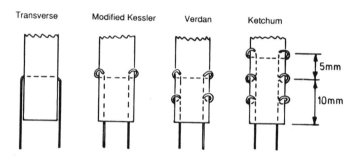

Figure 7. The transverse, modified Kessler, Verdan and Ketchum repairs.

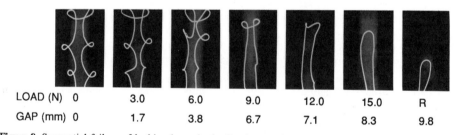

LOAD (N)	0	3.0	6.0	9.0	12.0	15.0	R
GAP (mm)	0	1.7	3.8	6.7	7.1	8.3	9.8

Figure 8. Sequential failure of locking loops in the Ketchum technique using 4/0 stainless steel suture.

Discussion

This experiment showed clearly that the use of complex suturing methods did not give any significant gain in gap resistance over the use of a simple transverse pass of suture material. Since complex sutures entail greater manipulation of the tendon during the repair operation, which will damage the epitenon layer more and thus make post-operative adhesions more likely,[39] there is a good clinical reason for utilising the simplest suture method possible.

It was obvious (Figure 9) that the addition of extra locking loops did not make the repairs stronger, because the loops were loaded sequentially rather than sharing the load, but that this strategy led to greater amounts of suture material being liberated into the repair gap. Thus, it was concluded that the best core stitch to resist gap formation should not incorporate locking loops, but simple features instead, such as the transverse stitch.

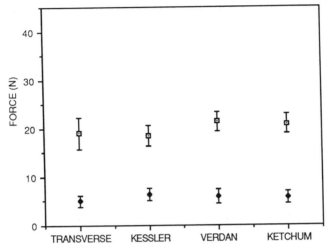

Figure 9. Comparison of the strength of four suture methods, 5/0 stainless steel.

DEVELOPMENT OF STRONGER PERIPHERAL SUTURE METHOD

The experiments of the previous section showed that it was transverse passes of suture material through the tendon fibres that gave security for the suture, resisting gap formation. It was noted that the peripheral stitches tested earlier had been based on passes through the epitenon layer; their placement in the superficial layer was intended to avoid eversion of the cut ends, rather than to provide strength. The handling qualities of the epitenon led surgeons to believe that this continuous tissue layer gave a secure hold for the sutures, since the underlying tendon fibre bundles tend to split apart easily. However, there was no data available as to the holding strength of the epitenon in comparison to that of the tendon fibres. It was decided to investigate this.

Material and Methods

Tensile tests of sutures placed into cadaveric human digital flexor tendons were used to examine the holding strength of the epitenon and the tendon fibres.[30] The fresh tendons were cut into 30 mm lengths and 5/0 multifilament stainless steel sutures were placed transversely through the epitenon layer 10 mm from one end. A thin steel template gauge ensured that each suture passed through the epitenon for a 6 mm transverse distance. Care was taken to keep the needle beneath the epitenon layer, but not in the tendon fibres - it was visible throughout the passage of the suture.

Fifty specimens were tested in extension to failure, using an Instron tensile test machine running at 50 mm/min. One end of the tendon was clamped into one jaw of the test machine, while the two ends of the suture were taken down to the other clamp. The specimens were kept moist with Ringer's solution throughout, to prevent changes to the epitenon due to dehydration. Tests were rejected if the suture broke in the clamp. After this test, sutures were introduced into the same specimens at the same places, with a 6 mm passage through the tendon fibres, and tested again.

Fingers from formalin embalmed cadavers were decalcified, dehydrated, wax embedded, cut in 5 μm transverse sections and stained using Azan trichrome.

Results

There was a significant difference ($p<0.001$, Student's paired t test) in the suture holding strength between the epitenon (10.6 + 4.3N, mean + s.d.) and the tendon fibres (19.0 + 4.1N). All sutures pulled out of the epitenon, while those in the tendon fibres were all ruptured. These results correlate with the histological appearance: the epitenon being a layer of loose connective tissue, variable in thickness and fragile looking when compared to the dense collagenous tissue of the tendon fibre bundles (Figure 10).

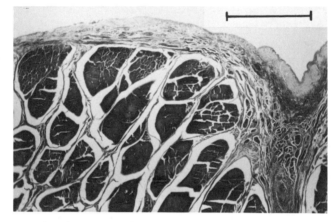

Figure 10. Transverse section of FDP tendon, with loose epitenon tissue cover and dense tendon fibre bundles. Stain: azan trichrome, bar = 0. 1 mm.

Discussion

With a mean strength increase of 83% if the sutures are placed into the tendon fibres rather than only below the epitenon, it seemed probable that a re-design of the peripheral stitches tested earlier would lead to a significant strength increase. Examination of the paths of the sutures in figure 4 shows that they all pass out of the tendon when making their transverse passes around it - the results shown above suggest that these configurations should be reversed, so that the transverse passes are into the tendon, thus gaining strength from passage through the tendon fibres.

These findings led to the design and testing of a new peripheral suture method.

Materials and methods

A new peripheral stitch was designed[30] with several aims in mind: to achieve good strength, to have a simple method that would minimise damage to the surface of the tendon, to avoid eversion of the edges of the tendon ends, to minimise the amount of suture material left on the surface of the tendon (which causes adhesions[33]), and to grasp the tendon away from the cut ends, since those zones are known to soften and lose their suture holding strength during the healing reaction.[32]

The new configuration is shown in figure 11. Stitches were first inserted parallel to the tendon fibres below the epitenon for 5mm from the cut end of the tendon. The suture was brought out of the tendon and then the transverse stitch was made, passing deeply into the tendon fibres, before returning to the tendon end in the epitenon. The suture was terminated by a knot on the tendon surface when the stitching had encircled the tendon.

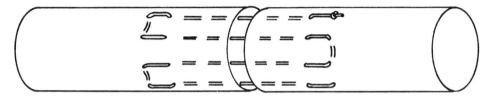

Figure 11. Configuration of the new peripheral suture technique.

The strength of the new peripheral stitch was found by tensile testing fifteen specimens, as described earlier, using the same 5/0 braided polyester (Ethibond), so that the strength could be compared to that found earlier for the commonly used over and over stitch and the stronger Halsted peripheral stitch. All of these methods were applied to tendons without a core stitch, so that the strength of the peripheral stitch alone could be examined, although this would not be used in isolation clinically. For each repair, the suturing was spaced out so that the tendon was encircled and the suture finished by knotting when eight strands of suture material linked the tendon ends.

Results

The forces necessary to cause 2 mm gap formation and the ultimate strengths are shown in figure 12. The new stitch was found to be significantly more resistant to both gap formation ($p < 0.03$) and rupture force ($p < 0.001$, Student's t test) than the Halsted stitch.

Discussion

This new technique is significantly stronger than other peripheral stitch configurations. The situation has now changed from one where the peripheral stitch was regarded as merely a 'tidying up' operation, to one where the peripheral stitch alone is now as strong as the complete Kleinert modification of the Kessler core stitch plus over and over peripheral, the most commonly-used method in clinical practice. Since the holding power of the tendon fibres with this technique is stronger than the sutures, which always snap before pulling out, further strength increases can only come from passing the suture across the repair site more times or by using a stronger suture material. Since the deep transverse passes are easily inserted and allow the running configuration to draw the tendon ends together, there is no reason to use complex locking features such as those of the Savage stitch.[43]

THE STRENGTH OF THE NEW SUTURE TECHNIQUE

A new suture technique was designed, based on the experience gained. The new peripheral suture method was added to a core stitch in which the suture material was taken along the tendon just below the epitenon for 12 mm, then brought out of the tendon and passed across the tendon diameter, before returning to the cut end to form a simple square stitch. This was made using 4/0 multifilament stainless steel, and the peripheral stitch was added using 5/0 braided polyester, as above. This repair was made in six fresh cadaveric tendons, as was the Halsted peripheral plus square core stitch described by Wade,[54] and also the Kleinert method, all using the same suture materials. The cross sectional area of the repair site was estimated by measuring the major and minor diameters at the site of anastomosis using calipers, then assuming an elliptical section.

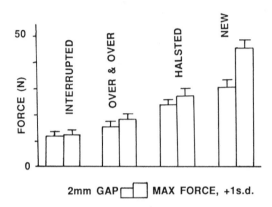

Figure 12. Strength of interrupted, over and over, Halsted and new peripheral suture methods, mean + s.d.

Results

The new suture method was significantly stronger than the Kleinert (p<0.001) and Wade (p<0.02) methods (Figure 13). The cross sections were not significantly different, the mean area with the new method being 82% of that with the Kleinert suture.

DISCUSSION AND CONCLUSIONS

The new suture technique has evolved logically from the results of the experiments performed. The individual details of the technique, such as grasping the tendon away from the cut ends, or of minimising the amount of suture material on the surface, have borne in mind the results of investigations of the biological reactions to sutures, published earlier. The new technique does not increase the bulk of the tendon more than accepted methods, so it should not jam in the tight sheath when the repaired finger is mobilised. The increased strength of the technique means that injured hands can be rehabilitated more boldly than with the accepted suture methods, perhaps introducing active movement of the fingers immediately after surgery.

Finally, figure 14 is a summary of all the work in this chapter. It is humbling to see how weak the repairs remain when compared to the strength of the intact tendons, which shows that there is much room for further work in this area, so that patients can rehabilitate their injured hands rapidly, without the fear of disrupting the work of the surgeon .

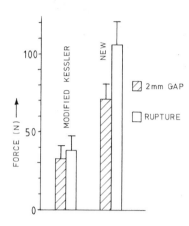

Figure 13. Forces required for 2mm gap and ultimate strength of the Kleinert ('modified Kessler') and new suture methods. Mean + s.d., N=6.

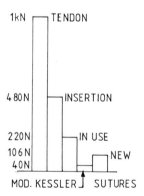

Figure 14. Breaking strength (N) of intact tendons, tendon insertions, loads in use, commonly used 'modified Kessler' sutures, and the new suture method.

Acknowledgements

The work in this chapter was carried out with Zahra Mashadi, David Pring, Warwick Radford, Peter Wade and Rory Wetherell, with financial support from Cyanamid Ltd and the Wellcome Trust. The author thanks them all. Figures 1,2,4,7-11 have appeared before in the Journal of Hand Surgery, and are reproduced with permission.

REFERENCES

1. M. Abrahams, Mechanical behaviour of tendon in vitro, a preliminary report, *Med. Biol. Eng.* 5:433 (1967).
2. A.A. Amis, Variation of finger forces in maximal isometric grasp tests on a range of cylinder diameters, *J. Biomed. Eng.* 9:313 (1987).
3. A.A. Amis, and M.M. Jones, The interior of the finger flexor tendon sheath and the functional significance of its structure, *J. Bone Joint Surg.* 70B:583 (1988).
4. K.N. An, E.Y. Chao, W.P.Cooney, and R.L. Linscheid, Normative model of human hand for biomechanical analysis, *J. Biomech.* 12:775 (1979).
5. J.V. Benedict, L.B. Walker, and E.H. Harris, Stress-strain characteristics and tensile strength of unembalmed human tendon, *J. Biomech.* 1:53 (1968).
6. N. Berme, J.P. Paul, and W.K. Purves, A biomechanical analysis of the metacarpophalangeal joint, *J. Biomech.* 10:409 (1977).
7. R.B. Bourne, H. Bitar, P.R. Andreae, L.M. Martin, J.B. Finlay, and F. Marquis, In-vivo comparison of four absorbable sutures: Vicryl, Dexon plus, Maxon and PDS, *Can. J. Surg.* 31:43 (1988).
8. J.H. Boyes, J.N. Wilson, and J.W. Smith, Flexor tendon ruptures in the forearm and hand, *J. Bone Joint Surg.* 42A:637 (1960).
9. S. Bunnel, "Surgery of the Hand," Fifth edition, Lippincott, Philadelphia, (1956).
10. R.E. Caroll, and R.M. Match, Common errors in the management of wrist laceration, *J. Trauma* 14:553 (1974).
11. E.Y. Chao, and K.N. An, Determination of internal forces in human hand, *J. Eng. Mech.*, Divn. ASCE 104:255 (1978).
12. E.Y. Chao, J.D. Opgrande, and F.E. Axmear, Three-dimensional force analysis of finger joints in selected isometric hand functions, *J. Biomech.* 9:387 (1976).
13. A.E. Cronkite, The tensile strength of human tendons, *Anat. Rec.* 64:173 (1936).
14. J.R. Doyle, and W. Blythe, The finger flexor sheath and pulleys: anatomy and function, in: "Symposium on Tendon Surgery in the Hand", American Academy of Orthopaedic Surgeons, C.V. Mosby, St. Louis, (1975).
15. A. Ejeskar, Flexor tendon repair in no-man's land: results of primary repair with controlled mobilisation, *J. Hand Surg.* 9A:171 (1984).
16. J.H. Evans, and J.C. Barbenel, Structural and mechanical properties of tendon related to function, *Equine Vet. J.* 7:1 (1975).
17. R.C. Folmar, C.L. Nelson, and G.S. Phalen, Ruptures of the flexor tendons in hands of non-rheumatoid patients, *J. Bone Joint Surg.* 54A:579 (1972).
18. R.H. Gelberman, J.S. Van de Berg, G.N. Lundborg, and W.H. Akeson, Flexor tendon healing and restoration of the gliding surface; an ultrastructural study in dogs, *J. Bone Joint Surg.* 65A:70 (1983).
19. J.P. Hallett, and G.R. Motta, Tendon ruptures in the hand with particular reference to attrition ruptures in the carpal tunnel, *The Hand* 14:283 (1982).
20. E.H. Harris, L.B. Walker, and R.B. Bennett, Stress-strain studies in cadaveric human tendon and an anomaly in the Young's modulus thereof, *Med. Biol. Eng.* 4:253 (1966).
21. W.C. Herrick, H.B. Kingsbury, and D.Y.S. Lou, A study of the normal range of strain, strain rate and stiffness of tendon, *J. Biomed. Mater. Res.* 12:877 (1978).
22. C.E.A. Holden, and M.D. Northmore-Ball, The strength of the profundus tendon insertion, *The Hand* 7:238 (1975).
23. J.M. Hunter, Anatomy of flexor tendons, pulleys, vincula, synovia, and vascular structures, in: "Kaplan's Functional and Surgical Anatomy of the Hand, 3rd edition", M. Spinner, ed., Lippincott, Philadelphia (1984).
24. I. Kessler, and F. Nissim, Primary repair without immobilisation of flexor tendon division within the digital sheath: an experimental and clinical study, *Acta Orthop. Scand.* 40:587 (1969).

25. L.D. Ketchum, N.L. Martin, and D.A. Kappel, Experimental evaluation of factors affecting the strength of tendon repairs, *Plast. Reconst. Surg.* 59:708 (1977).

26. H.E. Kleinert, S. Schepel, and T. Gill, Flexor tendon injuries, *Surg. Clin. N. Am.* 61:267 (1981).

27. G.D. Lister, Incision and closure of the flexor sheath during primary tendon repair, *The Hand* 15:123 (1983).

28. P.G. Lunn, and D.W. Lamb, "Rugby finger" - avulsion of profundus of ring finger, *J. Hand Surg.* 9B:69 (1984).

29. Z.B. Mashadi, and A.A. Amis, The effect of locking loops on the strength of tendon repair, *J. Hand Surg.* 16B:35 (1991).

30. Z.B. Mashadi, and A.A. Amis, Strength of the suture in the epitenon and within the tendon fibres: development of stronger peripheral suture technique, *J. Hand Surg.* 17B:171 (1992).

31. Z.B. Mashadi, and A.A. Amis, Variation of holding strength of synthetic absorbable flexor tendon sutures with time, *J. Hand Surg.* 17B, 278 (1992).

32. M.L. Mason, and H.S . Allen, The rate of healing of tendons, *Ann. Surg.* 113:424 (1941).

33. H.P. Matthews, H.J. Richards, Factors in adherance of flexor tendons after repair, *J. Bone Joint Surg.* 58B:230 (1976).

34. P.E. McMaster, Tendon and muscle ruptures - clinical and experimental studies of the causes and location of subcutaneous ruptures, *J. Bone Joint Surg.* 15:705 (1933).

35. G.A.W. Murray, and J.C. Semple, A review of work on artificial tendons, *J. Biomed. Eng.* 1:177 (1979).

36. T.F. Nealon, Fundamental Skills in Surgery, 3rd edition, W.B. Saunders, Philadelphia (1979).

37. E.E. Peacock, Fundamental aspects of wound healing relating to the restoration of gliding function after tendon repair, *Surg. Gynaec. Obstet.* 119:241 (1964).

38. D.G. Pennington, The locking loop tendon suture, *Plast. Reconstr. Surg.* 63:648 (1979).

39. A.D. Potenza, Critical evaluation of flexor tendon healing and adhesion formation within artificial digital sheaths; an experimental study, *J. Bone Joint Surg.* 45A:1217 (1963).

40. D.J. Pring, A.A. Amis, and R.R.H. Coombes, The mechanical properties of human flexor tendons in relation to artificial tendons, *J. Hand Surg.* 10B:331 (1985).

41. D.J. Riemersma, and H.C. Schamhardt, The cryojaw, a clamp designed for in vitro rheology studies of horse digital flexor tendon, *J. Biomech.* 15:619 (1982).

42. N.B. Rogers, A review of the use of prosthetic materials in tendon surgery, District Columbia, *Med. Ann.*, 39:411 (1970).

43. R. Savage, In vitro studies of a new method of flexor tendon repair, *J. Hand Surg.* 10B:135 (1985).

44. H. Seradge, Elongation of the repair configuration following flexor tendon repair, *J. Hand Surg.* 8A:182 (1983).

45. E.M. Smith, R.C. Juvinall, L.F. Bender, and J.R. Pearson, Role of the finger flexors in rheumatoid deformities of the metacarpophalangeal joints, *Anhritis Rheum.* 7:467 (1964).

46. I. Spar, Flexor tendon ruptures in the rheumatoid hand: bilateral flexor pollicis longus rupture, *Clin. Orthop.* 127:186 (1977).

47. P.J. Stern, Multiple flexor tendon ruptures following an old anterior dislocation of the lunate; a case report, *J. Bone Joint Surg.* 63A:489 (1981).

48. H. Tkaczuk, Tensile properties of human lumbar longitudinal ligaments, *Acta Orthop. Scand.*, Suppl. 115:5 (1968).

49. J.R. Urbaniak, J.D. Cahill, and R.A. Mortensen, An analysis of tensile strength of tendon anastomoses, *J. Bone Joint Surg.* 55A:884 (1973).

50. C.E. Verdan, Half century of flexor tendon surgery: current status and changing philosophies, *J. Bone Joint Surg.* 54A:472 (1972).

51. C.E. Verdan, Reparative surgery of flexor tendon in the digit, in:"Tendon Surgery in the Hand", C.E.Verdan, ed., Churchill Livingstone, New York (1979).

52. A. Viidik, L. Sandqvist, and M. Magi, Influence of post-mortal storage on tensile strength characteristics and histology of ligaments, *Acta Orthop. Scand.*, Suppl. 79:1 (1965).

53. P.J.F. Wade, I.F.K. Muir, and L.L. Hutcheon, Primary flexor tendon repair: the mechanical limitations of the modified Kessler technique, *J. Hand Surg.* 11B:71 (1986).

54. P.J.F. Wade, R.G. Wetherell, and A.A. Amis, Flexor tendon repair: significant gain in strength from the Halsted peripheral suture technique, *J. Hand Surg.* 14B:232 (1989).

55. P.S. Walker, and M.J. Erkman, Clinical evaluation of finger joints, in: "Human Joints and their Artificial Replacements," P.S. Walker, Thomas, Spingfield (1977).

56. B.O. Weightman, and A.A. Amis, Finger joint force predictions related to design of joint replacements, *J. Biomed. Eng.* 4:197 (1982).

BIOMECHANICS OF FLEXOR PULLEY RECONSTRUCTION

Kai-Nan An, Gan-Tyan Lin, Peter C. Amadio, William P. Cooney, III,
and Edmund Y.S. Chao

Orthopedic Biomechanics Laboratory
Mayo Clinic/Mayo Foundation
Rochester, Minnesota 55905, U.S.A.

INTRODUCTION

Many surgical techniques have been described for pulley reconstruction to restore hand function.[2,3,7,8] Biomechanically, a successfully reconstructed pulley should maintain adequate strength and similar compliance as those of the normal intact ligaments. In addition, it should provide the proper constraint of tendon to restore the functions of the digits. In this study, the efficacy of different types of pulley reconstruction on hand function were compared based on the joint motion and tendon excursion measurements. The mechanical properties of reconstructed pulleys were also tested to determine the strength and stiffness.

METHODS AND MATERIALS

Tendon Excursion

Tendon excursions and joint angle were measured by a method described previously.[1] Each finger was disarticulated through the carpo-metacarpal joint, with the tendon divided more proximally and mounted in a specially-designed jig (Figure 1). A fixed weight of 500 gms. was attached to the flexor digitorum profundus tendon. To assess the tendon excursion due to proximal interphalangeal joint (PIP) motion, the metacarpophalangeal (MP) and distal interphalangeal (DIP) joints were held fixed by Kirschner wires. Proximal interphalangeal joint motion and flexor tendon excursion were recorded simultaneously by an X-Y plotter or digital computer.

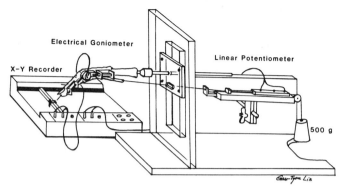

Figure 1. Experimental setup used to measure the tendon excursion during joint rotation of finger digit. Used with permission of C.V. Mosby Company (Figure 3 in Lin et al.[4]).

Mechanical Test

The mechanical properties of the normal and reconstructed pulley were obtained by using an Instron Series IX automated material testing machine (Instron Corporation, Canton, MA) integrated with a Vaxmate computer (Digital Equipment Corporation, Marlboro, MA) to operate the machine and to collect the data. The fingers were firmly held in place and prevented from sliding by a set of wedged pins. Tendon loading devices were made of metal rods to fit various configurations of pulley reconstruction. A distracting force was directly applied by an Instron materials testing machine until the pulley ruptured. The maximal breaking load, stiffness, displacement at maximal load and energy absorption of the reconstructed pulleys were all measured (Figure 2).

Procedures

Three methods of pulley reconstruction were evaluated (Figure 3): (a) The *"always present fibrous rim"* (described by Andreas Weilby[11] and reported by Kleinert and Bennett[3], 1978) was retained for weaving a tendon pulley reconstruction, (b) a

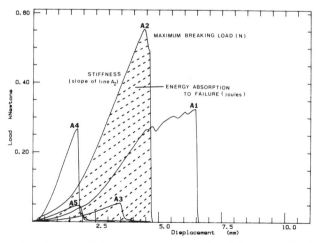

Figure 2. Parameters from the load-displacement curve are used to define the mechanical property of the normal and reconstructed pulley. Used with permission of the Churchill Livingstone (Figure 3 in Lin et al.[6]).

proximal interphalangeal (PIP) volar plate *"belt loop"* pulley, constructed as described by Karev et al.[2], (1987), and (c) a length of another tendon was passed around the proximal and middle phalanges as a single, double, or triple loop.[7] To standardize the effect of suture fixation on material properties of the reconstructed pulleys, each pulley was fixed with ten interrupted sutures of 4/0 polydioxanone. To assess the functional results of pulley reconstruction, we used the tendon-excursion to joint motion model previously reported.[1]

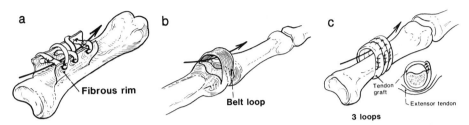

Figure 3. Three pulley reconstruction procedures are studied: (a) the "always present fibrous rim" (b) a PIP volar plate "belt loop" pulley (c) a length of another tendon was passed around the proximal and middle phalanges as a single, double, or triple loop. Used with permission of Churchill Livingstone (Figure 1 in Lin et al.[5]).

Mechanical Properties of Reconstructed Pulleys

The mechanical properties of five types of A2 pulley reconstructions were tested by using 30 cadaver digits. These include employing the fibrous rim of the previous pulley, volar plate *"belt loop"* pulley reconstruction, one loop around the proximal phalanx, double loop around the proximal phalanx, and triple loop around the proximal phalanx. The flexor digitorum superficialis tendon was split in half to be used with the fibrous rims of the previous pulley for pulley reconstruction in a portion of this experiment. The number of sutures was constant for specimens within groups.

Efficacy of Pulley Reconstruction on PIP Joint Function

Fifteen digits from fresh-frozen cadaver hands were used to study the effects of different types of pulley reconstruction on PIP joint function based on tendon excursion and joint rotation. Digits with intact pulleys were used as a normal reference and compared with digits in which individual pulleys were cut alone or in combination, and were left divided or repaired by pulley reconstruction. There were five combinations of pulley injuries studied. In Group I (comprising nine digits), the A3 pulley was excised leaving the A2 and A4 pulleys intact. In Group II (six digits), the A2 and A4 pulleys were cut but the A3 pulley was left intact. In Group III (fifteen digits), the remaining A2, A3, and A4 pulleys in the digits of Group I and Group II were excised, and the A2 and A4 pulleys were reconstructed. The method of reconstruction was type iii, a double or triple loop around the proximal and middle phalanges. In Group IV (the same digits as Group III), the volar plate[2] was used for A3 pulley reconstruction, but the A2 and A4 pulleys were not reconstructed. In Group V (eleven digits), the same digits were reevaluated, but the A2 and A4 pulleys were reconstructed as in Group III, and tin addition, belt loop reconstruction was used.

Four parameters were derived from the curves of tendon excursion and joint motion (Figure 4) and used for comparison. Tendon excursion was standardized to enable moving the PIP joint through an arc of 90° with an intact, uninjured tendon-pulley system. The range of movement of the joint in the operated fingers produced by this standardized excursion was called the *effective range of motion*.

Three different measures of excursion were made. The *absolute tendon excursion* was the excursion from full extension to 90° of flexion, as measured with the flexor tendon set at its normal length in the neutral position. After division of the pulley and as a result of bow-stringing, the tendon left at this length for the various injured and reconstructed sequences added a certain amount of slack to the tendon system, which needed to be taken up before any joint motion could occur. This amount of tendon slack was termed *"bow-stringing laxity."* By subtracting the bow-string laxity from the absolute tendon excursion, *"relative tendon excursion"* was derived. Maximum joint motion with unrestricted tendon excursion was also measured.

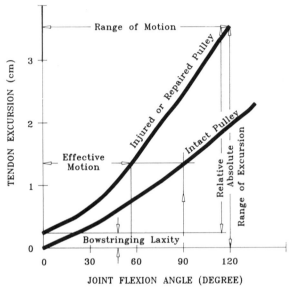

Figure 4. Definition of the parameters used for assessing tendon-excursion and joint displaced relationship. Used with permission of Churchill Livingstone (Figure 2 in Lin et al.[5]).

RESULTS

Mechanical properties of reconstructed pulleys

The maximum strength of the reconstructed pulleys using both fibrous rim and A3 belt loop techniques, was about half of that of the intact A2 pulley (Figure 5). The breaking strength of the loop-around-bone technique increased proportionately from 100 N to 520 N, with increasing numbers of loops. The stiffness of the reconstructed pulley systems (Figure 6) was low when compared to the normal A2 pulley. Although maximal strength and stiffness of the reconstructed pulleys were generally much less than normal, the energy to failure of the reconstructed pulleys was, except for the A3 belt loop and single loop-around-bone, greater than that of a normal intact A2 pulley (Figure 7).

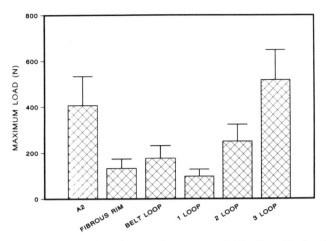

Figure 5. Maximum breaking load of the intact and reconstructed A2 pulley. Used with permission of Churchill Livingstone (Figure 9 in Lin et al.[5]).

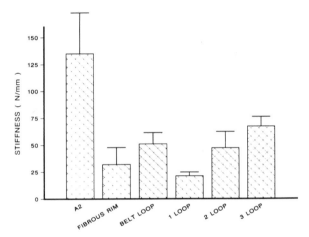

Figure 6. Stiffness of the intact and reconstructed A2 pulley. Used with permission of Churchill Livingstone (Figure 10 in Lin et al.[5]).

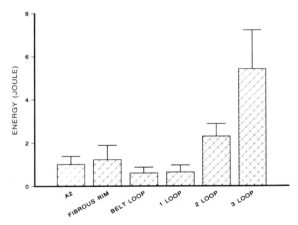

Figure 7. Energy absorption to failure of the intact and restructed A2 pulley. Used with permission of Churchill Livingstone (Figure 11 in Lin et al.[5]).

Efficacy of Pulley Reconstruction on PIP Joint Function

Little effect on the maximum range of PIP joint motion by pulley reconstruction was observed (Figure 8). The belt loop reconstruction did decrease this range of motion somewhat because the increased bulk anterior to the PIP joint tended to block flexion. The results for absolute tendon excursion are shown in Figure 9. Resection of the A3 pulley alone had little effect. However, if both A2 and A4 pulleys were divided, the absolute tendon excursion increased by 30% through the range of PIP joint motion. Reconstruction of the A2 and A4 pulleys still resulted in a requirement for 20% more excursion than normal. However, a reconstruction of the A3 pulley alone or in combination with A2 and A4 pulleys resulted in a requirement for tendon excursion than slightly less than or equal to normal to achieve maximum joint motion.

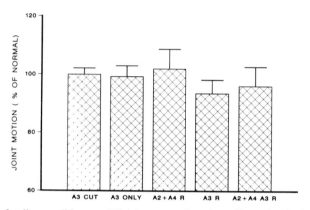

Figure 8. Effect of pulley conditions on PIP joint range of motion. Group 1, A3 alone was cut; Group 2, A2 and A4 were cut and only A3 was intact; Group 3, A2 and A4 were reconstructed; Group 4, A3 was reconstructed by belt loop; Group 5, All A2, A3, and A4 were reconstructed. Used with permission of Churchill Livingstone (Figure 3 in Lin et al.[5]).

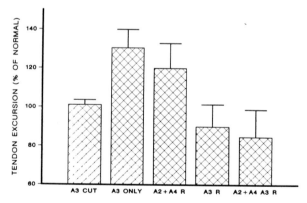

Figure 9. Effect of pulley conditions on PIP absolute tendon excursion. Used with permission of Churchill Livingstone (Figure 4 in Lin et al.[5]).

Disturbance of the normal pulley system altered the tendon excursion further by creating a bow-stringing effect. Division of the A2 and A4 pulleys caused the greatest laxity (Figure 10). The relative tendon excursions were calculated by subtracting this laxity from the absolute tendon excursion (Figure 11). For a given amount of available tendon excursion, the possible joint range of motion will depend upon the state of the pulleys (Figure 12). In this study, the effective joint motion for different pulley systems was calculated based on the amount of tendon excursions required to move the normal PIP joint through 90°. Laceration of the A2 and A4 pulleys increased bow-stringing, and thus, decreased greatly the effective motion. Reconstructed pulleys can restore an effective motion to at least 90% of normal.

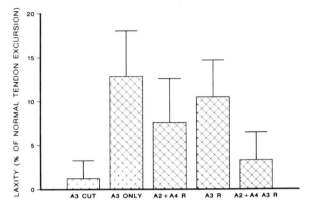

Figure 10. Effect of pulley conditions on PIP bow-stringing laxity. Used with permission of Churchill Livingstone (Figure 5 in Lin et al.[5]).

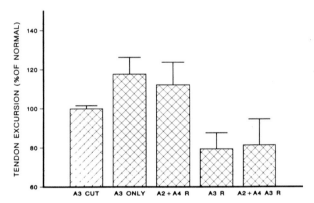

Figure 11. Effect of pulley conditions on PIP relative tendon excursion. Used with permission of Churchill Livingstone (Figure 6 in Lin et al.[5]).

DISCUSSION

Flexor tendon pulley reconstruction is a important challenge for the hand surgeon. Little comparative data is available on the potential strengths and weaknesses of the methods commonly employed. This study shows that reconstruction of both A2 and A4, or, alternating, reconstruction of the A3 pulley by the belt loop technique, is effective in restoring joint motion, but the A2/A4 reconstruction provided a more normal combination of tendon excursion and joint motion than either the belt loop alone or the belt loop combined with A2 and A4 reconstruction.

There are several factors which caused the differences in the tendon excursion of the reconstructed pulleys as compared with those of a normal intact system. The location of the reconstructed pulley will affect the moment arm and thus the excursion of the tendon. The stiffness of the reconstructed pulley will affect the constraint, and thus, will effect bow-stringing of the tendon. Physiologically, if the tendon-muscle structures are intact, any change in the absolute tendon excursion implies limitation of the active range of joint motion, both because of the quadriga effect,[10] where the flexor profundi share a common muscle origin and the necessary alteration of optimum length-tension relationships. Each of these can disturb normal finger function.

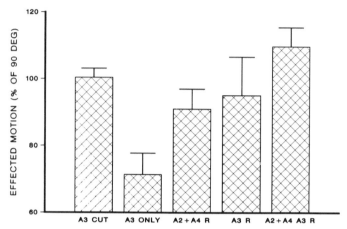

Figure 12. Effect of pulley conditions on PIP effective joint motion. Used with permission of Churchill Livingstone (Figure 7 in Lin et al.[5]).

Relatively low energy absorption and maximal breaking strength as compared to normal pulleys were observed for those reconstructions of repair at the fibrous rim and single loop pulley. Double, and particularly triple-loop reconstructions and belt loop pulleys have greater strength initially, but no method of pulley reconstruction using tendon provides normal characteristics of stiffness or displacement. The ability of several reconstructions to absorb more energy before failure than a normal A2 pulley is of questionable value, since the maximal load to failure remains quite low. The excess energy is absorbed by stretching out of the reconstruction, permitting tendon displacement (bow-stringing) and worsening the tendon excursion/joint motion relationship. *In vivo* studies will be necessary to test the properties of a healing pulley reconstruction over time. Pulley reconstruction in which there is a biomechanical feature of low stiffness may *"stretch out"* even if they do not rupture; the final effect in either case may be a failed pulley reconstruction.

Based on the data of tendon tension,[9] the initial strength and stiffness provided by double and triple loop pulley reconstruction may be sufficient to allow early protected active range of motion. Such techniques might be particularly appropriate when reconstructing pulleys around an active tendon prosthesis, a tendon graft for which early mobilization is desired, or when reconstructing pulleys following tenolysis. One should also keep in mind that the associated adhesion and friction between the tendon and the reconstructed pulley system are important parameters in the restoration of the functions of digits. This specific parameter should be used for the assessment of various techniques for pully reconstruction, as well.

ACKNOWLEDGEMENTS

This study was supported by National Institutes of Health Grant AR 17172.

REFERENCES

1. K.N. An, E.Y.S. Chao, W.P. Cooney, and R.L. Linscheid, Tendon excursion and moment arm of index finger muscles, *J. Biomech.* 16(6):419 (1983).
2. A. Karev, S. Stahl, and A. Taran, The mechanical efficiency of the pulley system in normal digits compared with a reconstructed system using the "belt loop" technique, *J. Hand Surg.,* 12A(4):596 (1987).
3. H.E. Kleinert, and J.B. Bennett, Digital pulley reconstruction employing the always present rim of the previous pulley, *J. Hand Surg.* 3:297 (1978).
4. G.T. Lin, P.C. Amadio, K.N. An, and W.P. Cooney, Functional anatomy of the human digital flexor pulley system, *J. Hand Surg.* 14A:949 (1989a).
5. G.T. Lin, P.C. Amadio, K.N. An, W.P. Cooney, and E.Y.S. Chao, Biomechanical analysis of finger flexor pulley reconstruction, *J. Hand Surg.* 14B:278 (1989b).
6. G.T. Lin, W.P.C. Cooney, P.C. Amadio, and K.N. An, Mechanical properties of human pulleys, *J. Hand Surg.* 15B:429 (1990).
7. I. Okutsu, S. Ninomiya, S. Hiraki, H. Inanami, and N. Kuroshima, Three-loop technique for A2 pulley reconstruction, *J. Hand Surg.* 12A:5(1):790 (1987).
8. L.H. Schneider, and J.M. Hunter, Flexor tendons - late reconstruction, in: "Operative Hand Surgery (2nd Edition)," D. P. Green, ed., Churchill Livingstone, New York, 3:2020 (1988).
9. F. Schuind, M. Garcia-Elias, W.P.C. Cooney, and K.N. An, Flexor tendon forces: *in vivo* measurements, *J. Hand Surg.* 17A:291 (1992).
10. C.E. Verdan, Syndrome of the quadriga, *Surg. Clin. of N. Am.* 40:425 (1960).
11. A. Weilby, Personal Communication, (Reference in paper of Kleinert and Bennett, 1978) (1973).

DEEP LIGAMENTS OF THE INTERPHALANGEAL JOINT OF THE THUMB

Ulrich Frank and Hans-Martin Schmidt

Department of Anatomy
University of Bonn
Nussallee 10
5300 Bonn 1, Germany

INTRODUCTION

Descriptions of ligamentous structures of the fingers go back to the 18th and 19th century.[1,11,31,32] At that time, the interest was focused on the collateral ligament and accessory collateral ligament of the metacarpophalangeal and interphalangeal joints of the fingers. Improved surgical procedures required a more detailed anatomical investigation. Landsmeer[15] drew attention to the retaining ligaments of the extensor mechanism which he called "retinacular ligaments" with "transverse" and "oblique" components. Further publications[10,18,28,30] confirmed Landsmeer's findings although Kaplan[13] pointed out that Weitbrecht had mentioned the same structure over 200 years before. In addition, in the 1950s two ligaments were described which did not appear to be in contact with capsular ligaments nor with the extensor or flexor tendon: Moerike[20] reported on a nail-halter which was fixed to the distal phalanx ; one year later, Flint[7] referred to a ligament he called "interosseous ligament" which is attached to the palmar aspect of the tuberosity of the distal phalanx. All these ligamentous structures may be summarized as deep ligaments of the fingers in contrast to the superficially located retaining skin ligaments mentioned by Cleland[4] and Grayson.[9] In the past 30 years, the deep ligaments of the fingers have been described in a number of comprehensive papers and monographs. However, little reference is made in these works to the ligamentous structures found at the interphalangeal joint of the thumb. Therefore, we present in this paper topographic data of this closely circumscribed region, based on dissections specifically performed for the study of the collateral ligament, accessory collateral ligament, nail-halter, proper phalangeal ligament and the oblique retinacular ligament. Furthermore, we analized the pulley system of the flexor tendon, dissecting the oblique pulley and the annular pulley A2. In addition, a thin ligament was demonstrated which Schmidt and Lanz[25] proposed to name "accessory retinacular ligament". It is part of the pulley system, spreading from the oblique pulley to the annular pulley A2 and located only at the ulnar border.

Advances in the Biomechanics of the Hand and Wrist
Edited by F. Schuind *et al.*, Plenum Press, New York, 1994

MATERIALS AND METHODS

This study is based on observations of 25 cadaver specimens of human thumbs. All of them originate from adults of both sexes and are preserved by a 10% formaldehyde solution. An off-centered longitudinal incision on the palmar and dorsal surface was performed and the skin carefully peeled off; thereby it was possible to avoid severing subcutaneous structures. Further dissection was carried out with a microscope with a magnification of 10 to 30. Applying the dividers, we got topographic data of the ligamentous structures which were analized in a statistical program. We determined the mean value, maximal value, minimal value and standart deviation. Finally, all ligaments were painted with superior coloured pencils in order to get more contrast in the continous-tone photographs.

RESULTS

Collateral ligament

The collateral ligament is nearly completely covered with thin fibers originating from three different structures (Figure 1). The most dorsally located fiber is spreading in the longitudinal axis from a small bony eminence at the distal third of the proximal phalanx to a bony attachment of the distal phalanx just below the insertion of the extensor tendon. It is inconstantly present and consists of 20 to 30 ligamentous bundles. The second structure continues the course of the oblique pulley; therefore, it can only be found at the radial border. These fibers are stronger than the more dorsally located ones and form a dense network with superficial fibers of the annular pulley A2; the latter builds up the third structure of these covering fibers. As shown in figure 4, the collateral ligament originates from a concave-shaped line at the head of the proximal phalanx. It crosses the interphalangeal joint with an angle of about 40° - depending on the joint position - and inserts into the convex-shaped line of the base of the distal phalanx. For an accurate measurement, we determined the proximal/distal length and the dorsal/palmar width (Figure 4); the results are shown in table 1.

115mm

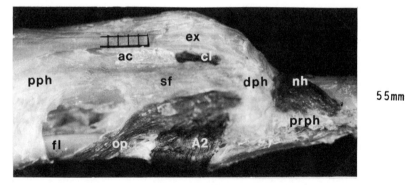

Figure 1. Radial aspect of a left thumb: ac: accessory collateral ligament; A2: annular pulley; cl: collateral ligament; dph: distal phalanx, ex: extensor tendon; fl: flexor tendon; nh: nail halter; op: oblique pulley; prph: proper phalangeal ligament; pph: proximal phalanx; sf: superficial fibers.

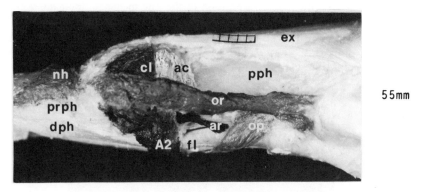

Figure 2. Ulnar aspect of a left thumb. ac: accessory collareal ligament; A2: annular pulley; ar: accessory retinacular ligament; cl: collateral ligament; dph: distal phalanx; ex: extensor tendon; fl: flexor tendon; nh: nail halter; op: oblique pulley; or: oblique retinacular ligament; prph: proper phalangeal ligament; ph: proximal phalanx.

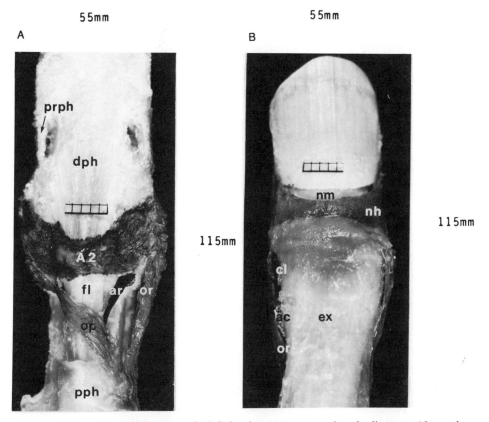

Figure 3. Figure 3-A. Palmar aspect of a left thumb: ar: accessory retinacular ligament; A2: annular pulley; dph: distal phalanx; fl: flexor tendon; op: oblique pulley; or: oblique retinacular ligament; prph: proper phalangeal ligament; pph: proximal phalanx. **Figure 3-B.** Dorsal aspect of a left thumb: ac: accessory collateral ligament; cl: collateral ligament; ex: extensor tendon; nh: nail-halter; nm: nail matrix; or: oblique retinacular ligament.

There is nearly no difference between the radial and ulnar collateral ligament. The distal length exceeds the proximal length by about 2 mm, the palmar width the dorsal width by 3 mm. So, in the neutral-0-position of the interphalangel joint the collateral ligament looks like an assymetrical trapezoid (Figure 5). The rectangular shape of the collateral ligament shown in figure 4 is caused by the hyperextension of the interphalangeal joint. Thus, it was possible to demonstrate the single fiber-bundles which compose the entire collateral ligament.

115mm

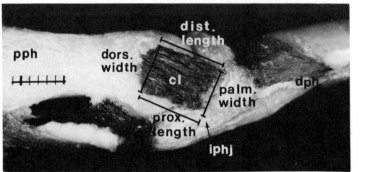

55mm

Figure 4. Detail: collateral ligament: cl: collateral ligament; dph: distal phalanx; iphj: interphalangeal joint; pph: proximal phalanx.

Table 1. Collateral ligament (all data in mm).

| | χ | | s | | χ_{max} | | χ_{min} | |
| | mean value | | standard deviation | | maximum | | minimum | |
	rad	uln	rad	uln	rad	uln	rad	uln
prox. length	7.6	7.5	1.2	1.1	10.0	10.0	6.0	6.0
dist. length	9.4	9.4	1.3	1.3	12.0	12.0	7.0	7.0
dors. width	3.2	2.9	0.7	0.5	5.5	4.0	2.0	2.0
palm. width	6.3	6.2	1.0	1.2	8.0	8.0	4.0	3.0

Accessory collateral ligament

The collateral ligament originates from the head of the proximal phalanx. As shown in figure 5, the fibers are diverging to the palmar side and get in contact with the annular pulley A2 and the palmar plate. The more proximally located part of the accessory collateral ligament was removed in order to make the free border of the palmar plate visible. The proximal/distal length and the dorsal/palmar width were

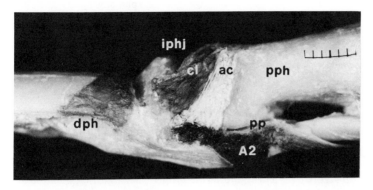

115mm

55mm

Figure 5. Detail: accessory collateral ligament: ac: acccessory collateral ligament; A2: annular pulley; cl: collateral ligament; dph: distal phalanx; iphj: interphalangeal joint; pp: palmar plate; pph: proximal phalanx.

determined in the same way as mentioned above (Figure 4). The data are shown in table 2.

As in the collateral ligament, there is nearly no difference between the radial and ulnar accessory collateral ligament. The distal length exceeds the proximal length by 1mm, the palmar width the dorsal width by 1.5-2.0 mm.

Table 2. Accessory collateral ligament (all data in mm).

	χ		s		χ_{max}		χ_{min}	
	rad	uln	rad	uln	rad	uln	rad	uln
prox. length	6.9	7.0	1.1	1.1	9.0	9.0	5.0	5.0
dist. length	7.9	8.0	1.3	1.2	10.5	10.5	6.0	6.0
dors. width	2.9	2.9	0.4	0.4	3.5	4.0	2.0	2.0
palm. width	4.7	4.6	0.7	0.7	6.5	6.5	4.0	3.5

Nail-halter

The total expansion of this dorsally located ligament is demonstrated in figure 3-B. It is only covered with a skin layer and therefore has to be prepared very carefully. The dumbbell-shape of the nail-halter is characterized by a proximal/distal and radial/ulnar border. The proximal border is attached to the bone over the entire width of the proximal third of the distal phalanx. As we were able to show, the radial and ulnar border insert into the proper phalangeal ligament (Figure 6). The distal border is turned inwards and builds up a pouch for the nail matrix. The topographic data are presented in table 3. The ulnar and radial width were measured at the attachment to the proper phalangeal ligament, the middle width in the center of the nail-halter.

Table 3. Nail-halter (all data in mm).

	χ	s	χ_{max}	χ_{min}
length	16.2	1.9	20.0	13.0
rad. width	8.2	0.9	10.0	7.0
uln. width	8.1	1.2	10.0	6.0
mid. width	2.3	0.4	3.0	1.5

Proper phalangeal ligament

As shown in figure 3-A and figure 6, the proper phalangeal ligament is situated at the palmar aspect of the distal phalanx. By means of the magnification in figure 6, it can be recognized that the proper phalangeal ligament is spreading from the basal tubercle to the tuberosity of the distal phalanx at the radial and ulnar border. Normally, blood vessels for the arterial supply to the nail bed run through the oval foramen between the ligament and the body of the distal phalanx, but they were removed in this dissection for a better demonstration of the ligament. The data of the proper phalangeal ligament of the radial and ulnar border are presented in table 4. There is no difference between the radial and ulnar proper phalangeal ligament. The range is remarkable : the smallest proper phalangeal ligament is only half the size of the biggest one.

Table 4. Proper phalangeal ligament (all data in mm).

	χ		s		χ_{max}		χ_{min}	
	rad	uln	rad	uln	rad	uln	rad	uln
length	6.6	6.6	1.0	1.0	8.0	8.0	4.5	4.5
width	1.3	1.3	0.3	0.3	2.0	2.0	1.0	1.0

Oblique retinacular ligament

In contrast to all the ligaments mentioned above, we were able to demonstrate the oblique retinacular ligament exclusively at the ulnar side, but it was absent in 16% of our specimens. As depicted in figure 2, it originates from the lateral wedge of the base of the proximal phalanx, runs parallel to the flexor tendon and inserts into the tubercle of the base of the distal phalanx. At the attachment zone of the extensor tendon, the more dorsally situated fibers of the oblique retinacular ligament are connected with fibers of the extensor tendon. The data of the ligament are presented in table 5.

Annular pulley A2

As shown in figure 3-A, the pulley is entirely transverse in orientation. It is located about 15 mm proximally of the insertion of the flexor tendon and is centered above the palmar plate of the interphalangeal joint. In table 6, the data of this ligament are demonstrated.

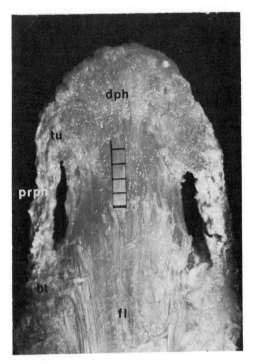

80mm

115mm

Figure 6. Detail: proper phalangeal ligament: bt: basal tubercle; dph: distal phalanx; fl: flexor tendon; prph: proper phalangeal ligament; tu: tuberosity.

Table 5. Oblique retinacular ligament (all data in mm).

	χ	S	χ_{max}	χ_{min}
length	23.6	2.3	27.0	20.0
width	2.3	0.6	3.0	1.0

Oblique pulley

The oblique pulley, shown in figure 2 and figure 3-A, originates from the ulnar border of the base of the proximal phalanx. It passes obliquely in a distal direction and inserts into the radial border of the head of the proximal phalanx closely to the interphalangeal joint. Some of its fibers continue the course and cover the accessory collateral ligament and the collateral ligament, as mentioned above. The length and width can be seen in table 7.

Accessory retinacular ligament

As shown in figure 7, this ligament is spreading longitudinally from the origin of the oblique pulley to the proximal border of the annular pulley A2. Therefore, it can only be dissected at the ulnar side. With an average length of 6.3 mm and a width of

Table 6. Annular pulley (all data in mm).

	χ	S	χ_{max}	χ_{min}
length	12.8	2.5	18.0	9.0
width	6.4	1.6	10.0	4.0

Table 7. Oblique pulley (all data in mm).

	χ	S	χ_{max}	χ_{min}
length	16.2	3.0	21.5	11.0
width	3.0	0.8	4.5	2.0

Table 8. Accessory retinacular ligament (all data in mm).

	χ	S	χ_{max}	χ_{min}
length	6.3	1.2	8.5	4.0
width	1.3	0.4	2.0	1.0

1.3 mm the accessory retinacular ligament can be easily overlooked. The entire statistical data are presented in table 8.

DISCUSSION

Collateral ligament

With regard to the reflection upon the collateral ligament, two problems must be considered: first, the nomenclature of this ligament was controversial until recently. Second, the interest was focused on the collateral ligament of the finger joints, particulary of the metacarpophalangeal joint. The joints of the thumb have not attracted a lot of attention. In table 9, we give a general survey of the different nomenclature used for the description of the collateral ligament in the past 150 years.

In the period from 1840 to the edition of the Baseler Nomina Anatomica (BNA, 1895) the term "lateral ligament" was used in general which was then slightly modified to "collateral ligament". Landsmeer[15] proposed the name "oblique lateral ligament" to accentuate the oblique course, but this recommendation was not accepted. Bock[2] mentioned that the ligamentum laterale is spreading from a rough lateral impression of the distal part of a phalanx to the lateral tubercle of the proximal part of the following phalanx. This arrangement corresponds to our findings. As far as we know, the only detailed anatomical description of the collateral ligament found at the interphalangeal joint of the thumb was presented by Pahnke.[22] Unfortunately, we could not confirm some of his findings. After having dissected 40 thumbs, he specified that the average proximal length of the radial collateral ligament is 12.2 mm. In contrast, we were able to show that the average proximal length is 7.6 mm (Table 1). In consideration of the fact that the total length of the proximal phalanx is about 30 mm,[25] it is difficult to understand that a closely adjacent region like the distal third of the proximal phalanx and the proximal third of the distal phalanx should be crossed by a ligament of a length as 12.2 mm. On the other hand, the comparison of the palmar width of the radial collateral ligament demonstrates identical findings in both investigations: the average width is 6.3 mm.

115mm

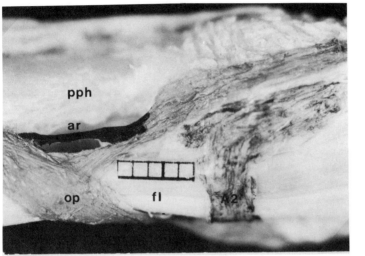

80mm

Figure 7. Detail: accessory retinacular ligament: ar: accessory retinacular ligament; A2: annular pulley; fl: flexor tendon; op: oblique pulley; pph: proximal phalanx.

Accessory collateral ligament

Fick[6] pointed out that there are usually a little bit proximally to the collateral ligament some fiber bundles, which are not glittering and generally separated from the collateral ligament by a fissure. He called them "ligamentum capituli arciforme volare" and considered this ligament to be a pecularity of the metacarpophalangeal joint. Surprisingly, a lot of papers confirm Fick's finding that the ligament can only be demonstrated at the metacarpophalangeal joint. Table 10 lists the authors and the different nomenclature. Moberg[19] maintained that "the JNA (Jenaer Nomina Anatomica; 1935) name for Fick's ligamentum capituli arciforme volare is ligamentum collaterale accessorium". This is not true. Neither the JNA (1936) nor the six editions of the PNA (Pariser Nomina Anatomica; 1955-1989) referred to this ligament. As for the collateral ligament, Pahnke[22] was the only person to describe the accessory collateral ligament of the interphalangeal joint of the thumb. According to his measurements, the average proximal length of the radial accessory collateral ligament is 8.1 mm, the average palmar width 6.6 mm. We cannot confirm his findings: in our dissection, the average proximal length is 6.9 mm, the average palmar width 4.7 mm (Table 2). The difference is difficult to explain; the scale bars integrated in Pahnke's and our figures (Figures 4 and 5) seem to establish our findings but the quality of Pahnke's continous-tone photographs does not make a definite conclusion possible.

Table 9. Nomenclature of the collateral ligament.

Bock (1840)	Lig. laterale s. accessorium phalangum
Arnold (1845)	Lig. laterale digitorum
Henle (1856)	Lig. accessorium rad. et uln.
Luschka (1865)	Lig. laterale
Quain/Hoffmann (1870)	Lig. accessiorium rad. et uln.
Gegenbauer (1883)	Lig. laterale
BNA (1895)	Lig. collaterale
Fick (1905)	Lig. collaterale
Landsmeer (1969)	oblique lateral ligament
PNA (1989)	collateral ligament

Table 10. Nomenclature of the accessory collateral ligament.

Fick (1904)	Lig. capituli arciforme volare
Rouvière (1943)	Fascieau glénoidien
Haines (1951)	collateral ligament of palmar pad
Moberg and Stener (1954)	accessory collateral ligament
Landsmeer (1955)	metacarpo-glenoid ligament
Simmers and de la Caffinière (1981)	accessory collateral ligament

Phalangoglenoid ligament

Haines[10] mentioned that "another fibrous band (= deep fibrous band) is sometimes found passing to the phalanx between the hood ligament and the deep insertion of the interosseous". In his paper, he presented a drawing of this ligament: it runs obliquely from the palmar plate of the metacarpophalangeal joint to the lateral border of the base of the proximal phalanx. Hakstian and Tubiana[11] considered this ligament to be a part of the capsular complex; therefore, they proposed the name "pars phalangoglenoidalis". Pahnke[22] was the first to present topographic data of the phalangoglenoid ligament of all finger joints. For the interphalangeal joint of the thumb, he described a phalangoglenoid ligament at the radial and ulnar border. In contrast to his findings we were able to show that the "phalangoglenoid ligament" of the interphalangeal joint is built up by superficially located fibers which cover the collateral ligament and continue the course of the oblique pulley at the radial border.

Nail-halter

Moerike[20,21] referred to ligamentous bundles of the corium running over the proximal end of the nail matrix to the lateral tubercles of the distal phalanx. Demonstrating some cross-sections he was in the position to state that these bundles pass above and below the matrix epithel, thus building up a longitudinally streched, narrow pouch. He held this ligament responsible for the pushing of cells from proximal to distal in a longitudinal direction, a phenomenon which is in contrast to the normally observed cell supply proceeding in a vertical direction from basal to superficial. The nail-halter was very often missed at the distal phalanx of the little toe; in all cases the nail was growing vertically. Moericke found the nail-halter at fingers and toes of humans but also at claws of cats, martens, hedgehogs, hens and lions. Thirty-six years later, Schmidt[24] was the first to present new anatomical details of the nail-halter. Performing his investigations at fingers and toes of humans, he confirmed the bony attachment zone of the proximal border of the nail-halter to the distal phalanx. In contrast, he pointed out that the ulnar and radial borders of the nail-halter insert into the phalangoglenoid ligament, annular pulley A5 and the proper phalangeal ligament. He stated the nail-halter is totally incorporated by this additional insertion into the functional unit of the distal interphalangeal joints. For the thumb, we were able to confirm Schmidt's findings taking into account the different ligamentous arrangements at the interphalangeal joint of the thumb. So, as shown in figure 8, the radial and ulnar borders of the nail-halter insert exclusively into the proper phalangeal ligament. In figure 9, the nail matrix was removed and thus it was possible to inspect the narrow cleft of the nail-halter built up by the fibers which run over the nail matrix.

Proper phalangeal ligament

In the 18th century, Weitbrecht[32] mentioned to various ligaments running to small bony projections jutting from the sides of both ends of the distal phalanges. Brooks[3] noted the ligament briefly. The most detailed description of this ligament is presented by Flint,[7] who considered himself erroneously as the explorer and proposed to name it "interosseous ligament". This term is not correct, since the ligament is not spreading between two adjacent bones as is expressed by "interosseous". For this reason Schmidt und Lanz[25] favored the term "proper phalangeal ligament". Soon,[27] who did not cite the findings of Flint in his paper, recommended the name "paraterminal ligament". Flint[7] described that "the ligament extends from the base of the terminal phalanx to the cap of bone at the distal pole. At the base of the terminal phalanx the ligament is continous with the collateral ligament of the terminal interphalangeal joint". The latter statement is qualified by Stack[28] who states that "the ligament is in part continous proximally with the collateral ligament". We agree with Stack that only single fibers of the collateral ligament are directly connected with fibers of the proper phalangeal ligament. We think that the proper phalangeal ligament has two functions: first, it builds up a fibro-osseous compartment together with the distal phalanx for the protection of dorsally passing vessels and nerves as described by Flint. Second, it represents the attachment zone of the radial and ulnar border of the nail-halter and participates indirectly in the fixation of the nail matrix.

Oblique retinacular ligament

The course and function of the oblique retinacular ligament dissected at the fingers was presented in a number of excellent publications.[10,15,18,29,30] The fibers arise from the distal fourth of the proximal phalanx, cross the proximal interphalangeal joint and join the lateral margin of the extensor tendon. The only publication considering the oblique retinacular ligament of the thumb was written by Milford.[18] He is not sure about the existence of the ligament and states that "in the thumb there seem to be transverse and oblique retinacular ligaments". We missed a description of the total number of oblique retinacular ligaments he prepared. Milford pointed out that "a small tendineous band originates from a discrete muscle bundle deep within the muscle mass of the abductor pollicis brevis. It is also seen in the adductor pollicis. The oblique retinacular ligament passes lateral over the metacarpophalangeal joint...".

80mm

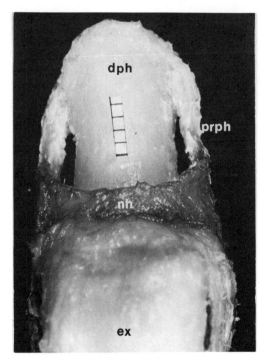

110mm

Figure 8. Detail: attachment zone of the nail-halter: dph: distal phalanx; ex: extensor tendon; nh: nail-halter; prph: proper phalangeal ligament.

We disagree with this description. As shown in figure 10, the oblique retinacular ligament originates distally from the metacarpophalangeal joint from the lateral border of the proximal phalanx. Some fibers are in contact with insertion fibers of the adductor pollicis, but most of them have a bony origin. Furthermore, Milford explained "that the oblique retinacular ligament inserts, along with the tendon, into the distal phalanx. The fibers of this band are indistinguishable from those of the tendon at their insertion". As presented in figure 11, we found a connection between the oblique retinacular ligament and the extensor tendon respectively the nail-halter, but we think the fibers are distinguishable in every case. To our knowledge, we are the first to present detailed topographic data of the oblique retinacular ligament of the thumb.

Oblique pulley, annular pulley A2, accessory retinacular ligament

Doyle and Blythe[5] presented a study of the pulley system of the thumb. Although they described the course of the oblique pulley and the annular pulley A2 in the same way as we do, their topographic data are totally different: according to their study, the width of the oblique pulley is 9-11 mm, the width of the annular pulley 8-10 mm. In

90mm

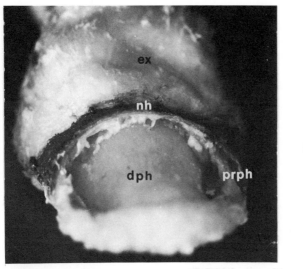

85mm

Figure 9. Detail: pouch of the nail-halter: dph: distal phalanx; ex: extensor tendon; nh: nail-halter; prph: proper phalangeal ligament.

contrast, we determined the average width of the oblique pulley with 3.0 mm, the average width of the annular pulley A2 with 6.4 mm. Unfortunately, Doyle and Blythe did not refer to the mode of technique they used for the measurements. Knott and Schmidt[14] presented data of the pulley system of the fingers and the thumbs: in this study, the width of the annular pulley A2 measures 6.73 mm; data of the oblique pulley were not published.

Doyle and Blythe stated that "in the fingers, two basic configurations are seen - annular and cruciform. In the thumb, annular and oblique pulleys are found". In our dissection, we were able to demonstrate in 28% of specimens fibers originating from the oblique pulley but spreading perpendicularly to the course of the oblique pulley. In figure 12, the most impressive "cruciform oblique pulley" is shown. Furthermore, a synovial cyst can be seen at the radial border of the oblique pulley. As far as we know, the accessory retinacular ligament was first described by Schmidt and Lanz.[25] It may be considered as a "link ligament" between the oblique pulley and the annular ligament A2, reinforcing the ulnar border of the flexor tendon sheath. Conditioned by the course of the oblique pulley, the ulnar space between the two pullies could be considered as a "locus minoris resistentiae" which is strengthened by the accessory retinacular ligament.

CONCLUSIONS

Our investigation of deep ligaments of the interphalangeal joint of the thumb establishes the contrast to the ligamentous systems of the finger joints. First, we were able to demonstrate the absence of the phalangoglenoid ligament; fiber bundles originating from three different structures cover the collateral ligament and the accessory collateral ligament superficially. Second, the radial and ulnar border of the nail-halter inserts exclusively into the proper phalangeal ligament; at the fingers, there is an additional connection to the phalangoglenoid ligament and the annular pulley A5. Third, the origin and attachment zone of the oblique retinacular ligament differ remarkably from that of the fingers: the fibers originate distally, not proximally from the metacarpophalangeal joint particularly from the lateral border of the proximal phalanx and attach to the lateral tubercle of the distal phalanx and to the extensor tendon. In the fingers, the ligamentous bundles are indistinguishably connected with the lateral border of the extensor tendon; a bony attachment zone is not described.

120mm

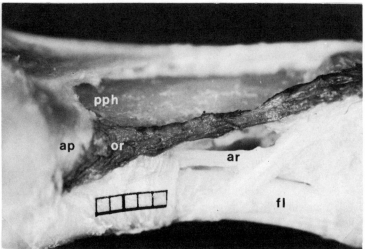

80mm

Figure 10. Detail: origin of the oblique retinacular ligament: ar: accessory retinacular ligament; ap: adductor pollicis (insertion); fl: flexor tendon; or: oblique retinacular ligament; pph: phalanx proximalis.

Fourth, we demonstrated in 28% of our specimens a cruciform shape of the oblique pulley. The additional fibers originate from the oblique pulley and run perpendicularly to the course of the ligament to the ulnar border. Finally, we are the first to present topographic data of a thin ligament which Schmidt and Lanz[25] proposed to name accessory retinacular ligament. It is part of the pulley system and can only be dissected at the thumb. For these pecularities of the deep ligaments of the interphalangeal joint of the thumb we consider that a profound knowledge of these structures is essential to perform surgical procedures successfully. Further investigations are necessary, especially of the functional significance of the oblique retinacular ligament of the thumb.

120mm

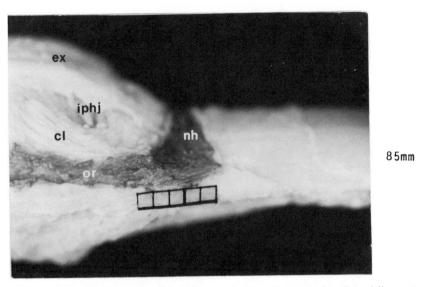

85mm

Figure 11. Detail: attachment zone of the oblique retinacular ligament: cl: collateral ligament; ex: extensor tendon; iphj: interphalangeal joint; nh: nail-halter; or: oblique retinacular ligament.

85mm

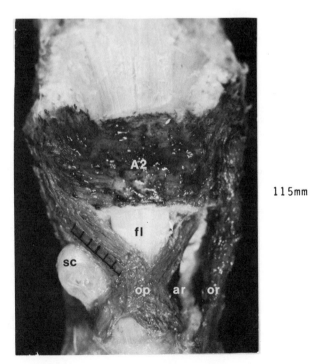

115mm

Figure 12. Detail: pulley-system of the flexor tendon: ar: accessory retinacular ligament; A2: annular pulley; fl: flexor tendon; op: oblique pulley; or: oblique retinacular ligament; sc: synovial cyst.

REFERENCES

1. F. Arnold, Handbuch der Anatomie des Menschen, Freiburg im Breisgau, *I. Band* 221, (1845).
2. C.E. Bock, Handbuch der Anatomie des Menschen, Leipzig, *I. Band* 222, (1840).
3. H.St.J. Brooks, On the distribution of the cutaneous nerves on the dorsum of the hand, *Trans. R. Acad. Med. Ireland*, 6:463 (1888).
4. J. Cleland, On the cutaneous ligaments of the phalanges, *J. Anat.Physiol.*, 12:526 (1878).
5. J.R. Doyle, and W.F. Blythe, Anatomy of the flexor tendon sheath and pulleys of the thumb, *J. Hand Surg.*, 2:149 (1977).
6. R. Fick, Handbuch der Anatomie und Mechanik der Gelenke, Jena, *I. Teil*, 278 (1904).
7. M.H. Flint, Some observation of the vascular supply of the nail bed and terminal segments of the fingers, *Brit. J. Plast. Surg.*, 8:186 (1956).
8. C. Gegenbauer, Lehrbuch der Anatomie des Menschen, Leipzig, 256 (1883).
9. J. Grayson, On the cutaneous ligaments of the phalanges, *J. Anat.*, 75:164 (1941).
10. R.W. Haines, The extensor apparatus of the fingers, *J. Anat.*, 85:251 (1951).
11. R.W. Hakstian, and R. Tubiana, Ulnar deviation of the fingers; the role of structure and function, *J. Bone Joint. Surg.*, 49A:299 (1967).
11. J. Henle, Handbuch der systematischen Anatomie des Menschen, Braunschweig, *I. Band, Zweite Abtheilung*, 108 (1856).
12. C.E.E. Hoffmann, Quains Lehrbuch der Anatomie, Erlangen, *I. Band.*, 197 (1870).
13. E.B. Kaplan, Functional and surgical anatomy of the hand, 2nd edition, J.B.Lippincott, Philadelphia (1965).
14. Ch. Knott, and H.M. Schmidt, Die bindegewebigen Verstärkungseinrichtungen der digitalen Sehnenscheiden an der menschlichen Hand, *Gegenbauers morph. Jahrbuch,*, 132:1 (1986).
15. J.M.F. Landsmeer, The anatomy of the dorsal aponeurosis of the human fingers and its functional significance, ...
16. J.M.F. Landsmeer, Anatomical and functional investigations on the articulation of the human fingers, *Acta Anat.*, 25: 1 (1955).
17. J.M.F. Landsmeer, Observation of the joints of the human finger, *Ann. Rheum. Dis.*, 28:11 (1969).
18. L. Milford, Retaining ligaments of the digits of the hand, *Saunders* (1968).
19. E. Moberg, and B. Stener, Injuries of the ligaments of the thumb and fingers, *Acta Chir. Scan.* 106:166 (1954).
20. K.D. Moerike, Unsere Finger-und Zehennägel, *Aus der Heimat*, 63:151 (1955).
21. K.D. Moerike, Ein bindegewebiges Halfter um das Matrixepithel des Nagels und der Kralle, *Z. Anat. Entwickl.-Gesch.*, 119:23 (1955).
22. J. Pahnke, Über die Articulationes metacarpophalangeales und interphalangeales der menschlichen Hand, Würzburg, *Med. Inaug-Diss.*, (1987).
23. H. Rouvière, Anatomie humaine, Tome 2e, Masson, Paris (1943).
24. H.M. Schmidt, et al., Das Moerike-Halfter: die bindegewebige Fixierung des Epithels der Nagelmatrix, *Verh. Anat. Ges.* 86. Vers. (Ergh.Anat. Anz. 172):286 (1991).
25. H.M. Schmidt, und U. Lanz, Chirurgische Anatomie der Hand, Hippokrates-Verlag, Stuttgart (1992).
26. B.P. Simmons, and J.Y. de la Caffinière, Physiology of flexion of the finger, In Tubiana: *The Hand*, Saunders (1981).
27. P.S.H. Soon, et al., Paraterminal ligaments of the distal phalanx, *Acta Anat.*, 142: 339 (1991).
28. H.G. Stack, Some details of the anatomy of the terminal segment of the finger, *Acta Orthop. Belg.* 24:113 (1958).
29. H.G. Stack, Muscle function of the fingers, *J. Bone Joint Surg.*, 44B:899 (1962).
30. R. Tubiana, and P. Valentin, The anatomy of the extensor apparatus of the fingers, *Surg. Clin. N. Amer.*, 44:897 (1964).
31. H. von Luschka, Die Anatomie des Menschen, Dritter Band, Tübingen, *I. Abteilung,* 145 (1865).
32. J. Weitbrecht, Syndesmologia sive historia ligamentorum corporis humani, quam secundum observationes anatomicas concinnavit, et figuria et objecta recentia adumbratia illustravit, Petropoli; ex typographica *Academiae Scientiarium* Anno MDCCXLII.

ANATOMY OF THE PALMAR PLATES: COMPARATIVE MORPHOLOGY

AND MOTION CHARACTERISTICS

Sylvain Gagnon, Michael J. Botte, Ephraim M. Zinberg,
and Steven N. Copp

Hand and Microvascular Surgery Service
Department of Orthopaedics and Rehabilitation
8-894 UCSD Medical Center
225 Dickinson Street
San Diego, CA 92101, USA

The metacarpophalangeal (MCP) joints and the proximal interphalangeal (PIP) joints have different tolerances for immobilization in a flexed position. Clinically the MCP joints usually tolerate immobilization in flexion with less loss of motion than the PIP joints (with joint contracture probably related to soft tissue stiffness involving the palmar plates, collateral ligaments, and joint capsule). The palmar plates and associated check-rein ligaments have been implicated in the formation of contracture of the PIP joint.[13,14] Although the anatomy of the PIP joint palmar plates has been studied,[1,4,9,13] little data is available concerning the anatomy of the MCP joint palmar plates and the anatomical differences between the palmar plates of these two joints. Watson[14] has hypothesized that certain differences in the anatomy and motion characteristics of the palmar plates of these joints explain their different degrees of tolerance to immobilization in flexion. The purpose of this study was to evaluate this hypothesis by examining in greater detail both anatomic differences between these palmar plates and differences in motion characteristics.

MATERIALS AND METHODS

Forty-eight MCP joints and 48 PIP joints were dissected from the index, long, ring and little finger of 12 fresh cadavers using 3.5x magnification. A digital micrometer was used to measure the lengths of the palmar plates in extension and 90 degrees of flexion (N=96). The change in length of the palmar plate when the joint was passively flexed from 0 degrees to 90 degrees of flexion, was termed the shortening of the palmar plate. Half the MCP joint palmar plates (N=24) and half the PIP joint palmar plates (N=24) were examined for gross morphologic characteristics and surrounding attachments. The palmar plate thicknesses were measured at the central portion.

Light microscopy (LM) with Haematoxylin and Eosin stain of long finger MCP palmar plates (N=6) and middle finger PIP palmar plates (N=6) were obtained for qualitative evaluation of fiber pattern. Surface scanning electron microscopy (SEM) with a 360 Cambridge Scanning Electron microscope, was performed on MCP joint and PIP joint palmar plates from 2 long fingers and 1 index finger (N=3) of two fresh cadavers (death occurring with 18 hours).

The other half of the specimens (24 MCP joint palmar plates and 24 PIP joint palmar plates) were studied in situ radiographically. A radiopaque marker (hemoclip) was placed on the proximal edge of each palmar plate. Lateral radiographs at a fixed distance were taken of each ray at zero degrees and 90 degrees of flexion. The distance traversed by the marker relative to the joint was measured in millimeters (mm), and was termed the migration. These data were adjusted for metacarpal length discrepancy.

Data concerning shortening and migration were reported as mean values as well as percentage of initial length. Statistical analysis was made using the pooled t-test. Individual ray data were analyzed but because of sample size, intra-hand and inter-hand results were not significant and data will be reported for all fingers assuming that they were comparable because of length standardization.

RESULTS

Gross Morphology

Compared to the PIP joints, the MCP joint palmar plates were less well-defined with blending attachments to multiple surrounding structures. The attached structures to the MCP palmar plates included the flexor tendon sheath palmarly, the transverse metacarpal ligament, lumbrical and interosseous muscle fascia laterally, and collateral ligaments and periosteum dorsally. In ten hands (83% of specimens), a sesamoid bone was present within the MCP palmar plate, located on either the radial side of the index and/or the ulnar side of the little finger palmar plates. In contrast, the PIP palmar plates had well-defined borders.

The surface of the MCP joint palmar plate appeared to consist of irregular transverse fibers. Compared to the smoother, more regular surface of the PIP joint palmar plate, with well defined borders, the proximal border of the MCP joint palmar plate was not usually distinguishable because of loose areolar tissue surroundings. However, the proximal border was identifiable by palpation with a probe. The deep or the dorsal aspect of the MCP palmar plate consisted of thin capsular tissue inserting near the articular border of the metacarpal head. In flexion, a very small recess or pouch was formed proximally by this loose tissue. The MCP palmar plate inserted distally on the palmar border of the proximal phalanx near the articular surface.

Length and thickness

The mean MCP joint palmar plate length was twice the length of the PIP joint palmar plate (p<0.01 - Table 1). The mean thickness of the central portion of the MCP joint palmar plate was 0.3mm thinner than the PIP joint palmar plate (p<0.01 - Table 1).

Table 1. Comparative Data of MCP and PIP Palmar Plate

Palmar Plate[+]

Measurements:	Length	N=48	MCP	11.2	(SD 1.62)*
Data in Extension	(mm)		PIP	5.6	(SD 1.35)
	Thickness	N= 6	MCP	1.34	(SD 0.34)*
	(mm)		PIP	1.65	(SD 0.39)
Motion Data at 90	Shortening	N=48	MCP	33.80	(SD 6.80)*
Degrees of Flexion	(% initial length)		PIP	26.60	(SD 10.20)
	Migration	N=24	MCP	7.85	(SD 1.72)**
	(adjusted for		PIP	6.39	(SD 1.37)
	metacarpal length)				
	Migration	N=24	MCP	79.00	(SD 18.00)*
	(% initial length)		PIP	139.00	(SD 24.00)

[+] Mean value for all digits

Statistically Significant Difference with PIP
*   P < 0.01
**   P < 0.05

mm = millimeters
N   = number of specimens measured
SD   = standard deviation

Palmar plate shortening with flexion

With passive joint flexion, the MCP joint palmar plate shortened to 66.2% of its initial length (33.8% shortening); the PIP joint palmar plate shortened to 73.4% of its initial length (26.6% shortening). A significant difference was found with the Student-t test ($p<0.01$) between the shortening of the two palmar plates during flexion to 90 degrees (Figure 1).

LM and SEM

On all specimens, with both light and scanning electron microscopic examination, the MCP joint palmar plate appeared to consist of loose connective tissue arranged in disorganized strands (Figures 2A, 3A, 4A). The PIP joint palmar plates appeared to consist of more dense, homogeneous connective tissue (Figures 2B, 3B, 4B). There was no significant difference in cellularity (sizes and numbers) between the two tissues; both appeared relatively avascular.

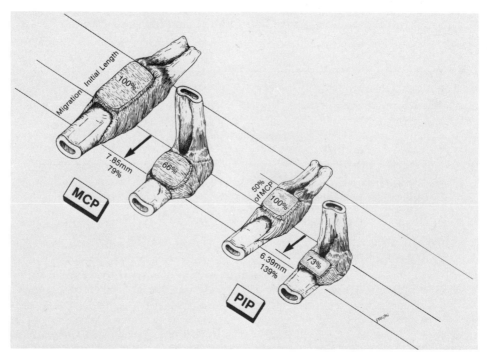

Figure 1. Illustration showing palmar plate shortening and migration at the MCP and PIP joint. The MCP joint palmar plate shortened to 66 % of initial length (33 % shortening) and migrated 7.85 mm or 79 % of its initial length. The PIP joint palmar plate shortened to 73 % of initial length (27 % shortening) and migrated 6.39 mm or 139 % of its initial length.

Plate Migration characteristics with flexion

The absolute distance of migration of the radiographic markers of the two joints (corrected for magnification and for metacarpal length) differed by only 1.5mm ($p < 0.05$). However, considering the initial size discrepancy of the respective palmar plates, the MCP joint palmar plate migrated a mean of 7.85 mm or 79% of its initial length whereas the PIP joint palmar plate migrated a mean of 6.39 mm or 139% of its initial length (Figure 1 and Table 1) which represented a significant difference ($p < 0.01$).

DISCUSSION

Clinically, the MCP joints seem to tolerate immobilization in flexion better than the PIP joints with less chance of forming contractures. This is substantiated by the recommendations for hand immobilization with the MCP joint in up to 70 degrees of flexion and the PIP joint in only 10 degrees of flexion.[5,6] When severely contracted, the PIP joint palmar plate often requires surgical release of its proximal border or neighboring attachments.[3,14] In contrast, few discussions address the surgical release of the palmar plates of MCP joints. When flexion contractures of this joint do occur, treatment is often accomplished by intrinsic or extrinsic muscle release or lengthening, or partial palmar fasciectomy.[7,8,10,11,12]

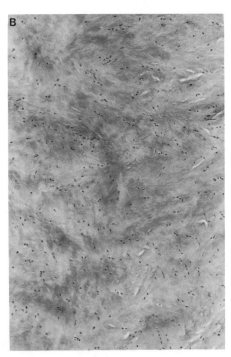

Figure 2. Figure 2A. H & E stain of MCP joint palmar plates with 25x magnification showing loose connective tissue arranges in strands. **Figure 2B.** H & E stain of PIP joint palmar plates with 25x magnification showing dense, homogenous connective tissue.

Though differences exist in collateral ligaments and osseous configuration of the joints, anatomic differences between the palmar plates of the MCP and PIP joints have been postulated to account, in part, for their differences in tolerance to immobilization flexion.[1,13,14] The MCP palmar plate has been hypothesized to be more elastic, with a "crisscross" fiber pattern that allows it to collapse (or shorten) with flexion and "open" on extension.[13,14] Watson has implied that the MCP palmar plate can shorten up to 40% during flexion, compared to only 9% shortening of the PIP joint palmar plate, which is composed of a more dense, less collapsible connective tissue.[14] In addition, the fibrous bands, that extend from the proximal phalanx to the proximal edge of the PIP joint palmar plate, thicken during prolonged flexion to form check-rein ligaments that contribute to fixed joint contracture.[1,13,14]

Our findings seem to be in agreement with the discussion of Watson, and support the collapsibility and elastic nature of the PIP joint palmar plates compared to the PIP joints. Although we were not able to demonstrate the "crisscross" fiber pattern of the MCP joint palmar plates compared to the PIP palmar plate, the histological and electron microscopic findings of our study do demonstrate that the MCP palmar plate is composed of more loosely arranged fibers which could conceivable allow expansion during joint extension and collapse during flexion. This may explain why the MCP palmar plates were found to shorten more than the PIP joints (34% shortening compared to 27% shortening, respectively). In addition, the migration findings indicate that the MCP joint palmar plate migrates less than the PIP palmar plate (79% compared to 139%), also in agreement with those of Watson. The greater migration of the thicker,

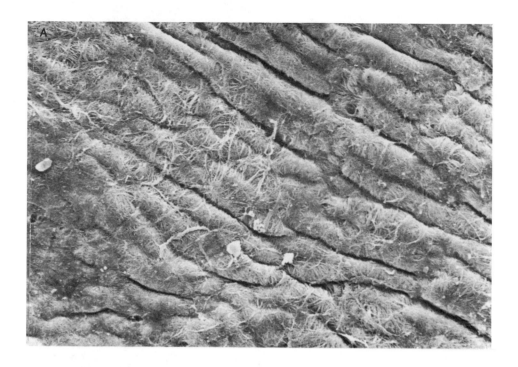

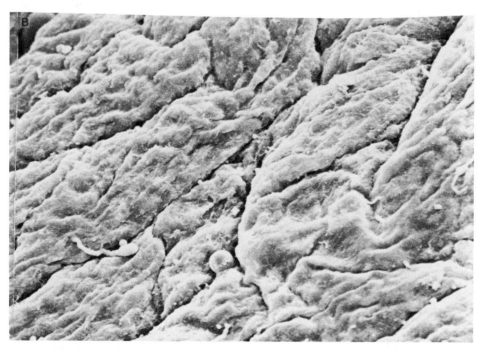

Figure 3. Figure 3A. Scanning electron microscopy of MCP palmar plate; magnification 600x. **Figure 3B.** Scanning electron microscopy of PIP palmar plate; magnification 600x.

Figure 4. Figure 4A. Scanning electron microscopy of MCP palmar plate (magnification 2.6 x 10³x) showing the looser appearing connective tissue. **Figure 4B.** Scanning electron microscopy of PIP palmar plates (magnification 2.6 x 10³x) showing more dense-appearing connective tissue.

more dense PIP palmar plate seemed to be due to its inability to collapse or shorten during flexion of the joint, it thus migrates proximally.

To summarize, several anatomic features and differences between the palmar plates of the MCP and PIP joints have been identified. The MCP palmar plates were found to be longer, thinner, composed of looser connective tissue that was more "collapsible" during joint motion. The PIP palmar plates were found to be shorter, thicker, and composed of a more dense-appearing connective tissue that was less collapsible but migrated more during passive joint flexion. These characteristics may contribute to the greater tendency of the PIP joint to develop palmar plate contracture, along with the presence of the proximal attachments (check-rein precursors).

This study does not address the other neighboring anatomic structures that may contribute to joint contracture such as the collateral ligaments, dorsal joint capsule, or joint osseous architecture. Studies isolating the relative contributions of these structures would be necessary to more fully understand the clinical behavioral differences of the MCP and PIP joints.

ACKNOWLEDGEMENTS

We wish to thank Mr. William R. Collins, Ms. Linda Kitabayashi and Mr. Tom Byrne for their technical assistance.

REFERENCES

1. W.H. Bowers, J.W. Wolf, J.L. Nehil, and S. Bittinger, The proximal interphalangeal joint volar plate: an anatomical and biomechanical study, *J. Hand Surg.* 5:79 (1980).

2. B.L. Chandrarajan, F.T. Chaykowski, W.J. Forrest, J.T. Bryant, V.P. Wyss, and T.D.V. Cooke, Volar plate insertions of the interosseous muscles of the hand, *J. Hand Surg.* 13A:309 (1988).

3. R.M. Curtis, Capsulectomy of the interphalangeal joints of the fingers, *J Bone Joint Surg.* 36A:1219 (1954).

4. P. Gad, The anatomy of the volar part of the capsules of the higher joints, *J. Bone Joint Surg.* 46B:362 (1967).

5. D.P. Green, General principles, (In operative hand surgery, 2nd. ed., Green D.P. ed.) Churchill Livingstone, New York, 1 (1988).

6. R.S. Idler, R.T. Manktelow, G. Lucas, W.H. Sietz, D.C. Bush, MP. Rosenwasser, F.M. Watson, and M.D. Putman, (Hand Surgery Manuals Committee, American Society for Surgery of the Hand), The hand: primary care of common hand problems, 3rd edition, Churchill Livingstone, New York, 37 (1990).

7. M.A.E. Keenan, R.A. Abrams, D.E. Garland, and R.L. Waters, Results of fractional lengthening of the finger flexor in adults with upper extremity spasticity, *J. Hand Surg.* 12:575 (1987).

8. M.A.E. Keenan, E.P. Todderud, R. Handerson, and M.J. Botte, Management of intrinsic spasticity in the hand with phenol injection or neurectomy of the motor branch of the ulnar nerve, *J. Hand Surg.* 12:734 (1987).

9. K. Kuczynski, The proximal interphalangeal joint: anatomy and causes of stiffness in the fingers, *J. Bone Joint Surg.* 50B:656 (1968).

10. J.D. Lubahn, G.D. Lister, and T. Wolfe, Fasciectomy and Dupuytren's disease: a comparison between the open-palm technique and wound closure, *J. Hand Surg.* 9A:53 (1984).

11. C.R. McCash, The open palm technique in Dupuytren's contracture, *Br. J. Plast. Surg.* 17:271 (1964).

12. R.J. Smith, Intrinsic muscles of the fingers: function, dysfunction, and surgical reconstruction, AAOS Instructional Course Lectures, C.V. Mosby, St. Louis, Mo. 24:200 (1975).

13. K.H. Watson, T.R. Light, and T.R. Johnson, Check-rein resection for flexion contracture of the middle joint, *J. Hand Surg.* 4:67 (1979).

14. K.H. Watson, and S.H. Turkellaub, Stiff joints, in: Green DP, ed. Operative hand Surgery, New York, Churchill Livingstone, 537 (1988).

Part II

FORCE ANALYSIS

ON A MODEL OF THE UPPER EXTREMITY

Bo Peterson

Centre for Biomechanics
Chalmers University of Technology
S-41296 Göteborg, Sweden

INTRODUCTION

The function of the upper extremity is an interesting and important issue. The function of the hand and the wrist can not easily be separated from the rest of the extremity. There is a mechanical and a mental coupling. However, it is not apriori clear how to distinguish the two effects. The whole system is an underdetermined mechanical system and it is difficult to make measurements both in vivo and in vitro. These complications make it difficult to evaluate a mechanical model.

A model will strongly contribute to the understanding of the whole situation. For example the mechanical equilibrium equations (for the force and torque), which are necessary ingredients in such a model, will select only physically feasible forces in the structure. Although this might look restricted it still allows for infinitely many solutions for the forces. It means that the forces can be redistributed normally. This can give insight in training or rehabilitation after injury. It is also clear that a geometrically correct model can be used to study the separation of the mechanical and mental coupling between subparts of the system.

A model is also useful for the study of the function of individual muscles, in particular those with wide attachments. It is of recent interest to study how parts of the wide muscles can be used independently.[12]

The only threedimensional models (up to the authors knowledge) are those described by the group around Christian Högfors[6,7,11] and those described by the group around Rients H. Rozendal.[15,16,17]

GEOMETRY, COORDINATE SYSTEMS AND EULER ANGLES

The upper extremity, as considered here, consists of the clavicula, the scapula, the humerus, the radius, the ulna and the "hand". The bones can be seen as rigid bodies. The first four of them are connected to each other with ball joints. The clavicula is

Advances in the Biomechanics of the Hand and Wrist
Edited by F. Schuind *et al.*, Plenum Press, New York, 1994

connected to the sternum with a ball joint. However, the distal end of the humerus is connected to the ulna with a hinge joint. The distal end of the radius has a concave part which is connected to the convex distal end of the ulna. In this way these two distal ends are kept a constant distance apart. As a consequence the radius can only rotate around an axis through the proximal end of the radius and the distal end of the ulna - referred to as the arm long axis. The joint at the hand is not yet further modelled.

Each ball joint introduces 3 degrees of freedom and each hinge joint (the ulna and the above mentioned radius rotation) introduces 1 degree of freedom. Together this makes 11 degrees of freedom. The scapula constraint from the thorax might eliminate some - dependent on the specific situation.

In the following we concentrate on the right shoulder. Bone fixed coordinate systems are introduced in the clavicula, the scapula and the humerus as Högfors et al. [6,7] with a slight modification (Figures 1,2). Here we consider a situation where the elbow joint axis is in the humerus 1-3 plane but rotated an angle $-\alpha$ (minus α) around the humerus 2-axis from being parallel with the humerus 3-axis (Figure 1-B). The ulna coordinate system is introduced by putting the origin Ω^u at the intersection of the humerus-ulna hinge axis and the humerus 1-axis. The distal ulna joint centre Ω^a is defined to be on the ulna 1-axis. The elbow joint axis is in the ulna 1-3 plane. Here the ulna 3-axis is rotated an angle $-\beta$ (minus β) around its own 2-axis from the hinge axis (Figure 2-A). The radius coordinate system with origin Ω^r in the centre of the humerus - radius ball joint has the distal end Ω^b on the radius 1-axis (Figure 2-A). The radius 3-axis is chosen in such a way that it is in the ulna 1-3 plane when the radius 1-axis also is in this plane. When the radius 1 and 3 axes are in this later plane its 3-axis is rotated an angle $-\eta$ (minus η) around the radius 2-axis from the hinge axis. This situation is shown in figure 2-A. A hand coordinate system is introduced with origin at the distal end Ω^a of the ulna and with the 1-axis along the ulna 1-axis. The 3-axis of the hand system is intersecting the radius 1-axis. The only way the hand is modelled is as a means for applying loads in a specific point as above.

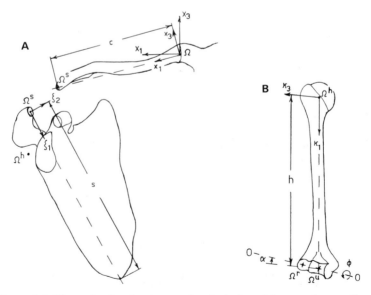

Figure 1. Figure 1-A. Figure showing the sternum, the clavicula and the scapula as well as respective coordinate systems. **Figure 1-B.** Figure showing the humerus and its coordinate system as well as the elbow joint axis 0-0. The points $\Omega^h,\Omega^u,\Omega^r$, are the origins of the humerus, the ulna and the radius coordinate systems.

The rotation ϕ around the elbow axis is defined to be 180° with straight arm when the ulna 1-axis is in the humerus 1-3 plane. With bent arm this angle is less than 180°. Please note that this angle of rotation is not the same as the angle between the humerus and the ulna. This is due to the slight tilt of the elbow joint axis relative both the humerus and the ulna. However, the difference is usually small.

The rotation γ around the arm long axis (i.e. an axis through the radius origin Ω^r and the distal end Ω^a of the ulna) is defined to be zero when the radius 3-axis is orthogonal to the ulna 1-3 plane. For a rotation $\gamma = 90°$ the radius 3-axis is in the ulna 1-3 plane.

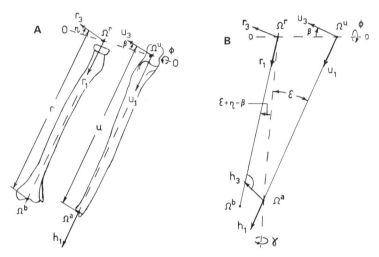

Figure 2. Figure 2-A Figure showing the ulna and the radius when the rotation angle γ around the axis Ω^r–Ω^a is 90°. **Figure 2-B.** Figure showing the lower arm geometry. The ulna 1-axis is in the humerus 1-3 plane when the elbow angle ϕ is 180°. The radius 1 and 3-axes are in the ulna 1-3 plane when the angle γ around the Ω^r– Ω^a axis (i.e. arm long axis) is 90°.

For clarity the situation may be put in operator from. Let R_i (θ) represent a rotation an angle θ in positive sense around the i-axis and let this operate on frames (not coordinates). The orientation of the ulna system is expressed as a rotation R_2 (-β) R_3 (180 - ϕ) R_2 (- α) of the humerus system. The orientation of the radius system is expressed as a rotation R_2 (-($\varepsilon + \eta - \beta$)) R_1 (-90 + γ) R_2 (ε) of the ulna system. Here the angle ε is the angle between the radius and the ulna origins (Ω^r resp Ω^u) as seen from the distal end Ω^a of the ulna (Figure 2-B).

The orientation of the humerus system, the clavicula system and the scapula system are given relative a body fixed coordinate system - the sternum system.[6,7,11] The body fixed system has the origin in the sternoclavicular joint with the 1-axis orthogonal to a sagittal plane and pointing towards the right side. The 2-axis is orthogonal to a frontal plane pointing forwards. In some of the other references the 2-axis of the sternum system is not orthogonal to a frontal plane. This has no practical influence on what follows here. The humerus, the clavicula and the scapula systems are expressed as a rotation of the type R_1 (γ) R_2 (-β) R_3 (α) of the sternum system. There are also angles to tilt the sternum system (i.e. the whole body relative a laboratory system).

In the above we have interchangeably used what is called Euler angles or Cardan angles to represent the orientation of the various bones. These two means of representation of the rotation group in three dimensions are essentially the same. A treatise on the subject can be found in Rimrott.[14]

THE SHOULDER RHYTHM

The shoulder rhythm is defined as the function for each of the Euler angles for the orientation of the clavicula and the scapula expressed by the Euler angles (Cardan angles etc.) for the humerus.[6] This concept is useful because knowing these functions one only needs to know the Euler angles for the humerus (instead of also the additional 6 for both the clavicula and the scapula). These functions can be determined by experimental methods.[6]

Knowledge of the shoulder rhythm functions allows us to compute synthetic motion patterns (not performed at the experiment) for the clavicula and the scapula.[6] Expressions for these functions, in terms of elementary functions, are given in Karlsson et al.[11] In this later work the data from Högfors el al.[6] are used. In fact the functions from Karlsson and Perterson[11] are valid for a larger range of motion than these from Högfors et al.[6] This extension was made after extensive trial and error experiments in combination with computer graphics. The visualization of a sequence of postures is a powerful tool in evaluating the rhythm functions - much better than comparing numbers.

THE MUSCULAR APPROXIMATION, EQUILIBRIUM EQUATIONS AND CONTROL QUANTITIES

The muscles are in this model approximated as strings running the shortest distance between the attachment points for the particular muscle.[7] The wide spread muscles have been divided into several strings. As far as the author understands there is a lack of knowledge about the extent of independence of such subparts of the muscles.[12] However, the final model suits as an instrument to make investigations of such matters. The muscles (both in reality and in our approximation) are restricted by various surfaces (lines, cylinders, spheres etc.) from running straight between their end points. As much as practically possible such minimum path length problems have been solved algebraically.

This work differs from Högfors et al.[7] and Karlsson and Peterson[11] as follows. Here we incorporate the brachialis and the brachioradialis. We also introduce the medial and lateral part of triceps as well as the real attachment points of the biceps and the triceps.

The method of free body diagram technique from mechanics has been applied to the problem in the following way. The exterior load consists of a force and a couple, both with arbitrary direction and size, applied at the hand. The intrinsic weight of the upper and lower arm have been taken account of. The clavicula, the scapula, the whole arm from the humerus origin and outwards together with the exterior and intrinsic loads are "cut out" and have been treated as free bodies. The forces are directed along unit vectors $\mathbf{e}_i$ tangent to the strings. Their points of application $\mathbf{r}_i$ are somewhere on the string (along the application line of the force) relative some arbitrary point (here the sternum origin). The 18 equilibrium equations can be set up in a straight forward way and leads to an equation of the type

$$\sum_{j=1}^{N} a_{ij} F_j = \begin{cases} Q_i^e & \text{for } i = 1\text{-}6 \\ 0 & \text{for } i = 7\text{-}18 \end{cases}$$

where: F_i are the unknown forces
          Q_i^e are the exterior forces (i = 1-3) resp. torques (i = 4-6)
          N is the amount of unknown forces.
          Typical elements in the coefficient matrix a_{ij} for the muscular forces are

$a_{k+i,j} = (\mathbf{e}_j)_i$ for the force equilibrium equations part
$a_{k+3+i,j} = (\mathbf{r}_j \times \mathbf{e}_j)_i$ for the momentum equilibrium equations part
        k = 0 for the entire arm
        k = 6 for the scapula
        k = 12 for the clavicula

The torque equilibrium around the elbow joint is used for a 19th equation. There are not yet enough muscles in the model to set up an equilibrium equation for the torque around the arm long axis. However, the total torque around this axis is computed as a control quantity. It remains to estimate if the pronators and subinators, which are not taken account of, can balance this torque. Of course one has to be very precise in the specification of the application point for the exterior load, relative the arm long axis, in order to make such studies meaningful.

FURTHER MODELLING AND THE MATHEMATICS FOR OBTAINING SOLUTIONS

The 19 equations obtained above are in general not sufficient to give a solution for the 50 unknown forces - including contact forces between the bones. The 19 equations are exact (within their approximations of the reality), seen as part of the classical theory of mechanics. In the following we are forced to rely on assumptions of much more vague nature.

We impose restrictions that the muscular forces have to be positive. Furthermore we assume that the muscular forces are bounded by a material constant σ N/m^2 times their physiological cross section A_i m^2, see for example Karlsson et al.,[11] Veeger et al.,[17] Dul et al.[4] and An et al.[1] which gives $0 \le F_i < \sigma \cdot A_i$ for the muscles. We allow small negative contact forces representing ligaments that are not really modelled in another way.

The cavitus glenoidalis is a relatively shallow surface. Thus the contact force in the glenohumeral joint can not deviate too much from the normal to this surface. We have included a nonlinear constraint for the angle θ between the force and the normal above.

$$0 \le \sin\theta \le \text{const.}$$

For the value 1 of this constant the constraint is fully relaxed. This constraint is difficult to interpret in a more physical way. This kind of nonlinear constraint can also lead to mathematical effects which is hard to believe that the human body can handle. Here we consider the possibility of holes or other cut outs in the solution space which makes it non convex.

What is needed next is a mathematical principle for the distribution of the forces over the different muscles. It is clear that the human body can choose among several different strategies which can be reflected in such a principle. One principle can be based on the endurance time T_i for muscle number i which can be expressed as

$$T_i = k_i \left(\frac{A_i}{F_i} \right)^{p_i}$$

Here k_i and p_i are constants dependent on the composition of the muscular fibre.[4] The principle states that the body distributes the forces so that the shortest endurance time is maximal which reads maximize $\min_i T_i$. Very little muscle fibre composition data is known. As a matter of fact the different muscles don´t have the same composition in a specific individual. The mathematics involved is also difficult to handle. An alternative to this discrete mathematics is to introduce a differentiable function (objective function) of the unknown forces which should be minimized. This leads to what is called a nonlinear programming problem.[18] The constant p_i in the endurance time is roughly equal to 2 and all constants k_i are roughly the same - for the nonexisting average man. One can use

$$\phi = \Sigma_i \frac{1}{T_i}$$

as an objective function which roughly means that the endurance times should be kept high. As the body might choose between several principles which are not clear to us it is not absolutely necessary to be extremely detailed about the particular choice of objective function. Experience of the use of different objective functions seems to support the idea that the most important ingredients are that [1] the forces divided with their respective physiological cross section are used, and [2] the objective function is convex.

It is clear that the solution space is convex as long as there are only linear constraints. The influence of the specific choice of objective function varies dependent on the particular situation under study. In the extreme case of maximum exterior loads the result is not at all dependent on the particular choice of objective function. It merely serves as a trick to handle the mathematics in order to obtain a solution.

THE COMPUTER PROGRAM

The theory as outlined above was turned into a computer program. The routines C05 NBF (for roots of transcendental equations) and E04 UCF (for minimizing a function with various constraints) were taken from the NAG fortran library. The following indata was provided from a separate file: bone length parameters, muscle attachment points, tendon attachment points, parameters for the muscular constraints, centre of gravity and mass parameters for upper and lower arm. Euler angles for the body position, the clavicula, the scapula, the humerus, the ulna and the radius. Exterior torque, force and point of action, muscular physiological cross sections and limiting forces. The indata are processed by the computer program as described in the flow chart in figure 3. The NAG routines requires a considerable amount of control quantities as indata and also provides still more as outdata on request. With convex optimization functions and "nice" nonlinear constraints the situation is manageable. However, in more complicated cases the situation easily gets out of control. The output from the computer program consists of the following quantities: the magnitude and direction of the forces as well as corresponding torques, muscular lengths.

CONSISTENCY OF THE MODEL

In order to establish external consistency the force and torque directions were checked against various books in anatomy and available skeletons. The torques were also compared with what can be found in Basset et al.[2] An assessment of the external consistency of the model has also been attempted in Karlsson et al.[10] where the maximum strength of the arm has been studied. This check has the merit that it is independent of the exact form of the minimization function. Internal consistency has been studied by small variations of the parameters in the model.[8] A more extensive treatise on the consistency of the model is given by Högfors et al.[5]

After these checks it seems reasonable to believe that the model with equilibrium equations, force magnitude constraints and possibly the contact force direction constraint can produce all possible physical solutions (infinitely many) for a given position. However, the modelling of the bodys strategy to distribute the forces over the muscles, possibly via objective functions, will require continued interest.

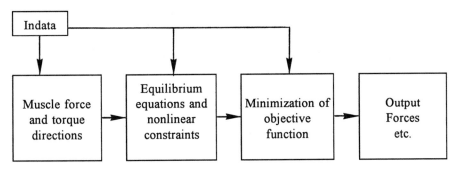

Figure 3. A flow chart for the computer program.

AN APPLICATION

As an application of the model we want to study the possibility to change the muscular force in two different muscles during some activity. First we study the trapezius muscle part number 3 in figure 4, compare with Högfors et al.[7] The orientation of the humerus relative a sternumsystem with the 3-axis vertical (compare with Högfors et al.[6,7]) and with Euler angles according to figure 4a in Högfors et al.[6] is $\alpha_h = 45.$, $\beta_h = -60.$, $\gamma_h = 32.3$. The elbow angle $\phi = 90.$ and the arm long axis angle $\gamma = 0$. The handload is a weight of 2 kg. The other parameter values are left out because they are not of particular interest. The muscle under study (the trapezius part no 3) is unrestricted in a first computation giving what we refer to as a normal case. After this the force value of this muscle is enforced to take the value from zero to two times the "normal" value with increments of 10N per computation. A solution was found for all these cases.

As the nonlinear contact force constraint is not in action (the limiting constant being too large) the case under study is a convex nonlinear programming problem. One consequence of this is that every value of the trapezius force between 0 and 200% of the "normal" can be assumed (all the values between were only computed as a check) in a hypothetical computation. The result of the study is presented in table 1.

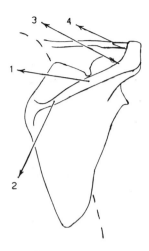

Figure 4. The trapezius muscle according to Högfors et al.[7]

After the trapezius we study a similar situation but now the supraspinatus force is changed in a similar way as the trapezius force. The result which is essentially the same is presented in table 1.

The surprising result that the two muscle forces can be varied as much as above has not been found in the literature. An in vivo experiment with EMG corresponding to the application above has been presented in Palmerud et al.[13] Here it was only possible to vary the trapezius force (i.e. EMG signal) voluntarily. Our computer model shows what is mechanically feasible but not what is necessarily feasible to perform mentally using the nerve system. The in vivo experiment essentially shows what the subjects under study could perform. It does not show what can be achieved by training. Last but not least the EMG signal is not easily related to the force.

DISCUSSION

Above we have presented a threedimensional model for the shoulder complex. We have explained how the model was developed and tested in a series of papers. During this process some important aspects have been demonstrated to the investigators.

Biomechanical systems with ball joints are fundamentally threedimensional. Concepts such as agonist and antagonist have no meaning (if one does not allow the agonist to also be antagonist at the same time). Twodimensional models are of little value because the human body will distribute the forces in such a way that they don´t fall in one and the same plane. This is the case almost regardless of distribution principle even before considering any kind of stability requirement.

It is impossible to treat the subparts independently by for example first solving the force problem for the arm with load (equilibrium equations with constraints and objective function) and then the same for the scapula and last for the clavicula. This will almost inevitable lead to an infeasable problem for the clavicula. This depends on the fact that the optimization procedure will select the wrong solutions by operating on a too restricted space (of forces) and in the end the solution space is exhausted.

Table 1. Forces, changes of forces and objective function. The normal case is where all forces are computed by the program. In the four other cases one force is given a prescribed value about + 100 % or - 100 % of the normal value.

			Normal case		Change in % F_i (or ϕ)			
					Trapezius variation		Supraspinatus variation	
			F_i	F_i	$F_{34}=$	$F_{34}=$	$F_{12}=$	$F_{12}=$
			N	%$F_{i\,max}$	-100%	+97%	-100%	+102%
1	1F	Latissimus dorsi 1 (upper)	0.	0.	-	-	-	-
2	1F2	Latissimus dorsi 2 (lower)	0.	0.	-	-	-	-
3	4F1	Pektoralis major (lower)	40.	6.	-81.	+51.	-15.	16.
4	13F1	Deltoideus med.	0.	0.	-	-	-	-
5	13F2	Deltoideus post.	0.	0.	-	-	-	-
6	14F1	Coracobrachialis	18.	6.	57.	-33.	20.	-21.
7	15F1	Infraspinatus 1 (upper)	102.	18.	-30.	21.	40.	-41.
8	15F2	Infraspinatus 2 (lower)	121.	20.	-14.	12.	37.	-37.
9	16F1	Subscapilaris 1 (upper)	11.	2.	223.	-100.	-24.	25.
10	16F2	Subscapilaris 2 (middle)	0.	0.	-	-	-	-
11	16F3	Subscapilaris 3 (lower)	0.	0.	-	-	-	-
12	17F1	Supraspinatus	124.	22.	-13.	7.	-100.	102.
13	18F1	Teres major	0.	0.	-	-	-	-
14	19F1	Teres minor	8.	3.	-18.	26.	42.	-43.
15	20Fc1	Biceps (long)	46.	15.	29.	-17.	11.	-11.
16	20Fc2	Biceps (short)	4.	17.	23.	-14.	15.	-16.
17	21F1	Triceps (long head)	0.	0.	-	-	-	-
18	24L	Coracohum. lig	99.	--	-100.	67.	54.	-55.
19	27.1	*Glenohum.contact force*	462.	--	-26	18.	13.	-13.
20	27.2	*Glenohum.contact force*	451.	--	4.	-2.	0.	0.
21	27.3	*Glenohum.contact force*	163.	--	-7.	7.	-4.	4.
22	4F2	Pektoralis major (upper)	8.	2.	30.	-20.	22.	-23.
23	13F3	Deltoideus ant.	181.	29.	26.	-19.	23.	-24.
24	2F1	Levatorscapulae	30.	11.	171.	-100.	-4.	4.
25	3F1	Omo-hyoideus	4.	4.	146.	-90.	-3.	3.
26	5F1	Pektoralis minor	0.	0.	-	-	-	-
27	6F1	Rhomboideus major	43.	13.	80.	-72.	-2.	2.
28	7F1	Rhomboideus minor	10.	6.	150.	-100.	-4.	4.
29	8F1	Serratus anterior (upper)	0.	0.	-	∞*	-	-
30	8F2	Serratus anterior (middle)	35.	9.	7.	2.	0.	0.

(Continued)

Table 1. (continued)

	Normal case		Change in % F_i (or ϕ)			
			Trapezius variation		Supraspinatus variation	
	F_i N	F_i %$F_{i\,max}$	$F_{34}=$ -100%	$F_{34}=$ +97%	$F_{12}=$ -100%	$F_{12}=$ +102%
31 8F3 Serratus anterior (lower)	111.	18.	20.	-27.	-2.	2.
32 12F1 Trapezius	46.	11.	-30.	-58.	-1.	2.
33 12F2 Trapezius (lower downwards)	4.	.1	-100.	836.	40.	-40.
34 12F3 Trapezius (upper)	137.	23.	-100.	97.	-4.	4
35 SN1 *Thorax constraint force*	92.	--	71.	-40.	-13.	14.
36 SN2 *Thorax constraint force*	0.	--	-	-	-	-
37 23.1 *Acromioclav. contact force*	251.	--	-43.	36.	7.	-7.
38 23.2 *Acromioclav. contact force*	44.	--	206.	-161.	66.	-67.
39 23.3 *Acromioclav. contact force*	-14.	--	8.	-9.	-16.	16.
40 9F1 Sterno cleido mastoideus	2.	3.	109.	-68.	3.	-3.
41 10F1 Sternohyoideus	0.	0.	-	-	-	-
42 11F1 Subclavius	0.	0.	-	∞*	-	-
43 12F4 Trapezius	0.	0.	-	-	-	-
44 22.1 *Sternoclav.contact force*	148.	--	-92.	75.	-4.	4.
45 22.2 *Sternoclav.contact force*	-110.	--	47.	-40.	-7	7.
46 22.3 *Sternoclav.contact force*	-2.	--	140.	-124.	50.	-51.
47 21F2 Triceps medial head	0.	0.	-	-	-	-
48 21F3 Triceps lateral head	0.	0.	-	-	-	-
49 28F1 Brachialis	98.	11.	-21.	13.	-11.	11.
50 29F1 Brachioradialis	13.	7.	-21.	13.	-11.	11.
Objective function ϕ	17.	--	32.	25.	15.	16.

- means unchanged zero
∞* means $F_{29} = 9N.F_{42} = 1N$

However, the solution for the first part might not be so bad. For example one can solve the 23 variable problem for the entire arm and find a reasonable force distribution.[11]

It is also noticed that in the simple model many muscles are not used. EMG studies often show that most muscles are engaged. The easiest way to engage more muscles - although quite senseless - is to apply a lower bound (different from zero) for the muscular forces. Another way is to study the stability of the force system in a similar way as in Crisco et al.[3] We can treat the muscles as springs and derive the total energy for the system. The second derivatives of the energy with respect to the parameters describing the orientation of the system can be seen as elements of a matrix. The eigenvalues of this matrix is a measure of the stability of the mechanical system. In particular the larger the lowest eigenvalue is the greater is the stability of the system.

If any eigenvalue is zero or negative the system is unstable. This scheme has been used for the humerus using the three Euler angles describing its orientation.[9] Analyzing the stability of the system as above does not answer the question of how to distribute the forces in order to increase the stability in a direct way. It merely serve as a means of evaluating the stability for given force distributions. However, as the eigenvalues (but not necessarily the eigenvectors) are continuous functions of the parameters it is in principle possible to optimize eigenvalues. The complications are: first it is not necessarily the same eigenvalue which is the lowest during the optimization, second there are large numerical problems involved due to the necessity of numerical derivation for obtaining the mixed partial derivatives. If one follows the ideas above it is likely that the most stable force distribution also involves more muscle forces different from zero than from the beginning. Of course there remains the determination of what level of stability a particular task demands.

Sometimes people are using the word stability in connection with luxation problems of a joint. Strictly this can not be done in the frame of our model because no specific modelling is made for the joint. One would have to introduce more parameters describing the relative motion between the different bones at the joint and a more detailed geometric description of the surfaces at the joint. Still using our model it is possible to study "unstable" behaviour of for example the scapula. One is lead to believe that such behaviour depend on lack of control of the muscles i.e. a problem in the nerve system.

In the future the model will be extended to include more pronators and supinators. The moment equilibrium around the arm long axis will be used as a 20th equation. It will then be possible to study the stability related to the rotation around the arm long axis and the elbow axis. The coupling between the other possible factors influencing the stability of the hand and the rest of the shoulder is a challanging field of study.

ACKNOWLEDGEMENTS

This work was supported by the Swedish Work Environment Fund. I am also grateful for valuable comments by Christian Högfors.

REFERENCES

1. K.N. An, F.C. Hui , B.F. Morrey, R.L. Linscheid and E.Y. Chao, Muscles across the elbow joint: a biomechanical analysis, *J. Biomech.* 14: 659 (1981).
2. R.W. Basset, A.O. Browne, B.F. Morrey, and K.N. An, Glenohumeral muscle force and moment mechanics in a position of shoulder instability, *J. Biomech.* 23:405 (1990).
3. J.J. Crisco, and M.M. Panjabi, Euler stability of the human ligamentous lumbar spine Part I: Theory, *Clin. Biomech.* 7:19 (1992).
4. J. Dul, G.E. Johnson, R. Shiavi, and M.A. Townsend, Muscular synergism - II. A minimum - fatigue criterion for load sharing between synergistic muscles, *J. Biomech.* 17:675 (1984).
5. C. Högfors, D. Karlsson, B. Peterson, and R. Kadefors, Biomechanical model of the human shoulder – III force predictions; preprint centre for biomechanics, Chalmers University of Technology, S-412 96 Göteborg, Sweden (1992).
6. C. Högfors, B. Peterson, G. Sigholm, and P. Herberts, Biomechanical model of the human shoulder Joint - II. The shoulder rhythm, *J. Biomech.* 24:699 (1991).
7. C. Högfors, G. Sigholm, and P. Herberts, Biomechanical model of the human shoulder Joint I – elements, *J. Biomech.* 20:157 (1987).
8. D. Karlsson, On the stability of a biomechanical shoulder model, biomechanics seminar p.87, Centre for Biomechanics, Chalmers University of Technology and Göteborg University (1989).
9. D. Karlsson, Internal shoulder forces in the case of an increased arm stability, biomechanics seminar, Centre for Biomechanics, Chalmers University of Technology and Göteborg University (1992).

10. D. Karlsson, C. Högfors, U. Järvholm, and B. Peterson, Strength of the shoulder, calculated theoretically, preprint Centre for Biomechanics, Chalmers University of Technology, S-412 96 Göteborg, Sweden (1992).

11. D. Karlsson, and B. Peterson, Towards a model for force predictions in the human shoulder, *J. Biomech.* 25:189 (1992).

12. S.E. Mathiassen, and J. Winkel, Electromyographic activity in the shoulder-neck region according to arm position and glenohumeral torque, *Eur. J. Appl. Phys.* 61:370 (1990).

13. G. Palmerud, H. Sporrong, P. Herberts, C. Högfors, U. Järvholm, R. Kadefors, and B. Peterson, Voluntary redistribution of muscle activity in human shoulder muscles, preprint Lindholmen Development, P.O. Box 8714, S-402 75 Göteborg, Sweden (1991).

14. F.P.J. Rimrott, Introductory Attitude Dynamics, Springer-Verlag (1989).

15. F.C.T. Van Der Helm, The Shoulder Mechanism, PhD thesis, Delft University of Technology, Delft, The Netherlands (1991).

16. F.C.T. Van Der Helm, H.E.J. Veeger, G.M. Pronk, L.H.V. Van Der Woude, and R.H. Rozendal, Geometry parameters for musculoskeletal modelling of the shoulder system, *J. Biomech.* 25:129 (1992).

17. H.E.J. Veeger, F.C.T. Van Der Helm, L.H.V. Van Der Woude, G.M. Pronk, and R.H. Rozendal, Inertia and muscle contraction parameters for musculoskeletal modelling of the shoulder mechanism, *J. Biomech.* 24:615 (1991).

18. W.I. Zangwill, Nonlinear Programming a Unified Approach, Prentice-Hall (1969).

A MODEL OF HUMAN HAND DYNAMICS

Elena V. Biryukova[1] and Vera Z. Yourovskaya[2]

[1]Institute of Higher Nervous Activity and
Neurophysiology of Russian Academy of Sciences
[2]Biological Department of Moscow State University

INTRODUCTION

Great variety and high adaptability of human hand movements are due to complex mechanical structure of the hand and subtle nervous control produced by CNS. As for the mechanical structure, a lot of studies have been devoted to the anatomy of hand joints,[6,21,32,35,37] hand tendons[2,3,29] and hand muscles.[4,16,19,20,24,31] These experimental data provide necessary basis for the elaboration of biomechanical models of the hand. In principle, these models allow to determine efforts produced by muscles in the course of hand movements. The latter problem, however, has been solved only for subsystems of the entire hand like a finger fixed in different static positions,[12] setting aside real dynamics of the movements of the entire hand.[6,13,14,15,30]

Below the attempt is made to develop the biomechanical model, which would include all principal muscles and degrees of freedom of the hand. Muscle efforts produced during the co-ordinated hand movements are then found by solving inverse dynamics problem.

MODELS OF LINKS

A biomechanical model of the human hand proposed here consists of 16 links, the links being rigid bodies connected by frictionless joints. These are carpal+metacarpal bones plus three phalanges for each finger which are considered to be links in the model (Figure 1).

Although the first link cannot be considered as a true rigid body, small movements in intercarpeal, carpometacarpeal and intermetacarpeal joints are nevertheless neglected. To be able to evaluate moments of inertia of links, the first link is modelled by rectangular parallelepiped, and the other links by right circular cylinders. Link sizes have been determined from special measurements and available anatomical data.[38] These sizes are given in table 1 in a normalized form. This means that length (a) of the first link as well as those of the other links (l) are divided by length of the hand (assumed to be equal to the distance between the center of radiocarpea joint and the third finger's tip); width and thickness of the first link (b and c) are divided by width and thickness of the forearm; the radii (r) of links are divided by their lengths (l).

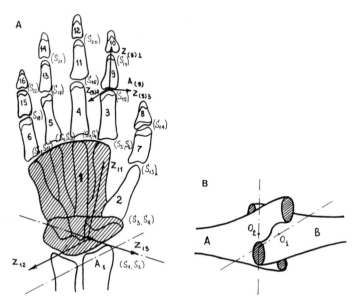

Figure 1. Figure 1-A. Links of the hand model with related systems of coordinates $A_n z_{nk}$ (n = 1,2, ... 16; k = 1,2,3); n is the link number, k is the coordinate number. **Figure 1-B.** Model of a joint with two degrees of freedom: fragments of two tori form joint surfaces. Link A rotates about the axe 0_1, link B rotates about the axe 0_2.

Masses of links are assumed to be equal to the masses of corresponding bones plus masses of all adjoining muscles, which have been determined using anatomical data.[33,34] Among hand muscles only those of thumb adjoin the bones. As for the others, only tendons with masses negligible in comparison with muscle masses adjoin the bones. Variations of link inertial characteristics, which may occur during muscle contractions are neglected. Masses of links divided by the mass of the whole hand are presented in table 1 as well.

MODELS OF JOINTS

Interphalangeal joints are modelled by cylindrical hinges having one degree of freedom each. Surfaces of bones forming radiocarpal, first carpometacarpal and 2-5 metacarpophalangeal joints are assumed to be fragments of tori[11,25] (Figure 1,b). As the rotations, which are possible in these joints, are around two orthogonal non-crossing axes, the joints are modelled by hinges with two degrees of freedom. The complete model has therefore 22 degrees of freedom. Positions of axes of rotation relative to the hand bones have been determined from anatomical measurements[11,17,25] (Figure 1,a). Exact positions of joint axes as well as the parameters of surfaces of two-degrees-of-freedom joints are given in Biryukova.[8]

Table 1. Sizes and masses of links in the model.

Link No	Linear dimensions			Masses
	a	b	c	
1	0.451	1.400	1.200	0.662
	l	r/l		
2	0.260	0.155		0.121
3	0.237	0.146		0.030
4	0.254	0.136		0.033
5	0.243	0.119		0.022
6	0.196	0.147		0.018
7	0.173	0.200		0.022
8	0.150	0.192		0.014
9	0.150	0.192		0.014
10	0.121	0.190		0.007
11	0.173	0.167		0.016
12	0.121	0.190		0.007
13	0.168	0.172		0.015
14	0.121	0.190		0.007
15	0.121	0.190		0.007
16	0.116	0.150		0.004

The systems of coordinates, which are necessary to describe kinematics and dynamics of the hand model are chosen to be attached to axes of rotation in the joints. Stationary system Ox_p is attached to elbow joint (Figure 2,a), the systems related to the links $A_n z_{nk}$ (n = 1,2, ... 16; k = 1,2,3), are attached to corresponding hand joints (Figure 1,a).

MODELS OF MUSCLES

Muscles included in the model are enlisted in table 2. They are modelled by weightless expandable threads.

3-D coordinates of thread origins and insertions viewed as points instead of areas, have been determined relative to the corresponding link-related coordinate systems from the special anatomical measurements of 52 specimen human arms.[39] These coordinates are presented in table 3. Forearm, which the stationary system is related to, has number zero. To denote muscle's origins and insertions letters from A to R are used. For the muscles with two caps (e.g. FDS or ECU) the letter is supplied with an index.

It should be noticed, that although in the strict mechanical sense muscle insertions and attachements must be the points of application of the resultant of all elementary efforts applied to the entire area, in the model they are reasonably chosen to lie inside the area. As far as we know, however, no experimental works exist concerning the distribution of elementary muscle efforts inside the areas of muscle attachment. Hence, the assumption made above is by the moment the only possibility for the "thread-muscle" model to be applicable in this case. This assumption seems to be the most crude one in the entire model, having been accepted, however, only for muscle origins located on forearm. Muscle insertions, located on wrist and fingers, are, in fact, those of their tendons, which are more thread-like looking than muscles themselves. The assumption above therefore seems sufficiently adequate. Directions of muscle efforts have been already found experimentally for some hand tendons, but not for muscles themselves.[2]

Table 2. Hand muscles included in the model.

Muscle No	Muscle name	Abbreviation
1	flexor carpi radialis	FCR
2	flexor carpi ulnaris	FCU
3	flexor digitorum superficalis	FDS
4	flexor digitorum profundus	FDP
5	flexor pollicis longus	FPL
6	extensor carpi radialis	ECR
7	extensor carpi ulnaris	ECU
8	extensor digitorum	ED
9	abductor pollicis longus	APL
10	extensor pollicis brevis	EPB
11	extensor pollicis longus	EPL
12	extensor indicis	EI
13	abductor pollicis brevis	APB
14	flexor pollicis brevis	FPB
15	opponens pollicis	OP
16	adductor pollicis	AP
17	abductor digiti minimi	ADM
18	flexor digiti minimi brevis	FDMB
19 - 22	lumbricales	LUMB
23 - 25	interossei palmares	IP
26 - 29	interossei dorsales	ID

Coordinates of muscle origins and insertions are given in the references related with corresponding links. The coordinates are given in the normalised form, i.e. each of them is divided either by corresponding link length (for the coordinate z_1) or link thickness (for the coordinate z_2) or link width (for the coordinate z_3). A coordinate value being greater than one means that the corresponding origin/insertion is located on a bone cap, which radius is greater than the average one of the bone itself.

Weightless non-expandable loops surrounding joints are also introduced into the model. These are necessary to describe "lines-of-actions" of muscles correctly. Schemes of some of hand muscles are presented in figure 2b illustrating the notion of "line-of-action".

KINEMATICS AND DYNAMICS OF THE MODEL

To describe kinematics and dynamics of the hand model described above, tensor formalism[22] is used. The movements of arm during writing,[23] eyes movements,[1] the movements of quadrupeds during postural adjustment[9,18] have been successfully studied using such formalism.

In terms of the formalism kinematics equations have the standard form, being the recurrent dependencies between base and generalized coordinates. The base coordinates are those of the centres of inertia of links and Eulerian angles of the links. All base coordinates are specified relative to the stationary coordinate system. The generalized coordinates are the angles of independent rotations in the joints.

To describe the dynamics of the system, Lagrange equations in the form of tensor convolution are used. To obtain the coefficients of these equations one needn't write down an expression for kinetic energy in an explicit form. Being compact enough due to index notation the equations have the same form for an arbitrary number of links in the model and any type of joint connections. Kinematics and dynamics equations in an explicit form are given in Biryukova.[8]

PROBLEMS OF REDUNDANCY

With the model above the inverse problem of dynamics was solved, i.e. given different hand movements, all possible groups of muscles capable of implementing these movements were found. A classification of motor redundancy suggested below (figure 3) allows to set up the problem of hand movements modelling unambiguously.

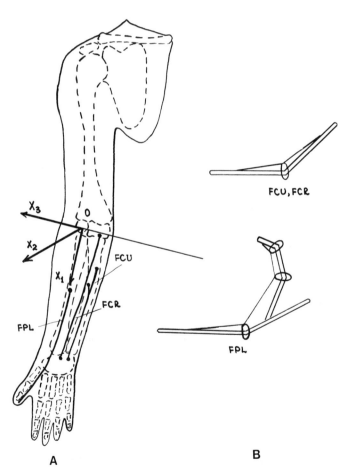

Figure 2. Figure 2-A. Stationary system Ox_p attached to elbow joint; threads corresponding to FCR, FCU and FPL. **Figure 2-B.** Schemes of "lines-of-actions" for these muscles.

Table 3. Coordinates of attachment points of the muscles.

Link No	Point of attachment	Muscle name	Coordinates of attachment point		
			z_1	z_2	z_3
0	A(1)	FCR	-0.049	0.000	-4.286
	$A_1(2)$	FCU	-0.049	0.000	-4.428
	$A_2(2)$	FCU	0.269	0.167	-3.071
	$A_1(3)$	FDS	0.054	0.143	-3.000
	$A_2(3)$	FDS	0.242	0.143	0.357
	A(4)	FDP	0.368	0.048	-0.714
	A(5)	FPL	0.628	0.048	-0.643
	A(6)	ECR	-0.273	-0.096	-1.428
	$A_1(7)$	ECU	-0.193	-0.048	-1.857
	$A_2(7)$	ECU	0.502	0.000	0.000
	A(8)	ED	-0.094	0.000	-2.000
	A(9)	APL	0.587	0.000	-0.214
	A(10)	EPB	0.722	-0.048	0.357
	A(11)	EPL	0.507	-0.071	0.500
	A(12)	EI	0.753	0.000	-0.928
1	B(1)	FCR	0.372	0.471	0.322
	B(2)	FCU	0.026	0.471	-0.322
	B(6)	ECR	0.282	-0.471	0.373
	B(7)	ECU	0.410	-0.353	-0.390
	B(13)	APB	0.231	0.471	0.203
	$B_1(14)$	FPB	0.282	0.471	-0.068
	$B_2(14)$	FPB	0.564	0.471	-0.085
	B(15)	OP	0.436	0.471	0.136
	$B_1(16)$	AP	0.538	0.471	-0.203
	$B_2(16)$	AP	0.782	0.353	-0.102
	B(17)	ADM	0.282	0.471	-0.356
	B(18)	FDMB	0.179	0.471	-0.508
	B(23) ⎤	2 os met	0.449	0.118	-0.085
	B(24) ⎥ IP	4 os met	0.423	0.059	-0.424
	B(25) ⎦	5 os met	0.449	0.059	-0.678
	B(26) ⎤	2 os met	0.385	0.000	0.068
	$B_1(27)$ ⎥	3 os met	0.372	0.000	-0.068
	$B_2(27)$ ⎥	- " -	0.397	0.000	-0.186
	$B_1(28)$ ⎥ ID	- " -	0.449	0.000	-0.322
	$B_2(28)$ ⎥	- " -	0.372	0.000	-0.492
	$B_1(29)$ ⎥	4 os met	0.436	0.000	-0.593
	$B_2(29)$ ⎦	- " -	0.410	0.000	-0.746
2	C(9)	APL	0.067	0.000	1.286
	C(15)	OP	0.400	0.000	0.857
	C(26)	ID	0.222	0.000	-1.000
7	D(10)	EPB	0.067	-1.000	0.000
	D(13)	APB	0.067	0.000	1.500
	D(14)	FPB	0.233	1.000	0.000
	D(16)	AP	0.167	1.000	-1.000
8	E(5)	FPL	0.308	0.800	0.000
	E(11)	EPL	0.077	-0.800	0.000

Table 3. (continued)

Link No	Point of attachment	Muscle name	Coordinates of attachment point		
			z_1	z_2	z_3
3	F(23)	IP	0.098	0.000	-1.500
	F(26)	ID	0.049	0.000	1.500
9	G(12)	EI	0.077	-1.000	0.000
	G(8)	ED	0.077	-1.000	0.000
	G(3)	FDS	0.308	1.000	0.000
10	H(4)	FDP	0.190	1.500	0.000
	H(8)	ED	0.095	-1.250	0.000
	H(12)	EI	0.095	-1.250	0.000
4	I(27)	ID	0.091	0.000	1.333
	I(28)	ID	0.091	0.000	-1.333
11	J(3)	FDS	0.267	1.200	0.000
	J(8)	ED	0.067	-1.000	0.000
12	K(4)	FDP	0.190	1.500	0.000
	K(8)	ED	0.095	-1.250	0.000
5	L(24)	IP	0.119	0.000	1.400
	L(29)	ID	0.095	0.000	-1.400
13	M(3)	FDS	0.276	1.000	0.000
	M(8)	ED	0.069	-1.000	0.000
14	N(4)	FDP	0.190	1.500	0.000
	N(8)	ED	0.095	-1.250	0.000
6	P(17)	ADM	0.029	0.000	-1.000
	P(18)	FDMB	0.029	0.000	-1.000
	P(25)	IP	0.176	0.000	1.400
15	Q(3)	FDS	0.381	1.000	0.000
	Q(8)	ED	0.095	-1.000	0.000
16	R(4)	FDP	0.200	1.677	0.000
	R(8)	ED	0.100	-1.333	0.000

The fact that different muscles exist, passing through the same joint is referred to as *anatomical redundancy*. The notion of *kinematic redundancy* which is also frequently encountered in the literature on the movement physiology, implies that the number of degrees of freedom of a system is greater than that necessary to reach the goal of a movement. In the process of movement learning some of the superfluous degrees of freedom are being frozen, which results in the synergies to arise.[7]

Let L be the number of degrees of freedom of a system, and M the total number of muscles. Since two muscles (agonist and antagonist) are necessary to implement a movement along each degree of freedom, the fact that M > 2L is referred to as *formal dynamic redundancy*, the value M-2L being its quantitative measure.

Note, that the hand model, having 16 degrees of freedom and 29 muscles is not formally redundant system. The number of unknowns in Lagrange equations is equal to M, while the number of equations is L. In our model M>L, and therefore to solve these equations we assume that the number of muscles implementing a movement is minimal and equals to L. To find all variants of muscular activity all possible combinations of M by L are considered. The efforts of (M-L) muscles, which are not included in the combination, are assumed to be equal to zero. This assumption implies that the sum of muscle efforts is minimal. This method of overcoming the redundancy was used in other works as well.[5] Lagrange equations are then being solved for each group of muscles included in the combination. Only positive solutions so obtained are physically meaningful, because the muscle can only pull not push.

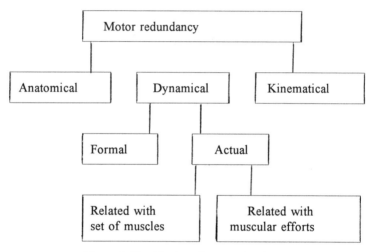

Figure 3. Classification of motor redundancy.

It is important to note, that the method above works only if a particular movement is given. Pattern of muscular activity depends essentially on the parameters of a movement (duration, amplitude etc.) as well as on the external conditions (e.g. external forces applied to the moving system). In other words, the parameters of a movement can affect not only the values of efforts generated by the muscles, but the fact that some muscles are presented in the pattern of muscular activity. In this context the existence of different combinations of muscles able to implement the given movement is referred to as the *actual redundancy*. The actual redundancy is related not only to the movement itself and external parameters, but also to the geometry of muscle apparatus. The descriptive geometrical method of actual redundancy investigation is presented in Biryukova and Platonov.[10]

The problem of hand movements modelling can therefore be formulated as follows. Assuming the number of muscles involved into the pattern of muscular activity be minimal, the actual redundancy of hand muscle apparatus is to be studied i.e. for the principal hand movements all possible patterns of muscular activities are to be found.

RESULTS OF THE MODELLING

Movement Modelling

Each hand movement can be viewed as a superposition of the movements in the individual joints: abduction/adduction (S^1) and flexion/extension (S^2) in radiocarpeal joint, abduction/adduction (S^3) and flexion/extension (S^4, S^{13}, S^{14}) of the thumb, finger flexion/extension (S^6, S^8, S^{10}, S^{12}) and abduction/adduction (S^5, S^7, S^9, S^{11}) in metacarpophalangeal joints, flexion/extension in the interphalangeal joints (S^{15} - S^{22}). Indicated in the parentheses are the generalized coordinates corresponding to the angles of rotation (Figure 2a).

The movements of a hand are given by the following formula:

$$S^1 = S_{inl} + (S_{fl} - S_{inl})*$$

$$(1 + \sin (p/T^1*(t - T^1/2)))/2, \tag{1}$$

where S_{inl} and S_{fl} denote the initial and final values of the angle respectively, and T^1 is the duration of the movement. The movement amplitude A is then equal to S_{fl} - S_{inl}. The initial and final velocities (but not the accelerations) are equal to zero. Equation (1) approximates reasonably such hand movements as finger flexion, as was validated by comparison with records taken from.[26,27]

Radiocarpal Joint Muscles Operation

Let's consider two modes of radiocarpal joint movement modelling. First we consider the system having two degrees of freedom (flexion/extension and abduction/adduction) and four muscles (FCR, FCU, ECR, ECU).

In Figure 4 muscle efforts produced during flexions with amplitudes A=10 grad (Figure 4a) and A=30 grad (Figure 4b) are shown. The acceleration and deceleration phases of the flexion are clearly distinguished, which is the consequence of how the movement is specified (1). In the case of A=10grad the deceleration phase can be executed in two ways: first, with the aid of two carpus extensors (ECU and ECR) and, second, with the aid of one extensor and one flexor (ECU and FCR). The only way to execute the acceleration phase here is with two flexors (FCR and FCU). On the contrary, for A=30grad this is the acceleration phase which can be executed differently (either with FCR and FCU or with ECR and FCU), while the deceleration one is unambiguously implemented with the aid of ECR and ECU. Then the movement amplitude affects not only the values of muscle efforts, but also the group of muscles, which could implement the movement.

Another conclusion, which can be made from the results above is that the co-operation of antagonists may be of pure mechanical nature and can be explained by the complexity of the joint with two degrees of freedom: the muscles have not only to flex the wrist but also to prevent it from abduction.

To demonstrate the effect the environmental parameters have on the pattern of muscular activity, let's consider the flexion of the hand in radiorarpeal joint under different directions of the gravity. If the gravity is orthogonal to the palm surface and its direction matches that of flexion, then the gravity "favours" the flexion. For fast movements the action of the gravity may be insufficient to implement the flexion, in which case wrist flexors are engaged. For slower movements the situation can occur when wrist extensors solely implement the flexion. The boundary between "fast" and

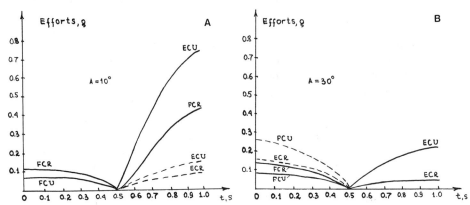

Figure 4. Figure 4-A. Time dependencies of muscle forces produced during flexion in radiocarpal joint with amplitude A=10 grad: joint angle is equal to zero at the beginning of the movement and is equal to 10 grad at the end of the movement; duration of motion is 1 sec. Two modes of deceleration phase implementation are represented by solid and dashed lines respectively. **Figure 4-B.** Time dependencies of muscle forces produced during flexion with amplitude A=30 grad: joint angle is equal to zero at the beginning of the movement and is equal to 30 grad at the end of the movement; duration of motion is 1 sec. Again, two modes of deceleration phase implementation are represented by solid and dashed lines respectively.

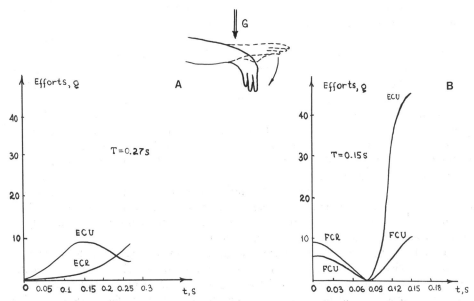

Figure 5. Time dependencies of muscle forces during the flexion of the hand in radiocarpal joint under the gravity; the gravity direction is denoted out by double arrow. **Figure 5-A.** Movement duration is equal to 0.27 sec. **Figure 5-B.** Movement duration is equal to 0.15 sec.

"slow" movements (in the sense above) for the maximal flexion amplitude (about 80 grad.) was found to be 230 ms. Examples of "fast" and "slow" movement implementations are shown in figure 5.

Patterns of muscular activity in the process of the same flexion under different directions of the gravity are presented in figure 6. Evidently, gravity direction is of crucial importance in muscle efforts as well as in which groups of muscles are involved in the movement.

With another mode of radiocarpal joint movement modelling we consider the system "wrist + finger" having six degrees of freedom (two in radiocarpal plus two in metacarpophalangeal plus two in interphalangeal joints) and nine muscles (FCR, FCU, ECR, ECU, ID, IP, FDS, FDP, EDC).

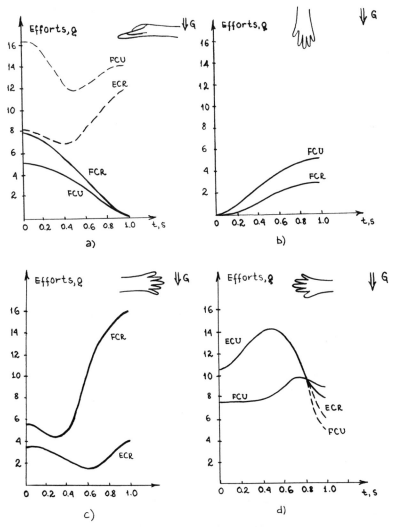

Figure 6. Time dependencies of muscle forces during the flexion in radiocarpal joint with the amplitude of 80 grad (joint angle is equal to zero at the beginning of the movement and to 80 grad at the end of the movement). Movement duration is 1 sec. The gravity direction is denoted by double arrow. Alternative modes of movement implementation are represented by dashed lines.

Two patterns of muscular activity were found when having performed the flexion in radiocarpal joint. In the first case the efforts of carpus flexors (FCU and FCR) and digit flexors (FDS and FDP) were of the same order (Figure 7a), in the second one digit flexors generated little efforts compared with those produced by carpus flexors (Figure 7b). We see, therefore, that the co-operation of wrist's and finger's muscles changes the pattern of muscular activity during the radiocarpal flexion. For example, this flexion cannot be implemented by two wrist flexors but only by the couple flexor-extensor. Finger flexors themselves cannot flex the wrist without wrist flexors.

The total muscle effort in the first case (Figure 7a) is greater than that in the second one (Figure 7b). These results being compared with experimental EMG recording in the process of training of hand skilled movements[28] allow to suppose that non-trained subject attempts to distribute efforts uniformly among the muscles while the well trained one minimizes the total muscle effort.

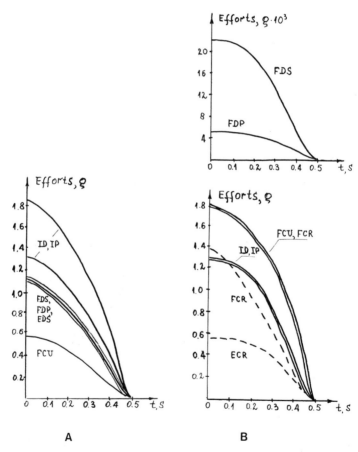

Figure 7. Time dependencies of muscle forces during flexion in radiocarpal joint. Movement amplitude is 80 grad (joint angle is equal to zero at the beginning of the movement and to 80 grad at the end of the movement). Movement duration 1 sec. **Figure 7-A.** In the first case forces produced by carpus and digit flexors are of the same order. **Figure 7-B.** In the second case digit flexors generated smaller forces as compared with those produced by carpus flexors.

Muscles' Operation During Finger Movements

Each finger has four degrees of freedom (two in metacarpophalangeal plus two in interphalangeal joints). Finger links are brought into action by six muscles (EDC, FDS, FDP, ID, IP, LUMB) (Figure 8). As finger extensor and flexors pass through several joints, the question arises, if independent movements in finger joints are still possible? The results of modelling allow us to positively answer the question.

The movements in finger joints were modelled according to the equation (1), with T=2.0 s, A=90 grad for all finger joints. The pattern of muscular activity during this movement is presented in figure 8(a,b). Having amplitudes and durations of joint movements been changed, the same muscles appeared to contribute to the movement, namely, EDC, FDP, FDS, IP, ID. The interossei produce the efforts of two orders greater than those of finger extensor and finger flexors. The fact that interossei do take part in finger flexion has been pointed out both in anatomical[2] and electrophysiological studies.[26] EMG recordings and the theoretical analysis of muscle efforts as well[14] show that these muscles are much more active during finger flexion than finger extensor and flexors.

Under certain values of amplitudes and durations lumbricales also take part in the movement not in place of any other muscles, but affecting only the latter's efforts. Therefore the finger muscle apparatus is not in fact redundant. It should be noted that

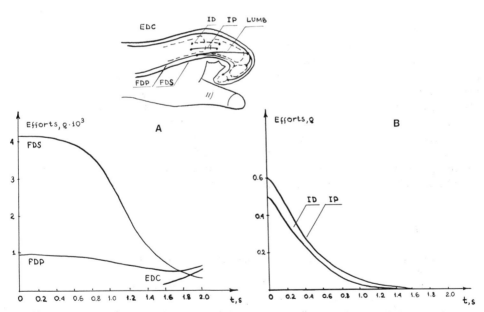

Figure 8. Time dependencies of muscle forces during finger flexion in metacarpophalangeal and two interphalangeal joints. Amplitude of flexion in each joint is equal to 90 grad (joint angles are equal to zero at the beginning of the movement and to 90 grad at the end of the movement), flexion duration is 2 sec. **Figure 8-A.** Forces produced by extrinsic muscles. **Figure 8-B.** Forces produced by intrinsic muscles.

the simultaneous activity of antagonists is the necessary condition for to implement the multijoint finger flexion.

The modelling of thumb movements shows that its principal movements can be implemented in many ways. At different stages of a movement different groups of muscles are brought into action. The principal thumb movements appeared to be impossible without intrinsic muscles, which produce much greater efforts than those produced by extrinsic muscles.

CONCLUSIONS

The model above only approximately describes real hand dynamics. The principal assumptions in the model are as follows: [1] muscles are considered to be threads with their origins/insertions being points; [2] such significant components of hand motor apparatus as ligaments are not dealt with;[12] [3] changes in muscle masses, which are due to muscle contractions, are ignored; [4] the method used to overcome the redundancy is the minimization of the total muscular effort in spite of some experimental data demonstrating that this is inadequate from the physiological point of view.[36]

Without these assumptions, however, it is hard to describe the dynamics of principal hand movements. We hope that more precise and detailed study based on new experimental data will give although more precise results, but not ones which would contradict to those below: [1] the problem of muscular apparatus redundancy should be considered as intimately related with the movement to be implemented; according to the parameters of the movement and external parameters, muscular apparatus, being redundant from the formal mechanical viewpoint, may implement the movement unambiguously or may give no acceptable implementation at all; hand muscular apparatus appeared to be actually redundant only for the movements in radiocarpal joint and those of thumb; as for the other fingers, the sole group of muscles responsible for their movements exists; [2] in spite of the fact that flexions/extensions of fingers are performed by multijoint muscles, independent movements in fingers' joints are still possible: it seems to be central nervous system not the structure of muscular apparatus what is responsible for the coordinated changes in the joint angles observed in practice; [3] the analysis of movements in complex joints and of coordinated movements in several joints assures that simultaneous activity of anatomical antagonists may be of pure mechanical nature; this fact must be taken into account in experimental data analysis to distinguish between pure mechanical effects and those related to nervous control; [4] no finger movements are possible without intrinsic muscles, which generate the efforts much greater than those produced by extrinsic muscles.

REFERENCES

1. E.V. Alexandrovich, Some results of the mathematical modeling of the human eyes muscular apparatus, *Biofizika*, 25:902 (1980).
2. K.N. An, E.Y. Chao, W.P. Cooney, and R.L. Linscheid, Normative model of human hand for biomechanical analysis, *J. Biomech.*, 12:775 (1979).
3. T. Armstrong, and D. Chaffin, An investigation of the relationship between displacements of the finger and wrist joints and the extrinsic finger flexion tendons, *J. Biomech.*, 11:119 (1978).
4. K.M. Backhouse, and W.T. Catton, An experimental study of the functions of the lumbrical muscles in the human hand, *J. Anatomy*, 88:133 (1954).
5. J.V. Basmajian, and M.A. McConaill, "Muscles and Movements", Williams & Wilkins, Baltimore, (1969).
6. N. Berme, J.D. Paul, and W.K. Purves, A biomechanical analysis of the metacarpophalangeal joint, *J. Biomech.*,10:409 (1977).

7. N.A. Bernstein, "Notes on Movement Physiology and Physiology of Activity", Medicina Publishers, Moscow, (in Russian), (1966).

8. E.V. Biryukova, "The Dynamics of Muscle-Bone Apparatus of the Human Hand", Ph.D. Thesis, Department of Mechanics and Mathematics, Moscow State University, (1986).

9. E.V. Biryukova, M. Dufosse, A.A. Frolov, M. Ioffe, and J. Massion, Biomechanical model of quadruped with flexible spine, "Electrophysiological Kinesiology" (W.Wallinga, H.B.K.Boom and J. de Vries eds.), Elsevier (Biomedical Division), 481 (1988).

10. E.V. Biryukova, and A.K. Platonov, On muscle apparatus excessiveness of the human hand, "Biomechanics in Medecine and Surgery", (V.K.Kalnberz ed.), Riga, 173, (in Russian), (1986).

11. B.M. Braude, "About the Mechanism of Radiocarpal Joint", St.Petersburg, 3, (in Russian), (1883).

12. H.J. Buchner, M.J. Hines, and H. Hemami, A dynamic model for finger interphalangeal coordination, J. Biomech., 21:459 (1988).

13. E.Y. Chao, and K.N. An, Determination of internal forces in human hand, J. Eng. Mech. Division, ASCE, 104:255 (1978).

14. E.Y. Chao, J.D. Opgrande, and F.F. Axmear, Three-dimensional force analysis of finger joints in selected isometric hand functions, J. Biomech., 9:387 (1976).

15. W.P. Cooney, and E.Y. Chao, Biomechanical analysis of static forces in the thumb during hand functions, J. Bone Joint Surg., 59-A:27 (1977).

16. D.L. Eyler, and J.E. Markee, The anatomy and function of the intrinsic musculature of the fingers, J. Bone Joint Surg., 36-A:61 (1954).

17. R. Fick, "Handbuch der Anatomie und Mechanik der Gelenke und Beruchsichtigung der Bewengenden Muskeln. Bd.3. Specielle Gelenks und Muskelmechanik", Jena, (1911).

18. A.A. Frolov, E.V. Biryukova, and M.E. Ioffe, On the influence of movement kinematics on the support pressure pattern during postural adjustment in quadrupeds, "Stance and Motion. Facts and Concepts", (V.S. Gurfinkel, M.E. Ioffe, J. Massion, J.-P. Roll eds.), Plenum Press, New York, 229 (1989).

19. R.W. Haines, The extensor apparatus of the finger, J. Anatomy, 85:251 (1951).

20. T. Hara, A study of mechanism of action of the extensor complex in human fingers, Fukuoka Acta Medica, 51:796 (1960).

21. D. Hirsch, D. Page, D. Miller, J.H. Dumbleton, and E.H. Miller, A biomechanical analysis of the meta-carpophalangeal joint of the thumb, J. Biomech., 7:343 (1974).

22. G.V. Korenev, "Goal-oriented mechanics of guided manipulators", Nauka Publishers, Moscow, (in Russian), (1979).

23. G.V. Korenev, and V.S. Pridvorov, About muscle efforts during goal-oriented movements, Izvestia Academii Nauk, Technical Cybernetics, 4:89 (1977).

24. J.M.F. Landsmeer, The anatomy of dorsal aponeurosis of the human finger and its functional significance, Anat. Record, 104:31 (1949).

25. P.F. Lesgaft, "About Bone Connection", St.Petersburg, (in Russian), (1866).

26. C. Long, and M.E. Brown, Electromyographic kinesiology of the hand: Part III. Lumbricales and flexor digitorum profundus of the long finger, Arch. Phys. Med. and Rehab., 43:450 (1962).

27. C. Long, M.E. Brown, and G. Weiss, Electromyographic kinesiology of the hand: Part II. Third dorsal interosseus and extensor digitorum of the long finger, Arch.Phys.Med.and Rehab., 42:559 (1961).

28. G.B. McFarland, U.L. Krusen, and H.T. Weathersby, Kinesiology of selected muscles acting on the wrist: electromyographic study, Arch. Phys. Med. and Rehab., 43:165 (1962).

29. R. Montant, and A. Baumann, Recherches anatomiques sur le système tendineux extenseur des doigts de la main, Ann. Anat. Pathol., 14:311 (1937).

30. C.W. Spoor, Balancing a force on the fingertip of a two-dimensional finger model without intrinsic muscles, J. Biomech., 16:497 (1983).

31. E. Sunderland, The action of the extensor digitorum communis, interosseus and lumbrical muscles, Am. J. Anatomy, 77:189 (1945).

32. R. Toft, and N. Berme, A biomechanical analysis of joints of the thumb, J. Biomech., 3:353 (1980).

33. I.I. Tzuran, "About the Relation of Muscle-Antagonists of the Human Body", St.Petersburg, (in Russian), (1882).

34. V.S. Varavin, "Materials for the Problem of Different Force Exibition by the Muscle of Upper and Lower Extremities", St.Petersburg, (in Russian), (1882).

35. H.J. Woltring, R. Huiskes, A. de Langer, and F.E. Veldpaus, Finite centroid and helical axis estimation from noisy landmark measurements in the study of human joint kinematics, J. Biomech., 18:379 (1985).

36. B.P. Yeo, Investigations concerning the principle of minimal total muscular force, *J. Biomech.*, 9:413 (1976).
37. Y. Youm, T.E. Gillespie, A.E. Flatt, and B.L. Sprague, Kinematic investigation of normal MCP joint, *J. Biomech.*, 2:109 (1978).
38. V.Z. Yourovskaya, "Investigation of Some Anatomical Parameters which affect the Arm Movement", Report of Biological Department of Moscow State University, Moscow, (in Russian), (1974).
39. V.Z. Yourovskaya, The pecularities of the fastening of the upper extremity muscles in the comparative aspects, *Voprosy Antropologii*, 70:38 (in Russian), (1982).

BIOMECHANICAL ANALYSIS OF THE WRIST JOINT IN ERGONOMICAL

ASPECTS

Krystyna Gielo-Perczak

Institute of Mechanics and Design
Warsaw University of Technology
02-524 Warsaw, Poland

INTRODUCTION

The mechanics of wrist movement is a current but relatively unexplored issue. Analysis of the current models and structure of the wrist joint shows that no general method of research exists for static analysis in the work place. In available references, the authors have made many special assumptions. In many respects, their modeling results did not confirm their experimental results.

We can assume that, based on biomechanical analysis, no general method of research exists in the modeling and defining of muscle and ligament stress servicing the wrist joint in statics conditions. Due to the complexity of the studied object, we need to make numerous simplifying assumptions and introduce into the equations varving data resulting from experiments. Numerous papers devoted to anatomic descriptions and functioning of the hand point to a number of important aspects that were not considered during previous biomechanical analysis of the joint: [1] structure of the hand which can be complicated and changing, depending upon the choice of the plane of motion; [2] the irregularity of the working joint surfaces at the radiocarpal level; [3] the dependence of the type of contact surface on the sex and the age of person under examination; [4] the change of the center of rotation, depending on the range of hand motion; [5] the length of ligaments; [6] the occurrence of joint laxity.

We need to formulate a new hypothesis of model to demonstrate the mechanism and structure of the wrist joint. This discussion analyses the wrist joint mechanism and structure. We consider in particular the geometry of the joint, specifically the geometry of the joints surfaces. A model of the wrist joint is created to define the forces developed by muscles during various positions assumed by the hand. First, the cooperation of muscles and ligaments servicing the wrist joint can be described by means of a statically indeterminate model of a variable structure that would include the influence of the loose joint as well as the geometry of the joint surfaces. Second, an appropriate choice of the acetabulum profile of the ulna and radius can help to match a person better with his or her working activity and to analyze the frequency of wrist joint overloading. Third, the analysis of jobs is necessary to identify and control risk factors of occupational cumulative trauma disorders.

PROBLEM OF MODELING OF WRIST JOINT

We propose a concept of so called "characteristic fibre", which we define as a set of points and segments joining these points that determine the shape of the muscle and the points of contact with other muscles (Figure 1). We introduce the notion of wrist joint load capacity into biomechanical analysis. The wrist joint load capacity is a quantity of the external loading that, if surpassed, disables the hand so that it can no longer carry the load because of damaging stresses that occur in the muscle, ligaments, and in the contact point of the radius and ulna with the proximal carpal bones.

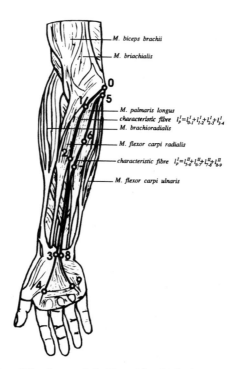

Figure 1. Presentation of the characteristic fibres 1_p^I and 1_p^{II} of the musculus palmaris longus.

Physical Model and Mathematical Description

The principal assumptions and limiting conditions for the geometry of the model of the wrist joint are as follows: [1] the joint in the said plane is assumed to have two degrees of freedom: theoretical research has been conducted during wrist deviation; [2] in the model, there is a cooperation of two muscles, the flexor carpi ulnaris and the flexor carpi radialis; [3] in the biomechanical analysis, a cooperation of the ligaments is proposed; the activity of both the ligament collaterale carpi ulnare and the ligament collaterale carpi radiale have been taken into consideration (Figure 2); [4] the analysis was carried on in the radial-ulnar deviation in the total range 60°; [5] the joint is loose; [6] while doing the analysis allowances have been made for a displacement of the instantaneous center of rotation; [7] in defining the forces occurring in ligaments and muscles, the equilibrium conditions, construction, and strength conditions have been considered; [8] analysis of the forces acting on the joint has been restricted to the conditions of statics; to define the forces in muscles and ligaments in the circumstances when the system is statically undefinable, the forces method has been used; [9] in calculations, the surface of the proximal carpal bones (PCB: scaphoid, lunate, and triquetral bones) is assumed to be ellipsoidal; [l0] the surfaces of the acetabulum of the ulna and radius (AUR) have been described, with a pencil, of straight lines N tangential to the acetabulum profile, bisecting the Z axis at an angle ϕ (N) at a distance z (N) from the beginning of the coordinate system, as well as with the value of the initial and final point of the acetabulum profile, measured along that the number of straight lines N= 7 renders the shape of the profile with a fair approximation (Figure 3); [11] the types of joints and their contact surfaces which in the modeling are (Figure 4): type A the surface of the PCB and the surface of the AUR have the same radius, type B: the surface of the PCB has a smaller radius than that of AUR; [12] the length of muscle acton is determined by the sum of the segments whose number depends on the degree of precise reconstruction of the line of a selected muscle fibre (characteristic fibre):

$$l_c\ (\alpha,\beta,\gamma\ ) = \sum_{i=1}^{n} l_i\ (\alpha,\beta,\gamma)$$

where:

(α,β,γ) are the angles measured in the lateral, the horizontal, and the frontal planes; acton - is the muscle or part of the muscle which muscle fibres displaying the force have the same or similar direction towards joints rotation axes.

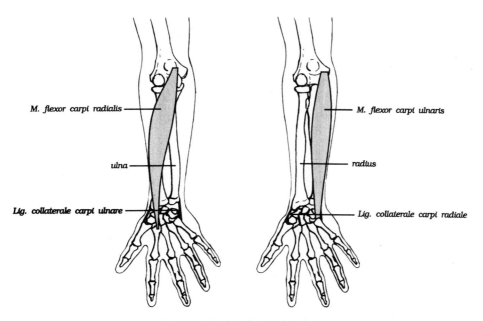

Figure 2. Model of the wrist joint.

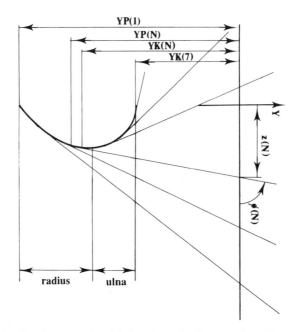

Figure 3. The surfaces of the acetabulum of the ulna and radius have been described with a pencil of straight lines N tangential to the acetabulum profile.

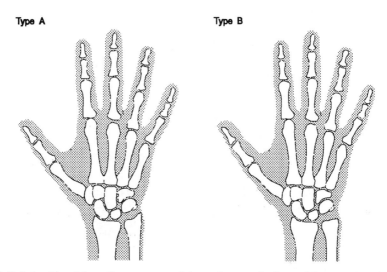

Figure 4. Relationship of the radius curvature of the surface proximal carpal bones and surfaces of the acetabulum of the ulna and radius.

The static equations for the assumed model of the wrist joint are the following (Figure 5):

$$M_A(N) = Pl_2 \sin\alpha_1 - FM_1(N)R + FM_2(N)h_{W3}(N) - FW_1(N)hw_1(N) + \\ + FW_2(N)hw_2(N) + M_2 \tag{1}$$

$$R_Z(N) = P\sin(\alpha+\alpha_1) + FM_1(N)\cos\kappa(N) + FM_2(N)\cos\kappa_1(N) + FW_1(N)\cos\Psi_1(N) + \\ + FW_2(N)\cos[\upsilon_3(N)+\upsilon_4(N)] - FS(N)\sin\phi(N) \tag{2}$$

126

$$R_Y(N) = P\cos(\alpha+\alpha_1) + FM_1(N)\sin\kappa(N) - FM_2(N)\sin\kappa_1(N) + FW_1(N)\sin\Psi(N) +$$
$$+ FW_2(N)\sin[\upsilon_3(N)+\upsilon_4(N)] + FS(N)\cos\phi(N) \qquad (3)$$

where:

$P, M2$	- reduced components of the external loading,
N	- a straight line tangent to the acetabulum of the radius and ulna profile,
$FM_1(N), FM_2(N)$	- forces developed in muscles M_1 and M_2, respectively,
$FW_1(N), FW_2(N)$	- forces developed in ligaments W_1 and W_2, respectively,
$FS(N)$	- forces of the contact of the proximal carpal bones with the acetabulum of the radius and ulna,
α	- joint angle measured in the radial and ulnar deviation,
$\alpha1$	- angle at which the external loading forces is applied in relation to hand,
$R, l_2, h_{w1}(N), h_{w2}(N), \kappa(N), \kappa1(N), \upsilon_3(N), \upsilon_4(N), \phi(N)$	
	- the quantities defining the geometry of the joint.

Whether the flexor carpi ulnaris, or the flexor carpi radialis is used depends on the direction of the motion. The ligaments can assume three possible states: the ligament collaterale carpi ulnare (LCGU) is tense and the ligament collaterale carpi radiale (LCCR) is loose, or the LCCU is loose and the LCCR is tense, or both are tense.

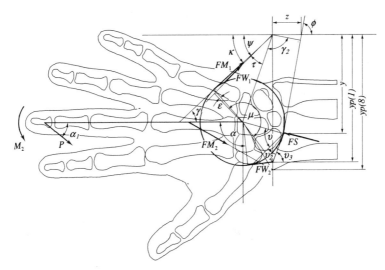

Figure 5. Two-dimensional mechanical model of the wrist joint.

To define the forces in muscles and ligaments under circumstances when the system is statically indeterminate the forces method has been employed. As a result, two zero-state sets of equations have been obtained:

$$M_{W1}{}^{(0)}(N) = Pl_2 \sin\alpha_1 - FM_1{}^{(0)}(N)R - FW_1{}^{(0)}(N)h_{W1}(N) + M_2 \qquad (4)$$

$$R_{W2}{}^{(0)}(N) = P\sin(\alpha+\alpha_1) + FM_1{}^{(0)}(N)\cos\kappa(N) + FW_1{}^{(0)}(N)\cos\Psi(N) +$$
$$- FS^{(0)}(N) \sin\phi(N) \qquad (5)$$

$$R_{W3}(N) = P\cos(\alpha+\alpha_1) + FM_1(N)\sin\kappa(N) + FW_1^{(0)}(N)\sin\Psi_1(N) + $$
$$- FS^{(0)}FS(N)\cos\phi(N) \qquad (6)$$

in the "1" state

$$M_{W1}^{(1)}(N) = FM_1^{(1)}(N)R + FW_1^{(1)}(N)h_{W1}(N) - FW_2^{(1)} h_{W2}(N) \qquad (7)$$

$$R_{W2}^{(1)}(N) = - FS^{(1)}(N)\sin\phi(N) + FM_1^{(1)}\cos\kappa(N) + FW_1^{(1)}(N)\cos\Psi(N) + $$
$$- FW_2^{(1)}(N)\cos[\upsilon_3(N)+\upsilon_4(N)] \qquad (8)$$

$$R_{W3}^{(1)}(N) = FS^{(1)}(N)\cos\phi(N) + FM_1^{(1)}\sin\kappa(N) + FW_1^{(1)}(N)\sin\Psi(N) + $$
$$- FW_2^{(1)}(N)\sin[\upsilon_3(N)+\upsilon_4(N)] \qquad (9)$$

where the forces in muscles $M1$ and $M2$ are as follows:

$$FM_1(N) = FM_1^{(0)}(N) + FM_1^{(1)}(N) \qquad (10)$$

$$FM_2(N) = FM_2^{(0)}(N) + FM_2^{(1)}(N) \qquad (11)$$

The forces in ligaments $W1$ and $W2$ are:

$$FW_1(N) = FW_1^{(0)}(N) + FW_1^{(1)}(N) \qquad (12)$$

$$FW_2(N) = X_1(N) \qquad (13)$$

The contact force of the proximal carpal bones with the acetabulum of the ulna and radius

$$FS(N) = FS^{(0)}(N) + FS^{(1)}(N) \qquad (14)$$

The value of $X_1(N)$ is calculated according to the dependency:

$$X_1(N) = - \left[FM_1^{(0)}(N) * FM_1^{(1)}(N) + FW_1^{(0)}(N) * FW_1^{(1)}(N) * \frac{lW1(N)EM1AM1}{lM1(N)EW1AW1} \right] /$$

$$* \left\{ [FM_1^{(1)}(N)]^2 + [FW_1^{(1)}(N)]^2 * \frac{lW1(N)EM1AM1}{lM1(N)EW1AW1} + \right.$$

$$\left. + \frac{lW2(N)EM1AM1}{lM1(N)EW2AW2} \right. \qquad (15)$$

The model of the wrist joint presented can be used to analyse the effect of the geometric parameters, characterizing the cooperating joint, on the value of the force in muscles and ligaments in the function of the position of the hand toward the forearm, and the loading conveyed by the hand.

Analysis of Joint Loading

According to the assumed load capacity definition in the biomechanical systems, we prove the thesis that the geometry of the joint surface is one of the factors that determine human physical strength.

The values of the load capacity, for example of the muscle Ml, are calculated from the following dependence:

$$NM_1(N) = \frac{P\,(\delta M1AM1 - \{M2\cos[\Psi(N) -\phi(N)]\} \,/\, [NM^{(0)}\,(N)]\,)}{FM\,1\,(N) - \{M2\cos[\Psi(N) - \phi(N)]\} \,/\, [NM^{(0)}\,(N)]} \tag{16}$$

RESULTS

We have analysed (1) the forces taking place in the elements of the joint that carry the load, and (2) the boundary load capacity for various admissable positions of the points of contact of the PCB in relation to the AUR, which are the functions of the hand position as well as the external loading. During the analysis of the forces in the carrying elements of the joint, the calculations for two shapes of the AUR have been performed. The analysis was carried out for two types of loads: [1] carrying a 2-kilogram object; [2] pushing a box on the table; the force applied to the hand is 10 N. We obtained sets of PCB positions admissible from the point of view of strength, and their values of the forces in muscles, ligaments, and the forces of the contact for the loads and both profiles. The results are presented in figures 6-9 where the value of the forces in ligaments 1 and 2, in muscle 1 and 2, and the value of the force of the contact between the PCB and AUR are shown. Analysis of the load of the joint with the acetabulum 1 is shown in figures 6 and 7; with the acetabulum 2 in figures 8 and 9 subjected, respectively to loads I and II. It can be seen in the figures that the position of the contact points of the PCB with the AUR, marked with arrows, and their number for the given load depends on the individual geometric features of the acetabulum. If we employ the notion of a maximum load capacity of the joint, it turns out that in the joint of an acetabulum profile 1, the PCB in the case of load I takes the position 5, and for load II, it takes also the position 5.

In the joint of an acetabulum profile 2 the PCB of in the cases of loads I and II take the position 4.

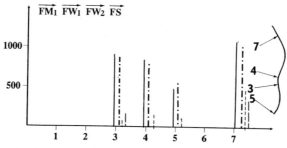

Figure 6. Analysis of loading of the wrist joint to load I with the profile 1 of the AUR. Forces in ligament 1 $FW1$(———), ligament 2 $FW2$ (– – –), musde $FM1$ (– • –), and of contact FS (– —).

The calculations proved that in the case of load I, smaller values of forces in muscles and ligaments occur in the joint with the acetabulum profile 1, whereas in the case of the load II, the values of the forces occur in the joint where the acetabulum has the profile 2. It is necessary to stress the fact that it is impossible to explicitly state which of the profiles is better, if we consider them from the point of view of those two loads. With the increase of the number of loads performed by the hand the calculations confirm the individual character of the carrying of the load by the hand according to the geometry of the wrist joint.

Along with the growth of the external force P, the values of the forces in the muscles and ligaments and the force of the contact increase. Analysis of the forces in muscles and ligaments and of the force of contact as well as the analysis of the boundary load capacity shows that their values are influenced by the geometry of the joint and by the type of loading carried by the hand. The calculations that have been carried out show that certain hand positions under given loads are out of the question for joints or certain definite acetabula. This conclusion finds its application in ergonomics, where there exists the need to define the so-called safe levels of manual forces. The results of such an analysis may be useful to traumatic surgery of the wrist joint.

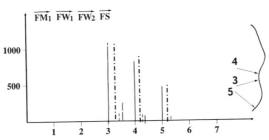

Figure 7. Analysis of loading of the wrist joint to load II with the profile 1 of the AUR. Forces in ligament 1 $FW1$ (———), ligament 2 $FW2$ (– – –), muscle $FM1$ (– • –), and of contact FS (– —).

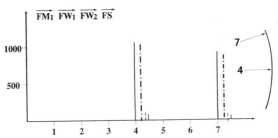

Figure 8. Analysis of loading of the joint to load I with the profile 2 of the AUR. Forces in ligament 1 $FW1$ (———), ligament 2 $FW2$ (– – –), musde $FM1$ (– • –), and of contact FS (– —).

METHOD OF STATIC LOAD EVALUATION OF THE WRIST JOINT ON WORKING STANDS

The paper discussed an application of the biomechanical model of the wrist joint in ergonomical activities, including the variety of joint types found in the population. An optimum shape AUR has been presented, achieved by means of a multicriterion optimization from among the specifically chosen samples of the wrist joint that had been subjected to various loads.

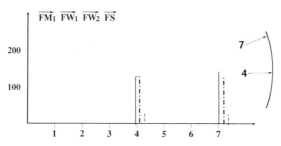

Figure 9. Analysis of loading of the joint to load II with the profile 2 of the AUR. Forces in ligament 1 $FW1$(——), ligament 2 $FW2$ (– – –), musde $FM1$ (– • –), and of contact FS (– —).

The essence of the proposed procedure resolves itself into the choice of such a worker for the working process, whose surfaces profile at the radiocarpal level of the wrist joint is known, to reduce the harmfulness of the joint loading to a minimum. In choosing the carpus of a maximum multicriterion functionality, the weight-correlation method was employed. This method solves the problem of adjusting the criteria, that is, characterizing the solution with one number established on the basis of partial utilities. The calculations proved that for certain types of occupations there exists an optimum shape of the joint surface for which the wrist joint reaches the maximum load capacity.

In the course of the procedure the following has been determined: [1] the set of working activities that cause considerable loading of the wrist joint; [2] the set of the acetabulum of ulna and radius profiles characteristic for the population was given; [3] the set of parameters for the evaluation of the joint loading during the realization of the production process in a definite period of time; [4] the interrelations between the set of joint profiles and the parameter of the working process.

In the above presented procedure, a certain multicriterion optimization method was used. The investigations were restricted to a certain class of chosen working process in a lamp factory. Ten typical kinds of activities were distinguished (Figure 10), performed at the assembly line.

A set of the AUR profiles were included in the analysis (Figure 11). Each type of hand activity was defined by giving [1] the angle position of the hand in relation to the forearm (α); [2] the direction and sense of the loading force (α_1); [3] the value of the force P applied in the middle of the palm.

The research was concerned with the movement of the hand in the frontal plane in the range of 20° (radial deviation) - 40° (ulnar deviation) as shown by Armstrong[1] and NASA[3] for different external loading values from 5 N to 30 N. The angle of the external load application in relation to the hand axis is contained within the range 60° - 150°. Using the values of the geometric parameters of the AUR as well as the data concerning the loading, the maximum load capacity for the respective acetabulum profiles during the realization of 10 different types activities has been calculated by means of an appropriate algorithm. The results are presented in table 1. In choosing the AUR of a maximum multicriterion functionality, the weight-correlation method was used because it solves the problem of adjusting the criteria, characterizing the solution with one number established on the basis of partial utilities.

The weight-correlation method is the method of scalarizing the utilities obtained, assuming that the issue is a multicriterion one, consisting in maximizing the weighted average coefficient of the resulting correlation between the compromise utility and the partial utilities as shown by Brzozowski and Pogorzelski.[2] For the instance of

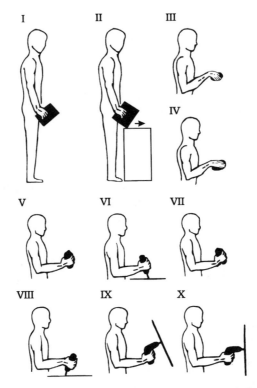

Figure 10. Ten typical of activities performed at the assembly line in lamp factory study.

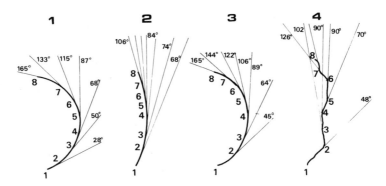

Figure 11. Acetabulum of ulna and radius profiles.

optimization, a weight-correlation criterion is offered, generated by the given finite set of creations representing the space in which the optimization is to be carried out. The optimization was conducted under conditions determined by the hand motion at the work stand on which the chosen 10 activities were realized. The maximum values of the joint load capacity during performing the 10 chosen types of activities, calculated for the 4 chosen radius and ulna acetabulum profiles have been regarded as partial criteria, where u^I_k, ...,u^X_k, $k = 1,...,4$. Partial criteria are treated as different evaluation aspects of the given acetabula.

Partial criteria are treated as different evaluation aspects of the given ulna and radius acetabula. A problem of adjusting q_1 ,..., q_{10} criteria arises by characterizing them with one number, determined on the basis of the calculated partial utilities $u1$,..., $u10$. This number is called a compromise utility u_o and it is the result of employing a certain compromise criterion q by the decision maker. The greater the value of u_o the better profile from the point view of the compromise criterion $q0$. It will enable the making of a ranking list of the profiles from the accepted set to assign the best of them for the given type of activity. As a result of the partial utilities proposed in the above presented way a ranking list of the profiles for the investigated workstand has been obtained. Table 2 shows that for the distinguished activities, the wrist joints with radius and ulna acetabulum profiles 4 is the best. The worst is the joint with profile 2.

Table 1. Maximum values of wrist joint capacity with the AUR profiles 1-4 during performance of activities I-X.

No. of profiles	Value for each activity									
	I	II	III	IV	V	VI	VII	VIII	IX	X
1	66	78	122	84	0	155	0	18	354	22
2	50	88	0	63	0	102	0	16	192	20
3	67	112	124	85	0	156	0	18	0	23
4	74	309	134	93	105	169	0	61	187	26

Table 2. Ranking list of the AUR profiles.

No.	Number of the profile	Compromise criterion
1	4	1.448
2	3	0.031
3	1	-0.103
4	2	-1376

On the basis of the knowledge of partial utilities the final form of the compromise criterion has been computed by means of the WAKOR program.

$$q0 = 0.957q1 + 0.851q2 + 0.842q3 + 0.945q4 + 0.836q5 + 0.906q6 +$$
$$+ 0.859q8 - 0.059q9 + 0.990q10 \qquad (17)$$

The values of wrist joint load capacity with every profile 1-4 for VII activity are zero so we assume that $q7 = 0$.

Table 3. Ranking list of activities.

No.	Number of the activities	Compromise criterion
1	IX	2.268
2	VI	0.875
3	II	0.830
4	IV	0.006
5	III	-0.147
6	I	-0.237
7	VIII	-0.769
8	X	-0.822
9	V	-0.847

Thus the participation of the individual criteria in the compromise criterion could have been fully observed. It implies not only the mutual meaning of the particular loadings but also that biomechanical capabilities of investigated joints have an influence on the final form of the criterion.

Another aspect of this problem is the analysis of the work stands to identify risk factors. Generally the hand posture in the activity VII is risky which brings the damaging stresses. Table 3 shows that for distinguished profiles, activity IX and VI are very dangerous for the wrist joint.

CONCLUSIONS AND SUGGESTIONS FOR FURTHER RESEARCH

We have proposed a proper biomechanical model of the wrist joint as a statically indeterminate system of a variable structure, where both muscles and ligaments cooperate. We defined a concept of a "characteristic fibre" of the muscle. Its choice depends on the muscle shape and the position of the muscle examined in relation to the muscles that are in contact with it.

An essential property of the assumed model of the wrist joint is the possibility of evaluating the influence of the geometric parameters, defining the cooperating joint surfaces, of looseness, the surface of the cross sections, and the lengths of muscles and ligaments on the quantity of their forces in the function of the hand position toward the forearm and the load conveyed by the hand.

We introduced a load capacity concept. The wrist load capacity is a quantity of the external loading, the surpassing of which brings about the disability of the hand to carry the load, owing to damaging stresses that occur in the muscle, ligaments, and in the contact point of the AUR with the proximal carpal bones.

The proposed optimization procedure can be introduced also to other types of working activities and other parameters of the wrist joint or the hand. On the basis of obtained results, a program of further research can be suggested. It would concern the modification and the extension of the model including a three-dimensional model and a greater number of muscles and ligaments or bones.

The concept of the characteristic fibre should be more extensively used when considering the contribution of muscles and their complex shape geometry. We also suggest expanding the range of practical applications for designing work place, and assigning employees to their work stands, examining the arduousness of the chosen activities for the human organism.

REFERENCES

1. T. J. Armstrong. "Biomechanical Aspects of the Upper Extremity in Work", Monograph, University of Michigan, Ann Arbor, (1985).
2. T. Brzozowski and W. Pogorzelski, "Natural values of quality indicators weight-correlation method", *Statistical News*, 42:44 (1992).
3. NASA. "Antropometric Source Book Volume 1: Antropometry for Designers", NASA Reference Publication 1024, National Aeronautics and Space Administration, Washington (1978).

CONTACT PRESSURES WITHIN WRIST JOINTS

Steven F. Viegas and Rita M. Patterson

University of Texas Medical Branch
Division of Orthopaedic Surgery
McCullough Bldg. Rm. 6.136 (G-92)
Galveston, Texas 77550, USA

INTRODUCTION

Normal carpal mechanics relies on the complex interplay between a sophisticated arrangement of carpal ligaments and carpal bone morphology.[25,32] A number of biomechanical studies have been performed to define the kinematics of the normal wrist and the changes in kinematics seen in degenerative states. Early workers studied x-ray films and cadaver specimens to observe relative motions in the wrist.[15,19,25] Investigators had also used quantified, three-dimensional techniques, such as sonic digitizers,[2,5] instrumented electromechanical linkages,[50] radiostereophotogrammetric techniques,[10] or three-dimensional computer imaging,[4] to define precise intercarpal motions.

Speculative patterns of radiocarpal joint contact areas and forces had been proposed, but the contact areas and forces of the wrist joint had only been partly documented [28,41] prior to the work done over the past six years.[42,57,58,59,60,61,62,63,64]

MATERIALS AND METHODS

This work has used a static positioning frame, pressure sensitive film (Fuji) and a microcomputer-based videodigitizing system, which was developed to characterize the biomechanics of the human wrist by measuring the contact areas and pressures in cadaver wrists in a variety of normal, simulated traumatic and surgically treated conditions.[63]

The method of testing involved the use of unembalmed fresh cadaver upper extremities of various ages that were free from any visible or radiographically identifiable deformities and/or degenerative changes. Each specimen was stripped of all its soft tissues except the joint capsule, the palmar and dorsal radiocarpal, ulnocarpal, and interosseous ligaments, the triangular fibrocartilage complex, and the radioulnar interosseous membrane. Each specimen was mounted on a loading jig (Figure 1)

Advances in the Biomechanics of the Hand and Wrist
Edited by F. Schuind *et al.*, Plenum Press, New York, 1994

permitting positioning of the wrist anywhere in its range of motion of flexion, extension and radio-ulnar deviation. The jig also allowed the forearm to be postured in pronation, neutral or supination. Axial loads of different weights, depending on the particular study being done, were applied by weights to a ball joint fixed to 4 mm diameter pins which were fixed into the medullary canals of the second and third metacarpals. The elbow was postured in 90 degrees of flexion and the humerus was fixed with two transverse threaded pins to the base plate of the loading jig.

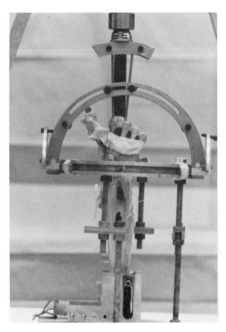

Figure 1. The wrist positioning and loading assembly set up to load the wrist through the second and third metacarpals.

Contact areas and pressures within the radioulnar carpal joint were measured in each position by Fuji prescale super low pressure sensitive paper[25] (C. Itoh, New York, N.Y.). The film was cut to match the surface of the distal radius and triangular fibrocartilage. The joint was distracted and the film inserted through a dorsal capsular incision. Spatial orientation of each area was determined by a "U" marker located externally to the distal radius and included on each print (Figure 2). The print image was then videodigitized and analyzed by a Better Basic program (Figure 3). The analysis calculated values based on pressure calibrations defining the intensity of color on the print. Contact areas, area centroids, and pressures were measured on each print for every position utilizing these color prints.

Tests were performed on the unaltered "normal" wrists and in every other study, depending on the particular issues being addressed, various conditions, injuries and/or procedures were performed on the wrists. The specimens could therefore be used as their own controls.

This research has provided information on the load transfer characteristics of the normal wrist joint.[61,63] It has also demonstrated the effects of ligamentous instabilities, [59,60,62,64] posttraumatic malalignments,[42] fractures[57] and various types of surgical procedures[58,59] on the wrist joint.

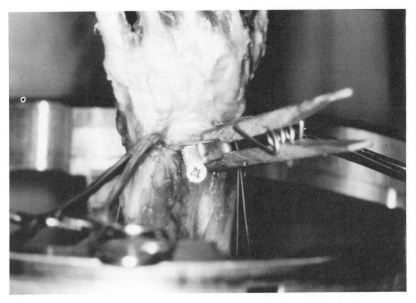

Figure 2. A specimen with a transducer clamped to the "U" marker, transfixed to the distal radius.

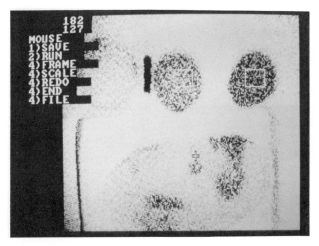

Figure 3. Computer screen showing the analysis of a digitized Fuji print displaying the calibration and scale markers on top.

RESULTS

Normal

Proximal wrist joint. The initial work done with a finite load of 23 lbs. in wrists which were free of any radiographic or gross abnormalities, demonstrated several previously undetermined facts. These included the finding that the scaphoid and lunate bones had separate, distinct areas of contact on the distal radius/triangular fibrocartilage complex surface (Figure 4).[19] The contact areas were found to be localized and accounted for a relatively small fraction of the joint surface (average = 0.206) regardless of wrist position. The contact areas shifted from a palmar location to a more dorsal location as the wrist moved from flexion to extension. This would suggest preferred positions of immobilization for different fractures, for example, in a wrist with a distal radius, volar lip fracture, placement in extension would unload that fracture fragment. Overall, the scaphoid contact area was 1.47 times that of the lunate. The scaphoid contact area was generally greatest with the wrist in ulnar deviation, i.e., with the scaphoid vertically oriented. The scapho-lunate contact area ratio generally increased as wrist position changed from radial to ulnar deviation and/or from flexion to extension. These increases in contact area are consistent with the position of the wrist when maximum grip strength is being obtained. Average high contact pressures varied with joint position; however, they were fairly low (average= 3.17 MPa, range 2 to 5.6 MPA) for this applied functional load of 103N (23 lbs). The intercentroid distance (scaphoid-lunate) averaged 14.91 mm and ranged from 10 to 20 mm with joint positon.

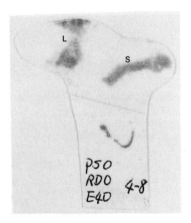

Figure 4. Photo of Fuji film print displaying the scaphoid (S) and lunate (L) contact areas as well as the reference "U" mark.

When normal wrists were loaded over a range of weights, a nonlinear relation was discovered between increasing loads and greater overall contact areas. The general distribution of the contact between the scaphoid and the lunate contact areas was consistent at all of the loads tested with 60% of the total contact area involving the scaphoid contact area and 40% involving the lunate contact area (Figure 5). Loads greater than 46 pounds were not found to significantly increase the overall contact areas implying that the cartilage of the wrist joint was maximally compressed at loads of this magnitude. At loads higher than 46 pounds it appeared that average high pressures increased in a more direct correlation with the increase in weight (Figure 6). The overall contact area even at the highest loads tested were not more than 40% of the available joint surface. The contact areas were not concentric or symmetric which is a characteristic of incongruent joints and demonstrates that the proximal wrist joint is an incongruent joint.[61] Some studies were also done which loaded the normal wrist through a variety of load paths (i.e. two metacarpals, five metacarpals and tendons). This work implied that the distal carpal row acts functionally as a single unit in load transfer,[61] which is consistent with its kinematic characteristics.[44]

Midcarpal joint. Contact area on the proximal side of the midcarpal joint was found to consist generally of four areas; the scaphoid-trapezium-trapezoid (STT), the scaphoid-capitate (SC), the lunate-capitate (LC), and the triquetrum-hamate (TH) (Figure

7). The contact areas accounted for less than 40% of the available joint surface even under loads of 118 lbs. The distribution of load through the midcarpal joint was STT 23.1%, SC 28.2%, LC 28.5%, and TH 20.2%. The midcarpal joint, like the radiocarpal joint, appears to transmit load through distinct areas and through a relatively small portion of the available joint surface. For all positions tested there was no significant difference in the contact area and/or the distribution of contact area in the midcarpal joint from one position to another of the twelve positions tested.

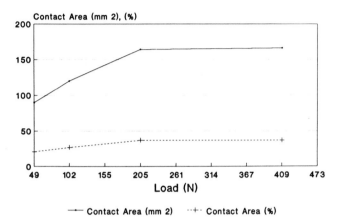

Figure 5. The nonlinear relationship between the contact area and the load. The contact area is normalized as a percentage of the available joint surface.

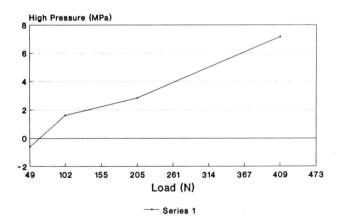

Figure 6. The average high pressure vs load of all wrists in the position of neutral pronation/supination, neutral flexion/extension, and neutral radio/ulnar deviation.

Five wrists were tested in the position of neutral pronation/supination, neutral radial/ulnar deviation and neutral flexion/extension (NNN) at 32, 46, 92, and 118 lb loads. The overall average contact area/joint area of the five wrists in the NNN position at 32 lb was 26.8% at 46 lb was 26.1% at 92 lb was 32.0% and at 118 lb was 36.0%.

The average distribution of the contact areas between the STT, SC, LC and TH in the five wrists was essentially the same at the different loads tested.

The average high pressures measured increased with each increase in load. The average high pressure at a load of 32 lbs was 0.50 MPa, at 46 lbs it was 2.55 MPa, at 92 lbs it was 5.18 MPa, and at 118 lbs it was 5.7 MPa.

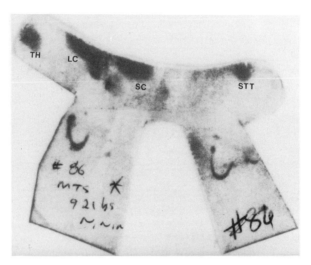

Figure 7. Print of the midcarpal joint with contact areas: scaphoid-trapezium-trapezoid (STT), scaphoid-capitate (SC), luno-capitate (LC), and triquetrium-hamate (TH).

Carpal Instabilities (Radial)

Cadaver wrists in which radial sided perilunate instabilities were simulated, developed areas of increased load. The location within the wrist where these increased loads developed correlated nicely with areas in which degenerative changes develop in patients with the same type of radial sided perilunate instability (i.e. scaphoid instability). Compared to a normal wrist, there is an overall decrease in load in the lunate fossa and a significant increase in load in the scaphoid fossa in the wrist with a Mayfield[33,34] stage III radial sided perilunate instability. In all `stages of perilunate instability, the contact areas remained a relatively small part of the overall joint surface. Pressures were significantly increased in wrists with stage III instability compared with normal wrists. The distance between the scaphoid and lunate contact areas changed little except when the wrist was placed in 20 degrees of extension, neutral radioulnar deviation, and 90 degrees of supination. In this position, there was a significant increase in the intercentroid distance. This correlates nicely with the fact that the stress view which best demonstrates the scapholunate gap on x-ray is the same position (i.e. AP, supinated, clenched fist view).

The effects of increasing perilunate instability on the load transfer characteristics of the wrist included a significant dorsal ulnar shift of the scaphoid centroid with increasing perilunate instability together with a less dramatic palmar ulnar shift of the lunate centroid. These changes in the centroids also correlate nicely with the changes in the carpal alignment with radial sided perilunate instability which has been called DISI.[29,11]

Knowledge gained from this phase of the studies was applied to the study of the comparative biomechanical effects of various surgical treatments for progressive radial sided perilunate instability.

There has been an increasing awareness over the recent years of the clinical entities known as carpal instabilities.[17,20,21,29,33,34,36,37,40,46,53,68] This increased awareness and the concern over these conditions are magnified by the belief that carpal instabilities are a precursor to degenerative changes in the wrist joint.[6,64,66,67,66,69] This phase of studies confirmed our earlier work, which has shown that the contact areas and pressures are concentrated more under the scaphoid and lessened under the lunate with progressive perilunate instability.[64] Watson [66,67,69] has demonstrated that the most common pattern of degenerative changes seen in the wrist is one which is consistent with changes seen in patients with scapho-lunate dissociation. Various limited fusions including the most popular for treatment of radial sided perilunate instability were simulated to attempt to assess the biomechanical efficacy of limited carpal fusions for the treatment of scapho-lunate dissociation.

Scaphoid-trapezium-trapezoid (STT) and scaphoid-capitate (SC) fusions transmitted almost all load through the scaphoid fossa. Scaphoid-lunate (SL), scaphoid-lunate-capitate (SLC) and capitate-lunate (CL) fusions all distributed load more proportionately, although not in exactly the same way as seen in the normal wrist, through both scaphoid and lunate fossae. The positioning of the carpal bones within a limited carpal fusion was also found to affect the load distribution in the wrist.

The SL, SLC or CL fusions, with attention to the relative carpal alignment within the limited fusion, appeared to offer more promise for treatment of perilunate instability biomechanically than the STT or SC fusions. On the other hand the significant loading of the scaphoid and unloading of the lunate seen in the study casts serious questions regarding the biomechanical efficacy of the SC and even the currently popular STT fusion as treatment for scapho-lunate instability.

Carpal Instabilities (Ulnar)

There is a great deal of confusion over carpal instabilities, particularly in the ulnar aspect of the wrist. This, in part, may arise from the duplicity and lack of uniformity in the use of terms such as dynamic and static, and the various classifications such as CID and CIND, and VISI, DISI and Midcarpal instability which are not always qouted or utilized within the defined parameters the original authors had intended.[16] Work was also done to further understand and categorize the biomechanics and etiology of ulnar sided perilunate instability known as VISI.

A staging system for ulnar sided perilunate instability was developed based on a series of cadaver dissections and load studies. The staging system includes the following: Stage I: partial or complete disruption of the lunotriquetral interosseous ligament, without clinical and/or radiographic evidence of dynamic or static VISI deformity; Stage II: complete disruption of the lunotriquetral interosseous ligament and disruption of the palmar lunotriquetral ligament, with clinical and/or radiographic evidence of dynamic VISI deformity; and Stage III: complete disruption of the lunotriquetral interosseous and the palmar lunotriquetral ligaments, attenuation or disruption of the dorsal radiocarpal ligament, with clinical and/or radiographic evidence of static VISI deformity (Figure 8). These three stages of instability would be classified as carpal instability dissociative (CID) type lesions because of the disruption between the lunate and the triquetrum. In addition to these instabilities there was also a static VISI deformity that would result from only attenuation or disruption of the dorsal radiocarpal ligament. This instability would be classified as a carpal instability non-

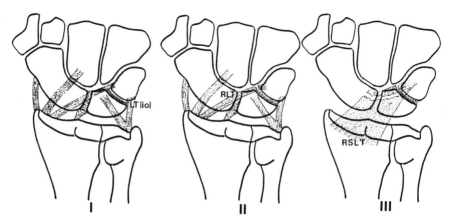

Figure 8. The three stages of ulnar sided perilunate instability as described by Viegas, et al.[60]

dissociative (CIND) type lesion because there is no disruption between the individual bones of the proximal or distal carpal row.

This work showed that increased motion developed between the lunate and the triquetrum with a tear of the lunotriquetral interosseous ligament. However, with a disruption of this ligament alone, an appreciable dynamic or static VISI deformity was not evident and could not be induced. This is compatible with the findings of the load studies, which overall, did not demonstrate significant differences in load distribution between the scaphoid and lunate fossa in the normal and a stage I instability. This would imply that the load distribution is not appreciably altered in cases where the lunotriquetral joint alone was disrupted. This is consistent with clinical findings that patients with incongruity between the lunate and the triquetrum generally had satisfactory clinical results.[35]

In a Stage II ulnar sided perilunate instability, a definite VISI deformity was evident during the application of a translational force on the dorsal aspect of the capitate and/or hamate, when the wrist was in some degree of neutral or flexion and radial, neutral or limited ulnar deviation. Sectioning the dorsal radiocarpal ligament and capsule at their attachment to the scaphoid and lunate, resulted in a Stage III ulnar sided perilunate instability and allowed a static VISI deformity to arise in the wrist. This static VISI deformity could only be demonstrated, however, in those same positions in which the dynamic VISI deformity could be attained. Whenever the wrist was brought into greater than 25 degrees of ulnar deviation or into extension, this would result in changing the VISI position to a position of normal carpal alignment. Reduction of the VISI deformity by ulnar deviation of the wrist has been described clinically. The symptomatic "clunk" that patients often describe can often be reproduced by this maneuver.[26] This is still consistent with the classic definition of a VISI deformity since it applies to the capitolunate angle in the lateral radiograph with the wrist in neutral flexion and neutral radioulnar deviation.

The information from these load studies would suggest the following: with the overall distribution between the scaphoid and the lunate remaining essentially the same throughout the progressive stages of ulnar sided perilunate instability, as defined in this paper, and the fact that even the high pressure centroids shifted to normal locations within portions of the functional range of motion of the wrist, one would not expect significant changes in the wear pattern in the radiocarpal joint. This expectation, coincides with the clinical observation of sparing of the radiocarpal joint in patients with VISI deformities.

There is clinical support for this concept of a sequential, progressive ligament disruption leading eventually to a static VISI deformity. Previous clinical studies of transscaphoid fracture dislocations have made an argument for the existence of a spectrum of lesions in those kinds of injuries.[55] Specifically, the palmar transscaphoid lunate fracture dislocation, is a more severe injury than the dorsal transscaphoid perilunate fracture dislocation. Anatomically, the difference between these two injuries is the disruption of the scaphoid and lunate attachment of the dorsal radiocarpal ligament and dorsal capsule. After fixation of the scaphoid fracture component of these injuries, the remaining pathology includes all the components of a Stage III ulnar sided perilunate instability, as defined in the anatomic model described in this paper. It is not surprising therefore, that in these cases, a static VISI can develop.[56]

The anatomic dissections in this study appear to support the concept that it is the flexion force exerted by the scaphoid upon the lunate, which is unopposed by the triquetrum but also the fact that the lunate and scaphoid are unrestrained by the disrupted dorsal radiocarpal ligament, that allows the carpus to posture in a static VISI alignment. The importance of this structure is emphasized by the fact that even disruption of the radiocarpal ligament alone allows the carpus to posture in a static VISI alignment.

Scaphoid Fracture

Scaphoid fractures and non-union of the scaphoid continue to be a problem. Various studies have suggested that the natural history of a scaphoid non-union is the development of radiocarpal degenerative arthritis.[30,45,54] One phase of our studies addressed the question of how the load transfer characteristics of the wrist were altered by a fracture of the scaphoid. A proximal pole scaphoid fracture model was studied since non-union and avascular necrosis of proximal pole scaphoid fractures are a proportionally greater problem in these than in other types of scaphoid fracture.

This work found that the amount of contact area born through the scaphoid fossa was essentially the same whether the scaphoid was intact, or following a simulated scaphoid fracture of its proximal pole or after resection of the proximal pole. The scaphoid contact area and pressure, although overall relatively constant, was redistributed following osteotomy, resulting in increased contact area under the distal fragment and no change or a slight decrease in the contact area under the proximal fragment of the scaphoid (Figure 9). After resection of the proximal fragment, all scaphoid contact area and pressure was born by the distal scaphoid fragment (Figure 10). The contact area and pressure characteristics of the lunate remained unchanged in all conditions compared to normal. There were no significant changes in the locations of the centroids of the scaphoid segments and the lunate in any of the conditions tested.

These load studies demonstrate that a scaphoid osteotomy does not significantly affect the load distribution through the lunate and suggests that the load through the proximal scaphoid fragment is also essentially unchanged or slightly decreased. It does, however, acutely increase and concentrate the load distribution through the distal fragment of the scaphoid. These changes of increased load through the distal scaphoid fragment and decreased load through the proximal fragment were even more dramatic when the wrists were tested using a greater force through the wrist. The load born by that distal scaphoid fracture fragment may also increase over time in an in vivo setting with chronic compromise of the associated ligamentous structures. The areas in which the load distribution is not changed or is lessened coincide with areas that are observed clinically to be spared from degenerative changes even after three decades.[30,45,54] The area in which the load distribution is increased, however, coincides with areas that are observed clinically to develop significant degenerative changes within one to two decades.[30,45,54] Some of our earlier studies have shown this kind of correlation, in models of scaphoid instability, between areas of increased and/or concentrated loads and areas of degenerative changes.[7,64]

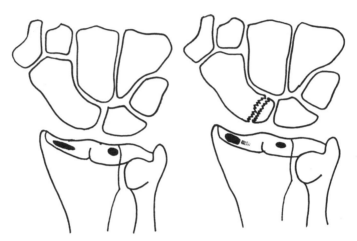

Figure 9. Display of the contact area between scaphoid and lunate and radius in the normal wrist and a wrist with a scaphoid fracture.

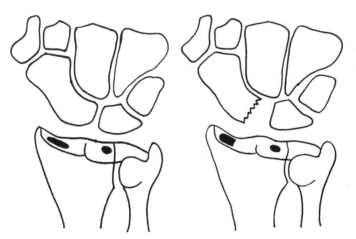

Figure 10. Display of the contact area between scaphoid and lunate and radius in the normal wrist and a wrist with the proximal pole of the scaphoid removed.

Scaphoid Silastic Implant

In clinical situations where significant degenerative changes have occured in the radioscaphoid or panscaphoid joints, one treatment alternative has been excision of the scaphoid and insertion of a silastic scaphoid. Over the years since silastic carpal implants first became available recommendations and implications of its load bearing characteristics, alone and with associated limited carpal fusions have been proposed with little or no proof of its load characteristics. With the increased concerns over silastic synovitis, a smaller implant has been used or a limited carpal fusion has been added to the procedure of silastic scaphoid implant by many surgeons with the assumption that these modifications would shield the scaphoid implant from bearing load. Another phase of our research was designed to attempt to understand how the scaphoid implant acts biomechanically by itself and when associated with a limited carpal fusion.

This study demonstrates that the scaphoid silastic (high performance) implant used as it is recommended did register significant although less contact area and pressure on the radius than the normal scaphoid. The scaphoid silastic implant also had different pressure centroids than the normal scaphoid it replaces. A smaller size scaphoid implant further decreased the load born by the implant (Figure 11). The decreased pressure/contact area (i.e. load) through the scaphoid implant was compensated by an increase in the lunate load. The addition of a limited carpal fusion did not appear to appreciably decrease the load born by a scaphoid implant. Therefore, the silastic scaphoid implant is a load bearing implant even when undersized or placed in association with a limited carpal fusion.

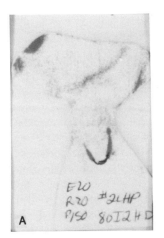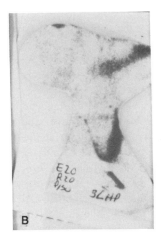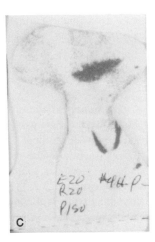

Figure 11. Fuji prints showing the increasing contact between scaphoid and radius for three sizes of silastic scaphoid implants, a) size 2, b) size 3, c) size 4.

Distal Radius Malunion

Distal radius fractures are the subject of much discussion.[1,3,8,9,12,13,14,22,23,27,31,38,39,43,47,48,49,51,52,63,65] They are relatively common fractures and have been reported to have a high complication rate.[1,3,9,23,43,52] Specific guidelines with respect to how much deformity can or should be accepted in treating distal radius fractures is lacking in the literature. This appears to be at least in part due to the difficulty in studying clinical outcomes of distal radius fractures where so many variables are involved. In an attempt to offer additional information on the effects of various components of distal radius malunion another phase of research was undertaken in which simulated radius malunions were analyzed. The components studied were radial shortening, radial inclination in the radio-ulnar plane and radial tilt in the dorso-palmar plane. In these simulated radius malunions, a decrease in the scaphoid contact area and an increase in the lunate contact area, were seen after decreasing the radial inclination from normal (19 to 30°) to 10° - a change of

147

only 9° to 20°. Variations in palmar inclination caused more concentrated contact areas, i.e. increased scaphoid and lunate high pressure contact areas without an appreciable change in the scaphoid/lunate load distribution, after angulating from normal (4° to 8° palmar) to 30° palmar or dorsal inclination - a change of 22° to 38°. While mild degrees of radial shortening caused a slight increase in the lunate load, more extreme shortening of 6 - 8 mm was noted to result in gross extraarticular ulnar impingement on the carpus.

This work suggests the following information and perhaps clinical guidelines for the reduction of fracture deformities and malunion. Fracture deformity or malunion comprised of less than 2mm of radial shortening, more than 10 degrees of radial inclination on the posteroanterior view and less than 20 degrees dorsal inclination on the lateral view, should not have any significant load changes compared to a normal wrist.

Additionally this work showed that an ulna styloid fracture increases the scaphoid/lunate area ratio only, and does not change the positions of the scaphoid and lunate high pressure areas or the scaphoid and lunate total contact areas. The fact that ulna styloid fractures change the load distribution so little may explain why they are often clinically asymptomatic. In simulating the various patterns and degrees of deformity, it was impossible to obtain displacements of the distal radius involving shortening greater than 4mm, or angulations greater than approximately 20° with the ulna styloid and TFCC intact. This suggests that fractures with this degree of displacement or more, not displaying an ulna styloid fracture, probably have a TFCC disruption.

Radio-ulnar Instabilities

Disruption of the distal radio-ulnar joint alone or in association with distal radius fractures is another wrist problem that we have studied. Three stages of radioulnar instability were studied: 1) an avulsion fracture at the base of the ulna styloid; 2) an avulsion fracture at the base of the ulna styloid plus disruption of the dorsal portion of the distal radioulnar joint capsule and; 3) an avulsion fracture at the base of the ulna styloid, disruption of the dorsal portion of the distal radioulnar joint capsule and disruption of the radioulnar interosseous membrane. All stages of radioulnar instability demonstrated a decrease in the lunate contact area in all wrist positions in which the forearm was in supination. In stage 3 instability there was also less lunate contact area in all wrist positions in which the forearm was in neutral pronation/supination. In stage 3 instability the lunate high pressure area centroid was abnormally palmar in all positions and the scaphoid high pressure area centroid was abnormally palmar in all wrist positions in which the forearm was in pronation or supination.

This would suggest that a position of neutral pronation/supination might be a preferred position for immobilization in acute stage 1 and 2 injuries and that immobilization alone in any position may not be adequate treatment for stage 3 injuries.

SUMMARY

This six years of research on the biomechanics of the human wrist has yielded a substantial amount of information. This information has increased our basic knowledge of carpal morphology and mechanics. It has also enhanced our understanding of the development of degenerative arthritis in certain types of clinical conditions. This additional information has already resulted in a better understanding of some of the current treatment methods and has raised questions regarding some others. Ideally the information which we have gathered will lead to more effective forms of treatment for various types of clinical problems in the wrist. The research is by no means a completed work, but is rather, the foundation upon which we base our continuing studies of other problems and the development of imaging, kinematic and mathematical modeling systems.

ACKNOWLEDGEMENTS

This work was supported in part by a grant from the Orthopaedic Research and Education Foundation # 87-459, and by a grant from the National Institute of Arthritis and Musculoskeletal and Skin Diseases # 5 R29 AR 38640-02.

REFERENCES

1. L. Ambrose, and M.A. Posner, Biplanar osteotomy for the treatment of malunited Colles' fractures, Presented at the 43rd Annual Meeting of the American Society for Surgery of the Hand, Baltimore, MD. Sept. 14-17, (1988).
2. J.G. Andrew, and Y.A. Youm, A biomechanical investigation of wrist kinematics, *J. Biomech.* 12:83 (1979).
3. R.W. Bacorn, and J.F. Kurtzke, Colles fracture. A study of two thousand cases from the New York state workman's compensation board, *J. Bone Joint Surg.* 35:643 (1953).
4. R.J. Belsole, D. Hilbelink, J.A. Llewellyn, M. Dale, S. Stenzler, and J.M. Raymack, Scaphoid orientation and location for computed three dimensional carpal models, *Orthop. Clin. N. Am.* 173:505 (1986).
5. R.A. Berger, R.D. Crowninshield, and A.E. Flatt, The three-dimensional rotational behavior of the carpal bones, *Clin. Orthop.* 167:303 (1982).
6. G. Blatt, Capsulodesis in reconstructive hand surgery: dorsal capsulodesis for the unstable scaphoid and volar capsulodesis following excision of the distal ulna, *Hand Clinics* 3:81 (1987).
7. A.D. Blevens, T.R. Light, W.S. Jablonsky, D.G. Smith, A.G. Patwardhan, M.E. Guay, and T.S. Woo, Radiocarpal articular contact characteristics with scaphoid instability, *J. Hand Surg.* 14A:781 (1989).
8. A. Colles, On the fractures of the carpal extremity of the radius, *Ed. Med. Surg. J.* 10:182 (1814).
9. W.P. Cooney, J.H. Dobyns, and R.L. Linscheid, Complications of Colles' fracture, *J. Bone Joint Surg.* 62:613 (1980).
10. A. de Lange, J.M.G. Kauer, and R. Huiskes, Kinematic behavior of the human wrist joint: a roentgen-stereophotogrammetric analysis, *J. Orthop. Res.* 3:56 (1985).
11. J.H. Dobyns, R.L. Linscheid, E.Y.S. Chad, E.R. Weber, and G.E. Swanson, Traumatic instability of the wrist, Instructional Course Lectures, AAOS, St. Louis: The C.V. Mosby Co., pp. 182 (1975).
12. D.L. Fernandez, Correction of post-traumatic wrist deformity in adults by osteotomy, bone-grafting, and internal fixation, *J. Bone Joint Surg.* 64: 1164 (1982).
13. D.L. Fernandez, Radial osteotomy and bowers arthroplasty for malunited fractures of the distal end of the radius, *J. Bone Joint Surg.* 70:1538 (1988).
14. J.J. Jr Gartland, and C.W. Werley, Evaluation of healed Colles' fractures, *J. Bone Joint Surg.* 33:895 (1951).
15. W.W. Gilford, R.H. Bolton, and C. Lambrinundri, The mechanism of the wrist joint with special reference to fractures of the scaphoid, *Guys Hosp. Rep.* 92:52 (1943).
16. L.A. Gilula, Ligament instability of the wrist: discussion of current classification systems, Wrist Investigators' Workshop, Paris, France (April 5, 1989).
17. W.D. Gordon, and M.D. Armstrong, Rotational subluxation of the scaphoid, *Canad. J. Surg.* 11:306 (1968).
18. E.N. Hanley, D.M. Boland, and H.K. Watson, Intercarpal arthrodesis for rotary subluxation of the scaphoid, *Orthopaedic Cons.* 41:1 (1983).
19. W. Henke, Die bewegungen der wandwurzel, *Z Rad Med III* 7:27 (1859).
20. F.M. Howard, T. Fahey, and E. Wojcik, Rotatory subluxation of the navicular, *Clin. Orthop.* 104:134 (1974).

21. W.T. Jackson, and J.M. Protas, Snapping scapholunate subluxation, *J. Hand Surg.* 6:590 (1981).

22. N.H. Jenkins, and W.J. Mintowt-Czyz, Mal-union and dysfunction in Colles' fracture, *J. Hand Surg.* 13:291 (1988).

23. J.B. Jupiter, and M. Masem, Reconstruction of post-traumatic deformity of the distal radius and ulna, *Hand Clinics* 4:377 (1988).

24. J.M. Kauer, The mechanism of the carpal joint, *Clin. Orthop.* 202:16 (1986).

25. J.M.B. Kauer, The interdependence of carpal articulation chains, *Acta Anat.* 88:481 (1974).

26. D.M. Lichtman, J.R. Schneider, A.R. Swafford, and G.R. Mack, Ulnar midcarpal instability-clinical and laboratory analysis, *J. Hand Surg.* 6:515 (1981).

27. A. Lindstrom, Fractures of the distal end of the radius, *Acta Orthop. Scand.* Suppl. 41, (1959).

28. R.L. Linscheid, Kinematic considerations of the wrist, *Clin. Orthop.* 202:27 (1986).

29. R.L. Linscheid, J.H. Dobyns, J.W. Beabout, and R.S. Bryan, Traumatic instability of the wrist, *J. Bone Joint Surg.* 54A:1612 (1972).

30. G.R. Mack, M.J. Bosse, R.H. Gelbermann, and E. Yu, The natural history of scaphoid non-union, *J. Bone Joint Surg.* 66A:504 (1984).

31. M. McQueen, and J. Caspers, Colles fracture: does the anatomical result affect the final function? *J. Bone Joint Surg.* 70:649 (1988).

32. J.K. Mayfield, R.P. Johnson, and R.F. Kilcoyne, The ligaments of the human wrist and their functional significance, *Anat. Rec.* 186:417 (1976).

33. J.K. Mayfield, R.P. Johnson, and R.K. Kilcoyne, Carpal dislocations: pathomechanics and progressive perilunar instability, *J. Hand Surg.* 5:226 (1980).

34. J.K. Mayfield, Patterns of Injury to Carpal Ligaments, *Clin. Orthop.* 187: 36 (1984).

35. A. Minami, T. Ogino, I. Ohshio, and M. Minami, Correlation between clinical results and carpal instabilities in patients after reduction of lunate and perilunar dislocations, *J. Hand Surg.* 11B:213 (1986).

36. L.G. Morawa, P.M. Ross, and C.C. Schock, Fractures and dislocations involving the navicular-lunate axis, *Clin. Orthop.* 118:48 (1976).

37. A. Mouchet, and J. Belot, Poignet à ressaut (subluxation médiocarpienne en avant), *Bull. Mem. Soc. Nat. Chir.* 60:1243, (1934).

38. T.M. Older, E.V. Stabler, and W.H. Cassebaum, Colles fracture: evaluation and selection of therapy, *J. Trauma* 5:469 (1965).

39. S. Overgaard, and S. Solgaard, Osteoarthritis after Colles' fracture, *Orthopedics* 12:413 (1989).

40. A.K. Palmer, J.H. Dobyns, and R.L. Linscheid, Management of post-traumatic instability of the wrist secondary to ligament rupture, *J. Hand Surg.* 3:507 (1978).

41. A.K. Palmer, and F.W. Werner, Biomechanics of the distal radioulnar joint, *Clin. Orthop.* 187:26 (1984).

42. D.J. Pogue, S.F. Viegas, R.M. Patterson, P.D. Peterson, D.K. Jenkins, T.D. Sweo, and J.A. Hokanson, The effects of distal radius fracture malunion on wrist joint mechanics, *J. Hand Surg.* 15:721 (1990).

43. R.M. Rubinovich, and W.R. Rennie, Colles' fracture: end results in relation to radiologic parameters, *Canad. J. Surg.* 26:361 (1983).

44. L.K. Ruby, W.P. Cooney III, K.N. An, R.L. Linscheid, and E.Y.S. Chao, Relative motion of selected carpal bones: a kinematic analysis of the normal wrist, *J. Hand Surg.* 13A:1 (1988).

45. L.K. Ruby, J. Stinson, and M.R. Belsky, The natural history of scaphoid non-union: a review of fifty-five cases, *J. Bone Joint Surg.* 67A:428 (1985).

46. T.B. Russell, Inter-carpal dislocations and fracture-dislocations, a review of fifty-nine cases, *J. Bone Joint Surg.* 31B:524 (1949).

47. W.H. Short, A.K. Palmer, F.W. Werner, and D.J. Murphy, A biomechanical study of distal radius fractures, *J. Hand Surg.* 12:529 (1987).

48. S. Solgaard, Classification of distal radius fractures, *Acta Orthop. Scand.* 56: 249 (1984).

49. S. Solgaard, Function after distal radius fracture, *Acta Orthop. Scand.* 59:39 (1988).

50. H.G. Sommer III, and N.R. Miller, A technique for kinematic modeling of anatomical joints, *J. Biomech. Eng.* 102:311 (1980).

51. H.D. Stewart, A.R. Innes, and F.D. Burke, Factors affecting the outcome of Colles' fracture: an anatomical and functional study. *Injury* 16:289, 1985.

52. J. Taleisnik, and H.K. Watson, Midcarpal instability caused by malunited fractures of the distal radius, *J. Hand Surg.* 9:350 (1984).

53. O.J. Vaughan-Jackson, A case of recurrent subluxation of the carpal scaphoid, *J. Bone Joint Surg.* 31B:532 (1949).

54. M.I. Vender, H.K. Watson, B.D. Wiener, and D.M. Black, Degenerative change in symptomatic scaphoid nonunion, *J. Hand Surg.* 12A:514 (1987).

55. S.F. Viegas, The lunohamate articulation of the midcarpal joint, arthroscopy: *J. Arthroscopic Rel. Surg.* 6:5 (1990).

56. S.F. Viegas, J.W. Bean, and R.A. Schram, Transscaphoid fracture/dislocations treated with open reduction and Herbert screw internal fixation, *J. Hand Surg.* 12:992 (1987).

57. S.F. Viegas, R.M. Patterson, G.R. Hillman, and P.D. Peterson, The simulated scaphoid proximal pole fracture: a biomechanical study, *J. Hand Surg.* 16A:495 (1991).

58. S.F. Viegas, R.M. Patterson, P.D. Peterson, M. Crossley, and R. Foster, The silastic scaphoid: a biomechanical study, *J. Hand Surg.* 16:91 (1991).

59. S.F. Viegas, R.M. Patterson, P.D. Peterson, D.J. Pogue, D.K., Jenkins, T.D. Sweo, and J.A. Hokanson, The evaluation of the biomechanical efficacy of limited intercarpal fusions for the treatment of scapho-lunate dissociation, *J. Hand Surg.* 15A:120 (1990).

60. S.F. Viegas, R.M. Patterson, P.D. Peterson, D.J. Pogue, D.K. Jenkins, T.D. Sweo, and J.A. Hokanson, Ulnar sided perilunate instability: an anatomic and biomechanic study, *J. Hand Surg.* 15A:268 (1990).

61. S.F. Viegas, R. Patterson, P. Peterson, J. Roefs, A. Tencer, and S. Choi, The effects of various load paths and different loads on the load transfer characteristics of the wrist, *J. Hand Surg.* 14A:458 (1989).

62. S.F. Viegas, D.J. Pogue, R.M. Patterson, and P.D. Peterson, The effects of radioulnar instability on the wrist: a biomechanical study, *J. Hand Surg.* 15:728 (1990).

63. S.F. Viegas, A.F. Tencer, J. Cantrell, M. Chang, P. Clegg, C. Hicks, C. O'Meara, and J.B. Williamson, Load transfer characteristics of the wrist: part I, the normal joint, *J. Hand Surg.* 12A:971 (1987).

64. S.F. Viegas, A.F. Tencer, J. Cantrell, M. Chang, P. Clegg, C. Hicks, C. O'Meara, and J.B. Williamson, Load transfer characteristics of the wrist: part II, perilunate instability, *J. Hand Surg.* 12A:978 (1987).

65. R.N. Villar, D. Marsh, N. Rushton, and R.A. Greatorex, Three years after Colles' fracture, *J. Bone Joint Surg.* 69:635 (1987).

66. H.K. Watson, and F.L. Ballet, The SLAC wrist: scapholunate advanced collapse pattern of degenerative arthritis, *J. Hand Surg.* 9A:358 (1984).

67. H.K. Watson, and L.H. Brenner, Degenerative disorders of the wrist, *J. Hand Surg.* 10A:1002 (1985).

68. H.K. Watson, and R.F. Hempton, Limited wrist arthrodesis I, the triscaphoid joint, *J. Hand Surg.* 5:320 (1980).

69. H.K. Watson, and J. Ryu, Evolution of arthritis of the wrist, *Clin. Orthop.* 202:57 (1986).

JOINT CONTACT PRESSURES WITHIN THE NORMAL WRIST:

EXPERIMENTAL VERSUS ANALYTICAL STUDIES

Frédéric A. Schuind,[1] Kai-Nan An,[2] Ronald L. Linscheid,[2] William P. Cooney,[2] and Edmund Y.S. Chao[2]

[1]Université libre de Bruxelles
Cliniques Universitaires de Bruxelles
808 route de Lennik
1070 Brussels, Belgium
[2]Orthopedic Biomechanics Laboratory
Mayo Clinic/Mayo Foundation
Rochester, MN 55905, USA

INTRODUCTION

The forces applied to the wrist joint are mainly the result of those externally applied, and of those exerted by the wrist and fingers flexor and extensor muscle-tendon units. The transmission of these forces through the 8 carpal bones to the radius and ulna is difficult to assess, due to the complexity [1] of the geometry of the carpus, which constantly changes with position and function, [2] of the mechanical properties of the cartilage of the various joints and of the triangular fibrocartilage (TFC), and [3] of the disposition of the numerous ligaments which spread the forces throughout the wrist articular complex. During the past years, there has been however a number of attempts to measure or to calculate the joint contact pressures within the wrist joints.

EXPERIMENTAL STUDIES

Two techniques are currently used to measure the joint contact pressures within the wrist: pressure sensitive film,[17,21,23,27-30,32-34] and intraarticular pressure-sensitive transducers.[8,34] Strain gauges have been used on the surface of carpal bones,[15,16,24] but do not provide a direct measure of joint contact pressures.

Pressure sensitive film

Details of the technique and main results at the wrist may be found in the chapter *Contact pressures within wrist joints*, by S.F. Viegas and R.M. Patterson. Fuji film is

Advances in the Biomechanics of the Hand and Wrist
Edited by F. Schuind *et al.*, Plenum Press, New York, 1994

thin (0.28 mm) and has an immediate observable pressure distribution pattern that is quantifiable.[5,9,18,22,27-30] The technique has however some limitations. At the present time, measurements have only been static,[19] performed on cadaveric specimens,[7] not on living subjects. The need to perform an arthrotomy in order to insert the film, and the thickness of the film itself may potentially alter the load transmission through the joint.[3] It is also believed that artifacts could occur, related to the procedure of calibration or to joint shear forces.[30] Nonetheless, experiments using pressure-sensitive film have provided a considerable amount of knowledge on the contact pressures of the normal and abnormal wrist.

Pressure-sensitive transducers

Hara et al. used pressure-sensitive conductive rubber transducers to measure the contact pressures within the radio-carpal joint.[8] The sensors were thin (0.9 mm) and flexible and did not have to be replaced for each loading configuration. However, most of the limitations already reported concerning the Fuji film apply to such transducers.

ANALYTICAL STUDIES

There are two analytical techniques that can be used for joint contact force prediction in anatomical structures with a complex geometrical shape such as the wrist: the finite element method[4] and the rigid body spring modeling technique (RBSM).

The finite element method is extremely time consuming in modeling and execution and has not yet, to our knowledge, been used at the wrist.

The RBSM, originally developed and validated in civil engineering by Kawai,[13] was later perfected in the Biomechanics Laboratory of the Mayo Clinic by Chao through the help of Himeno, Tsumura, Samson and Westreich.[1,10,12,26] The RBSM program calculates the force transmission and the displacement of non-deformable two-dimensional bodies, connected to one another by springs of known linear stiffness. The rigid bodies are considered to be in equilibrium with external loading conditions. Reaction forces between adjacent bodies are produced by the spring system, based on the minimal strain energy principle. The method has been previously used at the wrist by Garcia-Elias et al. to examine the stabilizing structures of the transverse carpal arch,[6] and by Horii et al. to study the biomechanics of five different treatment modalities for Kienböck's disease.[11] We have used the method to establish a normal data base of force transmission through the normal wrist according to age, gender and anatomical variability.[20]

One hundred twenty normal wrist X-rays of adults (evenly divided to represent both genders and two age groups) provided the anatomical data. Reaction forces between the 5 metacarpals, the carpal bones and the distal radius and ulna (the "rigid" two-dimensional bodies, whose geometry was obtained from digitization of the radiological contours) were modeled using a system of compression linear springs, representing cartilage and subchondral bone, and of tensile linear springs, representing the 28 ligaments of the wrist. The spring constants were determined based on the material properties of wrist cartilage and ligaments. Specific axial loads (total 142 N) were applied along the metacarpals, to simulate a grasp function of 10 N with active stabilization of the wrist in neutral position.

The average force transmission and the peak pressures at the radio-carpal joint are reported in Table 1. A normal data base according to age, gender and wrist morphology was constituted: none of these factors had an important influence on the magnitude and pattern of load distribution. Further details will be found in our publication.[20]

Table 1. Distribution of the resultant forces and peak pressures in the radio-ulno-carpal joint of the normal wrist (n = 120).

Joint	Total Force Transmitted (N)		Peak Pressure (N/mm)	
	Mean	S.D.	Mean	S.D.
Radio-scaphoid	88.2	13.3	0.90	0.27
Radio-lunate	56.5	10.1	0.56	0.23
Ulno-lunate	12.9	5.2	0.14	0.06
Ulno-triquetral	2.6	1.2	0.04	0.02

DISCUSSION

Our study was the first attempt to establish a normal data base of force transmission through all wrist joints. In the present state of technology, such a data base could not have been obtained experimentally as the techniques of joint pressure measurement are tedious, and the results are not always accurate.[22] Furthermore, the experimental techniques are not presently applied to living individuals. Only computer simulation can predict the theoretical load repartition in the wrist of a given living subject.

The rigid body spring model is well adapted to the analysis of force distribution of human joints with complex geometrical characteristics, since the bones can be modeled as rigid non-deformable bodies in view of the much higher elasticity of the cartilage and ligaments.[14]

There is an obvious need to validate the results of our analytical study. Trumble,[25] using load cells affixed to the distal diaphyses of the radius and ulna in cadaver specimens, found that in the neutral position an average of 17.0% (± 10.4%) of the applied axial load was borne by the ulna. Werner,[31] using a similar model, reported an average ulnar load transmission of 18.4%. We found a similar pattern of radio-ulnar repartition of the load (Table 1). There was, however, less load transmitted to the ulna in the analytical study than in the experimental studies. It should be noted that we reported the load transmission through the TFC only, while Werner[31] and Trumble[25] reported the total load transmitted to the diaphysis of the ulna, through the TFC, and through the distal radio-ulnar joint.

Blevens,[2] Tencer,[23] and Viegas,[29] using pressure-sensitive film, demonstrated that under application of axial load, about 56% of the total radio-carpal forces were transmitted through the scaphoid and 44% through the lunate; the average radio-carpal contact areas were in neutral position 53-60% for the scaphoid fossa and 40-47% for the lunate fossa. These findings correspond quite well with our results of 61% force transmission through the radio-scaphoid joint and of 39% force transmission through the radio-lunate joint.

Hara et al.,[8] using their pressure sensitive conductive rubber sensor, found in similar loading conditions that in neutral wrist position, the radio-ulno-carpal force transmission ratio was 50% through the scaphoid fossa, 35% throught the lunate fossa and 15% through the TFC. Thus, the results from our study compare reasonably with those based on experimental studies.

CONCLUSION

In view of the limitations of the experimental techniques, we believe that the theoretical calculation of joint contact forces using computer simulation holds considerable promise, particularly for the wrist joints. It should even be possible in the future to use the model as a clinical tool for the orthopedic surgeon, for example in a patient with a collapsed necrosed lunate, to predict the optimal leveling of the ulna, on the basis of the calculated unloading of the lunate. However, much work must first be done to assess the mechanical properties of the degenerated wrist.

REFERENCES

1. K.N. An, S. Himeno, H. Tsumura, T. Kawai, E.Y.S. Chao, Pressure distribution on articular surfaces: application to joint stability evaluation, *J. Biomech.* 23:1013 (1990).
2. A.D. Blevens, T.R. Light, W.S. Jablonsky, D.G. Smith, A.G. Patwardhan, M.E. Guay, T.S. Woo, Radiocarpal articular contact characteristics with scaphoid instability, *J. Hand Surg.* 14-A:781 (1989).
3. T.D. Brown, D.T. Shaw, In vitro contact stress distributions in the natural human hip, *J. Biomech.* 16:373 (1983).
4. T.D. Brown, A.M. Gigioia, A contact-coupled finite element analysis of the natural adult hip, *J. Biomech.* 17:437 (1984).
5. T. Fukubayashi, H. Kurosawa, The contact areas and pressure distribution pattern at the knee, *Acta Orthop. Scand.* 51:871 (1980).
6. M. Garcia-Elias, K.N. An, W.P. Cooney, R.L. Linscheid, E.Y.S. Chao, Transverse stability of the carpus. An analytical study, *J. Orthop. Res.* 7:738 (1989).
7. V.K. Goel, D. Singh, V. Bijlani, Contact areas in human elbow joint, *J. Biomech. Eng.* 104:169 (1982).
8. T. Hara, E. Horii, K.N. An, W.P. Cooney, R.L. Linscheid, E.Y.S. Chao, Force distribution across wrist joint. Application of pressure sensitive conductive rubber, *J. Hand Surg.* 17:339 (1992).
9. H.J. Hehne, H. Haberland, W. Hultzsch, W. Jantz W, Measurements of two dimensional pressure distributions and contact areas of a joint using a pressure sensitive foil, in "Biomechanics: Principles and Applications", Martinus Nijhoff, The Hague (1982).
10. S. Himeno, K.N. An, H. Tsumura, E.Y.S. Chao, Pressure distribution on articular surface: application to muscle force determination and joint stability evaluation, North American Congress on Biomechanics, Montreal, 97 (1986).
11. E. Horii, M. Garcia-Elias, K.N. An, A.T. Bishop, W.P. Cooney, R.L. Linscheid, E.Y.S. Chao, Effect on force transmission across the carpus in procedures used to treat Kienböck's disease, *J. Hand Surg.* 15-A:393 (1990).
12. R.W.W. Hsu, S. Himeno, M.B. Coventry, E.Y.S. Chao, Normal axial alignment of the lower extremity and load-bearing distribution at the knee, *Clin. Orthop.* 255:215 (1990).
13. T. Kawai, A new discrete model for analysis of solid mechanics problems, Seisan Kenkyu 29:208 (1977).
14. G.E. Kempson, C.J. Spivey, S.A.V. Swanson, M.A.R. Freeman, Patterns of cartilage stiffness on normal and degenerate human femoral heads, *J. Biomech.*, 4:597 (1971).
15. C. Kenesi, D. Gastambide, J.P. Lesage, Le syndrome de Kienböck. Etude biomécanique, *Rev. Chir. Orthop.* [Suppl.] 159:126 (1973).
16. V.R. Masear, E.G. Zook, D.R. Pichora, M. Krishnamurthy, R.C. Russell, J. Lemons, M.W. Bidez, Strain-gauge evaluation of lunate unloading procedures, *J. Hand Surg.*, 17-A:437 (1992).
17. A.K. Palmer, F.W. Werner, Biomechanics of the distal radioulnar joint, *Clin. Orthop.* 187:26 (1984).
18. B. Rieck, O. Paar, P. Bernett, Die intraartikuläre Druckmessung. Eine neue Methode zur Anwendung des Druckmessfilmes "Prescale", *Z. Orthop.* 122:841 (1984).
19. H. Schmotzer, C.L. Vaughan, I.D. Learmonth, Direct measurement of tibiofemoral contact pressures for various deformities of the knee, Proceedings, 7th Meeting of the European Society of Biomechanics, Aarhus (1990).
20. F. Schuind, W.P. Cooney, R.L. Linscheid, K.N. An, E.Y.S. Chao, Force and pressure transmission through the normal wrist, (submitted for publication), (1992).

21. W.H. Short, F.W. Werner, M.D. Fortino, A.K. Palmer, Distribution of pressures and forces on the wrist after simulated intercarpal fusion and Kienböck's disease, *J. Hand Surg.* 17-A:443 (1992).

22. T.J. Stormont, K.N. An, B.F. Morrey, E.Y. Chao, Elbow joint contact study: comparison of techniques, *J. Biomech.* 18:329 (1985).

23. A.F. Tencer, S.F. Viegas, J. Cantrell, M. Chang, P. Clegg, C. Hicks, C. O'Meara, J.B. Williamson, Pressure distribution in the wrist joint, *J. Orthop. Res.* 6:509 (1988).

24. T. Trumble, R.R. Glisson, A.V. Seaber, J.R. Urbaniak, A biomechanical comparison of the methods for treating Kienböck's disease, *J. Hand Surg.* 11-A:88 (1986).

25. T. Trumble, R.R. Glisson, A.V. Seaber, J.R. Urbaniak, Forearm force transmission after surgical treatment of distal radioulnar joint disorders, *J. Hand Surg.* 12-A:196 (1987).

26. H. Tsumura, S. Himeno, K.N. An, W.P. Cooney, E.Y.S. Chao, Biomechanical analysis of Kienböck's disease, *Orthop. Trans.* 11:327 (1987).

27. S.F. Viegas, A.F. Tencer, J. Cantrell, M. Chang, P. Clegg, C. Hicks, C. O'Meara, J.B. Williamson, Load transfer characteristics of the wrist. Part I. The normal joint, *J. Hand Surg.* 12-A:971 (1987).

28. S.F. Viegas, A.F. Tencer, J. Cantrell, M. Chang, P. Clegg, C. Hicks, C. O'Meara, J.B. Williamson, Load transfer characteristics of the wrist. Part II. Perilunate instability, *J. Hand Surg.* 12-A:978 (1987).

29. S.F. Viegas, R. Patterson, P. Peterson, J. Roefs, A. Tencer, S. Choi, The effects of various load paths and different loads on the load transfer characteristics of the wrist, *J. Hand Surg.* 14-A:458 (1989).

30. S.F. Viegas, R.M. Patterson, F.W. Werner, Joint contact area and pressure, in "Biomechanics of the Wrist Joint", K.N. An, R.A. Berger, W.P. Cooney ed., Springer-Verlag, New-York, 99 (1991).

31. F.W. Werner, A.K. Palmer, R.R. Glisson, Forearm load transmission: the effect of ulnar lengthening and shortening, Transactions, 28th Annual Orthopedic Research Society Meeting, New Orleans, 273 (1982).

32. F.W. Werner, A.K. Palmer, R.G. Utter, Distal radial osteotomy for the treatment of Kienböck's disease: a biomechanical study, Proceedings, 34th Ann. Meeting, *Orthop. Res. Soc.*, Atlanta, 411 (1988).

33. F.W. Werner, D.J. Murphy, A.K. Palmer, Pressures in the distal radioulnar joint: effect of surgical procedures used for Kienböck's disease, *J. Orthop. Res.* 7:445 (1989).

34. F.W. Werner, K.N. An, A.K. Palmer, E.Y.S. Chao, Force analysis, in "Biomechanics of the Wrist Joint", K.N. An, R.A. Berger, W.P. Cooney ed., Springer-Verlag, New-York, 77 (1991).

STRESS ANALYSIS OF THE RADIOCARPAL JOINT FROM A

DETERMINATION OF THE SUBCHONDRAL MINERALISATION PATTERN

Magdalena Müller-Gerbl,[1] Nicholas Löwer,[1] Klaus Wilhelm,[2] Rolf Kenn,[3] and Reinhard Putz[1]

[1] Anatomische Anstalt
[2] Department of Hand Surgery
[3] Department of Radiology
   Ludwig-Maximilians-Universität München
   Pettenkoferstr.11
   8000 München 2, Germany

INTRODUCTION

It is well known that there is a regularity of distribution of the subchondral bone density in the larger joints. This has been shown by Knief[1] and Konermann,[5] using x-ray densitometry. Some years ago, Pauwels[12] convincingly demonstrated that the distribution of the subchondral bone density reflects the localisation of the *long-term stresses* acting upon the articular surface of a joint. For a very long time no diagnostic method has been available by which the long-term stress to which joints are subjected can be accurately assessed *in vivo*. The conventional AP x-rays are summation pictures which can give no precise information about the actual area of distribution.

We have employed computed tomography to develop a method - CT osteo-absorptiometry[9,10,11] - which might allow one to estimate the area distribution of subchondral mineralisation in the radiocarpal joint in living people. This method enables the observer to watch the progress of normal adaptation of a joint to its mechanical circumstances, and to furnish information about its "loading history".

It is the purpose of this study to elucidate the relationship between mineralisation and the mechanical condition of the wrist-joint, in a sample of normal subjects and in patients by means of CT osteoabsorptiometry.

MATERIAL AND METHODS

CT data sets with a section thickness of 1 mm were obtained from both hands of 8 healthy persons (age range 24-25 years) and 16 patients from the Hand Surgery

Outpatients Department (age range 23-62 year). The diagnosis of the patients was: 2 distal radius fractures, 2 osteomalacia of the lunate, 3 scaphoid/lunatum fractures, 5 scaphoid pseudarthroses, 4 patients with pain, but no osseous changes). The CT scans were taken perpendicular to the articular surfaces of the carpus and distal radius.

The densitometric evaluation was achieved by means of CT Osteoabsorptiometry,[9,10] which is based upon the Hounsfield unit scale.

By means of a three-dimensional reconstruction program, which is integrated with the computer tomography itself, it is possible both to construct the entire articular surface and to indicate the separation of the various density levels (Figure 1). By employing an image-analysing system, the density grades may be demonstrated in false colors and projected one upon the other.

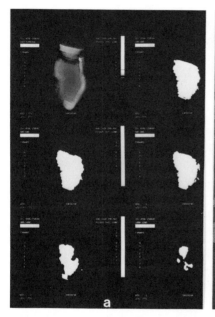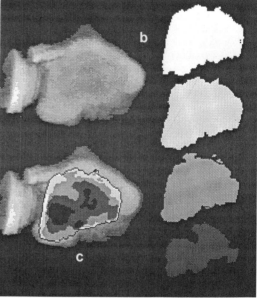

Figure 1. Three-dimensional CT osteoabsorptiometry of a right distal radius; a. Pictures of the whole articular surface of the distal radius and the different density zones resulting from the 3-D reconstruction (seen from distal); b. Method of producing density maps in a false-colour display; c. Resulting image.

The advantage of this procedure over earlier methods[9] is that the subchondral density can be projected exactly on to the *individual shape* of the articular surface of the joint in a much shorter time.

RESULTS

In normal cases the distal joint surface of the radius presents two density maxima. In 7 out of 17 hands, the maximum closer to the ulnar had a relatively higher density. In 7 hands, the radially located maximum showed a higher density (Figure 2), and in 3 hands, the same degree of mineralisation was recorded.

A comparison of the maxima on the opposed articular surfaces of the carpal bones revealed in 13 hands a higher degree of mineralisation in the scaphoid than in the lunate (Figure 2).

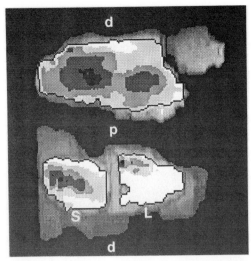

Figure 2. Density maps of the left wrist joint of a 25-year old healthy subject. The articular surface of the radius is seen from distal, the articular surfaces of the scaphoid and lunate are seen from proximal. The darkest colours always symbol the areas of highest density (d = dorsal, p = palmar, S = scaphoid, l = lunate).

Comparison of the two sides in right-handed people revealed, in most cases, a greater degree of mineralisation on the right than on the left.

In the presence of a badly reduced fracture of the distal radius, the density pattern was deviated from the normal. We found a displacement of the density maxima towards the dorsal part of the carpal surface (Figure 3).

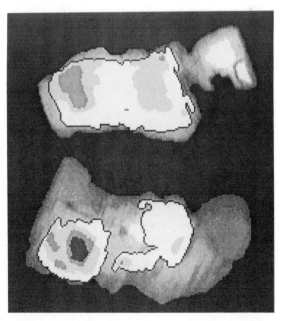

Figure 3. Density patterns of the right wrist joint of a 54-year old patient with a badly reduced distal radius fracture. The maximal density in the radial surface are displaced towards the dorsal parts.

161

Both patients with osteomalacia of the lunate showed a much higher bone density in the lunate than in the scaphoid. In the more severe case (grade two), the radial surface had a higher and the ulnar part exhibited an extremely low mineralisation (Figure 4)

In a 35-year-old man who had a pseudarthrosis of the right scaphoid and is now suffering from osteoarthrosis, the density pattern was completely different from the normal. Maximal density was found in the palmar periphery, that means in an area which normally exhibits only a low degree of mineralisation (Figure 5).

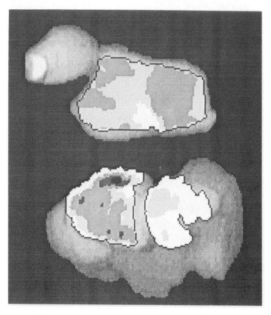

Figure 4. Density pattern of the right wrist joint of a 32-year old patient with lunatum osteomalacia grade II.

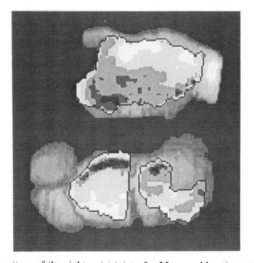

Figure 5. Density pattern of the right wrist-joint of a 35-year old patient with osteoarthrosis.

DISCUSSION

Bone is a living tissue which can specifically adapt itself, both in its architecture and in the distribution of the mineral salts. The latter results in increased density to the stress produced by external forces. It is apparent that the radiological density of subchondral bone is an expression of the long-term stress acting on a joint, as Pauwels[13] demonstrated in the acetabulum. The pressure brought about by the load on a joint reflects in its distribution and degree the stress to which its components have been subjected. Since it is possible with appropriate methods to determine the density of the material adjacent to the synovial cavity, a precise estimate of stress accepted by the joint can be determined.

The distribution of subchondral material in a normal radius shows a typical pattern with two central maxima. The most notable finding in healthy subjects is that, in the great majority of cases, the load received by the radio-scapho-lunate joint is transmitted through the scaphoid. These results are in agreement with those of Koebke[2,3] and Koebke and Mockenhaupt.[4] Equally impressive is the fact that the lunate surface of the radius shows, on average, nearly as pronounced a maximum as the scaphoid surface. Even so, this distribution of the load is not regularly reflected in the lunate itself. One must therefore conclude that, in spite of the suggestion that the lunate and its corresponding surface on the radius transmit the greater part of the load,[6,7,15,16] it is the scaphoid and scaphoid surface of the radius which are in fact most heavily involved. A similar conclusion was reached by Viegas,[17] whose experiments with pressure-sensitive film showed that "the general distribution of the contact areas between the scaphoid and lunate was consistent at all of the loads tested with 60% of the total contact area involving the scaphoid area and 40% involving the lunate contact area. Obviously it is impossible to compare single measurements taken from the wrist-joint with strain gauges or pressure-sensitive films directly with our own density patterns, since such measurements only record the momentary stress in each joint examined and not the amount of stress overtime. With pressure sensitive film, the strength and direction of the forces involved and the technique of implantation of the film all have an influence on the results. Such a procedure can provide no reliable information on the actual load to which a joint has been subjected for a period of time, since the precise contact between surfaces during movement, the frequency with which they come together during the course of a day and the strength of each individual force must all be known. In contrast to this, the subchondral density pattern can be used to assess the long-term stress acting on a joint, since it represents a summation of those parameters mentioned above.

In normal subjects this subchondral density pattern reveals a significant difference between the left and right sides of the body. In the right-handed people mineralisation is usually greater in the right wrist-joint, in left-handed people the reverse is true.

Patients with various disorders of the wrist joint usually present a density pattern which differs widely from the normal. Since, however, only a limited number of patients with the same condition have become available, only the general trend can be indicated.

In patients with a badly reduced Colles fracture, the maximal stresses are displaced dorsally. Like Möllers,[8] who examined the distribution of subchondral bone density at the distal end of the radius in dissecting-room specimens, we were able to show a typical pattern of two maxima in normal bones which distinguished them from pathological specimens. The abnormal pattern - particularly a dorsal displacement of the density maxima - must be regarded as the expression of an altered distribution of stress over the surface of the radius, and can account for the not infrequent pain and the danger of subsequent arthrosis developing.

It is remarkable that two patients with osteomalacia of the lunate bone showed a significant increase in the mineralisation of this bone in comparison with the scaphoid. When the disease is severe, the second normally present maximum in the radius is no longer demonstrable, probably owing to a failure in this region to transmit load.

One patient with severe arthrosis of the wrist-joint presented an entirely different density pattern. A single band-like density maximum appeared in the palmolateral part of the radial articular surface; that is to say, in a peripheral region in which only a low degree of mineralisation is normally found. This increased mineralisation occurred in the area of the articular cartilage where pathological changes were most marked. Apparently there is a connection between an increase in a region of normally low mineralisation, and changes in the adjacent articular cartilage. This has also been observed in the knee-joint. Further research is required to establish the connection beyond doubt.

In summary it can be stated that the regularly occurring, reproducible distribution patterns of subchondral mineralisation may be regarded as an expression of the individual long-term mechanical situation reflecting the loading-history on a joint. Obviously there is an adaptation to altered mechanical circumstances. The application of CT osteoabsorptiometry has enabled investigators to make individual analyses of the wrist-joints of living subjects to be made.

CT osteoabsorptiometry is a method of estimating the mineralisation pattern of a joint surface *in vivo*, which the conventional antero-posterior-radiograph (being a summation picture) naturally cannot provide. Thus a method is available which allows the adaptive capacity of an individual joint to be assessed, and enables progressive observations to be made after a joint has been subjected to abnormal or at least altered mechanical conditions.

Since this non-invasive technique makes it possible to monitor these changes in the living, it offers a variety of applications either in basic research on the skeletal system, or as a routine clinical procedure.

REFERENCES

1. J.J. Knief, Materialverteilung und Beanspruchungsverteilung im coxalen Femurende Densitometrische und spannungsoptische Untersuchungen, *Z. Anat Entwickl-Gesch* 126:81 (1967).
2. J. Koebke, Anatomie des Handgelenkes und der Handwurzel, *Unfallchirurgie* 14:74 (1988).
3. J. Koebke, Subchondral bone density as a key for normal and pathological wrist stress, presented at the Nato Advanced Research Workshop, Brussels, May 22-23 (1992).
4. J. Koebke, J. Mockenhaupt, and A. Lorbach, Die Subtraktionsäquidensitometrie als Methode zur Analyse der Gelenkbeanspruchung - dargestellt am Beispiel des Radiokarpalgelenkes, in: "Osteologie Interdisziplinär", eds Werner E. and Matthiass H.H., Springer, Berlin (1991).
5. H. Konermann, Quantitative Bestimmung der Materialverteilung nach Röntgenbildern des Knochens mit einer neuen photographischen Methode, *Z. Anat Entwickl-Gesch* 134:13 (1971).
6. E. Koob, Die Mondbeinnekrose, *Handchirurgie* 5:173 (1973).
7. M.A. McConnaill, and J.V. Basmajian, "Muscles and movements, a Basis for Human Kinesiology", William and Wilkins, Baltimore (1969).
8. N. Möllers, K. Lehmann, and J. Köbke, Die Verteilung des subchondralen Knochenmaterials an der distalen Gelenkfläche des Radius, *Anat Anz* 161:151 (1968).
9. M. Müller-Gerbl, R. Putz, N. Hodapp, E. Schulte, and B. Wimmer, Computed tomography-osteoabsorptiometry for assessing the density distribution of subchondral bone as a measure of long-term mechanical adaptation in individual joints, *Skeletal Radiol.* 18:507 (1989).
10. M. Müller-Gerbl, R. Putz, N. Hodapp, E. Schulte, and B. Wimmer, Computed tomography-osteoabsorptiometry: a method of assessing the mechanical condition in the major joints in a living subject, *Clin. Biomech.* 5:193 (1990).

11. M. Müller-Gerbl, R. Putz, and R. Kenn, Demonstration of subchondral bone density patterns by three-dimensional Ct osteoabsorptiometry (CT OAM) as a non-invasive method for in vivo assessment of individual long-term stress in joints, *J. Bone Mineral Research* (in press).

12. F. Pauwels, "Gesammelte Abhandlungen zur funktionellen Anatomie des Bewegungsapparates", Springer, Berlin (1965).

13. F. Pauwels, "Atlas zur Biomechanik der gesunden und kranken Hüfte", Springer, Berlin (1973).

14. F. Pauwels, "Biomechanics of the Locomotor Apparatus", Springer, New York (1980).

15. G. Segmüller, Zur Lunatum-Malazie (Morbus Kienböck), *Orthopäde* 10:47 (1981).

16. G. Sennwald, "Das Handgelenk", Springer, Berlin (1987).

17. S.F. Viegas, R. Patterson, R. Peterson, J. Roefs, A. Tencer, and S. Choi, The effects of various load paths and different loads on the load transfer characteristics of the wrist, *J. Hand Surg.* 14:458 (1989).

FORCE TRANSMISSION THROUGH THE PROXIMAL CARPAL ROW

Hans P. Kern,[1] Hilaire A.C. Jacob,[1] and Gontran R. Sennwald[1, 2]

[1]Biomechanics Unit
Department of Orthopaedic Surgery, Balgrist
University of Zürich
Forchstrasse 340
CH-800- Zürich, Switzerland
[2]Chirurgie St. Leonhard
Pestalozzistrasse 2
CH-9000 - St. Gallen, Switzerland

INTRODUCTION

Knowledge of force transmission through the wrist is of paramount importance in understanding the normal joint mechanics, and in finding an explanation for the pathogenesis of osteoarthritis and other degenerative deformations like, for example, Kienböck's disease. Also, the reconstruction of ligaments and the performance of partial intracarpal arthrodesis of the wrist bones often leads to unsatisfactory results. This has prompted us to study the intracarpal force transmission in the immediate environment of the proximal carpal row. In contrast to the fingers and the forearm, the determination of forces of several carpal bones is reflected in just a few studies.[6,9,11,12,13,14,15] We first used a simplified mathematical model and then carried out measurements on fresh autopsy specimens to confirm our findings.

Part I: A SIMPLIFIED ANALYTICAL MODEL

Material and methods

To obtain the force vectors acting on the first carpal row we first had to detemine the following parameters: the position of the hand during application of the external force, the centers of joint cartilage contact areas, and the force input distribution about the metacarpal bones.

The position of the hand has been chosen by the following reflections: most tool-grips have a diameter of about 25 mm (1 inch). We measured the spontaneous wrist positions of 50 persons holding forcefully a standardized cylinder with a diameter of 25 mm. These persons of different professions were aged between 23 and 57 years and showed clinically healthy wrists. They had been instructed to forcefully hold a standardized cylinder with their dominating hand several times. Radius and ulna were observed to adopt a neutral pronation-supination position (elbow in 90° of flexion). Further, the abduction of the hand and the ulnar variance (estimated manually) were noted. We found that a fairly constant wrist position was adopted. This force position showed a mean value of 25° of dorsiflexion at the wrist joint and was therefore chosen as the standard position for the force measurements and calculations undertaken in this study. The same position has been described by Bunnell[3] as the "position of function".

For the estimation of forces, the cartilage contact areas had to be known. This data was obtained through 3D-MRI of the articulations of healthy test subjects with their wrists in the force position. A new scanner generation allowed a sufficient resolution of the 3D-reconstructions (GE signa advantage 1,5T, service coil, spoil gross with 3D-volume acquisition). The contact areas were determined visually by regarding the cartilage surface shapes of two neighboring bones in a computed 3D view of the MRI-scan.

The contact areas determined in this manner were compared with the densitometry measurements of carpal bones undertaken by Koebke et al.[10] On the assumption that the force position is the most frequent position adopted in applying heavy forces, the pressure distribution at the contact surfaces would correlate with the bone density in this areas. This exactly corresponded to our findings that is, the areas of contact as determined by us were identical with areas of high bone density as determined by Koebke et al.[10]

The relative forces on the metacarpals in the force position have been calculated by Horii et al.[7] based on theoretical calculations reported by Cooney and Chao,[5] Chao et al.[4] and An et al.[1] These values for the applied loads along the midaxis of the five metacarpals were as follows: thumb 22.5 N, index 33.0 N, long 42.2 N, ring 25.6 N, and small 19.7 N. The loads chosen for each metacarpal are in the order of those that would be experienced in vivo while grasping with a 1 kg force.

For the estimation of forces across the joints of the wrist, the following assumptions have been made: [1] the bone contact areas have been reduced to contact points, [2] the forces acting on the metacarpal bones have been applied in the plane through the axes of the metacarpal bones II - V (vectors A, B, C and D in figure 1), [3] only that component of force acting on the saddle joint of the thumb that lies in the plane of the metacarpals was employed for this two-dimensional study (vector Ll in figure 1), [4] the fingers, wrist and the forearm bones are in static equilibrium, [5] no trussing action of the ligaments of the wrist in maintaining the position of the bones relative to each other is required. This has been purposely assumed in order to see whether under such conditions equilibrium can still be maintained.

Part II: FORCE MEASUREMENTS

Material and methods

The two fresh cadavers used for this study were prepared as follows: the upper extremities had been exarticulated in the humero-glenoidal articulation. The humerus and forearm were freed of skin and of all muscles. The elbow joint was completely retained, leaving the joint capsule intact. The hand was left completely unimpaired. The tendons of the following muscles were cut through at their muscular origin: extensor carpi radialis longus (ECRl), extensor carpi radialis brevis (ECRb), extensor carpi ulnaris (ECU), flexor carpi radialis (FCR) and flexor carpi ulnaris (FCU). These tendons were then prolonged with cords, stitched on with surgical suture material. The tendons of the ECRl and ECRb were fixed together.

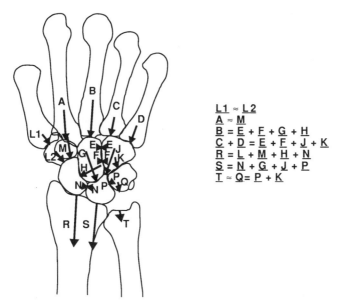

$$L1 \approx L2$$
$$A \approx M$$
$$B = E + F + G + H$$
$$C + D = E + F + J + K$$
$$R = L + M + H + N$$
$$S = N + G + J + P$$
$$T \approx Q = P + K$$

Figure 1. The force vector model in the plane of the metacarpal bones. The vectors show the distribution of force in the immediate environment of the first carpal row in the "force position".

To obtain reasonable measurements of force transmission through the carpal bones it is imperative to preserve the anatomical structures of the carpal articulations completely. Therefore, all methods which would entail partial or total destruction of the ligaments or joint capsules were precluded for this study. A possibility to measure forces by the application of dynamometers in the forearm only was considered. The experimental procedure was as follows.

The radius and ulna were rigidly fixed at their proximal and distal ends temporarily to a frame. Then 15 mm lengths of ulna and radius were removed from the diaphyseal region with an oscillating bonesaw (Figure 2-A). The position of the removed slices was chosen distal to the insertion of the membrana interossea. The reunion of the bones in their original position was effected by the implantation of electrical dynamometers of the strain-gauge type (Figure 2-B) after the temporarily used frame was removed. In this manner, the total force transmitted from the scaphoid and the lunate to the radius and the force exerted by the triquetrum on the TFCC and ulna were separately determined.

For determining the distribution of force between the scaphoid and the lunate the lateral part of the radius was separated from the rest of the bone by means of a slot after a strain-gauged bridge (dynamometer) had been attached rigidly to the bone components before separation (Figure 2-D).

By this means the force transmitted by the scaphoid alone would reach this lateral bone fragment and the strain-gauged bridge. To prevent dislocation from the original position of this bone fragment, the dynamometer was implanted in a strain-free state before the bone was cut (Figure 2-C). This was accomplished by first installing the dynamometer unit through over-sized holes in the bone. Once loosely installed, it was fixed to the bone by injecting bone cement into the spaces between the holes and the anchoring rods that passed loosely through the bone. After the bone cement hardened and the unit was fixed, the gap was made with an oscillating bonesaw under visual control using an X-ray image intensifier.

The distal radius dynamometer (bridge for radius fragment). The dynamometer used for measuring the force transmitted between scaphoid and radius consisted basically of two acrylic plates of 6 mm thickness, one on each side of the bone. Four strain gauges of 1.5 mm length were used on each of the two plates so that the axial and transversal forces could be calculated from the strain readings employing well-established methods of stress analysis and calibration tests.

The design was such that a maximum of bending rigidity together with a substantial amount of longitudinal strain sensitivity was obtained (Figure 3).

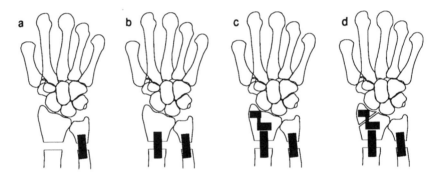

Figures 2. Implantation of the dynamometers (see text).

The proximal dynamometers for the radius and ulna. The two dynamometers for the radius and the ulna were similar. A central aluminum alloy tube (8 mm outer diameter, 0.5 mm wall thickness and 15 mm long) was furnished with strain gauges of 1.5 mm length that were so connected as to be sensitive to axial loads only. Heavy extension rods at each end of the dynamometers allowed them to be fixed into the medullary canal of the bones by means of bone cement.

Calibration. Calibration was performed with an Instron® material testing machine applying axial tensile force in steps of 20 N from 0 N to 200 N. As mentioned earlier in this section, 15 mm lengths of the ulna and radius were replaced through these dynamometers taking great care to fix the bones in their original positions. This was facilitated through use of a temporary frame and temporarily introduced 3 mm thick K-wires that passed transversally through the bones at their proximal and distal extremities (Figure 4).

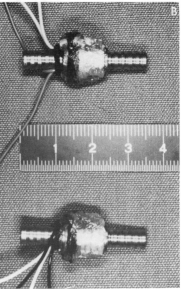

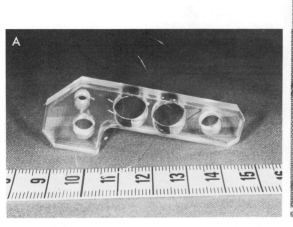

Figure 3. Figure 3-A. The distal radius dynamometer (bridge for radius fragment). **Figure 3-B.** The proximal dynamometers (radius and ulna).

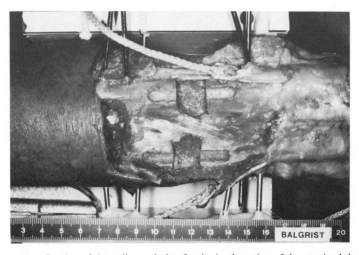

Figure 4. Temporary fixation of the radius and ulna for the implantation of the proximal dynamometers.

Experimental setup. The whole arm was supported in a loading frame, such that the humerus (flexed 90° at the elbow joint) was rigidly attached to the frame in a horizontal position, thus leaving the forearm project vertically upwards. The forearm and the hand were left unconstrained in this position (Figure 5).

The extensions of the wrist extensor tendons were fixed to a further dynamometer attached to the supporting framework so that the hand adopted the "force position" (25° dorsiflexion). The extensions of the flexor tendons were acted upon by weights. The weights were increased from 0 N to 100 N and then decreased from 100 N to 0 N in

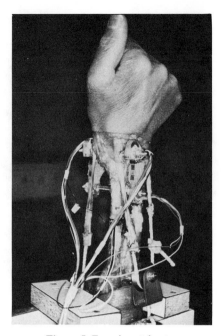

Figure 5. Experimental setup.

steps of 20 N. The force distribution in radial and ulnar deviations was measured with the application of weights of 100 N in the positions of 10° and 5° of radial deviation, and 17° and 32° of ulnar deviation.

RESULTS

It was always observed that the force measured by the dynamometers in series with the extensor tendons amounted to about 90% of the weight attached to the flexors, and therefore the total force acting across the joints of the wrist was about 190% the weight employed.

Figure 6 shows the results: the measured values were reproducible and were the same for increasing and decreasing external loads.

The relationship between the force transmitted through the scaphoid alone and the total amount passing through the radius is quite constant and is about 50%. This shows that the lunate must transmit the remaining 50% of the load on the radius.

Force distribution in radial and ulnar deviation is shown in Figure 7 with the application of a constant total force of 190 N.

With maximal radial deviation, the force transmitted to the radius becomes nearly 100% with about 76% transmitted through the scaphoid and the remaining 24% passing through the lunate.

Surprisingly in ulnar deviation, the scaphoid seems again to take more load than the lunate.

DISCUSSION AND CONCLUSIONS

Clinical experience with reconstruction of ligaments and the performance of partial intracarpal arthrodesis of the wrist bones still shows such unpredictable results that a closer investigation into the biomechanics of the complex wrist joint has indeed become

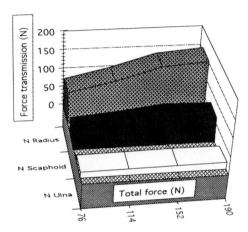

Figure 6. Force distribution on radius, scaphoid, lunate and ulna (see text).

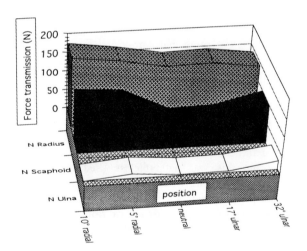

Figure 7. Force distribution in radial and ulnar deviations.

quite pressing. One of the many things deserving attention is the force transmission across the proximal carpal row.

Obviously, apart from external forces acting directly on the distal carpus, internal forces caused by the muscles would contribute to the major part of the joint load, this especially because muscle forces required to equilibrate external forces are usually several times larger than the latter. For the present investigation the every-day task of firmly gripping a handle was considered. This involved not only determining the position of the carpal bones in this particular working position, but also required an estimate of the muscle forces acting along the tendons that cross the carpal joints. As described earlier the most common position of the metacarpal bones in relation to the forearm was found to be in 25° dorsiflexion. A similar position had also been described by Bunnell,[3] who termed this the "position of function". Now, in gripping an object, all the flexor tendons that cross the wrist are involved, such that axial forces are transmitted through the bases of the metacarpals onto the distal row of carpal bones, and after passing the proximal row, enter the forearm. At this point the question also arises as to whether the ligaments of the wrist take any share of this force transmission, and if so, to what extent.

It was therefore decided to first consider the "pile" of carpal bones as being rather loose entities that could only transmit compressive forces across areas of contact. Also, negligible friction was assumed, and therefore the force vectors could only be directed normal to the joint surfaces in contact. This led us to determine the areas of contact as accurately as possible which was finally effected by in-vivo gripping exercises performed during magneto-resonance scanning.

Assuming load values acting along the axes of the five metacarpals as estimated by others,[1,4,5,7] we proceeded to carry out a rough analysis of the force transmission as shown in figure 1. It must be stressed that this is only a very crude estimate of the forces that act across the carpal joints, and only in one plane, too. One purpose of this investigation was, however, to obtain the information in a tentative manner as to whether the ligaments would be called upon to carry a significant share of the forces involved, or not.

The results obtained clearly show that hardly any ligament restraints are required to hold the "pile" of carpal bones together while transmitting force. It also appears that the lunate and scaphoid transmit approximately equal amounts of the greater part of total force to the radius.

Because of the various assumptions made in the above mentioned analysis, it was deemed necessary to verify the observations through an experimental test on fresh autopsy specimens. Of the two specimens at our disposal, preliminary radiographic examination showed a definite dissociation of the proximal carpal row between lunate and triquetrum in one but this was netherless used to rehearse the whole test procedure that included testing the feasibility of the special dynamometers, the arrangement used to load the specimen, and the best way to cut through the distal corner of the radius to separate the scaphoid bearing area from the lunate one. The second autopsy specimen was in very good condition with no pathology or irregularity relevant to the purpose of our investigation and therefore only the results obtained with this specimen have been presented here.

The results of the experimental investigation confirmed the findings of Palmer and Werner,[11,12] who convincingly showed that the greater part of the axial force is transmitted from the proximal carpal row to the radius, while the ulna takes only a small share of the total load. The investigation performed has furthermore shown the distribution of force between the lunate and the scaphoid, this in the neutral position of the wrist, as well as in ulnar and radial deviation (Figures 6 and 7). It might be remarked that Trumble et al.,[14,15] Rhomdane et al.,[13] and Horii et al.[6] also attempted to measure the force transmitted through the lunate but in our opinion the methods they used seem more prone to experimental errors than the arrangement described here.

The results of the experimental investigation totally confirm the prediction arrived at through the analytical estimation of force distribution presented in the first part of this presentation. The complexity of the load pattern of the first carpal row may help to explain the problems found in performing certain intracarpal arthrodesis. It probably is also true that the ligaments of the carpal bones are not significantly stressed while physiological loads are transmitted, as involved in gripping activities. Weber[16] for instance, found that tension on ligaments give rise to pain. As observed by Jacob et al.[8] in considering motion between the carpal bones, the topography of the joint surfaces might well be the major factor in enabling force transmission through this "pile" of bones without much need of any trussing elements like ligaments that perhaps, however, play an indirect but vital role in maintaining the overall structural form by virtue of their proprioceptive function.[2]

We wish to emphasize that in spite of the results of the experimental study coinciding so well with the predictions obtained from the theoretical model, it is absolutely necessary to carry out further investigations to confirm the observations mentioned above. In conclusion, we may summarize as follows: [1] in the neutral position the scaphoid/lunate ratio of force transmission is about 1:1, [2] the transmitted force relative to the total amount of force in this position is near 80% for the radius, and 20% for the ulna, [3] with increasing radial deviation, more force is transmitted through the radius; in maximal radial deviation we found nearly 100% transmitted through the radius, [4] the ratio of the force taken by the scaphoid relative to the lunate in radial deviation increases up to 4:1.

ACKNOWLEDGEMENTS

This work was supported by grants from "Stipendienfonds der Schweizenschen Gesellschaft für Orthopädie" and "Stiftung für Forschung, Weiter und Fortbildung in Handchirurgie der Chirurgie St. Leonard, St. Gallen".

REFERENCES

1. K.N. An, E.Y.S. Chao, W.P. Cooney, and R.L. Linscheid, Forces in the normal and abnormal hand, *J. Orthop. Res.* 3:202 (1985).
2. R.A. Berger, J.M.G. Kauer, and J.M.F. Landsmeer, Radioscapholunate ligament: a gross anatomic and histologic study of fetal and adult wrists, *J. Hand Surg.* 16A:350 (1991).
3. S. Bunnell, Surgery of the Hand, Edition 3, pp 42-48, JP Lippincott, Philadelphia (1956).
4. E.Y.S. Chao, J.D. Opgrande, and F.E. Axmear, Threedimensional force analysis of finger joints in selected isometric hand functions, *J. Biomech.* 9:387 (1976).
5. W.P. Cooney, and E.Y.S. Chao, Biomechanical analysis of static forces in the thumb during hand function, *J. Bone Joint Surg.* 59A:27 (1977).
6. T. Hara, E. Horii, K.N. Nan, W.P. Cooney, R. Linscheid, and E.Y.S. Chao, Force distribution across wrist joint: application of pressure-sensitive conductive rubber, *J. Hand Surg.* 17A:339 (1992).
7. E. Horii, M. Garcia-Elias, K.N. An, A.T. Bishop, W.P. Cooney, R.L. Linscheid, and E.Y.S. Chao, Effect on force transmission across the carpus in procedures used to treat Kienbock's disease, *J. Hand Surg.* 15A:393 (1990).
8. H.A.C. Jacob, C. Kunz, and G. Sennwald, Zur Biomechanik des Carpus - Funktionelle Anatomie und Bewegungsanalyse der Karpalknochen, *Orthopäde* 21:81 (1992).
9. C. Kenesi, D. Gastambide, and J.P. Lesage, Le syndrome de Kienböck; étude biomécanique, *Rev. Chir. Orthop.* 59:126 (1973).

10. J. Koebke, Ph. Fehrmann, and J. Mockenhaupt, Zur Beanspruchung des normalen und des pathologischen Handgelenks, *Handchir. Mikrochir. Plast. Chir.* 21:127 (1989).

11. A.K. Palmer, and F.W. Werner, The triangular fibrocartilage complex of the wrist - anatomy and function, *J. Hand Surg.* 6A:153 (1981).

12. A.K. Palmer, and F.W. Wemer, Biomechanics of the distal radioulnar joint, *Clin. Orthop.* 187:26 (1984).

13. L. Romdhane, L. Chidgey, G. Miller, and P. Dell, Experimental investigation of the scaphoid strain during wrist motion, *J. Biomech.* 23:1277 (1990).

14. T. Trumble, R.R. Glisson, A.V. Seaber, and J.R. Urbaniak, A biomechanical comparison of the methods for treating Kienböck's disease, *J. Hand Surg.* 11A:88 (1986).

15. T. Trumble, R.R. Glisson, A.V. Seaber, and J.R. Urbaniak, Forearm force transmission after surgical treatment of distal radioulnar joint disorders, *J. Hand Surg.* 12A:196 (1987).

16. E.R. Weber, Concepts governing the rotational shift of the intercalated segment of the carpus, *Orthop. Clin. North Am.* 15:193 (1984).

QUANTITATIVE FUNCTIONAL ANATOMY OF FINGER MUSCLES:

APPLICATION TO CONTROLLED GRASP

Joseph M. Mansour,[1,3] C. Rouvas,[1] J. Sarangapani,[1]
L. Hendrix,[1] and P.E. Crago[2,3]

[1]Department of Mechanical and Aerospace Engineering
[2]Department of Biomedical Engineering
[3]Department of Orthopaedics
Case Western Reserve University
Cleveland, OH 44106, USA

INTRODUCTION

The importance of the hand in basic activities of daily living has led to many investigations of the function of muscles which control hand movement. These studies have been both qualitative and quantitative. Qualitative studies have elucidated the role of particular muscles in obtaining a desired hand posture, and have helped to establish procedures for manual muscle testing. However, there are many applications where quantitative prediction of hand function is desirable.

Numerous investigations have been aimed at determining the forces in muscles and across joints for a particular function. Typically, this inverse approach leads to an indeterminate problem: there are more unknown muscle forces than equations to solve for these forces. One approach to circumventing this indeterminacy, which is not unique to the hand, has been to introduce some optimum criterion such as minimum total force or minimum muscle stress.[8,15,19] An alternative approach to the indeterminacy of muscle forces has been to specify sufficient muscle forces so that a determinate problem results. The specification of muscle forces could be based on electromyographic records for the particular grasp being studied, or muscle forces could be systematically eliminated and a series of determinate sets of equations are solved.[9] Using these computed muscle forces, joint contact forces can be found. Predicted muscle forces have ranged from less than one to approximately four times the applied external force at the distal phalanx of the index finger.

Studies of muscle and joint forces for a particular function do not, however, answer questions concerning the performance of the muscles in other tasks, the potential performance of a hand which is impaired due to muscle weakness or paralysis, or the

Advances in the Biomechanics of the Hand and Wrist
Edited by F. Schuind *et al.*, Plenum Press, New York, 1994

changes in hand function that might result from corrective surgery such as a tendon transfer. Some answers to these questions can be obtained from simulation of hand function.

Graphical-kinematic simulations have been used to illustrate tendon excursion for normal tendon anatomy and for transferred tendons.[6,7] These simulations incorporated detailed graphical representations of bones and tendon pulleys. They are kinematic in the sense that joint rotations are determined by the user, and the resulting tendon and pulley positions and tendon excursion are computed from the rotations. Recently, an extensive kinematic model which included the interaction of the finger with an object was developed.[4] Joint rotations were specified such that the fingers curled around the object. All of the above simulations did not include muscle forces nor were finger motions produced by any forces.

In contrast to these kinematic simulations, models driven by muscle-like actuators have also been developed.[5,20] These models, developed for the index finger, were driven by six muscles, flexor profundus, flexor superficialis, extensor communis, lumbricalis and palmar and dorsal interossei. Wells et al.[20] recognized the importance of passive moments at the joints in determining the equilibrium positions of the finger. This model simulated clinically relevant conditions. Buchholtz et al.[5] used this model to predict index finger grip strength as a function of the diameter of cylindrical objects.

The immediate purpose of this investigation was to determine the quantitative function of a selected group of intrinsic and extrinsic muscles of the index and long fingers and the thumb. A static simulation of joint position, driven by muscle model derived forces, was developed and used for this purpose. In the long term, the simulation will be developed to study both the potential utility of Functional Neuromuscular Simulation in restoring hand function in quadriplegic people and tendon transfer procedures.

METHODS

A model of hand function which included the index and long fingers and the thumb was constructed. The basic elements of the model were, rigid body links to represent the bones in each of the digits, a resultant passive moment at each joint and forces generated by extrinsic and intrinsic muscles (Table 1). These elements were assembled in a static model of finger position. The metacarpophalangeal joints of the index and long fingers were modeled as universal joints. The carpometacarpal joint was modeled with two independent rotations, flexion-extension and abduction-adduction and an axial rotation which was a function of the flexion-extension angle.[10] All of the more distal joints on the fingers and the thumb were modeled as hinges.

Equilibrium Equations

Equilibrium equations were written for the entire finger segment distal to each joint as opposed to the finger segment bounded between two adjacent joints.[17] This simplified the form of the equations. For example, considering only extrinsic muscles acting across the metacarpophalangeal joint of the index finger (Figure 1), we obtain for static moment equilibrium

$$(l_3 + l_2 + l_e) \times E + r^m_p \times F^m p + r^m_s \times F^m_s + r^m i \times F^m i + r^m_c \times F^m_c + p^m = 0 \quad (1)$$

and for force equilibrium

$$E + R^m + F^m_p + F^m_s + F^m_i + F^m_c = 0 \qquad (2)$$

where both Equations (1) and (2) are vector equations. The subscripts on muscle force (**F**) and moment arm (**r**) denote a specific muscle (Table 1), the superscript m refers to the metacarpophalangeal joint and **p** is the passive moment at the joint. The other symbols are described in Figure 1. Equations of the form of (1) and (2) were written for each of the joints of each of the digits; metacarpophalangeal, proximal interphalangeal and distal interphalangeal joints of the index and long fingers and the carpometacarpal, metacarpophalangeal and interphalangeal joints of the thumb. This led to a total of 18 nonlinear vector equations that were solved for joint angles resulting from the equilibrium of moments of active muscle force and passive moments.

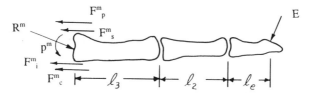

Figure 1. Free body diagram of the index finger distal to the metacarpophalangeal joint.

Table 1. Muscles included in the model and the symbol used to denote the muscle in equations such as (1) and (2).

Muscle	Subscript
Flexor Pollicis Longus	fl
Flexor Pollicis Brevis	fb
Extensor Pollicis Longus	el
Extensor Pollicis Brevis	eb
Abductor Pollicis Longus	al
Abductor Pollicis Brevis	ab
Flexor Digitorum Profundus	p
Flexor Digitorum Superficialis	s
Extensor Digitorum Communis	c
Extensor Indicis	i
Radial Interosseous	r
Ulnar Interosseous	u

Passive Moments

Relatively little data exists for the passive moments at the finger joints. We developed a device to measure the passive moments at the metacarpophalangeal, proximal interphalangeal and distal interphalangeal joint of the index and middle finger. The hand, wrist and forearm were secured in a splint leaving the finger free to rotate only about the joint being tested. At the metacarpophalangeal joint, passive moment was measured about both the flexion-extension and abduction-adduction axes. The subjects finger was rotated manually by an operator through a beam instrumented with strain gages. The strain gage output was calibrated to be proportional to the joint moment. The range of joint motion was determined for each joint and was limited by comfort of the subject. The joint was rotated through this range of motion in approximately three seconds. Five measurements of the passive moment were made at each joint.

Since the hand contains many multi-articular muscles it might be expected that the passive moment at one joint would be influenced by the position of other joints. The tenodesis grasp employed by some quadriplegics is an example of this coupling of passive moment with joint positions. A series of measurements were made with one joint of the finger free to rotate and the other two joints fixed in a series of positions with splints. While these tests showed some dependence of passive moment on the position of adjacent joints, the results obtained thus far are not conclusive. Since we were primarily interested in representative passive moments, the current model was implemented with moments that were measured at a given joint with the other joints held in the neutral position. Typical results for the metacarpophalangeal joint are shown in Figure 2. For the purposes of model simulation the passive moment was approximated by the midline curve of the loop in Figure 2 which was fit by an exponential function of joint angle Θ

$$P = A + Be^{C\Theta} - De^{F\Theta} \qquad (3)$$

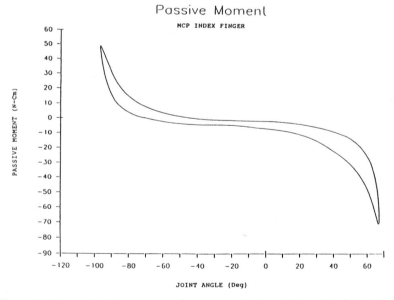

Figure 2. Measured passive moment at the metacarpophalangeal joint of the index finger.

Values of the coefficients A through F for the joints of the index and long finger are given in Table 2. The passive moment of the metacarpophalangeal joint in adduction (ulnar deviation) could not be measured due to the presence of the long finger; no data for D and F were available in this case. To prevent the finger from going into excessive adduction during the solution of the equilibrium equations, the term $-0.002\exp(50\Theta)$ was arbitrarily added to the equation to provide an increase in passive moment in adduction.

Table 2. Coefficients used to compute passive moments from Equation 3. Passive Moments at the joints of the long finger were assumed to be the same as those for the index finger.

Joint Finger	A	B	C	D	F
Index/Long					
DIP	2.23	0.04	-4.0	-10.74	4
PIP	0.84	0.06	-3.0	-8.42	3
MCP(a)	0.0	0.02	-19.0	-0.002	50
MCP(f)	0.01	0.51	-5.0	-0.05	5
Thumb					
ICP	0.05	0.88	-1.52	-2.23	18.44
MCP	0.19	0.20	-5.68	-2.37	4.30
CMC(a)	1.09	2.15	-8.28	-0.0001	22.80
CMC(f)	-1.24	0.01	-13.16	3.40	7.15

(a) abduction-adduction
(f) flexion-extension

At this time we are not able to measure the passive moments at the joints of the thumb. These moments were estimated based on joint range of motion and assuming an exponential relationship could also be applied to these joints. Coefficients for the passive moments at the thumb joints are also shown in Table 2.

Moments of Muscle Force

The moment of muscle force depends on both the moment arm of the muscle and the force in the muscle. Moment arms are a function of joint position. However, relatively little data is available on this functional dependency for all of the joint-tendon pairs the model. Furthermore, our own measurements have shown considerable subject to subject variation. We therefore used constant moment arms at each joint based on the data of An et al.[1]

It is tempting to represent muscle by a pure force generator which drives the simulation. However, the force in a muscle depends on its length, velocity and activation. For this static model of hand function, the force-velocity relationship has no influence on the predicted muscle force, which leaves length and activation as the major determinants of force. Muscle length depends on the distance between its origin and insertion and the compliance of the tendon in series with the muscle.

Under static conditions, muscle force, F_m, is given by an expression similar to that for a linear spring.[11]

$$F_m = m \ (l_m - L_o) \ \cos\varnothing \qquad (4)$$

where m is the activation, the length of the muscle, l_m is the distance from its origin to tendinous insertion and L_o (a constant) is the length at which no force is developed in the muscle and $\varnothing$ is the pennation angle of the muscle.

The force in the tendon, in series with the muscle, was modeled as a linear function of tendon length

$$F_t = k \ (l_t - l_{to}) \qquad (5)$$

where k is the tendon stiffness, l_t is the tendon length and l_{to} is the slack length of the tendon. Equating the muscle and tendon forces, and recognizing that the overall length of the muscle-tendon unit can be found from, $l = l_m + l_t$, leads to an expression for the muscle length

$$l_m = \frac{k \ . \ (l - l_{to}) + m \ . \ L_o \ . \ \cos\varnothing}{m \ . \ R \ . \ \cos\varnothing + k} \qquad (6)$$

The overall length, l, was determined as a function of joint angle from tendon excursion data.[12,1] The excursion data is essentially linear over a wide range of joint motion, which leads to an expression of the form

$$l = l_1 + l_2 \ \text{x} \ \Theta^{MCP} + l_3 \ \text{x} \ \Theta^{PIP} \qquad (7)$$

where l_1 is the length of the muscle tendon unit with the wrist and all finger joints in the neutral position, and l_2 and l_3 are constants obtained from the slope of the tendon excursion-joint angle curves and Θ^{MCP} and Θ^{PIP} are the joint angles at the metacarpophalangeal and proximal interphalangeal joints. Combining equations (4), (6) and (7) leads to an expression for muscle force

$$F_m = F_1 + F_2 \ \text{x} \ \Theta^{MCP} + F_3 \ \text{x} \ \Theta^{PIP} \qquad (8)$$

Where F_1, F_2 and F_3 are functions of the model parameters. Equation (8) was used to generate length dependent muscle forces in simulation equations such as (1) and (2). Details of incorporating the muscle model into the simulation are given in Sarangapani.[18]

Model Parameters

Use of the muscle and tendon models requires the estimation of several parameters. Tendon stiffness was determined from the data in Benedict.[2] Since the model relates forces to displacement, rather than stress to strain, the tendon stiffness was estimated from tendon areas (A), modulus of elasticity (E) and length l_{to} from

$$k = \frac{AE}{l_{to}} \qquad (9)$$

Tendon area was obtained from Blanton and Biggs.[3] The areas for embalmed tendons were used since limited data is available for unembalmed tendons. Tendon lengths were obtained from data given in Brand and Lieber.[13,14] Pennation angle was obtained from data given in Lieber.[13,14] The slack length of the muscle, L_o, was estimated based on the ratio of minimum to maximum force in a muscle. Taking the ratio of these forces from equation (4) leads to

$$L_o = \frac{l_{min} - r l_{max}}{1 - r} \qquad (10)$$

Where the ratio of r was chosen as 0.3.[11]

Solution Procedure

The simulation equations, such as those given in (1) and (2), are nonlinear algebraic equations. The system of equations describing finger position were solved numerically using Powell's method.[16] The use of Newton's method was also investigated, but it did not converge. Newton's method requires a good initial guess of the solution which was difficult to obtain. Powell's method is much less sensitive to the initial guess of the solution.

Applications

The primary motivation for developing this model of hand muscle function was to determine the muscles needed to perform basic manual tasks. As an example the requirements for tip pinch were investigated. Numerous hand positions were simulated using measured passive moments and phalanx lengths for one hand. The influence of individual muscles and groups of muscles were investigated by varying their activation.

The sensitivity of the model to tendon compliance was also investigated. Muscle force and joint positions were determined under constant activation of the flexor digitorum profundus (m = 0.100) against a constant 1N force perpendicular to the distal phalanx. Tendon compliance was varied from 300 MPa to 2000 MPa.

RESULTS

Tendon compliance was found to have little effect on the muscle force or joint position for the conditions investigated. Decreasing the Young's modulus of the tendon from 2000MPa to 300MPa resulted in a decrease of 0.2N in the force in the flexor digitorum profundus and the joint angles were changed by less than 2.5 deg. The more compliant tendon led to lower muscle force as would be expected. The shorter muscle produced less force for a given activation, and less joint flexion.

The approximate joint angles for the index finger in a tip pinch position are listed in Table 3. These were measured on a subject's hand using a goniometer, and represent target positions we wish to achieve with simulation.

A simulation with only passive moments resulted in an index finger position which was more flexed at the proximal interphalangeal joint and less flexed at the metacarpophalangeal and distal interphalangeal joints than the target position (Figure 3). To attempt to reach the target position, two simulation paths were followed, one in which only extrinsic muscles were used and the other which used both extrinsic and intrinsic muscles.

Table 3. Index finger joint angles for tip pinch.Negative values are flexion.

DIP	-31 deg.
PIP	-49 deg.
MCP(f)	-44 deg.
MCP(a)	0 deg.

(f) flexion-extension
(a) abduction-adduction

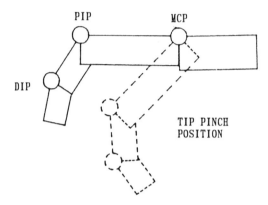

Figure 3. Simulated rest position of the index finger under the action of only passive moments (solid lines). The dashed line shows a target position for tip pinch.

When using only extrinsic muscles it was found that activation of the extensor digitorum communis (m = 0.080) drove the angle at the proximal interphalangeal joint close to the target value, but caused hyperextension of the metacarpophalangeal joint (Figure 4). The flexion of the distal joint was still somewhat less than the target. It is not possible to correct only the hyperextension at the metacarpophalangeal joint since most hand muscles are multiarticular. For example, activating the flexor digitorum profundus (m = 0.050) brings the metacarpophalangeal joint to approximately the correct position in flexion, but the proximal interphalangeal joint is now too flexed (Figure 5). This clawed posture can be somewhat reduced by activating multiple extrinsic muscles. Although these activations improve the finger positions, they do not give a good approximation of tip pinch.

The simulations shown in Figures 4 and 5 suggest that additional extension is needed at the proximal interphalangeal joint to achieve a desirable tip pinch position. Activating an extensor muscle, however, leads to excessive extension at the metacarpophalangeal joint. These observations, which are consistent with the results obtained from numerous other simulations and clinical experience, suggest that additional proximal interphalangeal joint extension and metacarpophalangeal joint flexion are needed. These motions cannot be obtained with extrinsic muscles, but they may be achieved with intrinsic muscles.

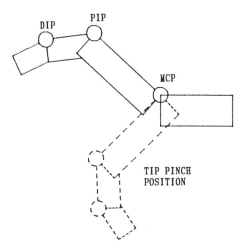

Figure 4. Simulated position of the index finger when only the extensor digitorum communis is activated (m = 0.080). The metacarpophalangeal joint is hyperextended, the proximal interphalangeal joint angle is close to that needed for tip pinch and the distal joint is somewhat extended relative to the tip pinch position.

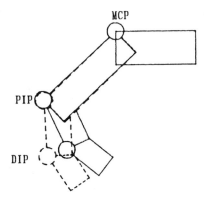

Figure 5. Simulated position of the index finger when only the flexor digitorum profundus is activated (m = 0.50). With respect to obtaining a tip pinch position, the metacarpophalangeal joint is well positioned while the proximal joint is over flexed and the distal joint angle is close to the target position.

The radial and ulnar interossei also known as the palmar and dorsal interossei, were used in an attempt to improve the tip pinch position of the index finger. An excellent approximation of the tip pinch position for the index finger could be achieved (Figure 6) by activating the ulnar interosseous (m = 0.320) and the radial interosseous (m = 0.640).

To position the thumb in opposition to the index finger, that is, to obtain a tip pinch position, the extensor pollicis brevis (m = 0.010) and the abductor policis longus (m = 0.200) were activated (Figure 7). In this simulation the index finger was slightly adducted (radially deviated) so that the medial portion of its finger pad contacted the thumb pad.

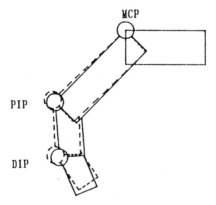

Figure 6. Simulated position of the index finger when only the radial and ulnar interossei are activated. Overall the finger is close to the tip pinch position.

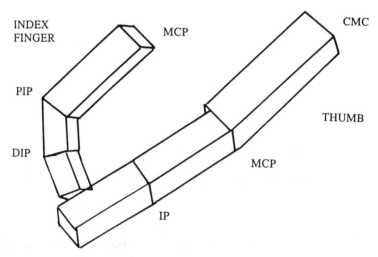

Figure 7. Simulated tip pinch position of the thumb and index finger. The index finger was positioned as in Figure 6, using only the interossei. The thumb was positioned by activating the extensor pollicis brevis and abductor pollicis longus.

DISCUSSION

This analysis of quantitative functional anatomy of muscles controlling hand function used only a subset of the muscles which act on the hand. The choice of muscles to include in this analysis was based on current and potential applications of Functional Neuromuscular Stimulations (FNS) which is used to restore hand function in quadriplegic people. Currently all of the thumb muscles and extrinsic muscles for the index and long finger listed in Table 1 are used in FNS implementations at our institution.

Intrinsic muscles acting on the index and long finger are generally not used in FNS implementations since they are difficult to stimulate, although electrodes were recently implanted in the interossei of one subject in our program. The actions of the radial and ulnar interossei were studied, since they are the most likely candidates for stimulation if it can be shown that there is a significant benefit to be gained by their use.

The current implementation of the model reproduces known hand behavior. Using only extrinsics it was not possible to produce an acceptable tip pinch position for the index and long fingers. All such attempts resulted either in hyperextension of the metacarpophalangeal joint or excessive flexion of the proximal interphalangeal joint. Using intrinsic muscles, the radial and ulnar interossei, resulted in an acceptable tip pinch position.

These simulations were performed without external forces. The fingers and thumb did not contact one another or any other object. The simulation is currently being modified to include contact with fixed cylindrical objects. Under these conditions it is expected that both intrinsic and extrinsic muscles will be needed to position the fingers and support the weight of an object. The small physiological cross section of the intrinsic muscles suggests they would not be able to generate the force needed to support any significant weight. Also, under conditions of grasp, joint rotations may be relatively small and muscle forces relatively high. The effect of tendon compliance on muscle length and force may then be greater than was observed thus far where we were only positioning the fingers without external forces.

ACKNOWLEDGEMENTS

This work was supported by the National Institute of Neurological Diseases and Stroke under contract NO1-NS-9-2356. We would like to thank Ali Esteki for his assistance in obtaining passive moment data.

REFERENCES

1. K.N. An, Y. Ueba, E.Y.S. Chao, W.P. Cooney, and R.L. Linscheid, Tendon excursion and moment arm of index finger muscles, *J. Biomech.* 6:419 (1983).
2. J.W. Benedict, L.B. Walker, and E.H. Harris, Stress-strain characteristics and tensile strength of unembalmed human tendons, *J. Biomech.* 1:53 (1968).
3. P.L Blanton and N.L. Biggs, Ultimate tensile strength of fetal and adult human tendons, *J. Biomech.* 3:181 (1970).
4. B. Buchholtz, and T.J. Armstrong, A kinematic model of the human hand to evaluate its prehensive capabilities, *J. Biomech.* 25:149 (1992).
5. B. Buchholtz, R. Wells, and T.J. Armstrong, The influenced object size on grasp strength: Results of a computer simulation of cylindrical grasp, 12th Meeting, Am. Soc. Biomech. Univ. of Illinois, (1987).
6. W.L. Buford, L.M. Myers, and A.M. Hollister, A modeling and simulation system for the human hand, *J. Clin. Eng.* 15:445 (1990).

7. W.L.Jr. Buford and P.E. Thompson, A system for 3-D interactive simulation of hand biomechanics, IEEE Trans. on Biomed. Engr. BME 34:434 (1987).

8. E.Y. Chao and K.N. An, Graphical interpretation of the solution to the redundant problem in biomechanics, ASME Paper No. 77-Bio-1, 1, (1977).

9. E.Y. Chao, J.D. Opgrande, and F.E. Axmear, Three-dimensional force analysis of finger joints in selected isometric hand functions, *J. Biomech.* 9:387 (1976).

10. W.P. Cooney, M.J. Lucca, E.Y.S. Chao, and R.L. Linscheid, The kinesiology of the thumb trapesiometacarpal joint, *J. Bone Joint Surg.* 63:1371 (1981).

11. P.E. Crago, Muscle input/output model: The static dependence of force on length, recruitment and firing period, in press, *IEEE Trans. Biomed. Eng.* (1992).

12. E.P. Frankenburg, Experimental evaluation of tendon moment arms at the MCP and PIP joints of the hand, M.S. thesis, Dept. of Mechanical and Aerospace Engr. Case Western Reserve Univ. (1990).

13. R.L. Lieber, B.M. Fazeli, and M.J. Botte, Architecture of selected wrist flexor and extensor muscles, *J. Hand Surg.* 15A:244 (1990).

14. R.L. Lieber, M.D. Jacobson, B.M. Fazeli, R.A. Abrams, and M.J. Botte, Architecture of selected muscles of the arm and forearm: Anatomy and implications for tendon transfer, personal communication manuscript in preparation, (1990).

15. A.G. Patriarco, R.W. Mann, S.R. Simon, and J.M. Mansour, An evaluation of the approaches of optimization models in the predictions of muscle forces during human gait, *J. Biomech.* 14:513 (1981).

16. M.J.D. Powell, A hybrid method for nonlinear equations, in: "Numerical Methods for Nonlinear Algebraic Equations," P. Rabinowitz, ed., Gordon and Breach Science Publishers. New York, NY, (1970).

17. C. Rouvas, The development of a mechanical model of the hand, M.S. thesis, Dept. of Mechanical and Aerospace Engr., Case Western Reserve Univ. (1989).

18. J. Sarangapani, A biomechanical model of the hand with applications to functional neuromuscular stimulation, M.S. thesis, Dept. of Mechanical and Aerospace Engr., Case Western Reserve Univ. (1991).

19. A. Seireg, and R.J. Arvikar, The prediction of muscle load sharing and joint forces in the lower extremity in walking, *J. Biomech.* 8:89 (1975).

20. R.P. Wells, D.A. Ranney, and A. Keeler, The interaction of muscular and passive elastic forces during unloaded finger movements: A computer graphics model, in: "Biomechanics Current and Interdisciplinary Research," S.M. Perren and E. Schneider, ed., Martinus Nighoff, Dordrecht, (1985).

ANATOMIC FORMS OF MALE AND FEMALE CARPOMETACARPAL JOINTS

G.A. Ateshian, M.P. Rosenwasser, and V.C. Mow

Orthopaedic Research Laboratory
Departments of Mechanical Engineering
and Orthopaedic Surgery
Columbia University
New York, NY 10032, USA

INTRODUCTION

The thumb carpometacarpal joint (CMC) is commonly afflicted with osteoarthrosis (OA), which is predominant in the female population over 45.[6] A common hypothesis attributes the prevalence of the development of this disease to the particular saddle-shaped anatomy of the articular surfaces of the CMC joint from female subjects. In early studies, MacConaill[8] postulated that the saddle configuration of the CMC joint is the most efficient for providing conjunct rotation during circumduction of the thumb, while Napier[9] described joint anatomy and congruence of the joint, and the stabilizing role of capsular ligaments. More recently, Smith and Kuczynski[11] classified the various shapes of the trapezium as sellar, triangular, ovoid and semi-cylindrical, and they noted that some of these CMC articular surface shapes predispose to, or accompany, OA changes. North and Rutledge[10] also found that the trapezium tended to be flatter in joints with early OA changes, and that CMC joints from female subjects had shallower surfaces than those from male subjects. Based on the work of Huiskes and co-workers,[5] we developed a close-range stereophotogrammetric (SPG) method to accurately quantify the anatomy of knee joint articular surfaces.[3] In the present investigation, we adopted our SPG method for a detailed anatomic study of the thumb carpometacarpal joint. Furthermore, using fundamental principles of differential geometry, each SPG determined 3-D surface was characterized by calculating its total surface area and principal curvatures. Congruence of each matching pair of metacarpal and trapezial surfaces was calculated, and comparisons were made between joints from female and male subjects.

Advances in the Biomechanics of the Hand and Wrist
Edited by F. Schuind *et al.*, Plenum Press, New York, 1994

MATERIALS AND METHODS

Eight female (aged 54 to 71, avg. 64) and five male (aged 61 to 80, avg. 70) fresh frozen human cadaver joints were used in this study. The articular surfaces of each joint were exposed by sharp dissection and placed within a calibration frame. A fine grid was optically projected on the surface and a pair of large format precision photographs (a stereogram) was obtained. The grid intersections were digitized with a high-accuracy (± 20μm) X-Y digitizer on each photograph. The three-dimensional coordinates of these intersection points (nodes) were reconstructed mathematically from the knowledge of their digitized photographic coordinates. The measurement uncertainty of these nodal coordinates was determined to be 20μm along each of the three spatial coordinate directions, using a calibrated flat plate.[2,3]

The nodal coordinates from each articular surface were fitted with the equation of a parametric biquintic spline, using a least-squares technique. The surface-fit error, which measures the difference between the predicted and actual coordinates of the surface nodes, was calculated for each specimen. Given the biquintic spline analytical representation of each articular surface, various surface differential properties were evaluated and graphically displayed.[2,4] Specifically, the principal curvatures κ_{min} and κ_{max} were calculated at all the surface nodes, and curvature maps were subsequently generated from these data. From these principal curvatures, the Gaussian curvature $K = \kappa_{min}\kappa_{max}$ and the rms curvature $\kappa_{rms} = [(\kappa^2_{min} + \kappa^2_{max})/2]^{1/2}$ were also evaluated and displayed. Finally, the principal directions of curvature were also obtained using a numerical integration scheme.[2] These quantities provide a precise description of the differential geometry of the articular surfaces of the thumb CMC joint.

A quantitative index of global joint congruence κ^e_{rms} was calculated for each joint, using surface averages of the principal curvatures of the trapezium and metacarpal.[2] The congruence of male and female joints were statistically compared using these values of κ^e_{rms}.

RESULTS

For the trapezium, the mean value and standard deviation of error over all specimens was 66.6 ± 17.1 μm, while for the metacarpal it was 65.5 ± 16.4 μm, indicating that the mathematical surface-fits are very faithful to the experimental data. Figure 1 shows elevation contour sections of the trapezium and metacarpal, which clearly exhibit the characteristic overall saddle shape of these surfaces.

A more detailed description of the topography of the trapezium and metacarpal is provided by various curvature maps. Thus, from the curvature calculations, we found that on the trapezium lines of minimum curvature were almost perfectly aligned with the radio-ulnar anatomic direction, while lines of maximum curvature followed the dorso-volar direction (Figure 2a). On the metacarpal, the opposite scheme occurred in the central aspect of the surface (Figure 2b). From the Gaussian curvature maps, it was observed that the trapezium is sellar over most of its surface, with narrow ovoid regions sometimes appearing around the periphery where the surface curls away. However, in four female trapeziums, ovoid regions extended in the form of dorso-volar bands on the ulnar and/or radial aspect of the joint (Figure 3a). The metacarpal was found to be sellar along a central band which extends from the radial to the ulnar aspect of the joint surface, and spans slightly more than half the width of the surface along the dorso-volar direction (Figure 3b). The dorsal and volar aspects of the metacarpal are ovoid in shape,

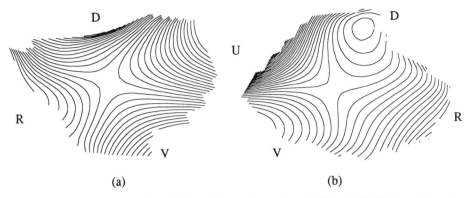

Figure 1. Elevation contour sections of the articular surfaces of a typical thumb CMC joint: a) trapezium, b) metacarpal.

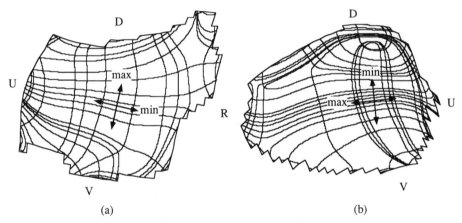

Figure 2. Lines of curvature for a typical pair of articular surfaces: a) trapezium, b) metacarpal. 'Min' and 'max' indicate the directions of minimum and maximum principal curvatures, respectively.

with the dorsal ovoid region being made up of one or sometimes two prominences, and the volar ovoid region having always one prominence.

From curvature maps of κ_{min}, it was found that the site of greatest concavity on the metacarpal occurs along a central band extending from the ulnar side and sometimes reaching the radial side of the joint (Figure 4). Similarly, curvature maps of κ_{max} on the trapezium demonstrate that the sites of greatest convexity occur on the ulnar aspect of the joint, with a narrow band extending dorsally toward the radial side (Figure 5).

As expected, female joint surface areas (trapezium: 1.05 ± 0.21 cm^2; metacarpal: 1.22 ± 0.36 cm^2) were statistically smaller than male joints (trapezium: 1.63 ± 0.18 cm^2; metacarpal: 1.74 ± 0.21 cm^2), and the surface area of the metacarpal from females was statistically greater than that of the trapezium. Statistical differences in maximum and minimum principal curvatures persisted between female and male trapezial surfaces normalized against the surface area, indicating fundamental differences in the anatomic shape of the trapezium between the sexes. Specifically, while all trapeziums were more curved in the dorso-volar plane than in the radio-ulnar plane, this characteristic was significantly more pronounced in female joints (Figure 6). As a consequence of the differences in size and shape, the congruence indices also differed between male and

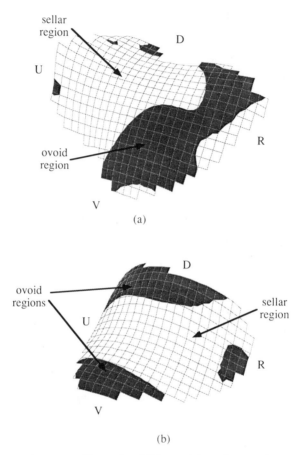

(a)

(b)

Figure 3. Gaussian curvature maps. **Figure 3-a.** While most trapezium specimens were sellar (saddle-shaped) over their entire surface, four female specimens displayed ovoid regions across a significant portion of their articular surfaces, as shown here. **Figure 3-b.** All metacarpal surfaces had ovoid prominences on their dorsal and volar aspects.

female joints. The mean and standard deviation of the congruence index for female joints was found to be $\kappa^e_{rms}= 129.3 \pm 33.3$ m^{-1} while that for male joints was $\kappa^e_{rms}= 63.5 \pm 24.7$ m^{-1}, indicating that male joints are more congruent ($\kappa^e_{rms}= 0$ indicates perfect congruence). The difference between female and male joint congruence was statistically significant ($p \leq 0.01$). Furthermore, in eleven out of thirteen specimens, the CMC joint was found to be more congruent in the radio-ulnar plane than the dorso-volar plane.

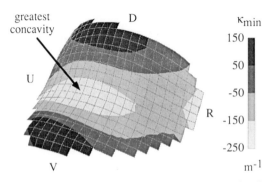

Figure 4. Minimum curvature map (κ_{min}) of a typical metacarpal articular surface.

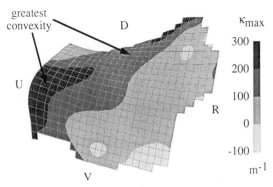

Figure 5. Maximum curvature map (κ_{max}) of a typical trapezium articular surface.

DISCUSSION

Our results on the shape of the trapezium and metacarpal are in general agreement with those of Kuczynski.[7] Similarly, these results agree with the observations of Napier[9] who had noted that the thumb CMC joint was more congruent in the radio-ulnar direction than the dorso-volar direction. Our study also provides new insight into the anatomy of the CMC joints in relation to gender. Significant differences were found between female and male joints for this sample of older specimens. Thus, we found that the trapezial surface has a different shape (i.e., curvature characteristics) in male and female CMC joints, even when the effects of size are eliminated. No such differences were found for the metacarpal surface. These results do not seem to agree with those of North and Rutledge[10] who found that the female trapezium is shallower than the male trapezium. However, this apparent contradiction may be attributed to the difference in measurement methods between these two studies. Furthermore, they included osteophytic spurs in their measurement of joint depth, while this study focused exclusively on the shape of the articular surfaces.

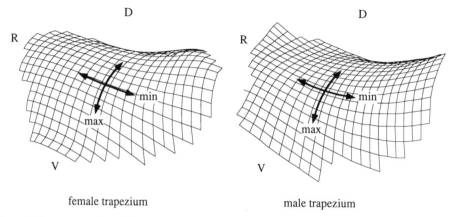

D   D

R   R

min   min

max   max

female trapezium   male trapezium

Figure 6. Female trapeziums demonstrated a higher ratio of maximum to minimum curvature than male trapeziums, indicating a gender-related difference in shape. No such differences were observed in the metacarpal.

Using quantitative indices of congruence, it was determined that male joints are significantly more congruent than female joints as a result of the different shape of the trapezial surfaces in the two sexes, as well as the difference in joint size. Consequently, contact areas in the female CMC joint will be smaller than in male joints under similar joint loading conditions, as can be assessed quantitatively from contact mechanics theories coupled with the curvature data of this study. Furthermore, joint incongruence leads to a quicker loss of load support from the interstitial fluid phase of articular cartilage.[1] This loss of load support, which occurs because water finds a shorter pathway to escape from the tissue through the incongruent joint surfaces, means that the solid collagen-proteoglycan phase of cartilage supports a larger share of the total applied load. Such larger loads may lead, over years of repeated cycles of loading, to fatigue damage of the collagen-proteoglycan network and degradation of the biomechanical performance of the joint. Thus, during similar activities of daily living which involve similar hand pinch or grasp load levels, female thumb CMC joints may experience higher stresses than male joints. These higher stresses may predispose females to degenerative disease of the thumb CMC joint.

While these results may provide a plausible explanation for the prevalence of OA in the female population, additional studies must be performed on a larger pool of specimens which are more representative of all age groups. For example, it would be of interest to determine whether gender-related differences exist from early adulthood, or whether these differences develop with age as a result of the loading environment of the male and female thumb CMC joints.

CONCLUSION

By applying basic principles of differential geometry, the curvature characteristics of the articular surfaces of the thumb CMC joint were quantified to provide a detailed description of the joint anatomy. From these quantitative data, accurate comparisons were made between the trapezial and metacarpal surfaces, as well as between female and male joints. Quantitative topographical measurements provide a useful tool to the study of joint biomechanics, as well as joint degenerative disease.

ACKNOWLEDGEMENTS

This work was supported in part by a grant from the National Institutes of Health (AR-38733) and a Center of Excellence for Orthopaedic Research Award from Bristol-Myers Squibbs/Zimmer.

REFERENCES

1. G.A. Ateshian, W.M. Lai, W.B. Zhu, and V.C. Mow, A biphasic model for contact in diarthrodial joints, ASME Adv. Bioeng, 22:191 (1992).
2. G.A. Ateshian, M.P. Rosenwasser, and V.C. Mow, Curvature characteristics and congruence of the thumb carpometacarpal joint: Differences between male and female joints, *J. Biomechanics*, 25:591 (1992).
3. G.A. Ateshian, L.J. Soslowsky, and V.C. Mow, Quantitation of articular surface topography and cartilage thickness in knee joints using stereophotogrammetry, *J. Biomechanics* 24:761 (1991).
4. J.M. Beck, R.T. Farouki, and J.K. Hinds, Surface analysis methods, IEEE CG&A December:18-36, (1986).
5. R. Huiskes, J. Kremers, A. de Lange, H.J. Woltring, G. Selvik, and Th.J.G. van Rens, Analytical stereophotogrammetric determination of three-dimensional knee-joint geometry, *J. Biomechanics* 18:559 (1985).
6. J.L. Kelsey, "Epidemiology of Musculoskeletal Disorders," Oxford University Press, New York, (1982).
7. K. Kuczynski, Carpometacarpal joint of the human thumb, *J. Anat.* 118:119 (1974).
8. M.A. MacConaill, Studies in the mechanics of synovial joints: II. Displacements on articular surfaces and the significance of saddle joints, *Ir. J. Med. Sci.* 6:223 (1946).
9. J.R. Napier, The form and function of the carpo-metacarpal joint of the thumb, *J. Anat.* 89:362 (1955).
10. E.R. North, and W.M. Rutledge, The trapezium-thumb metacarpal joint: The relationship of joint shape and degenerative joint disease, *Hand* 15:201 (1983).
11. S.A. Smith, and K. Kuczynski, Observations on the joints of the hand, *Hand* 10:226 (1978).

FINGER MATHEMATICAL MODELING AND REHABILITATION

Federico Casolo, and Vittorio Lorenzi

Politecnico di Milano, Instituto degli Azionamenti Meccanici
Piazza Leonardo da Vinci 32
I-20133, Milano, Italy

INTRODUCTION

Several pathologies or traumas (Figure 1) reduce the mechanical efficiency of the hand and are characterized by one of the following conditions: tendon rupture, tenolisis, tendons' adhesions, rupture of tendons interconnections, hand muscle deficiency or palsy, bones' deformity, reduced range of joint rotations and laxity or rupture of tendon sheaths.

Many kinds of interventions (Figures 2,3) on the musculo-tendon-skeletal complex can be proposed to recover the physiological functionality, when possible, or, at least, to restore few basic mechanical capabilities (e.g. grasping): those interventions can be surgical, physiotherapeutic or else they can involve the adoption of orthoses.

Surgical procedures can include: tendon transfer, tendon pulleys reconstruction, arthrodesis, tenodesis, tendon lengthening, etc; the selective muscular reinforcement is an example of rehabilitative therapy; moreover, splinting techniques can involve one or more joints of the hand.

OBJECTIVES OF FINGERS MATHEMATICAL MODELING

The development of mathematical models of the finger tendon complex allows to build simulation programs whose main purposes are: [1] to deepen the knowledge about mechanical causes and consequences of some hand pathologies, [2] to help the clinician in planning and optimizing their intervention for the best recovery of hand mechanical functions, [3] to develop ergonomical optimization of fingers' actions. As an example, the preliminary version of our simulation package, can handle the following tasks: [1] on the healthy finger: analysis of the range of motion of the finger chain, computation of the maximum force that the finger can exert in a given direction or selection of the best finger chain configuration to act against a given force or couple, evaluating, in the mean time, the force required by each tendon, [2] on the pathological finger:

Advances in the Biomechanics of the Hand and Wrist
Edited by F. Schuind *et al.*, Plenum Press, New York, 1994

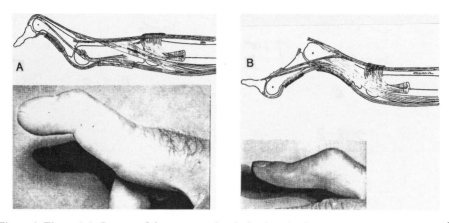

Figure 1. Figure 1-A. Rupture of the transverse band of retinacular ligament (swan neck deformity).[5]
Figure 1-B. Rupture of EC central slip at PIP joint (boutoniere deformity).[5]

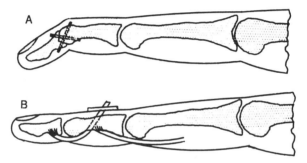

Figure 2. Figure 2-A. Arthrodesis of DIP joint. **Figure 2-B.** Tenodesis at DIP joint.

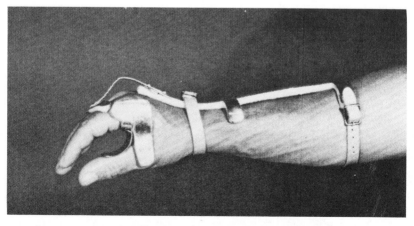

Figure 3. Splint preventing MCP joint hyperextension.[7]

evaluation of its "stability" (according to the definition given in a following section), and when this is not achievable, planning the recovery intervention as follows: in case of tendon transfer optimizing the location of the tendon insertions and checking the adequacy of the muscles to the new task; in case of arthrodesis or tenodesis, analyzing the adequacy of the technique and optimizing the angle of the joint to be fixed. Moreover, when the pathological finger is checked as stable, it allows all the analyses described for healthy fingers.

In the same way it is possible to analyze the effects of hypothetical hand-splinting systems or to select those muscles whose strengthening can be profitable for the functional recovery of the hand.

COMPONENTS OF A FINGER MODEL

Finger kinematics

The first and fundamental step in the development of the model is the choice of a set of anthropometric parameters, adequate to the description of the finger joints' kinematics. Joint surfaces, ligaments and capsules are the main constraints which drive the joints during the movements produced by the tendons or by the external forces; the complexity of these structures and the differences among subjects lead their modeling to a very critical question; in fact, to bypass the problem, most researchers refer only to the physiological motion of the joints, regardless to the constraining structure.

A useful analysis concerns the motion of the instantaneous screw axis. Some experimental measures demonstrate that the I.S.A. of finger joints are not fixed; however, for the interphalangeal joints, Youm[18] shows that the intersections of ISA with the transversal plane, lie in a 0.75 mm-radius circle. Consequently those joints can be adequately approximated by revolute pairs. Analogously the same author proves that "*functionally, the MCP joint is equivalent to a universal joint which has two degrees of freedom with negligible axial rotation*". Nevertheless the flexion axes of contiguous joints are not parallel (Figure 4-A).

The finger can thus be described as an open chain of 4 rigid bodies, where the metacarpal bone is the frame and, with four degrees of freedom (Figure 4-B). Therefore the parameters required to define this linkage are: segments length (l_0, l_1, l_2, l_3), axes orientations (defined, e.g. by Euler angles $\alpha_1, \beta_1, \gamma_1; \alpha_2, \beta_2, \gamma_2; \alpha_3, \beta_3, \gamma_3$), $\theta_1, \theta_2, \theta_3,$ θ_4 are the variables associated to the degrees of freedom: flexion-extension angles of MCP, PIP, DIP and the abdo-adduction angle of MCP, respectively.

Coordinate systems

Several coordinate systems can describe the position of the finger chain and that of the loads acting on it. For instance the following may be a convenient choice. Fixing a frame to each body of the finger chain, the relative and absolute position of those frames can be described by a 4x4 homogeneous matrix.[11] If we label each phalanx and joint, from proximal to distal ones, with $k = 1,2,3$ ("0" for the metacarpal bone) we can put the k frame on the center of k joint, with: X_k on the joint axis of rotation,[18] radially oriented, Y_k perpendicular to a plane defined by X_k and by $k+1$ joint center, dorsally oriented, and Z_k coherent with a right hand frame convention for right hand.

The position matrix Mk-1,k can be written in a general form as:

$$\mathbf{M}_{k\text{-}1,k} = \begin{vmatrix} Xx & Yx & Zx & Ox \\ Xy & Yy & Zy & Oy \\ Xz & Yz & Zz & Oz \\ 0 & 0 & 0 & 1 \end{vmatrix} = \begin{vmatrix} \mathbf{R}_{k\text{-}1,k} & \mathbf{T} \\ \hline 0 & 1 \end{vmatrix}$$

where the elements of R are the director cosines of the frame k with respect to the frame k-1, and the column matrix **T** contains the homogeneous coordinates of the origin of k with respect of k-1.

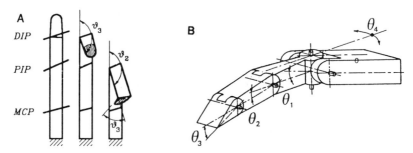

Figure 4. Figure 4-A. Joint axes of rotation are generally not parallel. **Figure 4-B.** Scheme of index finger joint kinematics.

Obviously $\upsilon1$, $\upsilon2$ and $\upsilon3$ are the phalanges rotation around X1, X2, X3 respectively, while $\upsilon4$ represents the proximal phalanges rotation around Y1. Using these conventions, when the relative position of the contiguous phalanges are known, the position matrix of the distal phalanx, with respect of the metacarpal frame, can be calculated by means of the simple expression:

$$\mathbf{M}_{0,3} = \mathbf{M}_{0,1} \cdot \mathbf{M}_{1,2} \cdot \mathbf{M}_{2,3}$$

Few others matrices with the same structure have been also recently introduced[9] in order to use the same notation for kinematics and dynamics.

Tendon-complex layout

Different kinds of tendon layout can be adopted for the finger model, according to the kind of the required simulation. In figure 5 a schematic layout of the tendon complex is displayed; it shows the tendons' actions at each joint: flexion or extension and, in the same time, adduction or abduction of the MCP joint.

With a more complex layout (Figure 6) it is possible to take also account of different branches of the same tendon and different tendons' interconnections. The adoption of this kind of layout is required in many applications, for example when the simulation takes account of lateral bands ruptures.

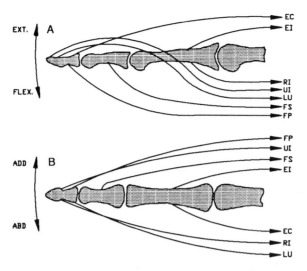

Figure 5. Functional layout of index finger tendons. **Figure 5-A.** flexors-extensors, **Figure 5-B.** adductors-abductors.

Tendon path

In order to insert the tendon model in the one of finger complex, it is required to define the tendon path and its insertion points to the bones; the path model must be described by a limited number of geometrical parameters. Most authors consider that the tendons' path along the central part of the phalangeal diaphysis does not vary during finger movements (Figure 7); consequently, there is no need to define the tendons' trajectory in this zone, when the evaluation of the tension within the sheaths located there is not required.

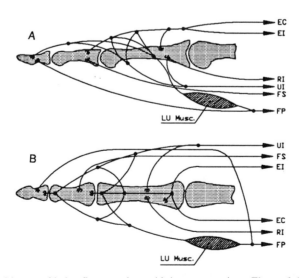

Figure 6. Functional layout of index finger tendons with inter-conncetions: **Figure 6-A.** flexors-extensors, **Figure 6-B.** adductors-abductors.

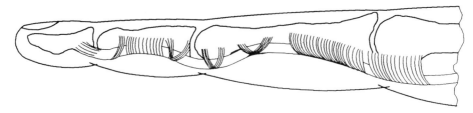

Figure 7. Sheaths along phalanx diaphysis prevents tendon bowstringing.

On the contrary, as the following chapters point out, an adequate analysis of the tendons' trajectory near the joints is required.

A small number of geometrical models for the tendons' path have been described in the literature; three of them have been introduced by Landsmeer,[14,15,16] and one by An and Chao.[1] In the following paragraphs we will refer to them as LI, LII, LIII and 2P respectively. All of these models assume that joints rotation does not affect the tendon path near the medial diaphysis.

According to LI, the tendons lean on the joint surfaces which are approximated to constant radius pulleys (Figure 8-A), thus, the tendon excursion is linearly correlated with the joint angle of rotation:

$$EXC_{LIij} = R_{ij} . \theta_j$$

where i is the tendon label (i=1,..,7) and j is the label for the d.o.f. (j = 1,..,4; d.o.f. = $\theta 1, \theta 2, \theta 3, \theta 4$).

The scheme corresponding to the LII model (Figure 8-B) is: a cable (tendon) passing through a ring free to rotate around the joint axis at a given distance **R.** Obvious considerations on this ring equilibrium allow to state that it will lie on the line bisecting the angle formed by the phalangeal long axis.

$$Exc_{LIIij} = 2 . R_{ij} . \sin(\theta_j/2); \quad i = 1,7 \quad j = 1,4$$

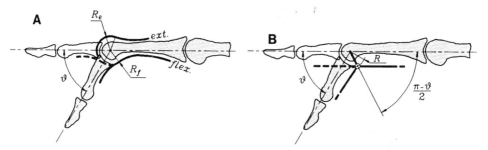

Figure 8. Figure 8-A. Landsmeer I model. **Figure 8-B.** Landsmeer II model.

According to LIII (Figure 9-A) the tendons' trajectory can be schematized by a circular arc which is delimited by the two straight lines located at distance y from the joint axis and orthogonal to the concurrent bones' long axes, and whose center is located at the intersection of the previous lines; moreover the arch is tangent to the straight lines located at distance d from the bone long axes and parallel to them. Consequently tendon path radius varies with joint rotation.

$$\text{Exc}_{\text{LIII}ij} = d_{ij} \cdot \theta j + yij \,((\theta_j/\tan(\theta_j/2)) - 2); \qquad i = 1,7 \qquad j = 1,4$$

On the other hand, the model 2P (Figure 9-B) assumes that the tendons' trajectory near the joints is a straight line connecting two points rigidly fixed to the bones. Therefore the six coordinates of those points are required.

$$\text{Exc}_{2Pi} = \Sigma_j \mid P^{(1)}_{2,ij} - P^{(1)}_{1,ij} \mid - \mid P^{(0)}_{2,ij} - P^{(0)}_{1,ij} \mid$$

where $P2 = P2(\theta j,0,0)$ for PIP and DIP joints, while $P2 = P2 \, (\theta 1,\theta 4,\theta 0)$ at MCP joint.

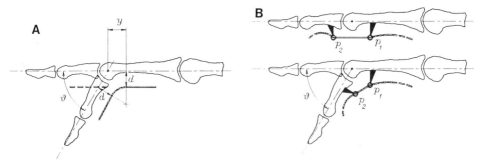

Figure 9. **Figure 9-A.** Landsmeer III model. **Figure 9-B.** Two points (2P) model.

A first qualitative analysis of the anatomical structure leads us to deduct that LI can be appropriate for the extensor tendons, except for the lateral bands of PIP, because those tendons lie on articular surfaces roughly circular; moreover, during joint hyperextension, the contact with the bone is assured by some ligamentous structures such as the transverse lamina (at MCP) e triangular ligament (at PIP).

Flexor tendons, on the contrary - and especially for big flexion angles - shift off the bones profile and run in strong sheaths that drive them to a path more similar to the LIII one. It is worth of noting that LIII model describes the links of the tendons to the bones (sheaths and ligaments) as deformable structures. 2P model can take into consideration also the lateral shifting of intrinsic tendons during flexion. Since metacarpo-phalangeal joint has two d.o.f. and the other joints' axes are generally not orthogonal to the bones' longitudinal axes, the development of a 3d tendon path model (as 2P is) seems to be profitable. The simplest way to take into account that tendon excursion at MCP depends both on flexo-extension and on abdo-adduction is to express this excursion as the sum of the excursion obtained by two Landsmeer models. I.E for LI at MCP:

$$EXC^* = Ri1 \cdot \theta1 + Ri4 \cdot \theta4;$$

for LIII at MCP:

$$EXC^* = EXC_{LIII1} + EXCL_{III4};$$

While these 3d extensions of Landsmeer models have not an evident geometrical meaning, other 3d path models can be developed taking into consideration the shape of the surfaces on which tendon lies. As an example, from an evolution of LI (Figure 10-A) for which extensors lie on a spherical surface, without friction, whose radius is Rs:P1 and P2 are, respectively, the starting point (fixed to the metacarpal bone) and the end point (fixed to the first phalanx) of the tendon path on the sphere. The coordinates of these points [Pix,Piy,Piz] are related to radius Rs by the obvious relationship:

$$R_s = \sqrt{P_{ix}^2 + P_{iy}^2 + P_{iz}^2}$$

The coordinates of P2, in the frame (0) rigid with the metacarpal bone, depend both on MCP angle of flexion ($\theta1$) and adduction ($\theta4$): $P2^{(0)} = P2(\theta1, \theta4)$. Tendon excursion is given by the simple expression:

$$Exc\,(\theta_1, \theta_4) = R_s \cdot \varphi_1 - R_s \cdot \varphi_0$$

where $\varphi1$ is the angle formed by P1 and P2 ($\theta1, \theta4$) and $\varphi0$ the angle between P1 and P2(0,0):

$$\varphi = a\cos \frac{|\ P_1^{(0)} \cdot P_2^{(0)}\ |}{|\ P_1^{(0)} \cdot P_2^{(0)}\ |}$$

$$P_2^{(0)} = \begin{vmatrix} 1 & 0 & 0 \\ 0 & c\theta1 & -s\theta1 \\ 0 & s\theta1 & c\theta1 \end{vmatrix} \cdot \begin{vmatrix} c\theta4 & 0 & -s\theta4 \\ 0 & 1 & 0 \\ s\theta4 & 0 & c\theta4 \end{vmatrix} \cdot \begin{vmatrix} P2x^{(1)} \\ P2y^{(1)} \\ P2z^{(1)} \end{vmatrix}$$

An other interesting model under test for the flexors starts from LII and is described in Figure 11, where points P1 and P2 are fixed on the phalanges and the segments AO1 and BO2 representing ligaments are free to rotate respectively around O1 and O2. A schema improving the results from 2P model, especially for extensor tendons, consists of the interposition of a sphere between the two points fixed to the bones, on which sphere the tendon can lie (Figure 10-B). The total excursion of a tendon (i) results:

$$EXC_i = EXC^*_i + EXC_{i2} + EXC_{i3}$$

where at different joints the excursion can be calculated by means of different path models.

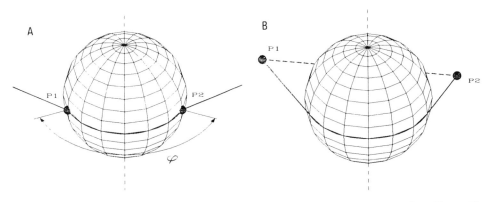

Figure 10. **Figure 10-A.** LI modified: the tendon lies without friction on a spherical surface. **Figure 10-B.** 2P model modified (a sphere of radius R is interposed between points P1 and P2).

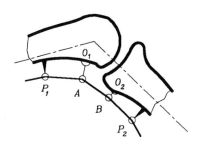

Figure 11. LII modified.

Tendons' moment arm

One of the most important parameters which are required to write the equilibrium equations for the finger is the lever arm of the tendons at each joint. This can be obtained from the path model geometry or, when tendon and sheath deformations are negligible, by differentiating the experimental diagrams $\text{EXC}_i(\theta_j)$.

Even if the lever arm, in most cases, can be calculated in the second way, the definition of the tendon geometrical path is anyway important: for example, to calculate the forces and couples acting on the joints' contact surfaces. If the sheaths deformation cannot be neglected, the moment arm calculation by differencing the experimental diagrams leads to inaccurate results: therefore it is necessary to add a few others terms to the derivative one, to take account of this deformation. This matter can be highlighted by the following example.

$d\text{EXC}/d\theta$ does not correspond to the moment arm of the tendons for LIII model (Figure 12-B); only for small angle of rotation these two values are quite near.

Looking to the tendon path, one can note that ligaments and sheaths linking the tendons to the bones vary their length with joint rotation; therefore the hypothesis of structures indeformability is not correct. The following passages show how the derivative method can be corrected. Let δe be the distance between the centers of curvature of the tendon paths corresponding respectively $t\theta$ and $\theta + \delta\theta$ (θ is the angle of joint flexion): $\delta e = R(\theta + \delta\theta) - R(\theta) = dR/d\theta \cdot \delta\theta$ that is $\delta e = -1/2(\theta/\sin^2(\theta/2))\delta\theta$. The radial displacement δh of the point of application of the sheath forces per unit of length (F/R) can be written as a function of the position on the sheath: $\delta h = \delta e(1-\cos\phi)$ (ϕ define the position on the sheaths and runs from 0 to $\theta/2$) The elementary work produced by sheath forces when the angle of flexion goes from θ to $\theta + \delta\theta$ is given by the following expression:

$$\delta W = 2 \int_0^{\theta/2} (F/R) \cdot \delta h \ R \ d\phi = 2 \int_0^{\theta/2} (F/R) \ R \cdot dR/d\theta \ (1-\cos\phi)\delta\theta \cdot d\phi =$$
$$= Fy(((\theta/2)/\sin^2(\theta y/2)) - (\sin(\theta/2)/\sin^2(\theta/2)))\delta\theta$$

Thus in this case (LIII model) it is not correct - at least from the theoretical point of view - to neglect the sheaths deformations (and their work) in writing equilibrium equation by means of the principle of virtual work.

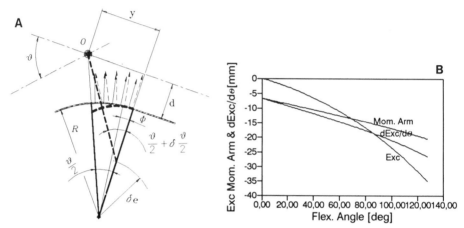

Figure 12. LIII model: **Figure 12-A.** moment arm calculation (continuous lines correspond to flexion angle θ, dashed lines to $\theta+\delta\theta$); **Figure 12-B.** Tendon Excursion, moment arm and dExc/dθ for FP at MCP joint calculated by means of LIII model.

In fact, if M is the couple required to maintain the joint equilibrium:

$$F \times \delta Exc \neq M \ \delta\theta$$

$$F \times \delta Exc = M \ \delta\theta + \delta W \qquad \text{i.e.:}$$

$$[F(d - \theta/\tan(\theta/2) + \theta/\sin(\theta/2) \cdot \theta/2/\sin(\theta/2))] \ \delta\theta = M \ \delta\theta + \delta W$$

that leads to the following equation:

$$F(d - y/\tan(\theta/2) + y/\sin(\theta/2)) = M$$

This means that in this case dExc/dθ is not the moment arm of the tendon; nonetheless as we have shown in this example, it is always possible to calculate the real moment arm by means of *V.W.P.* taking into account the path characteristics.

On the contrary, it is easy to demonstrate that for LI, LII and 2P models the relationship $dE_{XC}/d\theta = R$ is correct.

Muscle force evaluation

One important factor for the finger model is the choice of simple parameters adequate for a global description of the finger muscular capability. For the kind of models we are dealing with, it is sufficient to be able to evaluate approximately maximum muscle force and maximal shortening capability. A useful parameter for a first evaluation of muscular force is the muscle Physiological Cross Sectional Area (PCSA) that is the ratio muscle volume / fiber length. Some experimental data on this parameter for the upper arm have been reported in the literature.[6] Individual data to compute PCSA can be obtained in vivo by means of MRI or CT scan.

The adoption of this simple parameter is obviously a big simplification of the muscle behavior: the maximal force that it can provide depends in fact on other important factors such as shortening velocity, duration of the exercise, actual muscle length with respect to its reference length etc.

CHOICE AND OPTIMIZATION OF THE TENDON PATH MODELS

Generalities

In order to choose the best path model it is preliminarily required to evaluate the best geometrical parameters of each model, at each joint and for each tendon.

Two approaches can be adopted for this evaluation: the first one consists of the direct measure on the hand of those parameters; the second is based on numerical techniques.

An and Chao,[1] for example, provide the coordinates of the two characteristic points of 2P model by means of in-vitro biplanar radiographic recording. The accuracy of the measurements is affected by the technique used to highlight the natural tendon path on the x ray film: the insertion of radio opaque metallic wires, for instance, can vary the natural tendon path. The second approach, that we follow, consists of calculating the geometrical parameters that drive the theoretical diagrams "excursion versus joint rotation to best fit the the experimental ones.

Tendon excursion/joint rotation experimental curves

Very few diagrams "tendon excursion-joint rotation" have been published; the more recent and useful have been obtained by Chao-An[3] (Figure 14) and by Brand.[6]

The first authors report the excursion of the seven index tendons for the abdo-adduction and for the flexo-extension of MCP joint. The second work, does not report the results for all the tendons, but shows the diagrams "moment arm versus joint rotation" of some tendons at MCP, PIP and DIP obtained by deriving excursion experimental data.

All the above diagrams have been obtained in vitro by moving only one joint at a time, while the other ones were forced to maintain the 0 deg. angle (Figure 13).

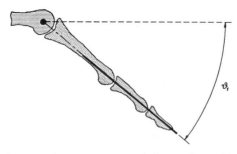

Figure 13. Usual tendon excursion measurement technique: only one "d.o.f." at once is moved.

Different techniques have been adopted by the two research groups. While the first one measured the excursions of cables connected to the proximal end of the tendons, the second group inserted inextensible metallic wires within the tendons in order to avoid the influences of tendon lengthening and of tendon interconnections.

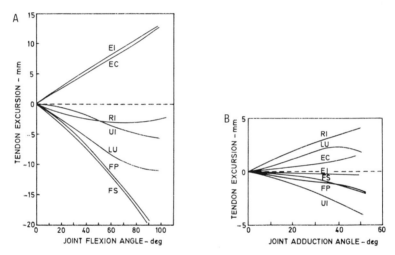

Figure 14 A,B. Experimental diagrams of tendon excursion versus joint rotation.[10]

These discrepancies and the specimen variability can probably explain the relevant differences of the diagrams obtained, especially for lumbricalis. In order to enlarge the experimental data bases and especially to fill some gaps, such as the lack of tendon excursion diagrams versus more than one joint angle at a time, new measures are required.

For this reason we also set up an experimental apparatus (Figure 16) which measures tendon excursions in finger specimens. It consists of: a finger holder; a set of three electrogoniometers, one of which is biplanar; a set of seven rotatory potentiometers for measuring the excursion of thin metallic cables connected to the tendon proximal end. All those gauges are connected to an a/d converter card of a personal computer for data storing and processing.

Due to the limited number of specimens tested (only two) it is not yet possible to deduce statistically relevant tendon behavior; however some new indications can be deduced (Figure 15).

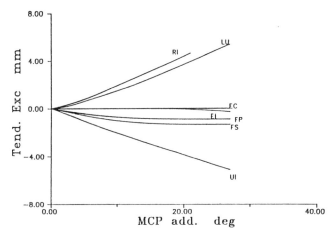

Figure 15. Experimental diagrams tendon excursion versus adduction while MCP flexion angle is θ1 = 40° (measured with the apparatus shown in Figure 16).

Geometrical parameters optimization

Since only an exiguous number of researches[1] reports on direct measures of the parameters for few specific geometrical models, and since the reproduction of this kind of experiments is quite complex, it is profitable to set up numerical methods for an indirect evaluation of these parameters starting from other available experimental data, such as the $EXC(\theta)$ ones described in the previous paragraph.

Our approach is based on the premise that in morphologically normal fingers the function $Exc(\theta_j)$ presents the same trend.

In order to choose the best model to represent the tendon course our first objective is to determine the geometrical parameters (Figures 8,9) allowing to the described path models the best reproduction of the experimental curves obtained by An-Chao or by Brand, according with the least square method.

Thus, for the main models, the optimal parameters satisfy respectively:

for LI and LII:

$$\Sigma k(y(\theta_{jk}) - f\ (R;\theta_{jk}))^2 = \varkappa^2 = min$$

for LIII:

$$\Sigma k(y\ (\theta_{jk}) - f(d,y;\theta_{jk}))^2 = \varkappa^2 = min$$

for 2P:

$$\Sigma k(y(\theta_{1k}) - f\ (P_1,P_2;\theta_{1k},0,0))^2 + \Sigma k(y(\theta_{4k}) - f\ (P_1,P_2;0,\theta_{4k},0))^2 = \varkappa^2 = min$$

at MCP and

$$\Sigma k(y(\theta_{jk}) - f(P_1,P_2;\theta_{jk},0,0))^2 = \varkappa^2 = min$$

at the other joints.

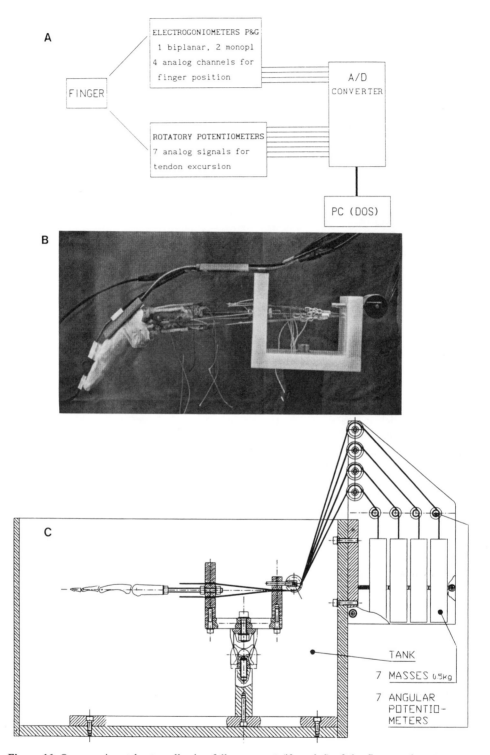

Figure 16. Our experimental setup allowing full movement (if needed) of the finger and contemporary measurement of tendon excursions. **Figure 16-A**. Scheme of data acquisition; **Figure 16-B.** finger holder with a mounted specimen (electrogoniometers P.&G. for joint rotations measurements); **Figure 16-C.** scheme of the experimental apparatus.

Where the summatories are extended to the n (k=1,n) experimental points of the jth d.o.f.

Unlike the Landsmeer models, the $EXC(\theta)$ from 2P is not linear with the parameters, therefore it requires the adoption of iterative optimization techniques such as the Marquardt method, that we have chosen. For a correct optimization of 2P parameters is important to input initial values not too far from the solution; for the other models this is not required; however, for all cases, acceptable initial values can be found from simple considerations on finger anatomy.

Limits on the geometrical parameters

An important step before the parameter optimization, is the evaluation of their range of variation suitable with the anatomical constraints.

Geometrical parameters values must assure that tendon path will lie within the volume delimited internally by bone surfaces and externally by the finger skin.

Consequently for LI and LII models the following condition must be verified:

$$|\text{Rbone}| \leq |\text{Rmod}| \leq |\text{Rskin}|$$

where Rbone and Rskin can be easily quantified by standard clinical examinations.

Analogously for LIII:

$$|\text{dbone}| \leq |\text{dLIII}| \leq |\text{dskin}| \quad (y \geq y\min > 0)$$

moreover, for LIII to prevent tendon loops:

$$\text{and, if } d \leq 0, |d| \leq y$$

(in our conventions for $D \leq 0$ the tendon is a flexor).

More complex is the numerical quantification of the limits for P1 and P2 coordinates in 2P model; since not only these points must lie within a feasible volume, but also the straight line connecting them (i. e. the tendon path) must not intersect bone and skin surfaces both with the finger extended (reference position) and during the movement.

Remarks on the optimization of 2P model

The optimization of 2P parameters leads to an indeterminacy even using the least squares method that allows to write a number of equation equal the number of unknowns. This means that some different combinations of the geometrical parameters converge on an optimal solution (with equal x). A simple planar example can clarify this concept: as shown in figure 17-A the distance between P1 and P2 (and consequently the tendon excursion), at a given joint angle, depends only on the length of the segments P1O, P2O and on the angle ß between them at the joint reference (extended) position. Thus, a generic function $EXC(\theta)$ can be obtained by infinite couples of points (a coordinate can arbitrarily be chosen within a certain dominion while the three others are univocally determined by the excursion function). Consequently the same value for x^2 can be obtained by an infinite number of parameters. This implies that, by developing x^2 in series, $\delta x^2 / \delta a \cdot \Delta a$ (where a is the vector of model parameters, i.e. it contains the six coordinates of P1 and P2) is 0 for a particular choice of Δa and also that **D.** Δa is 0 ([Dkl] = [$\delta^2\mathbf{x}^2$ /δakδae]); this may happen only if matrix **D** is singular and this explains why the Marquardt numerical method cannot find an optimal solution. Only three parameters, whose value depends on a fourth one arbitrarily chosen, can be obtained by an optimization technique. In a 3D case five parameters are independent. For this reason we chose to fix one of the parameter, more precisely that one for which was minimum the sensitivity of $\mathbf{x}^2$ ($\delta\mathbf{x}^2/\delta$ak = min) to the value determined by An's measures.

Choice of the tendon path models

The results obtained with the models described above, after the parameters' optimization, are generally well correlated to the experimental diagrams EXC(θ): in facts figure 17-B shows that the quadratic correlation index is generally over 0.9.

The histogram also reveals that 2P model can better approximate the tendon behavior for intrinsic muscles (Figures 17-B,18-B), while LIII (for flexors-figure 18-A) and LI (for extensors, where LIII degenerate in LI - figure 19-A) seems to be better for extrinsic. However, figure 19-B shows that the simplest model LI, except for RI, can approximate the excursion of all tendons fairly well, with a maximum error on the excursion of ± 1,5 mm.

On the contrary, for the tendons moment arm, we can find macroscopic differences on the results obtained with different models.

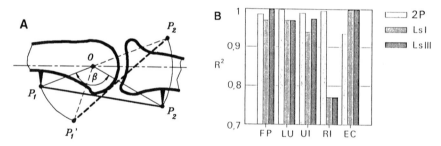

Figure 17. Figure 17-A. 2P model: the points P1-P2 and P'1-P'2 give the same Exc(y). **Figure 17-B.** Quadratic correlation index R2 of different tendon path models for some of index finger tendons.

Considering that: the moment arm can be estimated (for most of the possible tendon paths) by differentiating EXC(θ); this function presents relevant curvature for intrinsic muscles' tendons; also, (e.g. for RI), its derivative can change its sign, it follows that the adoption of the linear model (LI) seems to be inadequate for a precise evaluation of the path of those tendons. Only 2P is adequate for intrinsic because is the only model, out of the four reported in literature, that can give rise to both the excursion and the moment function, not monotonic.

In table 1 is summarized the preliminary choice of the path model to be adopted for each tendon at each joint, based on the results of their parameter optimization.

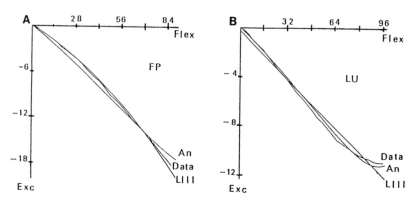

Figure 18. Comparison between experimental curves Exc(θ) (labeled Data) at MCP joint and the ones obtained from optimized tendon path models, for FP (**figure 18-A**), for LU (**figure 18-B**).

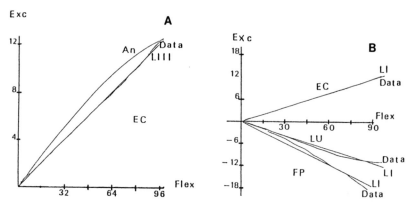

Figure 19. Comparison between experimental curves Exc(θ) (labeled Data) at MCP joint and the ones obtained from optimized tendon path models, for EC (**figure 19-A**), for LU EC and FP using LI model (**figure 19-B**).

Remarks on path models optimization at MCP

For MCP joint, tendon excursion is related to both joint flexion and to abduction. According to the simplest extension of Landsmeer's models to 3D, the total tendon excursion for a certain joint position (θ_1, θ_4) can be approximated to the sum of the excursion corresponding to two planar movements: $EXC(\theta_1, \theta_4) = EXC(\theta_1) + EXC(\theta_4)$.

Table 1. Preliminary choice of tendon path models for our finger model.

TENDON	EC			EI			RI			UI			LU			FS			FP		
JOINT	m	p	d	m	p	d	m	p	d	m	p	d	m	p	d	m	p	d	m	p	d
	c	i	i	c	i	i	c	i	i	c	i	i	c	i	i	c	i	i	c	i	i
	p	p	p	p	p	p	p	p	p	p	p	p	p	p	p	p	p	p	p	p	p
EXT/FLEX	E	E	E	E	E	-	F	-	-	F	E	E	F	E	E	F	F	-	F	F	F
ABD/ABD	A	-		A	-		A	-		A	-		A	-		A	-		A	-	
(MCP)	B			D			B			D			B			D			D		
MODEL	L	2	L	L	L	-	2	-	-	2	2	L	2	2	L	L	2	-	L	2	L
	1	P	1	1	1		P			P	P	1	P	P	1	3	P		3	P	1

Comparing the tendon excursion calculated in this way to the one obtained from the 2P model (Figure 20 for FP and EC) it is evident that the models have similar behavior on the planes $\theta_1 = 0$ and $\theta_4 = 0$ (on which they have been optimized in order to best fit the experimental curves $EXC(\theta_1)$ and $EXC(\theta_4)$) but they produce quite different results in other zones, especially for high values of θ_1 and θ_4. This is also evident in moment arm diagrams. Even if the finger physiological movements very seldom reach the extreme limits of two d.o.f. simultaneously, and the abdo-adduction active range decreases rapidly with flexion, for a more accurate optimization and choice of the path models it is profitable to refer to 3D excursion diagrams. They must be experimentally obtained varying more than a joint angle simultaneously (e.g. Excursion $= EXC(\theta_1, \theta_4)$.

Echographic verifications on the optimized tendon paths

Since an exiguous amount of data is available for the tendon path optimization and since anatomic differences among subjects are sometimes not negligible, it is important to develop a method for in vivo assessment of numerical results.

The very recent evolution of echographic systems allows to visualize the finger's tendon path with a not invasive technique. Measures can be carried out both in vivo and on anatomical specimens. Thus, it is possible to develop an experimental protocol to test the results of the path numerical optimization. So far the precision of the results obtained by processing the echographic images is not too high, but it is certainly sufficient for a qualitative check.

For this kind of analysis we use the finger testing apparatus previously described (Figure 16) in which the finger is maintained in water solution to simplify the positioning of the ultra sound source. This setup allows to scan the finger during the tendon excursion measurement. Echographic images are then digitally stored for further processing in order to extract the required geometrical information.

The following example (Figure 21) shows the echographic images (post processed) of FP at MCP and PIP joint and the path obtained from the optimization technique. Good agreement is generally found between echographic images and path optimization results.

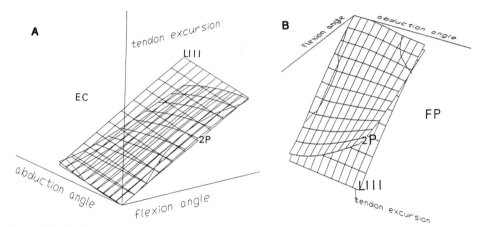

Figure 20. Exc($\theta 1,\theta 4$) calculated by means of 2P model and LIII model for Flexor Profundus (**figure 20-B**) and, for Extensor Communis (**figure 20-A**).

ASSEMBLY OF THE FINGER COMPLEX MODEL AND OF ITS EQUILIBRIUM EQUATIONS

Once the tendon paths have been individually settled, they can be used in the finger mathematical model. As previously described, the model can have different degrees of approximation according to the objectives of the simulation. The simplest model neglects tendons interconnection (Figure 5) and adopts only constant radius pulleys (LI). The most sophisticated one take into account finger interconnections (Figure 6) and define tendon path by means of different models (e.g. as shown in table 1).

Supposing to know the function $R(\theta i)$, for each tendon, at each joint i, it is possible to write the equilibrium equations for the whole finger: at each joint, in a given configuration υi, the torque Mi exerted by the n tendons spanning that joint is:

$$\Sigma j \; Rij(\theta i) \; Fj = Mi \qquad \text{where } j=1,n$$

Neglecting any torque exerted by the ligaments and by joint reactions, the joint equilibrium requires that:

$$Mi = Mext$$

Mext, that is the moment exerted by external forces, may be conveniently calculated by means of the virtual work principle:

$$\delta Wext = \Sigma l \; \mathbf{C}l \; . \; \delta \mathbf{P}l$$

where $\mathbf{C}$ are the external actions referred to the laboratory frame, and $\delta \mathbf{P}$ is the displacement of the points of action P.

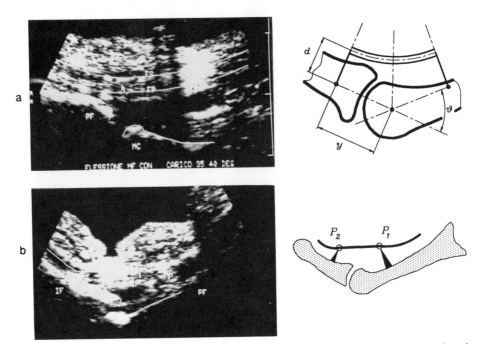

Figure 21. Echographic images of FP path at MCP (a) and PIP (b) joint compared with adopted tendon path model results.

If **C** is expressed with respect to a phalangeal frame k it can be reduced to the fixed frame (0) by means of the homogeneous transform matrix M:

$$\mathbf{C}^{(0)} = (\Pi_{j=1,k} \ \mathbf{M}_{j-1,j}) \cdot \mathbf{C}^{(k)}$$

Analogously if **P** is known in frame K, it can be reduced to the frame (0)

$$\mathbf{P}^{(0)} = (\Pi_{j=1,k} \ \mathbf{M}_{j-1,j}) \cdot \mathbf{P}^{(k)}$$

thus

$$\mathbf{P}^{(0)} = f \ (\mathbf{P}^{(k)}, \upsilon_1, \upsilon_2, \upsilon_3, \upsilon_4)$$

and

$$\delta\mathbf{P}^{(0)} = \Sigma_{j=1,k} \frac{\delta\mathbf{P}^{(0)}}{\delta\upsilon_j} \delta\upsilon_j$$

The moment arm L of the force k, with respect to the rotation axis for υ_j is:

$$L_{kj} = \frac{\delta\mathbf{P}^{(0)}}{\delta\upsilon_j} \cdot \frac{\mathbf{C}_k}{|\mathbf{C}_k|}$$

216

Note that the rotation axis can be a moving axis and must correspond to the one used in the calculations of tendon's moment arm.

While the joint stiffness is generally neglectable, the reaction torques that can be originated in the extreme positions of the joints, must be taken into account. Therefore equilibrium equations for each joint ($j=1,..,4$) can be written as:

$$\Sigma_i \, F_i \cdot R_{ij} = \Sigma_k \, L_{kj} \cdot |C_k| \pm M_j$$

where sign - is correct when $\upsilon j = \upsilon jmax$ and + when $\upsilon j = \upsilon jmin$, while in the intermediate positions M is null.

Since muscles (tendons) can exert only tractions it follows that $Fi \geq 0$ (in some positions $Fimin = F(EXC) > 0$ because of muscle passive stretching).

The upper limit for Fi can be empirically evaluated by means of the formula: $Fimax = k.PCSAi$ where $k = 3\text{-}5 \; 10^5$ Pa and PCSA is the Physiological Cross Sectional Area of the muscle, whose mean values can be found in the literature or can be clinically evaluated as above described.

Globally, if the system built with the described equations is used to calculate the muscle forces which are required to equilibrate the external actions , it contains a number of unknowns greater than the number of available equations. This kind of indeterminacy is rather common in biomechanics.

Different approaches are used in literature to solve this problem, mainly elimination and optimization techniques.

The first method is quite interesting because it provides all basic solutions (any other solution can be obtained as a linear combinations of these) thus it gives information on all possible ways for muscle load sharing, but, on the other hand, it does not help in the choice of these solutions that can be physiological.

For this reason the optimization methods are more diffuse: they compute the solution that minimize (or maximize) a function linear or not that can be provided by the user.

Related to the purpose of the research, the optimal functions can be very different in order to minimize the sum of joint reactions, or the muscle total work, or the greatest stress in muscles or in other anatomical structures such as tendons sheaths and ligaments, or maximize the force that can be exerted by the finger and so on.

For instance, to calculate the tension of each tendon when the finger is acting against a certain external system of forces (and moments) we chose to minimize ratio Fi/PCSA maximal among all muscles (i).

For this kind of situations the optimal criteria try to mimic the physiological behavior oriented to perform a given task in the "softer" and less expensive way.

On the contrary, if a maximal performance is required, as the request to exert the maximal force in a given direction, a probable natural behavior is to optimize the muscles action in order to reach the best result regardless of any other consideration; the physiological limits, such as the maximal muscle force or the maximal stress that all the involved anatomical structures can support, are of course, limiting factors. In this example the module of the applied load |C| is one unknown of the problem and in the meantime it is the function to be optimized; it is therefore generally possible to obtain a unique solution, for which only a tendon force distribution is feasible with the physiological limits.

Since in this kind of simulations the finger does not vary its geometry and position, internal forces and couples are linearly correlated with the external actions, and that also linear are the optimal functions; therefore the optimization can be performed by means of a linear method, such as the "simplex method".[4]

Finger "stability"

In order to study the normal or pathological finger mechanics an important matter is to check whether if a certain finger can maintain his configuration under any kind of external load compatible with the maximum tendon tension and with the maximal force that each muscle can exert. In this paper for sake of brevity, we refer to these properties as finger stability conditions.

In other words we define, in this contest, the finger as stable in a given position if the tendons (inextensible), fixing their excursions without laxity, are able to prevent any movement of the finger itself. It follows that muscles length must be the minimum compatible with that finger position. More into detail, it is possible to write a system of n relationships of this kind:

$$Exci \geq f(\theta 1, \theta 2, \theta 3, \theta 4) \qquad (*)$$

one for each tendon.

The stability (according to the previous definition) means that this system, for a set of excursions (EXCi) compatible with the tested finger position ($\theta 1, \theta 2, \theta 3, \theta 4$), must give as unique solution the same set of angles.

From a geometrical point of view: each unequality delimits an half space and all together a simplex; to have a unique solution this simplex must be closed (consequently at least dof+1 "planes" are required) and have zero volume.

This is only possible if an adequate number of tendons assume the minimum excursion compatible with a certain finger configuration; they must be at least equal to the number of finger degrees of freedom plus 1. In addition one has to verify the simplex zero volume, which can be done verifying that maximum and minimum value for υi are coincident.

If the system (*) is linear the simplex method is a suitable numerical technique to solve this optimization problem. When the function $EXC(\theta)$ is not linear, as it happens using tendon path models different from LI, the system is not linear. In this case the problem can be solved linearizing the system near the given configuration and then using the same numerical method; the finger geometry and the physiological limits to joints rotation seem to guarantee the convergence of the solution to the right results and prevents the linearized system from being instable when the nonlinear is stable.

SIMULATIONS EXAMPLES

As previously described, the main purpose of this work is to show a possible way to model some kinds of hand mechanical pathologies, in order to analyze their consequences on hand functionality and to choose and optimize appropriate rehabilitative interventions.

Thus, hereafter we describe how some pathologies can be introduced in the model and which kind of information can be provided by the simulation: [1] muscle palsy or damage: this case implies the reduction (up to zero) of the maximum muscular force; in the model this is equivalent to reduce the value of PCSA (or that of the constant k) of the injured muscles; as an example, the simulation of the "latent claw hand" clinic test, which is used to diagnose an intrinsic palsy, gives results which are consistent with the reality; it finds in fact that the finger, when subjected only to its own weight can maintain phalanges alignment but, if a dorsally directed little force is applied on the first phalanx it become unstable and assumes a claw position in which it can again support the applied load; [2] analogously, but by increasing muscles PCSA, it is also possible to simulate rehabilitative interventions, consisting of selective muscular reinforcements (which can be obtained or by means of either physiotherapy or muscle elettrostimulation); [3] in the same way other injuries such as tendon rupture or tendon adhesions can be simulated; [4] on the contrary, tendon subluxation, sheath rupture or sheath surgical dissection can be analyzed by changing the path model of the involved tendons or the values of their geometrical parameters; as an example, for partial dissection of FP sheath at MCP, it is convenient to substitute the model adopted for the healthy finger with 2P model (bowstringing model); [5] tendon transfer surgical interventions can be also simulated by composing the previous techniques; [6] the effects of joint deformity on tendon complex equilibrium can be approximately evaluated by changing the orientations of joint rotation axes; [7] an arthrodesis is simulated by suppressing the equilibrium equation which corresponds to the d.o.f eliminated, while a tenodesis or the adoption of a rigid orthosis require to limit the physiological range of motion of the involved joints.

Moreover, it is possible to evaluate how the maximum force that a finger can exert is related to the finger position (Figure 23), thus allowing to find the most convenient way, from the ergonomical point of view, to perform an action.

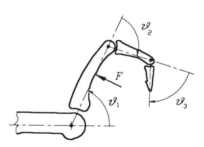

Figure 22. Latent clawing test: any little force applied on the first phalanx and dorsally directed causes the finger to assume a claw position.

Evaluation of a rehabilitative intervention

In order to evaluate the adequacy of a simulated intervention on the finger tendon complex, or to choose among different rehabilitative techniques, it is fundamental to carry on some tests on the results, with regard to the kind of medical problem that must be solved. As an example we refer to the simulation of interventions for restoring some mechanical functions in fingers lacking of some tendons contributions (e.g.because of muscle palsy).

Finger stability must be verified in a certain number of selected positions, characteristic of common physiological actions (key pinch, grasp etc.); then the maximum force that the finger, in those positions, can exert in appropriate directions can be evaluated. If the tendon course has been changed from the physiological one, the required excursions must be calculated in order to check if the muscles excursion capability is still adequate.

Moreover, in order to plan selective muscular reinforcements, it is profitable to compare the force required to each muscle to the maximal that they can exert. As an example figure 24 shows the tension required to each tendon in order to obtain from the finger the maximum force in a given direction: it is evident that FS is the only force that reaches the the maximum value for the muscle and consequently is the limiting factor for the maximum force of the finger; thus in this case, the reinforcement of flexor superficialis can alone produce a relevant increase in the finger performance.

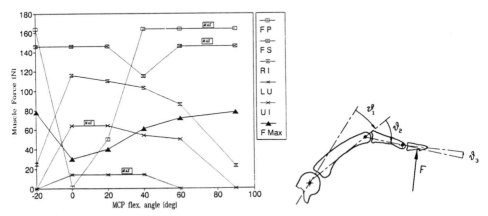

Figure 23. Finger maximal force and force distribution among muscles varying MCP flexion angle.

The evaluation of the level and distribution of the forces transmitted to the sheaths is another useful analysis for planning rehabilitative interventions; sheath stress is in fact another limiting factor for the maximum force of the finger. Neglecting friction, forces on sheaths can be easily calculated for a given position when tendon tension (T) and tendon path (R(s)=radius of curvature) are known (Figure 25), the forces exerted per unit of length (σ) can be calculated as: $\sigma=T/R(s)$. Consequently for this analysis it is fundamental to correctly define the finger path of tendons; along the finger diaphysis we can hypothesize that tendon (especially the flexors) runs parallel to the bone surface while, near the joints, we must also refer to the path models or to echographic measures. For instance, for 2P and LII models which have not a smooth path transmitted forces are concentrated in few points.

Most of the described analyses have been implemented in a computer simulation package that has been recently developed in our department and which preliminary version is now under test. It is intended as a tool to help the clinician, whose experience obviously could never be substituted, to evaluate the mechanical implications of finger pathologies and to plan interactively the rehabilitative intervention.

As an example figure 26 reproduces the screen output during the simulation of a tendon transfer intervention.

The diagrams on top displays the moment arm of each tendon at a given finger position with respect to two axis of rotation. When any semi-plane delimited by a straight line passing through the origin of one of the three diagram is empty, that is, does not contain any segment, the finger position is "unstable". However, if an external

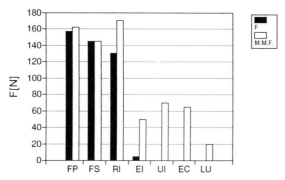

Figure 24. Finger maximal force and corresponding muscle forces compared with their maximal force.

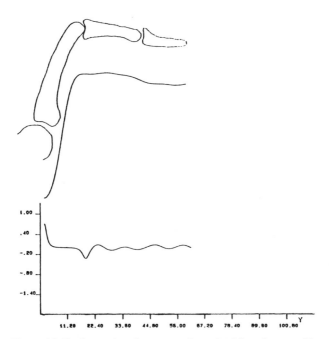

Figure 25. Tendon path and curvature determined from bone profile.

force whose moment arm (represented by a dashed line) does not lie in this semi-plane, it can be equilibtrated by the finger.

To replace the stability it is possible to vary the tendon course by: [1] choosing, on the finger skeletal drawing, the section where operate, which contains the tendon location; this will be displayed automatically; [2] picking on that section the tendon to be moved, and placing it at the new location; [3] choosing from the program menu the best model path for the new location of the tendon.

Thus the stability can be verified looking to the diagrams. After this preliminary simulation, other routines must check that the new finger geometry will satisfy all the conditions discussed in the previous paragraphs.

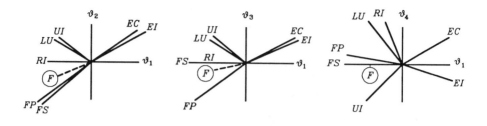

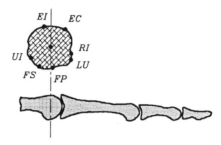

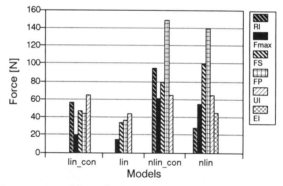

Figure 26. Output screen of our simulation program

Figure 27. Maximal force output and force distribution among muscles as obtained by means of different tendon path models: lin (only LI without tendon interconnections), lin_con (LI with interc.), nlin (nonlinear tendon path model without interc.) and nlin_con (nonlinear with interc.). Finger position is: $\theta_1=48°, \theta_2=50°, \theta_3=25°, \theta_4=0°$ (tip pinch).

CONCLUDING REMARKS

Some finger pathologies and recovery techniques can be simulated by means of mathematical models with different level of complexity.

A correct optimization of the geometrical parameters of the models requires to carry out new experimental tests, because the paper published on this matter are only few and are also lacking of some useful data (such as $EXC(\theta_1, \theta_2)$). Moreover the new echographic techniques are suitable for verifying the optimized parameters.

So far, very few experimental data are available in order to validate the finger model correctly .

However the few checks executed on the available experimental data, such as the ones on the maximum force that the finger can exert in some physiological positions, measured by Chao and An,[10] seem to be good for the non linear composite model (Figure 27).

The package developed in this research will not generate the only way to solve a mechanical problem of the finger, but it will offer to the clinician some helps in optimizing and testing interactively the intervention that his experience will suggest.

ACKNOWLEDGEMENTS

We must thank Dr.DePra (Legnano) for his important help in the echographic analyses.

We wish to thank publisher and authors listed below for their copyright permission to use their previously published figures: [5] Bonola et al. - La MANO (1981) Piccin Editore for figures 1a,b; [7] W.H.Bunch et al. - ATLAS OF ORTHOTICS (1985), The Mosby Company for figure 3; [10] E.Chao, K.N,AN et al - BIOMECHANIC OF THE HAND, World Scientific for figure 14.

This research is partially supported by CNR90.01106.ct07 and CNR92.03199.ct07 grants.

REFERENCES

1. K.N. An, E.Y. Chao, W.P. Cooney, and R.L. Linscheid, Normative model of human hand for biomechanical analysis, *J. Biomech.* 12:775 (1979).
2. K.N. An, K. Takahashi, T.P. Harrigan, and E.Y. Chao, Determination of muscle orientation and moment arms, *J. Biomech. Eng.* 106:280 (1984).
3. K.N. An, Y. Ueba, E.Y. Chao, W.P. Cooney, R.L. Linscheid, Tendon excursion and moment arm of index finger muscles, *J. Biomech.* 16:419 (1983).
4. M.S. Bazaraa, and J.J. Jarvis, "Linear programming and network flows", J. Wiley & Sons (1977).
5. A. Bonola, A. Caroli, and L. Celli, "La mano", Piccin ed., Padova (1981).
6. P.W. Brand, "Clinical mechanics of the hand", The C.V. Mosby Company, St. Louis (1985).
7. W.H. Bunch, R. Keagy, A.E. Kritter, L.M. Kruger, M. Letts, J.E. Lonstein, E. Byron Marsolais, J.G. Matthews, and L.R. Pedegana, "Atlas of orthotics", The C.V. Mosby Company, St. Louis (1985).
8. S. Bunnel, Surgery of the intrinsic muscles of the hand other than those producing opposition of the thumb, *J. Bone Joint Surg.* 24:1 (1942).
9. F. Casolo, G. Legnani, and R. Faglia, Industrial robots: application of a rationale solution for the direct and inverse dynamic problem, 2nd Int. Workshop on Advances in Robot Kinematics, Linz - Austria (1990).
10. E.Y.S. Chao, K.N. An, W.P. Cooney, R.L. Linscheid, "Biomechanics of the hand a basic research study", *World Scientific*, Singapore (1989).
11. J. Denavit, and R.S. Hartenberg, A kinematic notation for lower-pair mechanisms based on matrices, ASME Transactions, *J. Appl. Mech.* 6:215 (1955).
12. C. Harris, and G. Rutledge, The functional anatomy of the extensor mechanism of the finger, *J. Bone Joint Surg.* 54-A:713 (1972).
13. I.A. Kapandji, "Fisiologia articolare", Marrapese ed., Roma (1980).
14. J.M.F. Landsmeer, Anatomical and functional investigations on the articulation of the human fingers, *Acta Anatomica*, 25:1 (1955).
15. J.M.F. Landsmeer, The coordination of finger-joint motions, *J. Bone Joint Surg.* 45-A:1654 (1963).
16. J.M.F. Landsmeer, "Atlas of anatomy of the hand", Churchill Livingstone, New York (1976).
17. A. Storace, and B. Wolf, Functional analysis of the role of the finger tendons, *J.Biomech.* 12:575 (1979).
18. Y. Youm, TE.B. Gillespie, and L. Sprague, Kinematic investigation of normal mcp joint, *J. Biomech.* 11:109 (1978).

FORCE ANALYSIS OF THE THUMB

David J. Giurintano and Anne M. Hollister

Paul W. Brand Biomechanics Laboratory
Gillis W. Long Hansen's Disease Center
Carville, LA 70721, USA

INTRODUCTION

In this work, the thumb is modeled as a five-link manipulator with the links connected by hinge joints - one for each degree of freedom of the thumb.[11] Previous researchers modeled the thumb as three rigid links connected by a spherical joint (carpometacarpal joint), a universal joint (metacarpophalangeal joint), and a hinge joint (interphalangeal joint). Five static positions of hand function are analyzed - key pinch, tip pinch, screwdriver hold, wide grasp, and radial opposition. These positions will be modeled for both a median nerve paralyzed (MNP) and an ulnar nerve paralyzed (UNP) hand with a tendon transfer.

Median or ulnar nerve paralysis occurs as a result of trauma or disease. In Hansen's disease, the bacteria, *Mycobactrium leprae*, attacks the median nerve proximal to the carpal tunnel. The ulnar nerve is attacked proximal to the elbow. The loss of nerve function occurs as a result of swelling of the nerves, ischemia, and inflammation. Both sensory and motor losses occur as a result of the disease.

In order to restore the functional loss due to the paralysis, the flexor digitorum superficialis of the ring finger (FDSR) transfer will be used to restore the hand's balance with a Fink and Snow pulley[21] and an abductor insertion. Non-linear optimization techniques will be used to solve the redundancy problem associated with the over-constrained physical system. A computer graphics display is used to display the positions of the bones, locations of the muscle-tendon paths, and the directions of the applied load and the joint reactions.

BACKGROUND

Many researchers have studied the kinematics of the carpometacarpal joint of the thumb.[6,8,9,10,16,19] Motion was postulated to occur about two axes of rotation, perpendicular to each other and to the thumb metacarpal because of the saddle shaped

Advances in the Biomechanics of the Hand and Wrist
Edited by F. Schuind *et al.*, Plenum Press, New York, 1994

joint.[6,9,10,14] However, the position of the axes was not been verified. These authors modeled the carpometacarpal (CMC) joint as a simple universal joint, with the axes perpendicular to each other and to the bones.

With this model only flexion-extension and abduction-adduction can occur. Motion of the metacarpal on the trapezium includes flexion-extension, abduction-adduction and pronation-supination.[6,8,14] Cooney[6] showed that CMC joint motion occurred in all three anatomic planes. He further showed the CMC is a two degree of freedom joint - for a given position of flexion and abduction there is a set degree of pronation. Rudolph Fick[10] and Kapandji[14] postulated a third longitudinal axis to account for this pronation-supination motion. This is incompatible with the two degrees of freedom determined experimentally[6] and with the joint's saddle shape.

Fick[9] postulated that the metacarpophalangeal (MP) joint had two axes of rotation because of the two major curves on its surface. He did not locate them. He also postulated that the axes were perpendicular to the metacarpal and located in the metacarpal head. This model explains the two degrees of freedom observed in the joint. It accounts for flexion-extension and abduction-adduction but not the pronation-supination which occur.

Hollister et al.[12,13] determined the axes of the carpometacarpal joint, the metacarpophalangeal and the interphalangeal joint of the thumb. The flexion-extension axis of the CMC joint is located in the trapezium, and the abduction-adduction axis is near the base of the metacarpal. The axes do not intersect and are not perpendicular to each other or the trapezium or the metacarpal. The flexion-extension and abduction-adduction axes of the MP joint are located near the distal surface of the metacarpal. Likewise, these axes do not intersect and are not perpendicular to each other or the metacarpal. The axis of the interphalangeal (IP) joint runs parallel to the flexion crease of the joint and is not perpendicular to the phalanx.

Chao et al.[3] and Cooney and Chao[5] analysed the tendon forces and joint reactions in the fingers and the thumb. To reduce the number of unknowns in the thumb, they neglected the extensor pollicis brevis and extensor pollicis longus for pinch and grasp. They assumed the flexor pollicis brevis and the opponens pollicis to act as one muscle as a result of electromyogram studies used to determine muscle activity for function and by observation of the functional anatomy. The number of unknowns in the finger was reduced as a result of electromyogram studies.

Chao and An[4] used linear programming to solve for the unknown muscle forces and joint reactions of the hand for grasp and key pinch. The objective function was based on the assumption that the body attempts to minimize joint torque and maximize tendon force.

Crowninshield and Brand[7] used non-linear optimization to resolve the redundancy problem presented by the mechanical analysis of body mechanics. The objective function maximized was the endurance of musculoskeletal function. This was accomplished by minimizing the sum of the cube of the muscle stresses (muscle force divided by the physiological cross-sectional area).

An et al.[1] used several schemes to resolve the redundancy problem presented in elbow mechanics. The objective function minimized was the sum of the muscle stresses of the muscles that cross the elbow. This modified linear method was compared to objective functions that minimize the sum of the muscle forces or the sum of the squares of the muscle stresses using non-linear methods.

Myers et al.[18] developed a method for extracting data bones from CT scans for the creation of models for kinematic and kinetic analysis. Buford et al.[2] showed the usefulness of a computer graphics simulation used to resolve the locations of the axes of rotation of the thumb.

Giurintano et al.[11] used the Virtual Five-Link Thumb Model to simulate thumb mechanics. The thumb was assumed to be composed of five rigid links connected by hinge joints. These joints were not assumed to be orthogonal to the long axes of the bones or the anatomic planes. Non-linear optimization was used to solve for the muscle forces. Computer graphics was used to display the results of the simulation. CMC joint forces were predicted to be 65% greater in pinch than the model used by Cooney and Chao.[5]

Snow and Fink[21] proposed a procedure to restore the loss of function as a result of median nerve paralysis. It involved rerouting the FDSR through a window in the transverse carpal ligament and inserted into the abductor pollicis brevis.

MATHEMATICAL MODEL

The Virtual Five-Link Thumb Model was modified to allow removal of motors (a paralysis) and inclusion of a transfered motor to restore the function lost due to the paralysis. Homogeneous matrix transformations are used to move the bones to a functional position. The matrix used to transform the distal segment from a reference position, P, to a new position, P', in space is described mathematically as

$$P' = [T_1]P \tag{1}$$

The concatenated transformation to move a degree of freedom joint with non-orthogonal non-intersecting axes is

$$[T_1] = [Rotation\ about\ arbitrary\ axis\ X].[Rotation\ about\ arbitrary\ axis\ Y] \tag{2}$$

where

$$[Rotation\ about\ arbitrary\ axis\ X] = [TR\text{-}n1].[ROT\text{-}nz1].[ROT\text{-}ny1].[\Theta_1].$$
$$[ROT\text{-}py1].[ROT\text{-}pz1].[TR\text{-}p1]$$

The concatenated transformation to move a two degree of freedom joint with non-orthogonal non-intersecting axes is

$$[T_1] = [Rotation\ about\ arbitrary\ axis\ X].[Rotation\ about\ arbitrary\ axis\ Y] \tag{3}$$

where

$$[Rotation\ about\ arbitrary\ axis\ X] = [TR\text{-}n1].[ROT\text{-}nz1].[ROT\text{-}ny1].[\Theta_1].$$
$$[ROT\text{-}py1].[ROT\text{-}pz1].[TR\text{-}p1]$$

and

$$[Rotation\ about\ arbitrary\ axis\ Y] = [TR\text{-}n2].[ROT\text{-}nz2].[ROT\text{-}ny2].[\Theta_2].$$
$$[ROT\text{-}py2].[ROT\text{-}pz2].[TR\text{-}p2]$$

These transformations are for rotation of a object about an arbitrary axis in space. The CMC joint and the MP joint will be modeled as two degree of freedom joints, and the IP joint is modeled as a one degree of freedom joint. Either wire-frame or solid models of the bones can be used to visualize thumb motion in three-space.

Control points, which direct the line of action of the muscles or their long tendons of insertion, are all points logically associated with a bone segment. Thompson and Giurintano[22] used a building block approach to construct tendon models for the long flexors of the hand. Giurintano et al.[11] created two additional building blocks. Building block A simulates a tendon traveling in a straight line. The tendon control points are transformed by all of the translations and rotations of the bone segment to which the sheath is connected. Building block B simulates a tendon traveling in a circular path as it crosses a joint. Landsmeer's[15] third model is the basis of this simulation. Building block E is used to simulate an extensor tendon's dorsal travel over a joint in a circular arc with the radius of the arc equal to 1.3 times the distance between the two points on the circle. Building block F simulates an extensor tendon's dorsal travel over a joint in an arc modeled as a fourth order polynomial ($Y = A \cdot X^4 + B \cdot X^2 + C \cdot X + D$) with first derivative continuity at both the proximal and distal segments. Control points are used to determine the line of action of a tendon or a muscle as it crosses a joint. The information generated from the locations of the control points of the motors is also used to graphically display the tendon paths.

The flexor pollicis brevis (FPB) was modeled with 80% ulnar nerve innervation and 20% median nerve innervation. This was modeled by multiplying the physiologic cross-sectional area of the flexor pollicis brevis by 0.2 for the ulnar paralysis simulation and by multiplying the physiologic cross-sectional area of the flexor pollicis brevis by 0.8 for the for the median paralysis simulation. The transfered motor (FDSR) was modeled using tendon building block A[22] with a Fink and Snow[21] pulley and an insertion into the abductor pollicis brevis. Figures 1 and 2 depict the Virtual Five-Link Thumb Model modified to simulate median or ulnar nerve paralysis. The optimization problem to be solved is

$$\left.\begin{array}{l} \text{minimize } \sum \sigma_i^2 \\ \text{subject to } \sum M_{hinge\,axis(local\,coordinates)} = 0 \\ \qquad\qquad f_i \geq 0 \end{array}\right\} \tag{4}$$

where σ is the force of the muscle divided by its physiologic cross-sectional area and f_1 is the force of the muscle. This system is solved by the non-linear optimization techniques developed by McPhate[17] after Powell.[20] It can be descriptively represented by the method of parallel tangents in two dimensions. The constraint equations for solution of muscle forces for the Virtual Five-Link Thumb Model with a median nerve paralyzed hand and an FDSR transfer are

$$\sum M_{x1} = M_{x1}^{fpl} + M_{x1}^{epl} + M_{x1}^{fpb} + M_{x1}^{epb} + M_{x1}^{apl} + M_{x1}^{fdi} + M_{x1}^{adp} + M_{x1}^{fdsr} + M_{x1}^{al} = 0$$
$$\sum M_{y2} = M_{y2}^{fpl} + M_{y2}^{epl} + M_{y2}^{fpb} + M_{y2}^{epb} + M_{y2}^{apl} + M_{y2}^{fdi} + M_{y2}^{adp} + M_{y2}^{fdsr} + M_{y2}^{al} = 0$$
$$\sum M_{x3} = M_{x3}^{fpl} + M_{x3}^{epl} + M_{x3}^{fpb} + M_{x3}^{epb} + M_{x3}^{adp} + M_{x3}^{fdsr} + M_{x3}^{al} = 0$$
$$\sum M_{y4} = M_{y4}^{fpl} + M_{y4}^{epl} + M_{y4}^{fpb} + M_{y4}^{epb} + M_{y4}^{adp} + M_{y4}^{fdsr} + M_{y4}^{al} = 0$$
$$\sum M_{x5} = M_{x5}^{fpl} + M_{x5}^{epl} + M_{x5}^{al} = 0 \tag{5}$$

and constraint equations for solution of muscle forces for the Virtual Five-Link Thumb Model with an ulnar nerve paralyzed hand and FDSR transfer are

$$\sum M_{x1} = M_{x1}^{fpl} + M_{x1}^{epl} + M_{x1}^{fpb} + M_{x1}^{epb} + M_{x1}^{apl} + M_{x1}^{opp} + M_{x1}^{apb} + M_{x1}^{fdsr} + M_{x1}^{al} = 0$$

$$\sum M_{y2} = M_{y2}^{fpl} + M_{y2}^{epl} + M_{y2}^{fpb} + M_{y2}^{epb} + M_{y2}^{apl} + M_{y2}^{opp} + M_{y2}^{apb} + M_{y2}^{fdsr} + M_{y2}^{al} = 0$$

$$\sum M_{x3} = M_{x3}^{fpl} + M_{x3}^{epl} + M_{x3}^{fpb} + M_{x3}^{epb} + M_{x3}^{apb} + M_{x3}^{fdsr} + M_{x3}^{al} = 0$$

$$\sum M_{y4} = M_{y4}^{fpl} + M_{y4}^{epl} + M_{y4}^{fpb} + M_{y4}^{epb} + M_{y4}^{apb} + M_{y4}^{fdsr} + M_{y4}^{al} = 0$$

$$\sum M_{x5} = M_{x5}^{fpl} + M_{x5}^{epl} + M_{x5}^{al} = 0$$

(6)

The four positions analyzed in Giurintano *et al.*[11] were compared to the results of the paralyzed hand model. Another functional position was also defined - radial opposition. This is opposition of the thumb with a load applied to the radial side of the distal phalanx. This position was chosen because it most nearly imitates the function that is attempted to be restored in tendon transfer surgery - opposition.

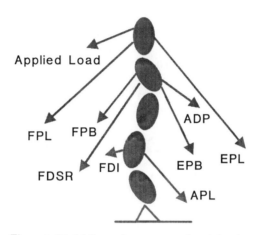

Figure 1. Model for median nerve paralyzed thumb.

RESULTS

Tables 1 and 2 list the results of the Virtual Five-Link Thumb Model applied to a median nerve paralyzed hand and an ulnar nerve paralyzed hand. Figures 3 through 7 depict the positions from the computer graphics simulation with the muscles' lines of action and the directions of the reactions at the joint surfaces.

The joint reactions are calculated for a point at the center of the surface of the trapezium (CMC joint), metacarpal (MP joint), and proximal phalanx (IP joint). For this simulation, the applied load was 147 N.

Figures 3 through 7 are of the computer graphics output of the Evans & Sutherland PS390 graphics terminal. The force vectors on the joint surfaces are represented with a single headed arrow and the moment vectors are represented with a double headed arrow.

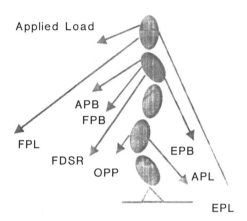

Figure 2. Model for ulnar nerve paralyzed thumb. (FPL = flexor pollicis longus. FPB = flexor pollicis brevis; FDSR = flexor digitorum superficialis of ring finger; FDI = first dorsal interosseus; ADP = adductor pollicis; EPB = extensor pollicis brevis; EPL = extensor pollicis longus; APL = abductor pollicis longus).

Table 1. Muscle forces (N) of Median Nerve Paralysis with an abductor insertion with 20% FPB palsy.

Muscle	Key Pinch	Screwdriver	Tip Pinch	Wide Grasp	Radial Opposition
FPL	980.1	980.1	355.7	0.0	0.0
FPB	441.1	739.9	0.0	-0.1	-0.1
EPL	27.0	223.2	0.0	0.4	-0.1
EPB	980.0	903.5	66.3	48.7	125.3
ADP	-2.0	-1.9	213.6	494.3	-1.3
APL	146.8	0.0	381.4	0.0	0.0
FDI	-1.3	-1.2	202.3	172.0	-0.3
FDSR	496.5	3.4	0.0	-0.1	953.0

Table 2. Muscle forces (N) of Ulnar Nerve Paralysis with an abductor insertion with 80% FPB palsy.

Muscle	Key Pinch	Screwdriver	Tip Pinch	Wide Grasp	Radial Opposition
FPL	980.1	980.2	94.5	-0.3	0.0
FPB	140.1	155.5	-0.2	-0.7	0.0
EPL	1.4	150.4	0.1	10.2	0.0
EPB	976.7	814.7	0.0	-0.2	321.9
OPP	-0.2	-0.1	980.0	978.6	980.0
APL	91.2	-0.1	184.7	0.0	-0.1
APB	84.8	605.9	-0.2	-0.7	10.4
FDSR	726.8	2.0	-0.3	-0.8	980.1

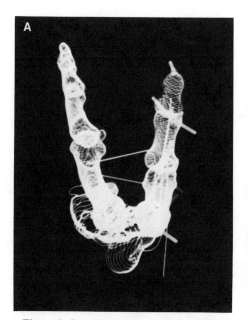
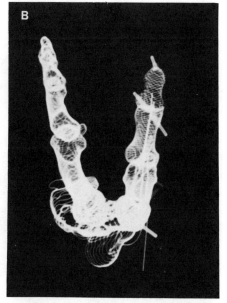

Figure 3. Computer simulation of key pinch for a) median nerve palsy, and b) ulnar nerve palsy.

231

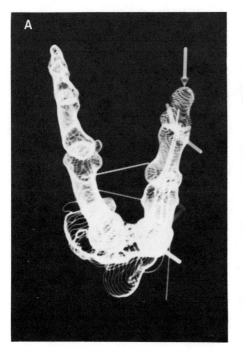
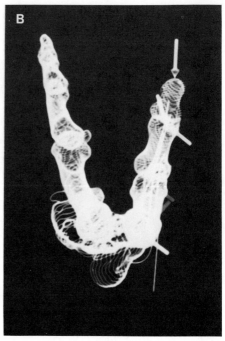

Figure 4. Computer simulation of screwdriver hold for a) median nerve palsy, and b) ulnar nerve palsy.

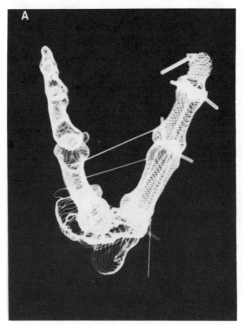
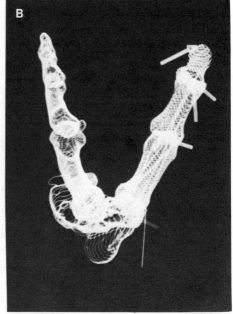

Figure 5. Computer simulation of tip pinch for a) median nerve palsy, and b) ulnar nerve palsy.

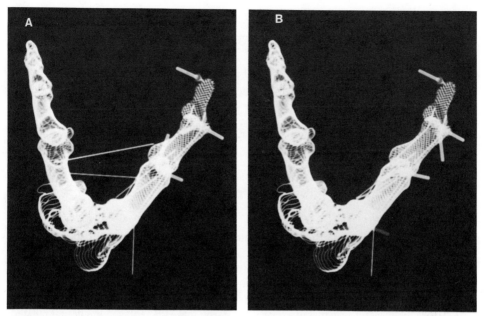

Figure 6. Computer simulation of wide grasp for a) median nerve palsy, and b) ulnar nerve palsy

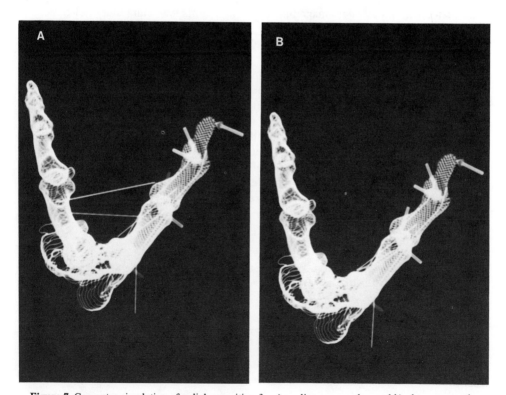

Figure 7. Computer simulation of radial opposition for a) median nerve palsy, and b) ulnar nerve palsy.

Table 3. Dominant Motors.

Posture	Normal Hand	Median Nerve Paralysis	Ulnar Nerve Paralysis
Key Pinch	FPL-FPB-EPB-APL	FPL-FPB-EPB-FDSR	FPL-FPB-EPB-FDSR
Screwdriver	FPL-FPB-EPL-EPB	FPL-FPB-EPL-EPB	FPL-FPB-EPL-EPB-APB
Tip Pinch	FPL-ADP-APL-FDI	FPL-ADP-APL-FDI	OPP-APL
Wide Grasp	EPB-ADP-OPP	ADP-FDI	OPP
Radial Opposition	FPL-EPB-OPP-APB	EPB-FDSR	EPB-OPP-FDSR

Table 3 is a comparison of the motors predicted as dominant in stabilizing the thumb for the four functional positions in Giurintano et al.[11] (normal hand) versus those motors predicted as dominant in the paralyzed hand.

The model predicted activity of the transfer (FDSR) in key pinch and radial opposition. In screwdriver hold the motors which stabilize the normal hand still exist; therefore the transfer was inactive. For ulnar palsy, the APB replaced the deficit created by the reduced capacity of the FPB, innervated in this example 80% by the ulnar nerve. The model predicted a reduced output of the FPL for tip pinch for median paralysis, and the OPP replaced the FDI for ulnar paralysis. Motor function for wide grasp was similar to the normal hand. For both median and ulnar paralysis in radial opposition, the transfer fired with an equal magnitude to the FPL in its maximum output of force.

CONCLUSIONS

The Virtual Five-Link Thumb Model has been modified to simulate either a median or ulnar nerve paralyzed hand. Five functional positions were modeled. The FDSR was used as a transfer with a Fink and Snow[21] pulley and an abductor insertion. The transfer was active for key pinch and radial opposition. In the other postures, a transfer routed with a more distal pulley and inserted into the adductor would have more closely approximated the replacement of adduction function. Any combination of percent median-ulnar innervation of the flexor pollicis brevis, transfered motor, pulley, and insertion can be simulated with the Virtual Five-Link Thumb Model.

ACKNOWLEDGEMENTS

The authors would like to thank Carol Langlois and Tanya Thomassie for their assistance in preparing this paper. Digital Diagnostics Incorporated provided the scan time for the CT images. Acromed Corporation provided travel funds.

REFERENCES

1. K.N. An, B.M. Kwak, E.Y. Chao, and B.F. Morrey, Determination of muscle and joint forces: a new technique to solve the indeterminate problem, *J. Biomech. Eng.* 106:364 (1984).

2. W.L. Buford, L.M. Myers, and A.M. Hollister, A modeling and simulation system for the human hand, *J. Clinical Engineering*, 15:445 (1991).

3. E.Y. Chao, J.D. Opgrande, and F.E. Axmear, Three dimensional force analysis of finger joints in selected isometric hand function, *J. Biomech.* 19:387 (1976).

4. E.Y. Chao and K.N. An, Determination of internal forces in human hand, *J. Engineering Mechanics Division, ASCE*, 104:225 (1978).

5. W.P.Cooney and E.Y. Chao, Biomechanical analysis of static forces in the thumb during hand function, *J. Bone Joint Surg.* 59-A:27 (1977).

6. W.P. Cooney M.J. Lucca E.Y. Chao and R.L. Linscheid, The kinesiology of the thumb trapeziometacarpal joint, *J. Bone Joint Surg.* 63-A:1371 (1981).

7. R.D. Crowninshield and R.A. Brand, A physiologically based criterion of muscle force prediction in locomotion, *J. Biomech.* 14:793 (1981).

8. B. Ebskov, De motibus motoribusque pollicis humani, sigil facul medicine schollhafnein, Copenhagen (1970).

9. A. Fick, ``Die Gelenke mit sattelformigen Flachen. Zeitschrift fur rationelle Medicin,'' Akademische Verlagshandlung von C. F. Winter, Heidelberg (1854).

10. R. Fick, ``Handbuch der Anatomie und Mechanik der Gelenke unter Berucksichtigung der bewegenden Muskeln, Vol II'' Verlag von Gustav Fisher, Jena (1908).

11. D.J. Giurintano, A.M. Hollister, W.L. Buford, D.E. Thompson, and L.M. Myers, A virtual five-link model of the thumb, *J. Biomech. (submitted)* (1992).

12. A.M. Hollister, W.L. Buford, L.M. Myers, D.J. Giurintano, and A. Novick, The axes of rotation of the thumb carpometacarpal joint, *J. Orthop. Res.* 10:454 (1992).

13. A.M. Hollister, D.J. Giurintano, W.L. Buford, L.M. Myers, and A. Novick, Axes of rotation of the thumb interphalangeal and metacarpophalangeal joints, *J. Orthop. Res. (submitted)* (1992).

14. I.A. Kapandji, Biomechanics of the thumb, in: The Hand, R. Tubiana, ed. WB Saunders, Philadelphia (1981).

15. J.M.F. Landsmeer, Studies in the anatomy of articulation, *Acta Morphol. Neerlando-Scandinavia*, 3:287 (1961).

16. M.A. MacConaill, Studies in the mechanics of synovial joints, *Irish Journal of Medical Science* :223 (1946).

17. A.J. McPhate, Mechanical design using nonderivative search and quadraric penalty functions, Oklahoma State Applied Mechanics Conference, paper 18 (1975).

18. L.M. Myers, W.L. Buford, D.J. Giurintano, and D.E. Thompson, Interactive segmentation of three-dimensional anatomical data for musculoskeletal modeling, Proceeding of 9th Annual National Computer Graphics Association III 143 (1988).

19. J.R. Napier, The form and function of the carpo-metacarpal Joint of the thumb, *J. Anatomy*, 89:362 (1955).

20. M.J.D. Powell, An efficient method for finding the minimum of a function of several variables without calculating derivatives, *Computing Journal*, 7:155 (1964).

21. J.W. Snow and G.H. Fink, Use of a transverse carpal ligament window for the pulley in tendon transfers for median nerve palsy, *Plast. Reconstr. Surg.* 48:238 (1971).

22. D.E Thompson, and D.J. Giurintano, A kinematic model of the flexor tendons of the hand, *J. Biomech.* 22:327 (1989).

STRESS ANALYSIS OF PIP JOINTS USING THE THREE-DIMENSIONAL

FINITE ELEMENT METHOD

Hiroyuki Hashizume, Takeshi Akagi, Hiroyoshi Watanabe, Hajime Inoue, and Takashi Ogura

Department of Orthopaedic Surgery
Okayama Universitv Medical School
Shikata-cho 2-5-1, Okayama City
Okayama 700, Japan

INTRODUCTION

Fracture dislocations of the proximal interphalangeal joint (PIPJ) are caused by indirect force from the finger tip running parallel to the longitudinal axis of the finger. However, onset mechanisms of the fracture dislocation are still obscure.

Biomechanical responses of bones and joints have been studied by various analytical techniques. The finite element method (FEM) has been successfully used for biomechanical studies in the field of the orthopaedic surgery.[3,4] The FEM has been developed originally for structural analysis in engineering,[5] however, its basic principle and procedure are also applicable for biological problems.

Clinically, we classified middle phalangeal base fractures,[1] which are the main intra-PIP joint fractures, into five types according to the first x-ray examination (Figure l). Types Ib,c, IIb,c and III are caused by axial force to the middle phalanx. These types of fractures are the object of the present study. In the present study stress distribution on a PIPJ model was analyzed using the three-dimensional finite element method. The results from FEM analysis were compared with those of experimentally produced fractures to study onset mechanisms of PIPJ fracture dislocations.

MATERIALS AND METHODS

To prepare FEM models, PIP joints taken from cadaveric fingers fixed with 10% formalin were used. Sagittally sectioned specimens of the PIP joints were made serially at intervals of 1 mm with a EXAKT BS-3000 bone sectioner. Structural shapes, obtained from microradiography of those specimens, were used to make two-dimensional models. Then, three-dimensional architectonic images were reproduced for three-dimensional FEM models. The external accuracy of the three-dimensional model was certificated by comparing it with a figure made with a non-contacted three-dimensional digitizer (Figure 2). The distance between each digitized section image was also 1mm according to the interval of the bone sectioner. The accurate distances of each superficial point were measured on these figures.

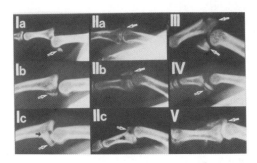

Figure 1. Classification of middle phalangeal base fractures.

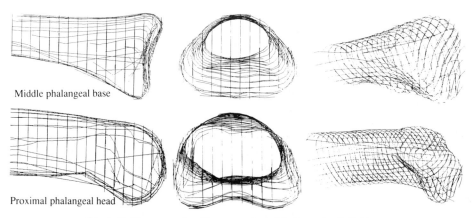

Figure 2. Figures made with non-contacted 3-dimensional digitizer.

The model was divided into four areas according to the following four different material properties: cortical bone, cancellous bone, subchondral bone and cartilage (Figure 3).

These material properties were assumed to be linearly elastic and isotropic. The thicknesses of cartilages, subchondral bones and cortical bones in the FEM models were set at 0.5 mm, 0.2 mm and 2 mm, respectively. Table 1 shows the material properties of these four areas.

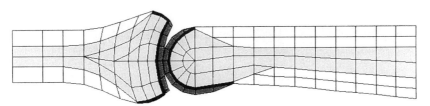

Figure 3. The area of four different materials in the FEM model.

Table 1. Material properties of the model.

	cortical bone	subchondral bone	cancellous bone	cartilage
Young's modulus	15,000MPa	1,000MPa	100MPa	5MPa
Poisson's ratio	0.3	0,2	0.2	0.49

(from Hayashi et al.[1]).

The total number of the elements is 180 triangles and quadrilaterals for two-dimensional FEM and 852 hexahedra for three-dimensional FEM (Figure 4). The number of nodes for two-dimensional FEM was 208 and that for three-dimensional FEM was 1246.

Loading data were obtained from fracture experiments using amputated fingers. PIPJ from cadaveric fingers were attached to a fixator of our own design with dental resin (Figure 5-A). Then, axial loadings were added to the middle phalanx, while the angles of the PIP joint were changed from hyperextension position to flexion position. In the computer simulation, the angles of 22.5 degrees hyperextension position, neutral position and 22.5 degrees flexion position were selected to analyze the stress distribution of the PIPJ. The maximum loading weight at the onset the fracture was from 30 kgf to 70 kgf (Figure 5-B). Therefore, axial loading used for the FEM model were set between 30 and 70 kgf. Loading conditions were determined by estimating the amount and position of axial pressure that might be added to the middle phalanx.

Restricting conditions were set by fixing both ends of the models at the same angles and depth as those of the fracture experiments. The length of the proximal phalanx remaining was 25 mm, and that of middle phalanx was 15 mm. The overall length of the model was decided to be 40 mm. General finite element program (MARC) was used for computer simulation analysis.

RESULTS

A two-dimensional model of the sagittal section showed that stress distribution changed according to the position of the PIPJ. Stress on the model was one kgf per square mm. The stress amount was divided into 20 ranks. Each rank was shown by a color display from dark blue on the lowest stress part to red on the highest. In the flexion position, highest concentration stress was found on the dorsal part. On the

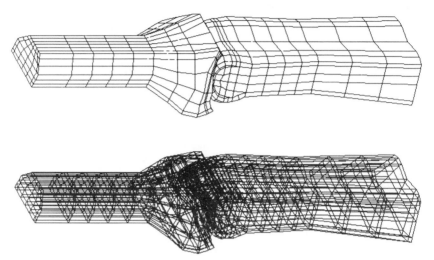

Figure 4. Three-dimensionl FEM models of PIPJ showing total number of the elements and nodes.

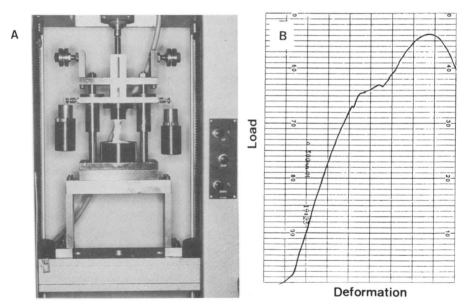

Figures 5. Figure 5A. Fracture experiments using amputated fingers. **Figure 5B.** Load-deformation curve.

contrary, in the hyperextended position it was found on the volar part. In the neutral position, it was observed on both volar and dorsal parts (Figure 6).

With higher magnification of the joint surface, deformations of four materials were easily seen. Articular cartilage was deformed from 0.5 mm to 0.1 mm in thickness. In the flexion position, higher axial load caused not only cartilaginous deformation but also sliding of the proximal phalangeal head to the dorsal. In the hyperextension position, sliding of the proximal phalangeal head occurred to the volar (Figure 7).

Reaction forces at fixed boundary conditions after a 1.3 mm movement in the neutral position are shown in table 2. The total amount of the stress force is 4.65 kgf per one square mm and the width of each phalanx is 7-10 mm. Therefore, the loading

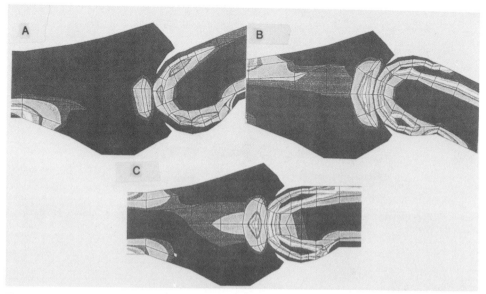

Figure 6. Stress distribution on two-dimensional model of sagittal section. A: hyperextension position. B: flexion position. C: neutral position.

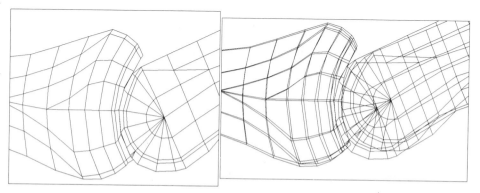

Figure 7. Articular cartilage deformation and sliding of the proximal phalangeal head, 22.5° hyperextension position.

weight becomes 32.6-46.5 kgf. These numbers are within the experimental boundaries of this study.

Three-dimensional analysis also showed superficially concentrated stress on the dorsal and volar aspect of the middle and proximal phalanx in the hyperextension, flexion and neutral position. Superficially concentrated stress are changing according to the position of the PIPJ (Figure 8).

Table 2. Reaction forces of the axial direction at fixed boundary conditions on the seven nodal points of proximal phalanx.

1.	-0.29622	2.	-0.87125	3.	-0.72783
4.	-1.0821	5.	-0.28398	6.	-0.78398
7.	-0.60319 (kgf/mm²)				

Figure 8. Superficially concentrated stress on three-dimensional model. A: hyperextension. B: flexion. C: neutral position.

Stress distribution on the articular surface of the middle phalangeal base was observable after removal of the proximal phalangeal head. There was no concentrated stress of the articular cartilage. In the view of the sagittally cut surface of the central portion, stress concentration was observed on the subchondral bone area.

Stress distribution on the subchondral bone in the neutral position was clearly shown after removal of the cartilage (Figure 9). In the neutral position, stress concentration was observed in both dorsal and volar areas. In the extension position, stress concentration moved volarlly. In the flexion position, it moved dorsally. The area of the stress distribution were similar to the experimentally produced fractures (Figure 10).

DISCUSSION

In the experiments using cadaveric fingers, Type Ic (volar split-depression fracture dislocation) was made in the hyperextension position of PIPJ. On the contrary, Type IIc (dorsal split-depression fracture dislocation) were made in the flexion position. Stress analysis using FEM models showed that concentrated stress distribution area changed according to the angles of the PIPJ. In the hyperextension position, concentrated stress was found on the volar side of the middle phalangeal base, and it was found on the dorsal side in the flexion position. In the neutral position, it was found on both sides.

Figure 9. Stress distribution on the subchondral bone and sagittally sectioned surface. A: hyperextension, B: flexion, C: neutral position.

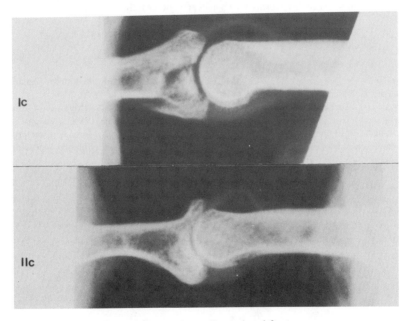

Figure 10. Experimentally produced fractures.

The forces on the finger tip are transmitted to the distal phalanx and then to the middle phalanx. When the interphalangeal joints are locked in the swan-neck position, the PIPJ enters the hyperextension position, causing volar fractures. On the contrary, when interphalangeal joints are locked in the button hole position, the PIPJ enters the flexion position, which makes dorsal fractures or fracture dislocations. Near the neutral position, axial force makes split depression fractures and impacted force in the complete neutral position may produce Type III fractures.

Our study are still qualitative, however, FEM analysis combined with the study of experimentally produced fractures are useful for the solution of onset mechanism of these fractures.

CONCLUSION

[1] Stress distribution changed as the angles of the PIPJ changed, [2] in the hyperextension position, concentrated stress was found on the volar side of the middle phalangeal base while it was found on the dorsal side in the flexion position; in the neutral position, concentrated stress was found on both sides, and [3] axial stress on the middle phalanx makes three different fractures, volar, dorsal, and both depending upon the angle of the PIPJ.

ACKNOWLEDGEMENTS

We are grateful to Mr. N. Nagayama for his co-operation of FEM analysis.

REFERENCES

1. H. Hashizume, F. Nagasawa, H. Wake, S. Hara, and O. Akahori, Study of middle phalangeal base fracture, in Japanese with English abstract, *J. Jpn. Soc. Surg. Hand* 8:709 (1991).
2. A. Hayashi, S. Tanaka, and S. Tsutsumi, Simulation analysis about development of osteoarthritis using finite element method, in Japanese, *J. Jpn. Orthop. Assoc.* 65:s1178 (1991).
3. M.C. Hobatho, R. Darmana, P. Paster, I.J. Barrau, S. Laroze, and J.P. Morucci. Development of a three-dimensional finite element model of a human tibia using experimental modal analysis, *J. Biomech.* 24:371 (1991).
4. T.E. Orr., and D.R. Carter, Stress analysis of joint arthroplasty in the proximal humerus, *J. Orthop. Res.* 3:360 (1985).
5. M.J. Turner, R.W. Clough, H.C. Martin, and L.J. Topp, Stiffness and deflection analysis of complex structures, *J. Aero. Sci.* 23:805 (1956).

Part III

MOTION ANALYSIS

NORMAL AND ABNORMAL CARPAL KINEMATICS

Marc Garcia-Elias,[1] Douglas K. Smith,[2] Leonard K. Ruby,[3]
Emiko Horii,[4] Kai-Nan An,[5] Ronald L. Linscheid,[5] Edmund
Y.S. Chao,[5] and William P. Cooney III[5]

[1]Hospital General de Catalunya
08190 Sant Cugat
Barcelona, Spain
[2]Mallinckrodt Institute of Radiology
Washington University School of Medicine
510 S. Kingshighway Blvd
St Louis, MO 63110, USA
[3]Tufts University School of Medicine
Boston, MA 02111-1854, USA
[4]Nagoya University School of Medicine
1-1-20 Daikominami, Higashiku
Nagoya, 461, Japan
[5]Orthopedic Biomechanics Laboratory
Mayo Clinic
Rochester, MN 55905, USA

INTRODUCTION

Most wrist problems (fractures, dislocations, ligament disruptions, bony malunions, dysplasias, etc) generate substantial alteration of individual carpal bone kinematics.[1,2,7,8,9,12] Unless properly corrected, the resulting abnormal intercarpal motion results in the development of late articular degeneration.[13]

From a basic science perspective, understanding carpal bone kinematics is essential as a database for future developments in wrist modelling, not to mention for the design of new prosthetic devices.[4] Kinematic knowledge is also important from a clinical point of view since it allows a more accurate evaluation of carpal disorders. Better results are likely to be obtained not only if the static geometrical carpal problem is solved, but also if the kinematic disorder is addressed when treating these injuries.

Stereoradiography has proved to be one of the most reliable methods for providing quantitative data about joint motion.[4,6,7,8,9,10,11,12] It is specially useful in composite joints

Advances in the Biomechanics of the Hand and Wrist
Edited by F. Schuind *et al.*, Plenum Press, New York, 1994

where irregularly shaped bones move around separate, and yet interdependent, axes of motion.[4] The possibility of accurately measuring individual carpal bone motion has represented an enormous contribution to the understanding of such a complex joint.

In the same category as the stereometric method, the biplanar radiographic method of kinematic analysis has also proved to be very useful in orthopedics. This method was initially introduced by Chao and Morrey in 1978 to study elbow motion.[5] Since then it has been utilized, and proven useful, in many other areas.

At the Mayo Clinic Orthopedic Biomechanics laboratory, several studies addressing the kinematics of different abnormal conditions of the wrist have been performed using the biplanar radiographic method of kinematic analysis.[7,8,9,10,12] In this chapter, the experimental methodology is presented, and a review of the results obtained for the normal and simulated abnormal conditions is analyzed.

MATERIAL AND METHODS

Twenty fresh human cadaver wrist specimens were used in these experiments. Four metal markers (tantalum spheres) were inserted into selected carpal bones, through small arthrotomies, with care to avoid injuring any intrinsic or extrinsic ligaments. The specimens were then secured to a plexiglas testing frame, by means of two Steinmann pins drilled across the two forearm bones, in neutral axial rotation.

The five muscles mainly involved in wrist motion (extensor carpi ulnaris, extensor carpi radialis longus and brevis, flexor carpi radialis, and flexor carpi ulnaris) were connected to calibrated springs, controlled by a turnscrew. By adjusting the individual spring tensions to simulate the physiologic tendon tension the wrist specimen could be balanced and placed into any wrist position. These tension values were based on the physiologic cross-sectional areas, and the relative electromyographic activity of each tendon in each wrist position.

A biplanar radiographic apparatus, consisting in a cassette holder maintaining two X-rays in an exact orthogonal relationship to two fixed X-ray tubes, was utilized in these experiments. Simultaneous anteroposterior and lateral radiographs were obtained in different wrist positions with the wrist in the loaded condition.

The position of the metal markers on each X-ray projection was digitized and stored in a PC computer with specially designed software programs allowing calculation of: [1] the spatial location of the markers with respect to a reference coordinate system, [2] the absolute motion of the selected bones, and [3] the relative motion between each rigid body pair, described as a rotation around, and translation along a unique axis, following the "screw displacement axis" concept.[4] This kinematic analysis system had shown an accuracy to within 0.4 mm displacement and 2 degrees angular rotation.[5]

Once the wrists had been tested for their normal kinematics, incisions were reopened and four conditions of carpal osteoligamentous integrity were simulated and studied using the biplanar radiographic method: [1] unstable scaphoid waist fractures (5 wrists),[12] [2] scapho-lunate dissociations (4 wrists),[9] [3] luno-triquetral dissociations (5 wrists),[8] and [4] scapho-trapezial-trapezoidal (STT) and scapho-capitate (SC) limited intercarpal fusions (6 wrists).[7] Dissociations were simulated by selectively sectioning specific ligaments. Unstable fractures of the scaphoid were made with an osteotome. Finally, limited fusions were simulated by means of two non-parallel Herbert screws inserted through selected articulations. Further details about these carpal kinematic studies can be found in previous publications.[7,8,9,10,12]

RESULTS

I: NORMAL CARPAL KINEMATICS

Unless constrained by external forces, the flexion-extension and/or radial-ulnar deviation moments generated by contraction of any one of the different wrist motor tendons result in wrist motion always starting at the distal carpal row. The bones of the proximal row do not move until the tautness of the ligaments crossing the midcarpal joint reaches a certain level.

Table 1 summarizes the average magnitude and direction of rotation of the carpal bones with respect to the radius, as found by testing the last five consecutive fresh human cadaver wrists. Orientation of the helical (screw) axes is described on an XYZ axis system, such that X lies along the radius, Y is transverse across the wrist, and Z is in the sagittal plane perpendicular to the other two axes. A positive rotation about the X axis means pronation, while negative values correspond to supination. Rotations about the Y axis mean flexion if negative, and extension if positive. Motion about the Z axis represents radial deviation (positive) or ulnar deviation (negative).

In normal wrists, very little intracarpal motion exists between the bones of the distal carpal row. From full flexion to extension, no more than 9 degrees (dgs) angular rotation at the hamate-capitate joint, no more than 6 dgs at the capitate-trapezoid joint, or than 12 dgs at the trapezoid-trapezium joint, have been recorded. The bones of the distal carpal row can not be thought of as a rigid structure but still as a single functional unit. In flexion of the wrist, they all follow a rotation about an axis which obviously implies flexion, but also some degree of ulnar deviation. In extension the tendency of all distal carpal bones is to rotate into extension and a slight radial deviation.

The bones of the distal carpal row also move synergistically (in about the same direction) in lateral deviations of the wrist. In radial deviation of the wrist, they all deviate radially, extend and supinate. In ulnar deviation, they flex, deviate ulnarly and pronate.

The proximal carpal row bones appear to be less tightly bound to one another than the bones of the distal carpal row. Despite differences in angular rotation, however, all proximal carpal row bones move synergistically with wrist motion. From full flexion to full extension of the wrist, the scaphoid rotates an average of 110 dgs with respect to the radius, while the lunate rotates only 76 dgs, and the triquetrum 88 dgs (Table 1). The different radii of curvature of the proximal aspects of these bones may explain such differences in angular rotation.

There is more relative motion at the scapholunate interval than at the lunotriquetral joint (34 dgs and 12 dgs, respectively). Differences in fiber length and orientation of the respective interosseous ligaments may account for such differences. Despite differences in absolute values of angular rotation, however, the direction of motion found by all the authors is quite consistent: during wrist flexion, the bones of the proximal carpal row go into flexion and ulnar deviation, while during wrist extension they extend and deviate radially.

The contribution of the radiocarpal and midcarpal joints to normal wrist flexion-extension is equally divided in only one-third of the wrists. In the remaining two-thirds, approximately 60% of the global flexion occurs at the lunocapitate interval while 66% of extension is radiocarpal dependent. However, if motion is recorded on the radius-scaphoid-trapezium linkage, more than two-thirds of the global arc of movement occurs at the radio-scaphoid interval. According to that, an arthrodesis of the luno-capitate joint theoretically would result in a larger restriction of the global wrist motion than a fusion of the scapho-trapezial-trapezoidal joint.

Table 1. Normal carpal bone motions with respect to the radius, as found by Horii et al.[8] (Average values of five consecutive specimens)

I: From neutral to extension

Moving bone	Orientation of screw axis (*)			Rotation (dgs) (mean ± SD)
	X	Y	Z	
Scaphoid	0.10	0.98	0.05	56.1 ± 9.2
Lunate	-0.09	0.98	0.15	31.2 ± 4.9
Triquetrum	-0.12	0.98	0.05	41.6 ± 5.1
Capitate	0.00	0.99	0.03	64.2 ± 6.8
Hamate	-0.01	0.99	0.00	65.0 ± 5.7

II: From neutral to flexion

Moving bone	Orientation of screw axis (*)			Rotation (dgs) (mean SD)
	X	Y	Z	
Scaphoid	-0.04	-0.96	-0.23	55.4 ± 16.9
Lunate	0.01	-0.90	-0.35	45.1 ± 19.8
Triquetrum	-0.01	-0.92	-0.30	47.7 ± 21.2
Capitate	0.02	-0.99	-0.10	77.0 ± 18.7
Hamate	0.04	-0.99	-0.10	72.2 ± 17.0

III: From neutral to radial deviation

Moving bone	Orientation of screw axis (*)			Rotation (dgs) (mean SD)
	X	Y	Z	
Scaphoid	0.24	-0.72	0.62	12.8 ± 6.9
Lunate	0.14	-0.78	0.63	13.0 ± 6.9
Triquetrum	0.16	-0.49	0.83	11.8 ± 7.2
Capitate	0.10	0.06	0.97	24.2 ± 8.3
Hamate	0.15	0.15	0.97	23.9 ± 7.3

IV: From neutral to ulnar deviation

Moving bone	Orientation of screw axis (*)			Rotation (dgs) (mean SD)
	X	Y	Z	
Scaphoid	-0.24	0.74	-0.55	22.7 ± 6.0
Lunate	-0.13	0.80	-0.54	25.4 ± 7.3
Triquetrum	0.01	0.64	-0.84	23.3 ± 7.3
Capitate	-0.21	-0.15	-0.95	28.7 ± 8.8
Hamate	-0.16	-0.08	-0.96	28.0 ± 9.7

(*) X-axis: + supination   - pronation
     Y-axis: + extension   - flexion
     Z-axis: + radial dev.   - ulnar dev.

From full radial deviation to full ulnar deviation, the scaphoid, the lunate and the triquetrum rotate an average of 36 dgs, 38 dgs, and 35 dgs respectively about quite similarly oriented axes, implying radial deviation but also extension. The three proximal carpal bones move synergistically from a flexed position in radial deviation to an extended position in ulnar deviation. This flexion-extension adaptive mechanism, present in normal wrists, allows a constant spatial congruity between the distal carpal row and the radius regardless of the wrist position.

II: ABNORMAL CARPAL KINEMATICS

All four simulated conditions of carpal integrity showed substantial changes in the kinematic behavior. All of them showed a gross alteration of the mechanisms by which the proximal carpal row is able to fill the changing space between the distal carpal row and the two forearm bones with motion.

Unstable fractures of the waist of the scaphoid generated significant changes in the overall carpal kinematics. The two scaphoid fragments not only moved independently, but their relative motion was more complex than previously recognized. After simulation of scaphoid fracture, the overall motion of the distal scaphoid fragment increased significantly, while the proximal scaphoid showed less rotation. Such differences in angular rotation are likely to promote the development of a dorsal and radial angulation at the fracture site ("humpback" deformity), a common finding in chronic scaphoid nonunions.

Disruption of the dorsal portion of the scapholunate interosseous ligament caused a significant change in the normal alignment of the scaphoid relative to the lunate. In all the four specimens studied, the scaphoid was observed to collapse primarily in flexion and pronation. Conversely, the lunate rotated an average of five degrees towards extension, which has been known as a "Dorsal Intercalated Segment Instability" (DISI). If both the palmar and dorsal portions of the scapholunate interosseous membrane were divided, the presence of the radio-scapho-lunate ligament was found not to prevent a dorsal subluxation of the scaphoid with scapholunate gap formation.

The effects of sectioning the supporting ligaments of the lunotriquetral joint were also analyzed. If only the dorsal and palmar lunotriquetral ligaments were sectioned, increased triquetrum motion was noted, but no static "Volar Intercalated Segment Instability" (VISI) was promoted. However, further sectioning of both the dorsal radio-triquetral and dorsal scapho-triquetral ligaments produced a gross alteration of the overall carpal kinematics, with the lunate adopting a palmar-flexed position, and the triquetrum rotating into supination.

Limited intercarpal fusions, especially those crossing the midcarpal joint, were also found to alter very significantly intracarpal kinematics (Figure 1). After simulating either an STT or an SC fusion, the flexion-extension component of scaphoid motion during radio-ulnar deviations was significantly neutralized. When fused to the distal row, the scaphoid no longer can control the dynamic midcarpal joint relationships, but behaves as a distal carpal row bone. Consequently, stress in the radioscaphoid joint is likely to increase in radial deviation, as well as tensions in the scapholunate ligaments in ulnar deviation.

DISCUSSION

Idiopathic degenerative osteoarthritis of the wrist is rare. When present, it usually is the consequence of an unresolved traumatic or inflammatory problem.[13] A chronic alteration

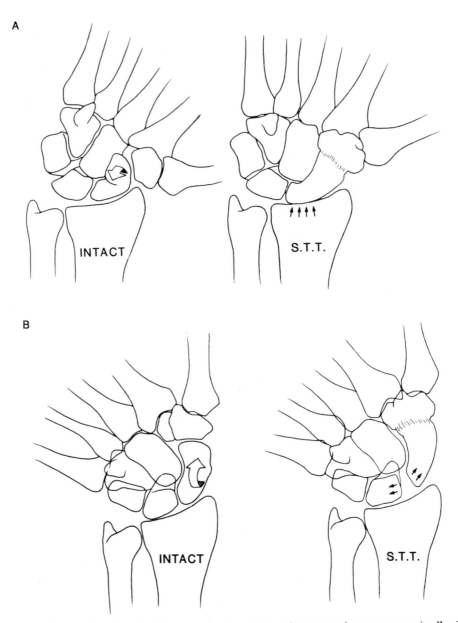

Figure 1. During ulnar deviation the scaphoid of an intact wrist rotates about a screw axis allowing extension, ulnar deviation, and pronation. After an STT fusion, the scaphoid behaves as another distal carpal row bone. Therefore, increasing pressure at the radioscaphoid joint is likely to appear during radial deviation (A), while increasing tension on the scapho-lunate ligaments during ulnar deviation is to be expected (B). [reprinted with permission of Garcia-Elias et al.[7]].

of the geometry of either one of the bones involved in the wrist may result in articular incongruity.[12] Rupture of specific ligaments may also generate articular dysfunction.[9] Our investigations tend to confirm the hypothesis that late degenerative osteoarthritis secondary to articular incongruity and/or ligamentous dysfunction can be explained not only by an abnormal distribution of forces within the wrist, as found by others,[3] but also by a substantial change in the intracarpal kinematics. In effect, an alteration in the force distribution may increase pressure on small areas of the joint, resulting in cartilage deterioration. Abnormal kinematic behavior may as well increase shear forces in areas not well prepared to resist them, thus promoting further degeneration, progressive radiocarpal and intercarpal ligament disruption, and therefore, further carpal instability. Studies such as the ones reviewed in this chapter are of importance not only because they may help in explaining the different stages of wrist degeneration after specific unresolved wrist problems, but also because they increase our attention to these apparently benign lesions. Undoubtedly, the higher the awareness of the problems created by overlooking these injuries, the better the chances of obtaining more durable results.

REFERENCES

1. R. Arkless R, Cineradiography in normal and abnormal wrists, *Amer. J. Roentg.* 96:837 (1966).
2. R.A. Berger, R.D. Crowninshield, and A.E. Flatt, The three-dimensional rotational behaviors of the carpal bones, *Clin. Orthop.* 167:303 (1982).
3. A.D. Blevens, T.R. Light, W.S. Jablonsky, D.G. Smith, A.G. Patwardhan, M.E. Guay, and T.S. Woo, Radiocarpal articular contact characteristics with scaphoid instability, *J. Hand Surg.* 14A:781 (1989).
4. E.Y.S. Chao, and K.N. An, Perspectives in measurements and modeling of musculoskeletal joint dynamics, *in*: "Biomechanics: Principles and applications," R. Huiskes, D. Van Campen, and J. Wijn, eds., Martinus Nijhoff, Der Hague (1982).
5. E.Y.S. Chao, and B.F. Morrey, Three-dimensional rotation of the elbow, *J. Biomech.* 11:57 (1978).
6. A. de Lange, J.M.G. Kauer, and R. Huiskes, Kinematic behavior of the human wrist joint: a roentgen-stereophotogrammetric analysis, *J. Orthop. Res.* 3:56 (1985).
7. M. Garcia-Elias, W.P. Cooney, K.N. An, R.L. Linscheid, and E.Y.S. Chao, Wrist kinematics after limited intercarpal arthrodesis, *J. Hand Surg.* 14A:791 (1989).
8. E. Horii, M. Garcia-Elias, K.N. An, A.T. Bishop, W.P. Cooney, R.L. Linscheid, and E.Y.S. Chao, A kinematic study of luno-triquetral dissociations, *J. Hand Surg.* 16A:355 (1991).
9. L.K. Ruby, K.N. An, R.L. Linscheid, W.P. Cooney, and E.Y.S. Chao, The effect of scapholunate ligament section on scapholunate motion, *J. Hand Surg.* 12A(2Pt1):767 (1987).
10. L.K. Ruby, W.P. Cooney, K.N. An, R.L. Linscheid, and E.Y.S. Chao, Relative motion of selected carpal bones: A kinematic analysis of the normal wrist, *J. Hand Surg.* 13A:1 (1988).
11. H.H.C.M. Savelberg, J.G.M. Kooloos, A. de Lange, R. Huiskes, and J.M.G. Kauer, Human carpal ligament recruitment and three-dimensional carpal motion, *J. Orthop. Res.* 9:693 (1991).
12. D.K. Smith, W.P. Cooney, K.N. An, R.L. Linscheid, and E.Y.S. Chao, The effects of simulated unstable scaphoid fractures on carpal motion, *J. Hand Surg.* 14A:283 (1989).
13. H.K. Watson, and J. Ryu, Evolution of arthritis of the wrist, *Clin. Orthop.* 202:57 (1986).

FUNCTIONAL ANATOMY OF THE CARPUS IN FLEXION AND EXTENSION AND IN RADIAL AND ULNAR DEVIATIONS: AN IN VIVO TWO- AND THREE-DIMENSIONAL CT STUDY

Véronique Feipel,[1] Marcel Rooze,[1] Stéphane Louryan,[2] and Marc Lemort[2]

[1]Laboratory for Functional Anatomy
University of Brussels
Route de Lennik, 808 (CP 619)
B-1070 Brussels, Belgium
[2]C.R.E.A.R.I.M.
Rue Héger-Bordet, 1
B-1000 Brussels, Belgium

INTRODUCTION

The complexity of carpal functional anatomy necessitates sophisticated experimental devices for assessment. Moreover many points in our knowledge and comprehension of the wrist in physiological and pathological conditions remain obscure. It is thus a challenge to develop an accessible investigation method, that is easy to handle and constitutes an educational as well as a diagnostic means to study normal and pathological carpal biomechanics.

Since the beginning of the Twentieth Century, various authors have tried to explain the wrist's functional behavior using carpal functional models (Figures 1,2,3). These models can be subdivided into roughly two categories: the "horizontal carpus" concept in two transverse rows and the "vertical carpus" concept in three longitudinal columns. Sennwald[23] considers a semi-annular structure surrounding the lunate, a model that cannot be included in any of these categories.

The two-row model of Lichtmann et al.[12] classifies the carpal bones in two rows whose composition is analogous to the classical descriptive anatomy (Figure 1a). The two rows, and especially the distal one, behave more or less like fixed entities.

Littler[14], in a study of functional hand architecture, considers a fixed element made up of the distal carpal row and the metacarpals II and III. These bones form the rigid part of the transverse carpal and metacarpal arches. Ruby et al.[18] stress the importance of intracarpal mobility, great in the proximal row, lesser in the distal one. Volz et al. [28]

Advances in the Biomechanics of the Hand and Wrist
Edited by F. Schuind *et al.*, Plenum Press, New York, 1994

include the trapezium in the thumb column, and consider the pisiform as a sesamoid bone formed within the flexor carpi ulnaris tendon.

Other authors[13,15,16] consider that the scaphoid bridges the two carpal rows and acts as a connecting rod between the inherently unstable proximal row and the distal one (Figure 1b).

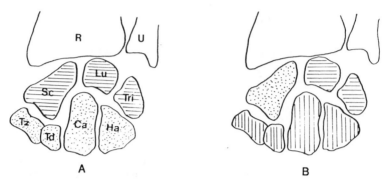

Figure 1. Horizontal carpal models. **Figure 1-A**. Lichtmann's model. **Figure 1-B.** Scaphoid bridging the two rows. R: radius, U: ulna, Sc: scaphoid, Lu: lunate, Tri: triquetrum, Tz: trapezium, Td: trapezoid, Ca: capitate, Ha: hamate.

The model of Kuhlmann et al.[11] is intermediate between the horizontal and vertical concepts, and takes into account the transverse osseous cohesion assumed by two horizontal rows, and the longitudinal cohesion, essential for muscular force transmission, by three vertical columns.

Gilford et al.[9] and later Kauer[10] propose a longitudinal model of the wrist, composed of three longitudinal chains. The element of the proximal row is an intercalated segment in each of these chains. Kauer stresses the importance of the scaphoid stabilizing the central column, the eccentricity of the scapho-lunate interosseous ligament, and the proximal curvature of these two bones in relation with their mutual displacement.

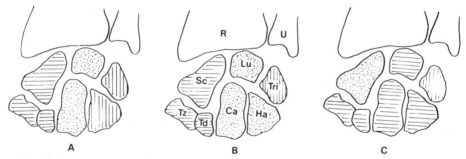

Figure 2. Vertical carpal models. **Figure 2-A.** Destot. **Figure 2-B.** Navarro. **Figure 2-C.** Taleisnik.

Concerning the vertical carpal models, the one proposed by Destot (cited by Taleisnik)[25,26] is certainly the simplest (Figure 2a). The internal, or rotational column is made up of the triquetrum and the hamate. This rotational column continues the axis of forearm pronation and supination. The central, flexion-extension column consists of the lunate and capitate, and the external column consists of the scaphoid, trapezium and

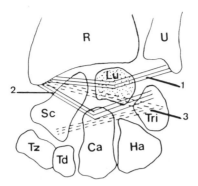

Figure 3. Sennwald's functional model: (1) proximal, and (2) distal palmar "V" ligaments, (3) dorsal "V" ligament.

trapezoid. Navarro (cited by Taleisnik)[25,26] considers that the triquetrum is the pivot of wrist and hand longitudinal rotations (Figure 2b). The central column is thus extended to the hamate.

Taleisnik[25,26] presents a "T"-shaped model, the lateral columns being reduced to the triquetrum and scaphoid, respectively (Figure 2c). This concept takes into account the special functions of these two bones, the scaphoid bridging the lunate-capitate joint space, and the triquetrum as an internal pivot.

Sennwald[23], studying carpal ligamentous anatomy, considers that the lunate is more firmly anchored to the forearm than the remaining carpals (Figure 3). Indeed, the palmar and dorsal "V"-shaped ligaments give a special support to the lunate, which is thus surrounded by a semi-annular structure formed by the remaining carpals. The radiographic observation of the wrist under longitudinal traction confirms this model, as the lunate remains closer to the radius than the other carpal bones.

In recent years many three-dimensional investigational methods have been applied to the study of carpal biomechanics. These techniques include sonic digitizing systems,[3] roentgen stereophotogrammetric analysis,[5,6,20,21,22] and similar methods using implanted pins[11,24] or reflecting markers[27] in combination with X-ray or video cameras. Because of their extreme precision, these are undeniably methods of choice in biomechanics. Unfortunately they cannot be applied in vivo.

Electrogoniometry (with three or six degrees of freedom) is applicable to living subjects, but is limited to the study of global motion and does not permit the assessment of individual carpal bone motion.

In the present study in vivo 3D CT, in combination with the mechanical and computational helical axis and rigid body concepts, is used in a trial to analyze qualitatively and quantitatively the displacements of the individual carpal bones in normal volunteers.

It is *a priori* obvious that this method lacks accuracy compared to the techniques cited above. However the aim of this study is to develop an investigational method that can be used for in vivo research purposes and, possibly, as a diagnostic tool.

MATERIAL AND METHODS

A CT study of the right wrist of 20 healthy volunteers was carried out in neutral position, radial (15°) and ulnar (30°) deviation. A second sample of 20 volunteers' right wrists was examined in neutral position, flexion (45°) and extension (45°). Five anatomical preparations were studied in radioulnar deviations. All examinations were executed on a Philips Tomoscan LX; the main scanning characteristics were 100 or 200 mA, 125 kV, filter 4, cervico-facial program, transaxial 3-mm thick slices with a 2-mm table increment, zoom 1.6.

A homemade fixation device (Figure 4) was used to avoid forearm motion and to allow positioning of the hand in the desired position.

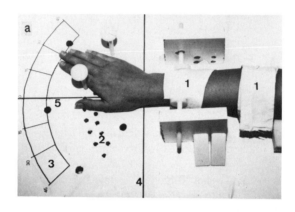

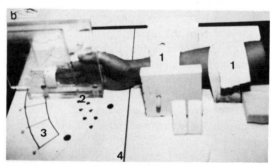

Figure 4. Fixation device. **Figure 4-a.** radioulnar deviations. **Figure 4-b.** flexion-extension. (1) forearm straps, (2) hand positioning device, (3) graduated scale, (4) transverse, and (5) longitudinal axes.

CT data were printed onto radiographic film and stored on floppy disks for further manipulation. Three-dimensional reconstructions were achieved using the I.S.G. Technologies C.A.M.R.A. S 200 computer facility. This interactive and user-friendly system offers wide possibilities of 2D data manipulation, density or gray level volume-based surface reconstruction, and positional and morphometric manipulation of 3D objects. Reconstructions were obtained for each position of the wrist. Direct photography of the 3D objects from the screen was used.

For the qualitative study, views of each wrist in the chosen positions were compared, and the positional variations of the carpal bones regarding neutral position were described.

The quantitative aspects of this study were preliminary ones, and were limited to the study of scaphoid, triquetrum, and hamate motion. The 3D coordinates of 3 points of each of these bones were determined using the 3D images along with the CT slices. Corrections for different zooming (CT scan, C.A.M.R.A., photograph) were easily obtained knowing the number of slices filmed, and measuring object size on the photographic picture. Figure 5 shows the landmark determination procedure for the scaphoid.

The bones were supposed to behave like rigid bodies during their displacements. Three-dimensional motion of a rigid body can be expressed in terms of a rotation Θ about and a translation $\overline{d}_p$ of the center of gravity along a line in space (motion axis), according to:

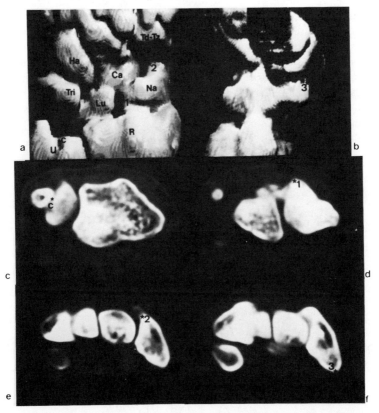

Figure 5. Landmark coordinate determination for the scaphoid. **Figure 5-a.** 3D reconstruction (dorsal view). **Figure 5-b.** 3D reconstruction (radial view); transverse CT slices. **Figure 5-c.** Level of the ulnar reference point c. **Figure 5-d.** Level of scaphoid landmark 1 (S1). **Figure 5-e.** Level of S2. **Figure 5-f.** Level of S3.

$$\overline{r} = M\overline{r}_o + \overline{d} \qquad (1)$$

where $\overline{r}_o$ is the position vector of the rigid body, r the position vector of the body after a motion step, M represents the rotation matrix, and d is the translation vector.

According to this concept, a moving rigid body is characterized by a six degrees of freedom (dof) mobility: [1] the determination of a point on a line by the intersection of the screw axis with one coordinate plane (2 dof = the 2D coordinates in this plane); [2] the direction of the line given by a vector, i.e., 2 independent components, for example, the length of the vector (2 dof); [3] the rotation angle about the axis (1 dof); [4] and the translation along the axis (1 dof).

The rotation matrix M is the matrix product of the matrices expressing the rotations about the ox, oy, and oz axes respectively, be it in a space-fixed or in a body-fixed reference system. According to Woltring[30] the choice of the rotation sequence, i.e. the matrix multiplication sequence, influences the results if the rotation about more than one axis is large, and if the rotation about the second chosen axis is large. In the case of the carpal bones, out-of-plane motions are important, especially for the proximal carpals in radioulnar deviation of the wrist.[21] The choice of the sequences is therefore difficult. In this study the procedure used by de Lange[6] and Savelberg[21] was followed.

For easier interpretation, the global rotation and translation were decomposed into rotations and translations along each of the axes of the space-fixed right-handed reference system. It was assumed that the forearm bones did not move during wrist motion, except for pronation/supination motion (which was taken into account).

The axis system was centered on the most proximal point on the distal surface of the ulnar head (Figure 6). The oy-axis was transverse and crossed the distal external vertex of the ulnar head, with an external positive direction. The longitudinal oz-axis was perpendicular to the oy-axis, and its positive direction was distal. The ox-axis was antero-posterior, with a dorsal positive direction (Figure 6).

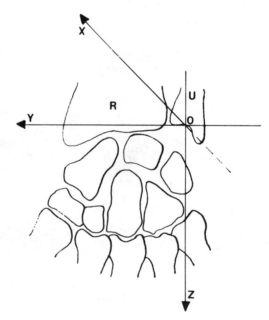

Figure 6. Reference system. U: ulna, R: radius, O: origin of the axes system. Arrows point in a positive axis direction (Oz: dorsal positive direction).

RESULTS

Qualitative Observations

Table 1 describes the observations made on the 3D images. Only some of these, the most relevant ones, will be discussed.

Lunate motion was remarkably less than that of its proximal row neighbors (Figures 7,9). If we observe the aspect of this bone in ulnar deviation of the wrist, for example, we see that a palmar translation of the lunate seemed to have occurred with respect to the neutral position. This apparent palmar translation is simply a lunate extension, due to the curved shape of its proximal (convex) and distal (concave) articular facets and the relationships of this bone with the radius proximally, and the capitate distally. Despite the functional importance of the scapho-lunate interosseous ligament, it seemed that the lunate was related to the movements of a rather poorly mobile entity that was composed of the capitate, the metacarpals II and III, the trapezium and trapezoid, and the lunate.

The scaphoid acted as a connecting rod at the external aspect of the wrist. Its stabilizing role was fulfilled mostly by flexion-extension motion and palmar dorsal translations (Figures 7, 9).

Table 1. Movements of the carpal bones during wrist motion.

Bone	U.D.	R.D.	Flex.	Ext.
Scaphoid	UD+E+DT	RD+F+PT	F+RD+PT	E+UD
Lunate	UD+E	RD+F	F+RD	E+UD
Triquetrum	UD+P+E	RD+S+F	F+S+RD	E+P+UD
Pisiform	PrT	DiT	PrT	DiT
Trapezium Trapezoid	UD+F	RD+E	F	E
Capitate	UD	RD	F	E
Hamate	UD+P	RD+S	F+S	E+P

Abbreviations: U.D. = UD = ulnar deviation; R.D. = RD = radial deviation; Flex.= F = flexion; Ext. = E = extension; P = pronation; S = supination; DT = dorsal translation; PT = palmar translation; PrT = proximal translation; DiT = distal translation.

The triquetrum and hamate displayed important rotations about longitudinal axes (intracarpal pronation and supination). Both bones seemed governed by the same motion scheme (Figures 7,8,9).

Beyond these global observations, comparison of the volunteers' wrists and of the anatomical preparations (upper limbs with sectioned wrist and hand muscles) revealed the importance of the flexor and extensor carpi ulnaris muscles in the stability of the internal carpus. Indeed, in the anatomical preparation sample, the triquetrum and the hamate displayed greater mobility than in the volunteers' wrists. Moreover, in the volunteers, sample, pisiform motion was relatively independent of triquetrum motion; the pisiform followed the excursion of the flexor carpi ulnaris tendon. However, in anatomical preparations, the triquetrum and the pisiform moved more uniformly. In the volunteers' sample, in ulnar deviation and flexion, the pisiform adopted a more proximal position with regard to the triquetrum, whereas in radial deviation and extension it was more distally located (Figure 9). These observations reinforce the concept advanced by Volz et al.[28] that the pisiform is a sesamoid bone in the distal flexor carpi ulnaris tendon.

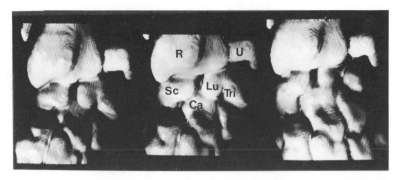

Figure 7. Wrist 3D reconstructions. Dorsal view. Radial deviation (left), neutral position (middle), ulnar deviation (right).

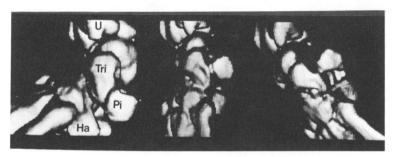

Figure 8. Wrist 3D reconstructions. Ulnar view. Extension (left), neutral position (middle), flexion (right).

Quantitative study

This part of our study was limited to the quantitative *in vivo* 3D analysis of scaphoid, triquetrum, and hamate motion during wrist flexion, extension and radioulnar deviation.

As large interindividual variations were observed, we had to adopt a stratagem to interpret these results. The results of the qualitative observations were considered as the expected behavior of the bones. Two-dimensional plots were constructed: [1] the percentage of expected patterns was determined for each motion component; [2] for each wrist movement and for each of the three bones, the 6 motion components (3 rotations and 3 translations) were graphically plotted against each other; [3] for each wrist displacement, the rotational components of each bone were plotted against the ones of the two remaining bones.

In each of these three approaches a percentage of 50% or more was considered relevant. This choice can be justified by the fact that this quantitative study is a preliminary approach to the in vivo study of carpal kinematics, based on a methodology that is liable to be improved in accuracy. A majority of the population can thus be found to display the described features. This approach does, however, not exclude the possibility of subdividing the population into several subgroups, in case of a more precise approach and of a larger sample.

Scaphoid motion (Table 2). Extension of the wrist resulted in scaphoid extension in 70 % of the wrists (Figure 10a), and ulnar deviation in 50 %. Sixty percent of the population was found in the extension-radial shift quadrant (Figure 10a), when these two components were plotted. No other plots appeared to show significative correlations.

Wrist flexion caused scaphoid flexion (75 %), radial deviation (50 %), and pronation (60 %). The graphic confrontations revealed that 50 % of the wrists displayed pronation and flexion simultaneously.

Scaphoid motion in ulnar deviation of the wrist was composed of ulnar deviation (75 %), extension and pronation (both 60 %). Two-dimensional plots showed that 60 % of the scaphoids shifted distally and palmarly, 50 % were found in the radial and palmar shift as well as pronation-extension quadrants. Fifty-five percent of the wrists were located in the scaphoid ulnar deviation and distal shift quadrant.

Wrist radial deviation caused scaphoid radial deviation in 65 % of the cases. None of the relationships was found to be significant.

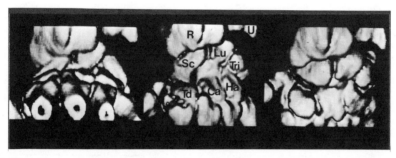

Figure 9. Wrist 3D reconstructions. Dorsal view. Extension (left), neutral position (middle), flexion (right).

Triquetrum motion (Table 3). The most relevant result in wrist extension was that 90 % of the wrists displayed triquetrum extension (Figure 10b). Triquetrum extension was combined with translations in radial, distal and dorsal directions, each in 60 % of the cases. Fifty percent of the triquetrums displayed simultaneous extension and ulnar deviation during wrist extension. Pronation-supination movements had no significant trend.

In wrist flexion triquetrum flexion occurred in 75 % of the cases. Many relevant correlations between motion components were observed. Triquetrum flexion was accompanied by radial deviation (70 %) (Figure 10c), palmar (65 %) and ulnar (60 %) shift, and/or supination (55 %). Moreover radial deviation of the triquetrum in wrist flexion was combined with palmar (65 %), proximal (60 %) and ulnar (65 %) translation, as well as with supination (60 %). Sixty-five percent of the wrists displayed simultaneous ulnar and palmar translation of the triquetrum in wrist flexion.

Component graphs were not as relevant as component percentage when triquetrum motion was analyzed in wrist radioulnar deviation. Ulnar deviation of the wrist induced triquetrum supination (70 %), flexion (60 %), and radial deviation (55 %), as well as proximal (70 %), palmar (65 %) and ulnar (75 %) shift.

In radial deviation of the wrist the triquetrum executed extension (75 %), radial deviation (65 %), supination (70 %), and palmar translation (70 %). The results of triquetrum motion related to the two latter wrist movements are interesting for the fact that three components (radial deviation, supination, and palmar shift) were present in most wrists during both (opposite) displacements of the wrist.

Table 2. Scaphoid motion. Proportions of individual and correlated components.

WRIST EXTENSION	%	WRIST FLEXION	%
E	70	F	75
UD	50	RD	50
E + RaT	60	P	60
		F + P	50

WRIST ULNAR DEVIATION	%	WRIST RADIAL DEVIATION	%
UD	75	RD	65
E	60		
P	60		
PT + DiT	60		
RaT + PT	50		
E + P	50		
UD + DiT	55		

Abbreviations: RaT = radial translation; UlT = ulnar translation (refer to table 1 for the remaining abbreviations).

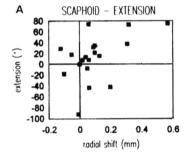

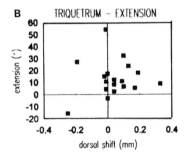

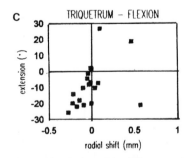

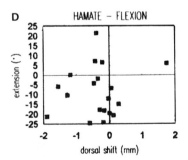

Figure 10. Graphic 2D plots of the different motion components of one bone. **Figure 10-A.** Wrist extension - scaphoid extension versus radial translation. **Figure 10-B.** Wrist extension - triquetrum extension versus dorsal translation. **Figure 10-C.** Wrist flexion - triquetrum extension versus ulnar deviation. **Figure 10-D.** Wrist flexion - hamate extension versus dorsal translation.

Table 3. Triquetrum motion. Proportions of individual and correlated components.

WRIST EXTENSION	%	WRIST FLEXION	%
E	90	F	75
E + RaT	60	F + RD	70
E + DiT	60	F + PT	65
E + DT	60	F + UIT	60
E + UD	50	F + S	55
		RD + PT	65
		RD + PrT	60
		RD + UIT	65
		RD + S	60
		UIT + PT	65

WRIST ULNAR DEVIATION	%	WRIST RADIAL DEVIATION	%
S	70	E	75
F	60	RD	65
RD	55	S	70
PrT	70	PT	70
PT	65		
UIT	75		

Hamate motion (Table 4). The most relevant components of hamate motion in wrist extension were radial deviation (75 %), pronation (75 %), extension (70 %), and ulnar translation (60 %). Mutual plots of the three rotational components showed in about 60 % of the cases simultaneous occurrence of these components.

Eighty percent of the hamates flexed during wrist flexion (Figure 10d). Palmar and proximal shift also occurred in 80 % of the cases. Another interesting component was ulnar deviation (70 %). Correlated motion components ranged between 50 and 60% (ulnar deviation - flexion, ulnar deviation - palmar shift, radial and proximal shift, radial and palmar shift, flexion - palmar shift (Figure 10d)).

No significant interaction was found for the hamate in wrist ulnar deviation. However, this wrist movement induced hamate ulnar deviation (55 %), pronation (65 %), palmar (75 %) and proximal (55 %) translations.

Wrist radial deviation caused hamate radial deviation in 70 % of the cases. Supination (60 %), and proximal (65 %) and radial shift (70 %) of the hamate were also relevant.

The behavior of the remaining wrists or bones, those that displayed behavior that differed from the ones described above, could be classified as marginal if the plot in the 2D confrontation was close to the expected or principal quadrant, or as unexpected if one component or both were opposite to the expected behavior. As sometimes the proportion of unexpected patterns was relatively important (up to 50 %), a trial was made to correlate the behavior of the different bones to explain the large interindividual variations. Relevant correlations are summarized in Table 5.

Wrist extension. Some interesting features confirmed the results of other authors concerning wrist extension. Simultaneous extension of the bones analyzed occurred in extension of the wrist (60 - 65 %). Moreover relevant correlations between, on the one hand, triquetrum extension and, on the other hand, scaphoid ulnar deviation (55 %) and supination (65 %), as well as hamate radial deviation (55 %) and pronation (65 %) were noted. Hamate pronation was accompanied by scaphoid extension (60 %) and supination (55 %). Triquetrum supination and hamate radial deviation were also significant correlated (55 %).

Table 4. Hamate motion. Proportions of individual and correlated components.

WRIST EXTENSION	%	WRIST FLEXION	%
RD	75	F	80
P	75	PT	80
E	70	PrT	80
UIT	60	UD	70
RD + P	65	UD + F	60
RD + E	55	UD + PT	50
P + E	60	RaT + PrT	55
		RaT + PT	60
		F + PT	60

WRIST ULNAR DEVIATION	%	WRIST RADIAL DEVIATION	%
UD	55	RD	70
P	65	S	60
PT	75	PrT	65
PrT	55	RaT	70

Wrist flexion. This wrist movement showed interesting correlations between flexion of the three bones (65 - 75%) (Figure 11a, 11b). The simultaneous movements of these bones during wrist flexion could be summarized as follows: the scaphoid flexed (Figure 11c); the triquetrum executed flexion, supination and radial deviation (Figure 11c,11d); and hamate motion was made up of flexion and ulnar deviation.

Wrist radial deviation. The only plots that showed relevant correlations were those between hamate radial deviation and triquetrum radial deviation (55%), and, respectively, extension (60%).

Wrist ulnar deviation. Except simultaneous hamate pronation and scaphoid ulnar deviation (55%), no significant correlations could be found between the individual motion components of the three bones in ulnar deviation of the wrist.

DISCUSSION

At the time of determining the experimental protocol, a compromise had to be found between data quality and number of positions to analyze, on the one hand, and ethical considerations, on the other hand. Radiation dose, examination time, and subject comfort had to be taken into account. The prone position, with maximal shoulder flexion, was indeed not easy to maintain for a long period of time, and patient movements could thus influence data quality. These considerations are especially important in view of the clinical application of this protocol. Thus only 3 wrist positions were analyzed for each subject.

Table 5. Wrist motion: correlated rotations of the scaphoid, triquetrum, and hamate.

EXTENSION	%	FLEXION	%
Tri-E + Sc-UD	55	Sc-F + Tri-RD	70
Tri-E + Sc-E	65	Sc-F + Tri-F	70
Tri-E + Sc-S	65	Sc-F + Tri-S	70
Tri-E + Ha-RD	55	Sc-F + Ha-UD	55
Tri-E + Ha-E	65	Sc-F + Ha-F	65
Tri-E + Ha-P	65	Ha-UD + Tri-RD	60
Tri-S + Ha-RD	55	Ha-UD + Tri-F	65
Ha-E + Sc-E	60	Ha-UD + Tri-S	55
Ha-P + Sc-E	60	Ha-F + Tri-RD	70
Ha-P + Sc-S	55	Ha-F + Tri-F	75
		Ha-F + Tri-S	60

ULNAR DEVIATION	%	RADIAL DEVIATION	%
Ha-P + Sc-UD	55	Ha-RD + Tri-RD	55
		Ha-RD + Tri-E	60

Abbreviations: Sc = scaphoid; Tri = triquetrum; Ha = hamate.

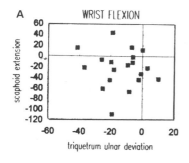

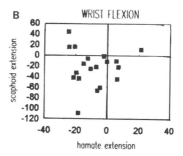

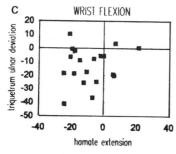

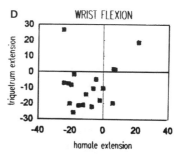

Figure 11. Graphic 2D "inter-bone" relationship in wrist flexion. **Figure 11-A.** Scaphoid extension versus hamate extension. **Figure 11-B.** Hamate extension versus triquetrum extension. **Figure 11-C.** Scaphoid extension versus triquetrum ulnar deviation. **Figure 11-D.** Hamate extension versus triquetrum ulnar deviation.

The results presented here thus do not describe sequential or instantaneous displacements of the carpal bones during wrist motion. We describe the variations of the carpal bone positions between neutral position and the chosen end position. It is nevertheless obvious that wrist motion consists of combined movements of the individual carpal bones about palmar dorsal, transverse, and longitudinal axes.

The results of this 2D/3D CT study are, in general, consistent with the observations of most authors such as Belsole et al.,[2] Berger et al.,[3] De Lange et al.,[5] Garcia-Elias et al.,[8] Kauer,[10] Linscheid,[13] Ruby et al.,[18] Sennwald,[23] and Taleisnik.[25,26]

However, the surprising mobility of the hamate, and especially the existence of its longitudinal rotations, prompted us to doubt the fixed element theories defended by Littler[14] and Kuhlmann et al..[11] It seemed that the hamate cannot be included in this element made up of the distal carpal row and the metacarpals II and III.

The lunate, on the other hand, displayed relatively poor mobility with regard to the scpahoid and the triquetrum. It is thus more likely that the lunate, the capitate, the trapezium, and the trapezoid form, together with the metacarpals II and III, a semi-rigid entity.

Our qualitative study confirmed that the pisiform is a sesamoid bone of the distal flexor carpi ulnaris tendon, as proposed by Volz,[28] and that this muscle and the extensor carpi ulnaris are important stabilizers of the internal carpus.

Quantitative results gave rise to large interindividual variations, so that the existence of one single functional carpal model is questionable in vivo. However, the principal behavior of the scaphoid, triquetrum, and hamate could be determined. This behavior cannot be considered as the only normal aspect of carpal kinematics as the percentage of "unexpected" motion patterns was very large (10 - 50%).

De Lange[6] and Savelberg,[21] among others, have described large interindividual variations in carpal bone motion and ligament behavior, which were confirmed by this in vivo study. The question that arises is how to explain these differences in carpal behavior. No correlation could be observed between these differences and sex, age, or level of physical activity. It seems likely that differences in carpal bone morphology could be responsible for different carpal motion schemes. Indeed, 3D object observations revealed important variations in the shape of the analyzed bones. The structure of the surrounding bones is also a factor that influences the mechanical behavior of a bone. Present biomechanical concepts allow us to conclude that, for instance, a morphological variation of the proximal articular surface of the hamate could interfere with triquetrum motion. Differences in the shape, strength, and insertions of wrist ligaments are also liable to induce different behaviors of the carpal bones.

The existence of compensatory mechanisms is also possible. Poor mobility of the scaphoid, for instance, could be offset by qualitatively or quantitatively different behavior of other bones. The global displacement could thus result from different combinations of individual motion of the carpal bones, which could explain that some motion components, classically considered as obvious, do not occur in every wrist.

In this respect, it is possible that some of these different motion patterns can be predisposing factors for wrist pathologies, such as DISI or VISI carpal instabilities.

These hypotheses have to be confirmed by a broader 3D CT study, by carpal bone shape analysis and by an anatomical study of wrist ligaments. Only a broader study can help to define the limits between normal and pathological carpal functional anatomy in living subjects.

A possible explanation for some divergent results with respect to previous studies may be found in different in vivo motion patterns, compared to carpal bone behavior in cadaver preparations, where freezing or other conservation methods, skin, muscle and ligament sectioning, as well as marker insertion have to be used.

Most of the studies dealing with carpal functional anatomy have been in vitro studies, involving a relatively small number of specimens. In contrast, the present study, due to the number of cases analyzed, enables us to show that the existence of different motion patterns is possible.

CONCLUSIONS

In this study an attempt was made to study the in vivo behavior of the carpal bones as a function of wrist motion. Our results do not describe sequential or instantaneous motion of the carpal bones. They are limited to the analysis of the variations of the positions of these bones between neutral position and the chosen end position.

The qualitative part of this 2D/3D CT analysis of carpal functional anatomy leads to the following results: [1] wrist motion consists of combined movements of the individual carpal bones about palmar, dorsal, transverse, and longitudinal axes; [2] the pisiform is a sesamoid bone in the distal flexor carpi ulnaris tendon; this muscle and the extensor carpi ulnaris are important stabilizers of the internal carpus.

The motion components of the scaphoid, triquetrum, and hamate in different wrist movements were quantified, and the principal behavior of these bones was described.

These quantitative results cast doubt on the existence of one universal model of carpal behavior. It seems indeed more likely that the important dispersion of carpal bone morphology and the existence of compensatory mechanisms influence the biomechanics of this region. Perhaps the analysis of the shape of a carpal bone, in relation with the morphology of the surrounding structures, will enable us to predict its functional behavior.

ACKNOWLEDGMENTS

This work was supported by the grant R&D/BFR/91/004 of the Ministère des Affaires Culturelles, Luxembourg. We thank the team of the CT department of the Institut J. Bordet, Brussels for their understanding and technical help. We express our gratitude to P. Salvia for his computational help.

REFERENCES

1. J.G. Andrews, and Y. Youm, A biomechanical investigation of wrist kinematics, *J. Biomech.* 12: 83 (1979).
2. R.J. Belsole, D.R. Hilbelink, A. Llewellyn, S. Stenzler, T.L. Greene, and M. Dale, Mathematical analysis of computed carpal models, *J. Orthop. Res.* 6: 116 (1988).
3. R.A. Berger, R.D. Crowninshield, and A.E. Flatt, The three-dimensional rotational behaviors of the carpal bones, *Clin. Orthop.* 167: 303 (1982).
4. R.P. Biondetti, M.W. Vannier, L.A. Gilula, and R.H. Knapp, Three-dimensional surface reconstruction of the carpal bones from CT scans: transaxial versus coronal technique, *Comput. Med. Imag. Graph.* 12: 67 (1988).
5. A. De Lange, J.M.G. Kauer, and R. Huiskes, Kinematic behavior of the human wrist joint: a roentgenstereo-photogrammetric analysis. J. Orthop. Res. 3: 56 (1985).
6. A. De Lange, A kinematical study of the human wrist joint, Thesis, Katholieke Universiteit te Nijmegen (1987).
7. K.H. Englmeier, A. Wieber, K.A. Milachowski, and C. Hamburger, Methods and applications of three-dimensional imaging in orthopedics, *Arch. Orthop. Trauma. Surg.* 109: 186 (1990).
8. M. Garcia-Elias, W.P. Cooney, K.N. An, R.L. Linscheid, and E.Y.S. Chao, Wrist kinematics after limited intercarpal arthrodesis, *J. Hand. Surg.* 14-A: 791(1989).
9. W.W. Gilford, R.H. Bolton, and C. Lambrinudi, The mechanism of the wrist joint with special reference to fractures of the scaphoid, Guy's Hosp. Rep. 92: 52 (1943).
10. J.M.G. Kauer, The mechanism of the carpal joint, *Clin. Orthop.* 202: 16 (1986).

11. J.N. Kuhlmann, and R. Tubiana, Mécanisme du poignet normal, *in*:"Le Poignet," J.P. Razemon and G.R. Fisk, ed., Expansion Scientifique Française, Paris: 62 (1983).

12. D.M. Lichtmann, J.R. Schneider, A.R. Swafford, and G.R. Mack, Ulnar midcarpal instability - clinical and laboratory analysis, *J. Hand Surg.* 6: 515 (1981).

13. R.L. Linscheid, Kinematic considerations of the wrist, *Clin. Orthop.* 202: 27 (1986).

14. J.W. Littler, Les principes architecturaux et fonctionnels de l'anatomie de la main, *Rev. Chir. Orthop.* 46:131 (1960).

15. M.A. MacConaill, The mechanical anatomy of the carpus and its bearings on some surgical problems, *J. Anat.* 75: 166 (1941).

16. J.K. Mayfield, R.P. Johnson, and R.F. Kilcoyne, The ligaments of the wrist and their functional significance, *Anat. Rec.* 186: 417 (1976).

17. F. Moutet, A. Chapel, P. Cinquin, and L. Rose-Pitet, Imagerie du carpe en trois dimensions, *Ann. Radiol.* 33: 128 (1990).

18. L.K. Ruby, W.P. Cooney, K.N. An, R.L. Linscheid, and E.Y.S. Chao, Relative motion of selected carpal bones: a kinematic analysis of the normal wrist, *J. Hand Surg.* 13-A: 1 (1988).

19. J. Ryu, W.P. Cooney, LJ. Askew, K.N. An, and E.Y.S. Chao, Functional ranges of motion of the wrist joint, *J. Hand Surg.* 16-A: 409 (1991).

20. H.H.C.M. Savelberg, J.G.M. Kooloos, R. Huiskes, and J.M.G. Kauer, A functional analysis of the ligaments of the human wrist joint (abstr.), *Acta Orthop. Scand.* 61:41 (1990).

21. H.H.C.M. Savelberg, Wrist joint kinematics and ligament behaviour. Thesis, Katholieke Universiteit te Nijmegen (1992).

22. G. Selvik, Roentgen stereophotogrammetry: a method for the study of the kinematics of the skeletal system, *Acta Orthop. Scand.* 60 (1989).

23. G. Sennwald, "L'Entité Radius-Carpe," Springer Verlag, Heidelberg Berlin (1987).

24. H. Seradge, P.T. Sterbank, E. Seradge, and W. Owens, Segmental motion of the proximal carpal row: their global effect on the wrist motion, *J. Hand Surg.* 15-A: 236 (1990).

25. J. Taleisnik, The ligaments of the wrist, *J. Hand Surg.* 1: 110 (1976).

26. J. Taleisnik, Post-traumatic carpal instability, *Clin. Orthop.* 149: 73 (1980).

27. T.E. Trumble, C.J. Bour, R.J. Smith, and R.R. Glisson, Kinematics of the ulnar carpus related to the volar intercalated segment instability pattern, *J. Hand Surg.* 15-A: 384 (1990).

28. R.G. Volz, M. Lieb, and J. Benjamin, Biomechanics of the wrist, *Clin. Orthop.* 149: 112 (1980).

29. P.M. Weeks, M.W. Vannier, W.G. Stevens, D. Gayou, and L.A. Gilula, Three-dimensional imaging of the wrist, *J. Hand Surg.* 10-A: 32 (1985).

30. H.J. Woltring, R. Huiskes, and A. De Lange, Finite centroid and helical axis estimation from noisy landmark measurements in the study of human joint kinematics, *J. Biomech.* 18: 379 (1985).

31. Y. Youm, and A.E. Flatt, Kinematics of the wrist, *Clin. Orthop.* 149:21 (1980).

ROLE OF THE WRIST LIGAMENTS WITH RESPECT TO CARPAL

KINEMATICS AND CARPAL MECHANISM

John M.G. Kauer,[1] Hans H.C.M. Savelberg,[1] Rik Huiskes,[2]
and Jan G.M. Kooloos[1]

[1]Department of Anatomy and Embryology
[2]Institute of Orthopaedics
Khatolieke University Nijmegen
P.O. Box 9101
6500 HB Nijmegen, The Netherlands

INTRODUCTION

Diagnosis and treatment of carpal joint disturbances require detailed insight into the function of the structures involved in the carpal mechanism. Kinematic analyses of carpal joint motions[1,3,5] have given evidence for the specificity of the carpal bone motion patterns in the various movements of the hand to the forearm. Destot[6] already offered an important contribution to the understanding of the disturbed carpal joint mechanism, stressing the functional interrelationship between carpal bone geometries and the ligamentous interconnections of these bones. In his view ligaments act as constraints as to intercarpal and radiocarpal mobility and direct the movements of the carpals during flexion and deviation of the hand. This does not alter the fact that in the recent litterature[2,4,18,19,21,22,23] the role of the ligaments with respect to the carpal mechanism has been interpreted differently. As a result we encounter different classifications of ligament trauma where the concept is followed that ligamentous interconnections can be depicted as separable entities that restrict passively mutual displacements of the carpals one on another and with respect to the bones of the forearm.

Structurally wrist joint ligaments are parts of a system that consists of interwoven fibre bundles differing in length, direction, texture and mechanical properties.[2,18,19,21,22,23,25] In order to get a better understanding of the function of this system and of the mechanism of ligament injury, morphologic and mechanical data of the wrist joint have to be viewed in relation to each other and in relation to bone geometries and specific shaped intercarpal contacts.[16]

Advances in the Biomechanics of the Hand and Wrist
Edited by F. Schuind *et al.*, Plenum Press, New York, 1994

271

RADIOCARPAL AND INTERCARPAL MOBILITY

For the analysis of the normal and the disturbed carpal mechanism several concepts are proposed.[7,8,9,10,13,14,15,17,27] In our concept the proximal carpal bones act as intercalated bones in longitudinal articulation chains that function with radiocarpal and intercarpal interdependent motions, the chains being interdependent in their motion patterns as well.[11,12,13] The interdependency of radiocarpal and intercarpal movements results in a specific carpal bone motion pattern for each movement of the hand to the forearm. Kinematic analyses of the carpal joint motions have provided evidence for this concept.[1,3,5]

The evaluation of carpal joint disturbances shows that the interdependency of positions and movements of the carpals essentially depends on the integrity of the ligamentous system.[19,20,26] Therefore, the ligamentous interconnections of the bones can be regarded as a system that restrics movements of the carpals one to another and to the radius. Besides, the ligaments play a role in the linkage of the longitudinal articulation chains and in the stabilizing function of one chain to the other.

The Articulation Chains

Gilford et al.[9] were the first to propose a functional concept for the wrist in order to explain certain conditions of disturbed carpal equilibrium. In this concept the central chain (capitate, lunate, radius) functions under the stabilizing influence of the radial chain (trapezium and trapezoid, scaphoid, radius). In this concept the ulnar chain (hamate, triquetrum, ulnar articular disc) is left out of consideration. In each chain the proximal carpal acts as an intercalated segment each of them influenced in their movements by the action of the corresponding distal carpal to the radius. In this way the central and radial chains viewed on their own act as two three-bar linkages with movements at two levels, the radiocarpal and midcarpal levels, that are totally independent. As a result the direction in which the parts of the chain will move is unpredictable where is accepted further that the proximal and distal carpals move around fixed pivot points related to both joint levels (Figure 1).

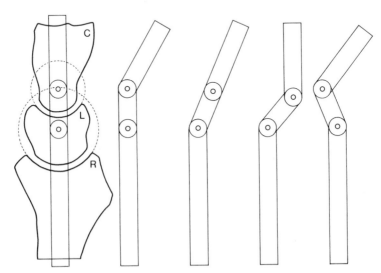

Figure 1. The central chain, radius (R), lunate (L), capitate (C), modelled as a three-bar linkage.[9] The proximal and distal carpals move around fixed pivot points related to the curvatures of the joints. Lateral view (From Kauer et al.,[15] with permission).

272

The Intercalated Bones

An important point in the mechanism of the carpal joint is the specific geometry of the intercalated bones giving them a well defined tendency to move. The lunate has a dorsopalmar wedge-shape, having a larger proximodistal dimension at the palmar aspect than at the dorsal aspect by which the lunate tends to move into a dorsal rotation.[13] This tendency is independent of any position of the capitate to the radius (Figure 2).

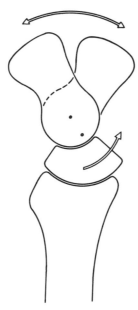

Figure 2. The lunate tends to move according to its wedge-shape. The tendency to move is independent of the position of the distal carpal. Lateral view (From Kauer,[12] with permission).

The geometry and the position of the scaphoid have to be viewed in the same way. The position of the scaphoid is different from that of the lunate being intercalated between capitate and radius as well as between trapezium and trapezoid and the radius. The proximal part of the scaphoid between capitate and radius has a wedge-shape in the same direction as the lunate and has therefore the same tendency to move, viz. into a dorsal rotation.[16] The contact of the scaphoid with the trapezium and trapezoid (STT) counteracts this tendency to move (Figure 3). The STT contact serves as a stop for the dorsal rotation of the scaphoid and with the scaphoid for the dorsal rotation of the proximal carpal row as a whole.[13] This mechanism includes that when the scaphoid is forced into a palmar rotation (palmar flexion and radial deviation of the hand) the lunate has to follow, against its tendency to move. In dorsal flexion of the hand and in ulnar deviation, the scaphoid can come again into the erect position and will come into this position by a dorsal rotation, the movement according to its wedge-shaped proximal part. This movement is facilitated by the moving tendency of the lunate into the same direction. The dorsal rotation of scaphoid and lunate will be executed only as far as is permitted by the STT contacts. The pure bone geometry-based antagonism of the trapezium-trapezoid and capitate actions at the moving tendency of the scaphoid can give an explanation for the "humpback" deformity seen in scaphoid fracture malunions.[24]

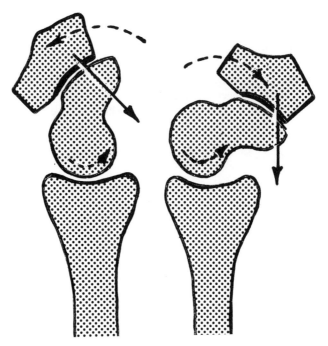

Figure 3. Position of the scaphoid between trapezium and trapezoid and radius. Lateral view.

Scapholunate Interaction

A comparison of lateral X-rays of the wrist in the endpositions of hand flexion and deviation demonstrates characteristic mutual displacements of the scaphoid and lunate (Figure 4). These displacements are the result of differences in curvature of the scaphoid and lunate articular facets at the radiocarpal joint.[12] The scaphoid rotates on the radius with a stronger curved facet than the lunate and will rotate faster than the lunate. As a result, the scaphoid shifts proximally with respect to the lunate and with respect to the capitate when the scaphoid rotates palmarly and shifts distally to lunate and capitate when rotating dorsally. It can be noted that this scapholunate shift will lead to differences in the angulations of the radial and central chains related to the dorsopalmar rotations of both bones.

The way scaphoid and lunate will shift one to another is essentially determined by their interosseous connection.[12] The scapholunate interosseous ligament has a very short dorsal part and a longer palmar part (Figure 5). The mutual mobility of the bones at the palmar aspects is greater than at the dorsal aspects. The shorter dorsal part of the interosseous ligament serves more or less as a turning point while the palmar part can bridge a changing mutual position of the palmar aspects of both bones. In the dorsiflexed hand the palmar part spans the cleft between the two bones and when the scaphoid and lunate come into their palmar rotated positions the palmar part of the interosseous ligament bridges an increasing difference of level between the palmar aspects of the scaphoid and lunate. As a result, the palmar cleft will be closed (Figure 6). This mechanism gives the scaphoid and lunate very specific positions during the movements of the hand to the forearm, as during the dorsopalmar rotation of these bones the rotation will be associated with pronation-supination along longitudinal axes.

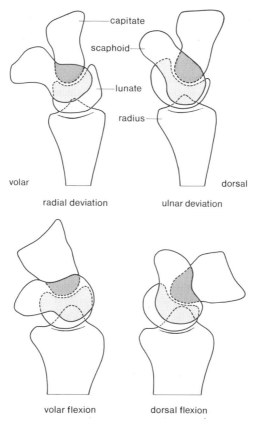

capitate

scaphoid

lunate

radius

volar   dorsal

radial deviation   ulnar deviation

volar flexion   dorsal flexion

Figure 4. Lateral projections of the radioscapholunocapitate complex in the four endpositions of the hand (From Kauer,[13] with permission).

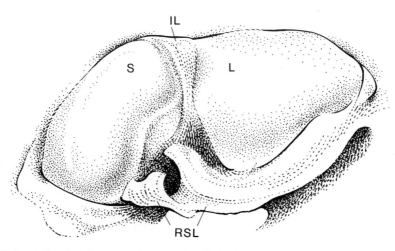

IL

S   L

RSL

Figure 5. Scapholunate interosseous ligament. Proximal view. Scaphoid (S), lunate (L), interosseous ligament (IL), connection with palmar radiocarpal ligament (RSL) (From Kauer al.,[15] with permission).

Lunotriquetrum Interaction

During the scapholunate interaction the triquetrum moves with the lunate. The interosseous interconnection between lunate and triquetrum consists of short fibres. In combination with the shape of the lunotriquetral interface which is directed obliquely from proximal and ulnar to distal and radial, the triquetrum can only move to the lunate into a proximodistal shift.[13] This shift can be observed when the hand is moved into a radioulnar deviation (Figure 7).

The scapholunate and lunotriquetral interosseous ligaments are important elements in the linkage of the intercalated proximal carpals one to another. The scapholunate ligament links the function of scaphoid and lunate in their respective chains while the lunotriquetral interosseous ligament links lunate to triquetrum. As the triquetrum can be depicted as an anchorage for most extrinsic (the longer capsular) wrist joint ligaments the lunotriquetral interosseous serves as a closure of the kinematical chain of the wrist joint.

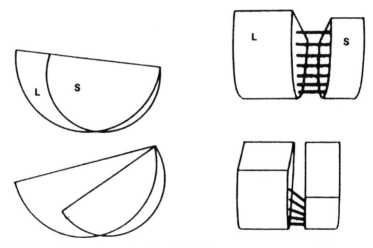

Figure 6. Model of scapholunate interaction. Scaphoid (S), lunate (L) (From Kauer,[12] with permission).

Extrinsic Ligament Function

The extrinsic, capsular, ligaments of the wrist joint are all part of a system of interwoven fibre bundles. The parts of the system differ in length, texture, direction and in mechanical properties.[21,22,23] Descriptions of these ligaments are made based on dissections of the system and on more or less artificial separations. Therefore, we encounter different classifications and different interpretations of the function of parts of the system.[4,18,19,21,22,23]

In general, ligaments of the carpal joint are regarded as the elements that counteract the mutual movements of the carpals and the movements of the carpus with respect to the radius and the ulna. In contrast with the intrinsic ligaments where a role as controls in the linkage of the movements of the proximal carpals could be demonstrated, a role that is consistent with the specific carpal bone geometry, the extrinsic ligaments can only be approached in their function by systemizing the system as bone to bone connections.[23]

distal

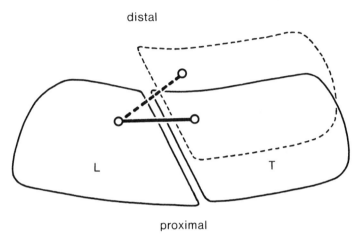

proximal

Figure 7. Model of lunotriquetrum interaction. Lunate (L), triquetrum (T) (From Kauer,[13] with permission).

The radiocarpal and ulnocarpal connections restrict the axial rotation of the hand to the forearm, so keeping the proximal carpals into contact with the articular facets on the distal radius. Cutting one of these connections results immediately in loss of stability at the radiocarpal joint. The function of the triquetrum as a keystone in the carpal complex has to be emphasized, with respect to this phenomenon, as a number of these ligaments are inserted into this bone. However, it has to be aknowledged at the same time that the stability of the wrist joint depends on the integrity of the ligamentous system as a whole. This implies that experimental procedures in the investigation of carpal ligament function in which selected ligamentous interconnections are cut sequentially, will not offer the data for the evaluation of ligament trauma.[19,21,22] When ligament function is inserted into an overall functional concept for the carpal joint it has to be recognized that the extrinsic ligaments consist of poorly stretchable material and that concepts in which ligaments function under the condition of a substantial elongation do not match their mechanical properties. In vitro experimentation on carpal ligament behaviour has shown that there are differences in stiffness between parts of the ligamentous system (Figure 8).

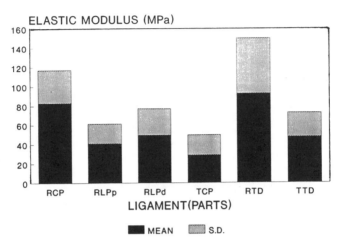

Figure 8. Elastic moduli of some (parts of) carpal ligaments. RCP, palmar radiocapitate; RLPp and RLPd, proximal and distal part of the palmar radiolunate; TCP, palmar triquetrocapitate; RTD, dorsal radiotriquetral; TTD, dorsal triquetrotrapezial. RCP and RTD are significantly stiffer than the other parts.

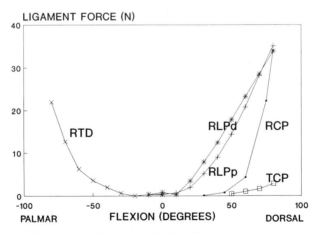

Figure 9. Ligament forces in the dorsopalmar flexion of the hand. Abbreviations as in Figure 8.

When data concerning ligament length changes and force-elongation relationships are combined the generated forces in the ligaments can be estimated. The calculated forces in some parts of the system,[22] give the suggestion that at least under normal conditions, the loading of the ligamentous system is relatively low. Only at the endpositions of flexion and of deviation of the hand some parts of the ligament system become strained (Figure 9). In how far these experiments match the in vitro situation has to be further investigated. Based on the available data the function of the carpal ligamentous system can be depicted as a very delicate support of the motion patterns of the carpal bones. Ligamentous interconnections are essential linkages with respect to the interdependency of the carpal articulation chains, the interdependency being based first of all on mutually attuned specific carpal bone geometries.

REFERENCES

1. R.A. Berger, Analysis of Carpal Bone Kinematics, Thesis, University of Iowa, Iowa (1980).
2. P. Bonjean, J.L. Houton, R. Linarte, and J. Vignes, Anatomical bases for the dynamic exploration of the wrist joint, *Anatomia Clinica* 3:73 (1981).
3. A. de Lange, A kinematical Study of the Human Wrist Joint, Thesis, University of Nijmegen, The Netherlands (1987).
4. A. de Lange, R. Huiskes, and J.M.G. Kauer, Wrist joint ligament length changes in flexion and deviation of the hand: An experimental study, *J. Orthop. Res.*, 8:722 (1990).
5. A. de Lange, J.M.G. Kauer, and R. Huiskes, Kinematic behavior of the human wrist joint: A roentgen stereophotogrammetric analysis, *J. Orthop. Res.* 3:56 (1985).
6. E. Destot, Anatomie et physiologie du poignet, in: "Traumatismes du poignet et rayons X", Destot E, (ed.) Masson, Paris (1923).
7. O. Fischer, Ueber Gelenke von zwei Graden der Freiheit, *Arch. Anat. Entwickelungsgesch*, Suppl. Bd.:242 (1897).
8. G. Forssel, Ueber die Bewegungen im Handgelenke des Menschen, *Scand. Arch. Physiol.* 12:168 (1902).
9. W.W. Gilford, R.H. Bolton, and C. Lambrinudi, The mechanism of the wrist joint with special reference to fractures of the scaphoid, Guy's Hosp. Rep. 92:52 (1943).
10. G.B. Günter, Das Handgelenk in mechanischer, anatomischer und chirurgischer Beziehung, Meissner, Hamburg (1850).
11. J.M.G. Kauer, The interdependence of carpal articulation chains, *Acta Anat.* 88:481 (1974).
12. J.M.G. Kauer, Functional anatomy of the wrist, *Clin. Orthop.* 149:9 (1980).
13. J.M.G. Kauer, The mechanism of the carpal joint, *Clin. Orthop.* 202:16 (1986).
14. J.M.G. Kauer, The longitudinal carpal chain, a model of carpal function, *Ann. R. Coll. Surg. Engl.* 70:166 (1988).
15. J.M.G. Kauer, and A. de Lange, The carpal joint, in: "Management of Wrist Problems", Taleisnik J. (ed.), *Hand Clin.* 3(1):23 (1987).
16. J.M.G. Kauer, A. de Lange, H.H.C.M. Savelberg, and J.G.M. Kooloos, The wrist joint; functional analysis and experimental approach, in: "Wrist Disorders; current concepts and challenges", R. Nakamura, R.L. Linscheid and T. Miura, Springer, Tokyo (1992).
17. R.L. Linscheid, J.H. Dobyns, J.W. Beabout, and R.S. Bryan, Traumatic instability of the wrist, *J. Bone Joint Surg.* 54:1612 (1972).
18. S.E. Logan, M.D. Nowak, Ph.L. Gould, and P.M. Weeks, Biomechanical behavior of the scapholunate ligament, *Biomed. Sci. Instr.* 22:81 (1986).
19. J.K. Mayfield, Mechanism of carpal injuries, *Clin. Orthop.* 149:45 (1980).
20. J.K. Mayfield, R.P. Johnson, and R.G. Kilcoyne, The ligaments of the human wrist and their functional significance, *Anat. Rec.* 186:417 (1976).
21. M.D. Nowak, and S.E. Logan, Distinguishing biomechanical properties of intrinsic and extrinsic human wrist ligaments, *J. Biomech. Eng.* 113:85 (1991).
22. H.H.C.M. Savelberg, J.G.M. Kooloos, A. de Lange, J.M.G. Kauer, and R. Huiskes, Human carpal ligament recruitment and three-dimensional carpal motion, *J. Orthop. Res.* 9: 693 (1991)
23. H.H.C.M. Savelberg, J.G.M. Kooloos, and J.M.G. Kauer, Kinematics of the human carpal bone-ligament complex, *Ann. Soc. R. Zool. Belg.* 119:65 (1989).

24. D.K. Smith, K.N. An, W.P. Cooney, R.L. Linscheid, and E.Y. Chao, Effects of a scaphoid waist osteotomy on carpal kinematics, *J. Orthop. Res.* 7(4):590 (1989).
25. J. Taleisnik, The ligaments of the wrist, *J. Hand Surg.* 1:110 (1976).
26. J. Taleisnik, Current concepts of the anatomy of the wrist, in: "Kaplan's Functional and Surgical Anatomy of the Hand", Spinner M. (ed.) Lippincott, Philidelphia (1984).
27. H. Virchow, Ueber Einzelmechanismen am Handgelenk, *Anat. Anz.* 21:369 (1902).

KINEMATIC GEOMETRY OF THE WRIST: PRELIMINARY REPORT

C. L. Nicodemus,[1] Steven F. Viegas,[1] and Karin Elder[2]

[1]Orthopaedic Surgery Department
[2]Office of Academic Computing
University of Texas Medical Branch
Galveston, Texas 77555, USA

BACKGROUND

This research significantly advances the state of the art in our ability to visualize the dynamic kinematics of the wrist. Beginning 150 years ago, investigations into the kinematics of the wrist were anatomical descriptions[5,6,8,10,20,21] with observations made by the eye and including the earliest use of the x-ray after its discovery in 1895. It was not until much later that McConail[12] began to apply basic mechanical laws to the curiosities and movement of the wrist and its complex of eight carpal bones. Motion studies began with cineradiography[1,26] and moved parallel with technological development of optical and electronic methods of measuring minute distances accurately while the joint is moving. Use of LED path generation on photographic plates was used in conjunction with both cineradiographic and cinematographic (movie film)[26] methods in an attempt to improve and automate the laborious and error prone requirement of hand digitizing data. With the introduction of the spark gap, or sonic digitizer, 3-D positional data could be taken automatically from moving joint components.[3,4,27] Experimental error associated with the sonic digitizing equipment, its dependency on acoustic related environmental conditions (temperature, humidity, air movement) and the physical size of the spark gap apparatus itself still left much to be desired. Because of these concerns, recent investigators have attempted to return to basic approaches (manual analysis of biplanar radiographs) to increase precision even at the expense of fewer data points and increase in digitization error.[7,9,14] CT scan imaging as a tool for evaluating moving objects or joint mechanics is not yet available owing to its requirement for a stationary object. However, its use in imaging the individual bone 3-D shapes and computing volumes, centroids and principal axes (of inertia) are definitely at hand.[2,16]

In this research, the CT scan imaging process developed at UTMB[15,16] replaces the biplanar radiographic method of localizing embedded marker with relation to anatomic land marks. Indeed, with the precision obtained in reproducing physical dimensions and volumes in the UTMB CT imaging process, the logical next step is to reproduce actual bone movement using the imaged objects. To our knowledge this has not been done before our pilot study effort.

MATERIALS AND METHOD

A fresh frozen cadaver arm was x-rayed to confirm absence of diseases or abnormalities that would affect normal range of motion or kinematic function of the carpal bones. The dorsal surface of the wrist was prepared for imbedding the triad target (TT) pins. The TT pins are plastic tapered pins approximately 5.0 cm in length and tapered up from 1 mm at the point to 3mm at the top. The top of the pin was modified to include three noncolinear 3.0 mm diameter spheres fixed to the pin head approximately 2 cm apart. The spheres were coated with 3M photoreflective paint to enhance their reflectivity to the video cameras. Initially, two of the TT pins were placed into holes drilled through the skin into the distal radius and proximal third of the 3rd metacarpal, such that the pins extended outward in a dorsal direction. They were offset along a longitudinal axis from radius to metacarpal so that they would not come into contact during full extension of the wrist. Prior to placing the pins, the narrow end was cut to a length such that the pin diameter slid freely into the 2.0-2.5 mm drilled hole. The tapered pin was then forced tightly into the hole such that the bone actually deformed the plastic. The pin design was selected to be rigid and not to flex or vibrate independently of the bone into which it was placed. Care was taken not to place the pin through tendons or other tissue that are involved in the wrist movement so as not to inhibit or influence the movement. A clearance area was cut in the skin around the pins to account for and permit skin movement. The implantation of these pins provided the gross wrist motion for which relative carpal bone motion was measured and reproduced.

In a similar fashion, TT pins were placed into the carpal bones in the order of most difficulty. These pins were placed using a Siemens C-arm fluoroscope for assistance in positioning. Because of its shape and limited dorsal surface exposure, the lunate was pinned first, followed by the scaphoid and capitate. Care was taken in each placement to eliminate interference between pins throughout the wrist's full range of motion. The hamate, trapezoid, trapezium and triquetrum were not pinned. Final pin placement is shown on Figure 1. Once the TT pins were placed, the forearm was stabilized in a pronated position and the hand was moved through its natural range of motion.

Four black and white NEC TI-23 video cameras (with 8.0 mm lenses) were arranged around and above the wrist so as to keep the track of the metacarpal and other TT pins in the same video frame space as the radius pins. Each camera had a co-axial mounted light to illuminate the pins' retroreflective surfaces. The video processor (Motion Analysis Corp. VP110) edge detection threshold was adjusted to ensure recording of the disk of reflected light for each TT pin's three spheres. All VCR's were triggered simultaneously and the motion recorded at 60 frames per second.

The computer (Sun SPARC-IPC workstation) software (Expert Vision TM Motion Analysis Corp.) tracked each of the 3mm diameter TT pin spheres (a total of 15) during the motion of the wrist. Each set of three pin points on the metacarpal and other carpal TT pins were converted into a transformation matrix relating the relative translating and rotating coordinate systems of each bone as the hand is moved. The 3x3 homogeneous Dual Screw Transformation Matrix (DSTM)[22,23] was calculated for each carpal bone, for each frame and stored on disk in an ASCII file for export into the Imaging Lab computer systems.

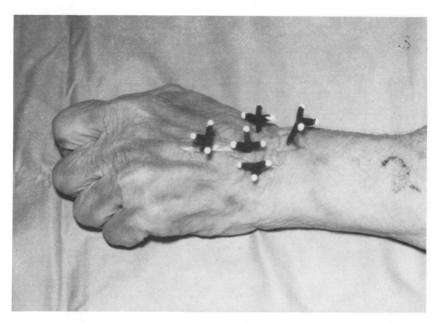

Figure 1. Cadaver arm with the triad pins surgically embedded in the radius, lunate, scaphoid, capitate and 3rd metacarpal.

Motion analysis of the wrist is accomplished by optoeletronic stereo cinephotogrammetry (OSCP). That is, multiple high speed (1/60 sec) non-interlaced video recording tracks the movement of small (3 mm) diameter triad target pins placed into the bone and projecting out 2.5-3.0 cm beyong the skin surface on the dorsal aspect of the hand and wrist. The system records the path of all pin heads during the ranging of the hand and automatically digitizes the 3-D positional data, saving the results in a computer file. At this time sampling speed, continuous or instantaneous kinematics can be used rather than the displacement or static position techniques. Thus, the process emulates the mathematical basis for the calculus wherein functions are continuous within the limit as changes are made very small. The continuous expressions describe the path, whereas the displacement expressions can only describe points along the path. The data are then analyzed to calculate positional vectors and matrix transformations necessary to compute the Instantaneous Screw Axes (ISA) of the joint at each frame (0.017 sec). The wrist is then CT scanned with the identical pins in place, and the triad pin coordinate transformations computed with reference to the CT scanned bone. The CT information is then combined with the motion analysis transformations and ISA relationships to recreate carpal bone movement. Further analysis and 3-D graphical imaging of the ISA data produces an ISA vector generated ruled surface which fully characterizes the kinematic geometry of the carpus.

This dynamic visualization and measurement greatly improves our understanding compared to process investigations done with planar or less precise methods. The wrist was CT scanned with the TT pins intact in the host bones. The tape output from the scan was then analyzed according to the process developed at UTMB.[16] Since the CT scan slice thickness is 1.5mm with a 1mm index space, the 3mm spheres were well imaged. The total length of the scan was approximately 15cm, centered near the proximal pole of the capitate. This provided adequate length of radius and third metacarpal to include their TT pins so as to establish the global wrist motion. A 3-D reconstruction of the wrist and pin image data was produced, and position of the TT pin coordinate system calculated with respect to the bone coordinate system. Figure 2 shows the reconstructed bone and pins set. This information was used with the motion data to reproduce the motion of the bones, compute the ISA, and generate its surface.

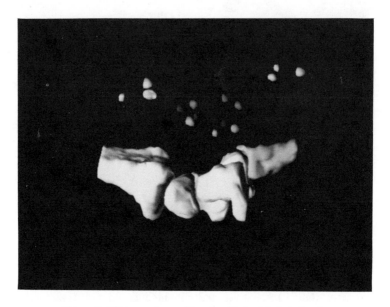

Figure 2. CT imaged carpal bones with the triad pin heads.

RESULTS

Figures 3 and 4 show sequenced (and expanded) positions for the lunate and capitate respectively. The motion of each of the pinned carpal bones (lunate, capitate and scaphoid) was visualized with 900 frames for a flexion extension movement of the wrist. By utilizing the motion analysis transformation matricies (the DSTM described earlier) it is possible to efficiently compute the ISA for each position by a matrix multiplication process. More importantly, use of the Dual Number notation yields a 3x3 orthogonal DSTM which means that inverse relationships are also possible through the same set of software.[13,22,23,24,25] That is, all kinematic geometry derived from this investigation can rapidly be referred to any bone or point within the wrist complex that was pinned and scanned. In turn, this approach allows for determination of direct intra and intercarpal motion analysis with freedom to select the reference bone.

DISCUSSION

Figures 3A & 3B represent the tip of the iceberg of available data. This progress report was prepared to inform as to the process utilized in this research and illustrate its earliest results. Much additional work is required in order to document precision, evolve a method of presenting the vast amount of relational kinematic data contained in these dynamic images, and relate them clinically. The process described in this paper also lays the foundation for analysis of the ISA surface characteristics of these carpal bone pairs in order to synthesize the kinematic motion for more realistic modeling.

The technique applied here for the first time opens the door for continued studies that will lead to the determination of relative motion (rolling and sliding) between bones. These data, combined with the load path investigations conducted at UTMB[17,18,19] will lead to considerable additional insight into aging, arthritic evolution and other disease states that appear to result from abnormal wear patterns on bone joint surfaces caused by abnormal load biomechanics and presumably abnormal kinematics. In addition, this approach will provide a means of quantifying for the first time the relative motion including displacement, velocities and accelerations of normal clinically described occurrances such as "clunks", "snaps" and "clicks". These occur clinically in abnormal wrists on a regular basis and here-to-fore have not been well defined. X-rays are static, two dimensional representations, inadequate for this sort of investigation. Fluoroscopy provides a dynamic representation but also only 2-D. CT and MRI imaging provide 3-D information, but are unfortunately static and do not offer insight into these dynamic events.

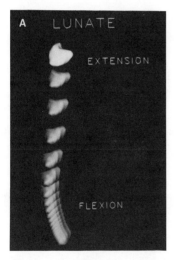 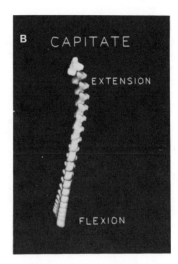

Figure 3. Figure 3A. Sequenced positions of the lunate as derived by motion analysis data. **Figure 3B.** Sequenced positions of the capitate as derived by motion analysis data. (Note: position distances are expanded for clarity).

This project is the first practical and clinically relevant application of the combined technologies of three-dimensional reconstruction of CT images and motion analysis. The successful marriage of these two technologies allows a more accurate and previously never seen animated example of the actual three-dimensional dynamic kinematics of the carpal bones of the wrist. This approach can be applied to any and all joints in the body and have a tremendous potential to impact both the basic research and clinical understanding of joint motion, both in normal and abnormal states.

The specific goals of this research project will offer a better understanding of the normal dynamic carpal kinematics of the wrist and the abnormal kinematics resulting from different types of ligament disruptions. Furthermore, the previously described snaps, clicks, and clunks of the wrist associated with carpal instabilities which have only been able to be described from what is believed to cause these clinically evident findings, or have been only in part described statically or two-dimensionally by radiographs or fluoroscopic evaluation, can now be viewed and three-dimensionally described and measured with this technology. This not only will afford a better understanding of the injury, but will also facilitate the understanding and the specific pathology that is being assessed in a clinical examination. Furthermore, this knowledge and this technology will afford the opportunity to assess the efficacy of various types of surgical treatments. This research offers an exciting and unique way to study this and many other problems in Orthopaedics.

ACKNOWLEDGEMENTS

The authors wish to acknowledge the support provided for this research effort by the John Sealy Memorial Endowment Fund. The motion analysis system was supported in part by the Moody Foundation.

REFERENCES

1. R. Arkless, Cineradiography in normal and abnormal wrists, *Am. J. Roentgenology* 96:837 (1966).
2. R.J. Belsole, D.R. Hilbelink, A. Llewellyn, M. Dale, and J.A. Ogden, Carpal orientation from computed reference axes, *J. Hand Surg.* 16A:1,82 (1991).
3. R.A. Berger, R.D. Crowninshield, and A.E. Flatt, The three-dimensional rotational behaviors of the carpal bones, *Clin. Orthop. Rel. Res.* 167:303 (1982).
4. R.B. Brumbaugh, R.D. Crowninshield, W.F. Blair, and J.G. Andrews, An in-vivo study of normal wrist kinematics, *J. Biomech. Eng.* 104:176 (1982).
5. T.N. Bryce, On certain points in the anatomy and mechanism of the wrist joint reviewed in the light of a series of roentgen ray photographs of the living hand, *J. Anat. Phys.* 31:59 (1896).
6. E.F. Cyriax, On the rotary movements of the wrist, *J. Anatomy* 60:199 (1926).
7. A. de Lange, J.M.G. Kauer, and R. Huiskes, Kinematic behavior of the human wrist joint: a roentgen-stereophotogrammetric analysis, *J. Orthop. Res.* 3:56 (1985).
8. R. Fick, Anatomic and meckanic de gelenke, *Jena* 3:357 (1911).
9. M. Garcia-Elias, W.P. Cooney, K.N. An, R.L. Linscheid, and E.Y.S. Chao, Wrist kinematics after limited intercarpal arthrodesis, *J. Hand Surg.* 14A:791 (1989).
10. J. Goodsir, Anatomical memoirs of John Goodsir, Edinburgh (1858).
11. G.L. Kinzel, A.S. Hall Jr., and B.. Hillberry, Measurement of the total motion between two body segments--I, analytical development, *J. Biomech.* 5:93 (1972).
12. M.A. MacConail, The mechanical anatomy of the carpus and its bearing on some surgical problems, *J. Anatomy* 75:166 (1941).
13. G.R. Pennock, and K.J. Clark, An anatomy based coordinate system for the description of the kinematic displacements in the Human Knee, *J. Biomech.* 23,1:1209 (1990).
14. L.K. Ruby, W.P. Cooney, K.N. An, R.L. Linscheid, and E.Y.S. Chao, Relative motion of selected carpal bones: a kinematic analysis of the normal wrist, *J. Hand Surg.* 13A:1 (1988).
15. H.D. Tagare, D.M. Stoner, K.W. Elder, C.L. Nicodemus, S.F. Viegas, and G.R. Hillman, Location and geometric description of carpal bones in CT images, in press (1992).
16. S.F. Viegas, G. Hillman, K. Elder, D. Stoner, and R. Patterson, Measurement of carpal bone geometry by computer analysis of 3-D CT images, *J. Hand Surg.*, in press (1992).
17. S.F. Viegas, R. Patterson, P. Peterson, J. Roefs, A.F. Tencer, and S. Choi, The effects of various load paths and different loads on the transfer characteristics of the wrist, *J. Hand Surg.* 14A:458 (1989).
18. S.F. Viegas, A.F. Tencer, J. Cantrell, M. Chang, P. Clegg, C. Hicks, C. O'Meara, and J.B. Williamson, Load transfer characteristics of the wrist: Part I the normal joint, *J. Hand Surg.* 12A:971 (1987).
19. S.F. Viegas, A.F. Tencer, J. Cantrell, M. Chang, P. Clegg, C. Hicks, C. O'Meara, and J.B. Williamson, Load transfer characteristics of the wrist: Part II perilunate instability, *J. Hand Surg.* 12A:978 (1987).
20. G. Von Bonin, A note of the kinematics of the wrist joint, *J. Anatomy* 63:259 (1929).

21. R.D. Wright, A detailed story of movement of the wrist joint, *J. Anatomy* 70:137 (1935).

22. A.T. Yang, Displacement analysis of spatial five-link mechanisms using (3X3) matrices with dual-number elements, *J. Engineering for Industry*, Trans. ASMS, 9,1: 152 (1969).

23. A.T. Yang, Inertia force analysis of spatial mechanisms, *J. Engineering for Industry*, Trans. ASME, 93,1:27 (1971).

24. A.T. Yang, and B. Roth, Higher-order path curvature in spherical kinematics, *J. Engineering for Industry*, Trans. ASME, 95,2:612 (1973).

25. A.T. Yang, and F. Freudenstein, Application of dual-number quaternion algebra to the analysis of spatial mechanisms, *J. Applied Mechanics*, Trans. ASME, 86,2:300 (1964).

26. Y. Youm, R.Y. McMurtry, A.E. Flatt, and T.E. Gillespe, Kinematics of the wrist-I, an experimental study of radial-ulnar deviation and flexion-extension, *J. Bone Joint Surg.* 60:423 (1978).

27. Y. Youm, and T.S. Yoon, Analytical development in investigation of wrist kinematics, *J. Biomech.* 12:613 (1979).

A COMPUTERGRAPHICS MODEL OF THE WRIST JOINT:

DESIGN AND APPLICATION

Jan G.M. Kooloos, Marc A.T.M Vorstenbosch, and Hans H.C.M Savelberg

Department of Anatomy/Embryology
University of Nijmegen
PO box 9101
6500 HB Nijmegen, The Netherlands

INTRODUCTION

In order to understand the morphology of the wrist joint the functions of both bone movements and ligament behaviour have been studied extensively. These studies have resulted into: [1] possible explanations of the joint kinematics based on a selection of parts of the joint and based on simple geometrical analogons of the bones,[12,13,14] [2] concepts of ligament behaviour that simplify ligament dimensions to single line-elements or to straight lines[22] and [3] explanations of ligament bone interaction that are based on assumptions about joint motion that exclude out-of-plane motions.[3,17] So, functional anatomical approaches have not led to an overall concept of wrist joint behaviour.

At the best we are able to conclude that the geometries of the carpal bones are responsible for the characteristic articulations in the wrist joint, that the ligaments assist to keep the bones in their courses and serve as check reins at the limits of the range of motion of the joint and that fibre-bundles within a carpal ligament behave differently.[6,20] In order to be able to predict therapeutical interventions (partial arthrodeses; ligament sectioning), a concept of the wrist joint is necessary and such a concept should include force transmission through the carpus. Already a 2D rigid spring model is available,[7,9] but 3D models are lacking.

In order to contribute to the shortcomings in our understanding of the wrist joint sophisticated techniques have been developed: accurate wrist joint kinematics can be assessed in cadaver experiments. Röntgenstereophotogrammetry is the main tool in this field. These experiments result into a description of the mutual carpal bone displacements in mathematical terms, viz. helical axes, Euler angles, translations etc.[5,9,18,19,20]

These experiments have indeed enlarged the knowledge of the joint but there is a growing need to interpret these mathematical kinematics in anatomical terms. To

Advances in the Biomechanics of the Hand and Wrist
Edited by F. Schuind *et al.*, Plenum Press, New York, 1994

that purpose anatomical models of different joint positions are built using non-invasive image techniques (CT/MRI). These 3D images of the joint should be matched with accurate kinematic parameters of the joint in order to study wrist joint problems. This is still work in progress.[24]

The cadaver studies and the imaging studies just mentioned could benefit from each other but there are two major problems that prevent the connection of these lines of research. First, stereophotogrammetry needs the introduction of artificial landmarkers (pellets) in the joint and this restricts the technique to in vitro experiments. Secondly, the anatomical models image only a selection of the living tissues in the joint: either bone or cartilage or ligaments are sharply pictured. We have established the connection between anatomy and kinematics. It is accomplished by removal of the pellets after the cadaver experiment and by reconstruction of the pellet positions in a 3D model of the wrist anatomy. This connection leads to a computergraphics model (CGM) of the carpus, a term used by Garg and Walker[8] for a similar model of the knee joint.

The nature of the CGM is to derive both kinematical and anatomical data of the same specimen; the linkage between kinematical and anatomical data is to determine the positions of the same set of artificial markers twice, once in the kinematical experiment and once in relation to the anatomical data. The first experiment delivers the quantitative rotations and translations of the markers for subsequent positions of the hand, the second experiment results in the positions of the markers and the geometrical data in spatial relation to each other. Then, the kinematical data are used to reposition the geometrical data. In this way the in vitro experiment is mimicked in a computer simulation.

The simulations of the carpal bone positions firstly enable the visualisation of the phenomena that have taken place during the cadaver experiment: How have the bones moved in respect to each other and how do the complex rotational and translational movements look like. Secondly, calculations can be made of the kinematical phenomena in the wrist that are not or difficult to measure in an experiment: how do the insertions sites of the larger extrinsic ligaments move; What is the behaviour of the short interosseus ligaments as derived from the shifts of their respective insertion sites to the bones; what is the extent of the contact areas between the bones and how do these shift during hand movements. Any question about the relative movements between parts in the joint can be posed to the model as long as the anatomy of these parts is included in the three-dimensional reconstruction. So far we have included: outer and inner contours of the cartilage, insertion sites of the ligaments and also the contours of the bones. Only histological material ensures the detection of all these tissues at the same time. This justifies our method of 3D reconstructions based on microsections of the cadaver hands. Other methods of 3D reconstructions such as CT based ones exclude, for instance, cartilage and ligament insertions sites.[1]

MATERIALS AND METHODS

A. Design

The method to come to a CGM exists of four subsequent steps: [1] the assessment of kinematical data from a cadaver experiment, [2] assessment of the anatomy: sectioning and three-dimensional reconstruction of the joint subsequent to the cadaver experiment, [3] the reconstruction of the pellet positions which is the linkage between the kinematical and the anatomical data and [4] the establishment of the simulation of the different wrist positions. Additionally, calculations of relative change in position of the elements that are taken up in the anatomical model can be made.

Kinematics. The most accurate way to assess the kinematics is to apply the Selvik method of röntgenstereophotogrammetric analysis (RSA[21]). This method has been made applicable to the wrist joint by de Lange[5] and comprises, in short, the following.

The radius of a cadaver arm is attached firmly into a fixed position; flexor and extensor tendons are loaded by means of constant force springs (20N; six in total), and the hand can be put in any position by moving a metal pin that is positioned into the third metacarpal. In every position of the hand the wrist bones are free to take in their characteristic positions. An arbitrary neutral position of the hand is taken as the starting-point for the movements. Then, the hand is either deviated or flexed or moved into a combination of flexion/deviation. Usually motion steps of 10 degrees are imposed to the hand. The method includes the insertion of artificial markers into the bones during the in vitro cadaver experiment. The markers in each bone form a polygon that serves as a rigid body for the further calculations. Results of the kinematics are based upon the rigid body mathematics and include Euler angles and helical axes.[10]

The only adaptation to the de Lange method we have made is that we place the markers in superficially drilled holes in the cortical bone in stead of shooting these into the trabecular bone. These markers are removed after the experiment. The holes are cone shaped and the diameter at the surface of the bone equals the diameter of the used marker (round pellets of 1.0 or 0.8 mm in diameter). These markers are fixed to the bones with a histoacrylate. The locations where the markers are placed are carefully cleared from fibrous matter. All bones can be reached without damage to the capsule, except the lunate bone. Palmarly, parts of the capsule are prepared to find the cortical bone. Care is taken not to violate the ligaments.

Anatomy. After the experiment markers are removed from the joint without damaging the holes. Following fixation, embedding, sectioning and a staining procedure, a series of histological microsections of 20 μm is yielded.[15] One out of five sections is used to reconstruct the drilled pellet holes and one out of twenty to reconstruct the other joint elements. Distances between subsequent sections are 0.1 and 0.4 mm, respectively. The selected sections are digitized by means of an image analyser (1024 x 1024 pixels; one pixel equals about 70 μm) and home-made software is available to distract the needed anatomical information from the images. This includes semi-automatic contour detection based on only a limited number of control points. The following joint elements are selected: bone contour, outer contour of cartilage, inner contour of cartilage, insertion sites into the bone, marker holes and external reference points. These reference points are used to reconstruct the joint.

Pellet reconstruction. The reconstructed holes are imported into a CAD program (AutoCAD, AutoDesk, Ltd). Virtual pellets on the same scale are generated, carefully moved into postion in the holes, and the centre point is registered. This is baptized the "pit-procedure". By this time, there is an original marker set (RSA-marker set) and a reconstructed marker set (CGM-marker set) for every marked carpal bone.

The polygons formed by the two marker sets should be rigid in respect to each other, i.e. the distances between the markers in the RSA-marker set and the respective markers in the CGM-marker set should be equal. These distances are calculated and when the mean error of the distances of these two respective marker sets is smaller than 0.2 mm, it is assessed that the polygons are rigid.[4] In order to find the best match between the RSA-marker set and the CGM-marker set, several observers perform the pit-procedure and this results into several possible CGM-marker sets per bone. The mean errors of these CGM-marker sets and the RSA-marker set of the respective bone in the neutral, starting-point position are calculated. The coordinates belonging to that CGM-marker set that shows the smallest error are taken as the actual coordinates of the respective bone.

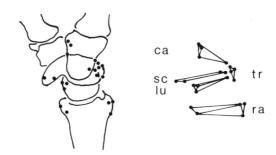

Figure 1. Contours and the markers of the radius (ra), lunate (lu), scaphoid (sc), triquetrum (tr) and capitate (ca), and further contours of the trapezium and metacarpals 1 and 3, drawn from a lateral X-ray from the cadaver experiment. On the right the markers are drawn again and connected to form the RSA-marker sets of the respective bones.

Simulations. Reconstructed anatomical data are also imported in the CAD program, and positions of pellets and bones are matched. The full surface reconstruction of the bones includes a triangularisation between adjacent contours. Home-made software accounts for bifurcations that occur, for instance, when the top of the lunate bone, consisting of a dorsal and a palmar horn, is sliced. See for an example figure 2 for the match between the best CGM-marker set, formed by the four introduced pellets, and the rebuilded surface anatomy of the lunate bone.

In order to estimate the sensitivity of the simulations for small remaining mismatches between RSA-marker set and CGM-marker set, test-simulations of the lunate bone are carried out. This implies that the CGM-marker set of the lunate is transposed in space in such a way that the rigid body tests equals the allowed limit of a mean error of 0.2 mm. Then, simulations of the lunate bone using these transposed CGM marker sets are performed. The differences in positions of the RSA-marker set and the test CGM-marker set during maximally radial and ulnar deviated handpositions are calculated.

Before simulations can be carried out, the positions of the bones during the time of sectioning have to be transposed to the starting-point of the RSA experiment. From this starting-point the translations and rotations as calculated from the RSA-experiment and that are needed to mimick the wrist positions from this RSA-experiment can be introduced to the anatomical data.

B. Application

As an example, the data of a wrist (female; 87 years; no radiographic abnormalities) are given. Since the distal bones move almost completely as one

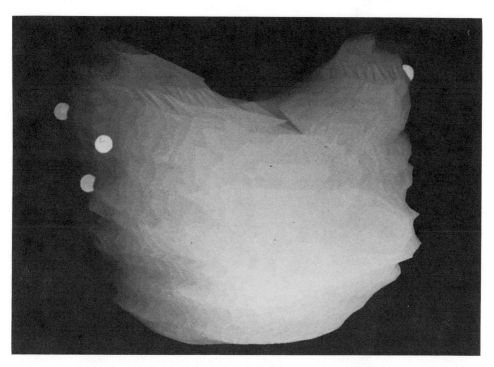

Figure 2. A photograph of the screen showing the reconstructed surface anatomy of the lunate in conjunction with the reconstructed CGM-marker set.

block,[20] only the capitate bone is taken along; other bones included are the radius, lunate, scaphoid and triquetrum. The surfaces of the insertion sites are divided in triangles and a line can be drawn from any triangle on one site to any triangle on the opposite site. In this way a pattern of fibre-bundles is constructed. The pattern is designed to resemble actual fibre-bundle directions as much as possible. Simulation of the various wrist positions delivers a range of length changes of the selected fibre-bundles. These length changes are the result of dissociations or approximations of the insertion sites and given as percentages change of the length in the neutral hand position. Positive changes in this length are connected to a dissociation of the insertion sites; a negative change to an approximation of the sites. Deviation data only are calculated and further processed in the CGM. The division of the insertion sites is discussed elsewhere;[15] they have been given figures which are also used here. A scheme is given in table 1.

Table 1. Scheme of ligaments and insertion sites as used in the simulations

origo bone	origo site	ligament (acronym)	insertion site	inserting bone
palmar				
radius	A1	radio-triquetrum palmare (RTP)	D	triquetrum
radius	A2	radio-lunatum (RL)	C	lunate
radius	A3	radio-scaphoideo-capitatum (RSC)	G	capitate
radius	A3	radio-scaphoid (RS)	H	scaphoid
scaphoid	H	scapoideo-capitatum (SC)	G	capitate
lunate	E	lunato-triquetrum (LT)	D	triquetrum
triquetrum	D	triquetro-capitatum (TC)	G	capitate
dorsal				
radius	I	radio-triquetrum dorsale (RTD)	K	triquetrum

RESULTS

Kinematics. Passive motion in the investigated joint is large. The movement space for deviation runs from 70 degrees ulnarly to 30 degrees radially (Figure 3). The ulnar deviation of the capitate bone does not follow the hand completely; about 55 degrees ulnar deviation is reached maximally. The capitate shows very little flexion and pronation during ulnar deviation. The lunate bone shows a smaller ulnar deviation than the capitate and some dorsal flexion and pronation during ulnar deviation. The triquetrum and the lunate bone follow each other shortly. Ulnar deviation of the scaphoid is relative large. Radial deviation is coupled to palmar flexion and supination for all bones except the capitate which shows a small dorsal flexion instead.

Reconstructions, linkage and simulation. The three-dimensional reconstructions establish the needed anatomical data and also the pellet holes. The resulting anatomical data of bone contours and insertion sites have been pictured elsewhere.[15,16] Actual simulation implies rotations about the x, y and z axis using the body-fixed axes and three translations along these axes. This finally gives new positions of the reconstructed bones. This has been pictured in figure 4 for the bones in the proximal row and for two positions, maximal ulnar deviation and maximal radial deviation.

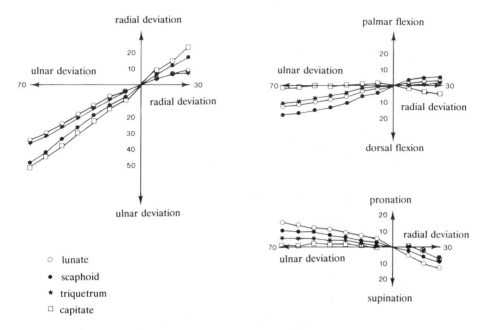

Figure 3. Movements of the marked bones during the ulnar and radial deviations. The overall rotation of the respective bones is split into rotations about three anatomical axes of the joint, viz. radioulnar, palmodorsal and proximodistal, the latter representing pronation and supination.

During the pit-procedure the markers are replaced as accurate as possible by positioning their midplanes equal to the opening of the hole. The influence of mismatches is tested as described in the materials and methods section. These tests show that the differences in rotations between the RSA-experiment and the CGM-simulations are maximally 0.6 degrees and 0.05 mm, respectively. It is concluded that the reconstruction of the marker positions by means of 3D-reconstruction propagates only minor errors in the simulated positions of the bones.

The study of ligament behaviour. The CGM model is designed to open the ability to study the movements of the elements that are not or only partly measurable in experimental set-ups. A future application will be the study of contact areas. This paper deals further with ligament behaviour by means of studying the position changes of the insertion sites.

All insertion sites could be found in this joint except the distal insertion of the lig. triquetro-trapezium dorsale. The palmar insertion sites are shown in figure 5. The line-elements are nominated by the ligaments they represent (Figure 5). Examples of a simulation of the RTP is given in figure 6. This ligament is fixed to

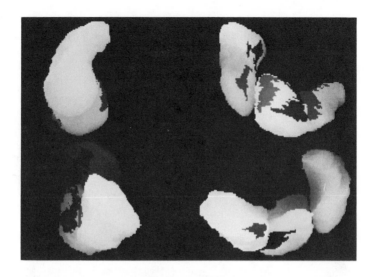

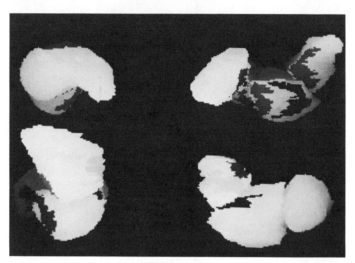

Figure 4. Photographs of the screen of three bones (lunate, triquetrum, scaphoid) in four views (upper left: radial view; upper right: palmar view; lower left: ulnar view and lower right: dorsal view) and in two positions of the hand (TOP: maximal ulnar deviation; BOTTOM: maximal radial deviation). The spots on the bone surfaces represent the insertion sites of various ligaments. Note that the top of the photographs is distal.

the radius and runs to the moving triquetrum and is simulated in six steps, from 70 degrees ulnarly to 30 degrees radially.

The simulated behaviour of all selected ligaments are given in figure 7. The simulation is executed with line-elements. In the following description these line-elements are named by the ligament they represent. The lengths of the proximal and distal line-element in the neutral, starting-point is given in table 2.

1) The sites of the RTP approach each other. The proximal parts of this ligament shortens during ulnar deviation for about 20%; the distal parts show only minor changes. During radial deviation, the distal parts show about 5% shortening.

2) The sites of the RL dissociate during ulnar deviation, showing an increase in length of the distal fibers of about 10%. There is hardly any change during radial deviation.

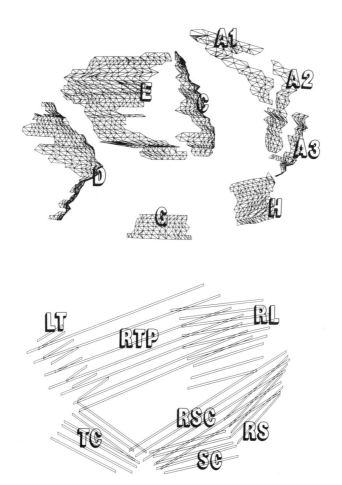

Figure 5. Reconstructed insertion sites on the palmar side of the joint (TOP) and the configured patterns of fibre-bundles that represent the different joint ligaments (BOTTOM). The figures and acronyms are explained in Table 1. Note that the lower figure can be placed on top of the upper one.

3) The insertion sites of the RSC dissociate and the distal line-elements elongate up to 15% during ulnar deviation. The line-elements shorten for 5% during radial deviation. The proximal line-elements show little change during the complete deviation movements.

4) The sites of both the proximal and distal parts of the RS dissociate largely and an increase in length to 25-30% is calculated.

5) The sites of the SC retain their orientation during the simulations. As a result the line-elements do not change much in length.

6) The proximal parts of the LT shorten largely up to 20% during ulnar deviation, and during radial deviation a large elongation is noted. The distal line-elements show an increase in length up to 10% during radial deviation.

7) The sites of the TC approach somewhat only during radial deviation.

8) The sites of the RTD approach largely during ulnar deviation and a shortening of the line-elements of 15-25% is calculated. Only during maximal ulnar deviation a substantial difference in proximal en distal fibres is found.

Table 2. The lengths of the line-elements, representing the selected ligaments, in the neutral position of the hand in millimeters.

ligament	proximal line-element	distal line-element
RTP	19.82	22.92
RL	11.48	13.60
RSC	22.93	19.27
RS	14.75	15.29
SC	8.11	9.64
LT	3.12	3.69
TC	9.39	10.46
RTD	14.42	14.94

Testing another line-element configuration. The line-elements are carefully developed to resemble the actual fibre-bundle distribution as much as possible. But variations to the actual courses are introduced. To estimate this, one ligament (RTD), is taken and is configured two times: once the line-elements resemble the actual course of the fibre-bundles as much as possible (Figure 5) and once the line-elements run to the more distal parts of the triquetrum. The line-elements in the latter configuration show an angle of about 20 degrees with the former configuration. It is shown that dissociations of the insertion sites of the RTD enlarge greatly due to the change in fibre-bundle configuration and increases in length up to 40 % can be calculated (Figure 8). It is concluded that a change in the orientation of the elements has a large effect upon the calculations. Interpretation of the translations is difficult, since these figures depent on the position of the origin of the body-fixed axes in the bones. This origin is placed at the position of an arbitrarily selected (the first coded) marker. Nevertheless the data are gathered and needed in the final simulations.

DISCUSSION

The establishment between kinematics and anatomy using the CGM can mimick a cadaver in vitro experiment. The CGM enables the study of those phenomena that are hard to measure in cadaver experiments. The method shows only minor errors between actual and simulated movements.

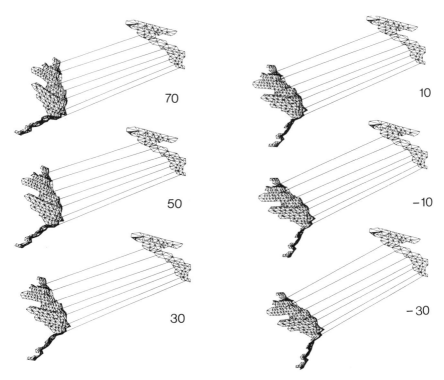

Figure 6. An example of a simulation of the RTD in six steps from 70 degrees ulnar deviated to 30 degrees radial deviated (= -30). Note the fixed insertion site on the radius on the upper right and the moving site on the triquetrum on the lower left, and also the resulting changing pattern of line-elements.

The calculated kinematics of this particular wrist are in agreement with the quantitative measurements of other authors.[2,5,19,20,23] Some of the ligament behaviour can be explained by connecting the kinematical data to the simulated length changes. For instance, the deviation movements of the scaphoid and capitate are about equal (Figure 2). This is related to the fact that the SC-part of the RSC does not do much, while the RS-part shows a large dissociation. Also, the triquetrum and the lunate show the same amount of both deviation and flexion. A small absolute elongation (about 1 mm) of the LT is the consequence, resulting nevertheless into a relatively large elongation.

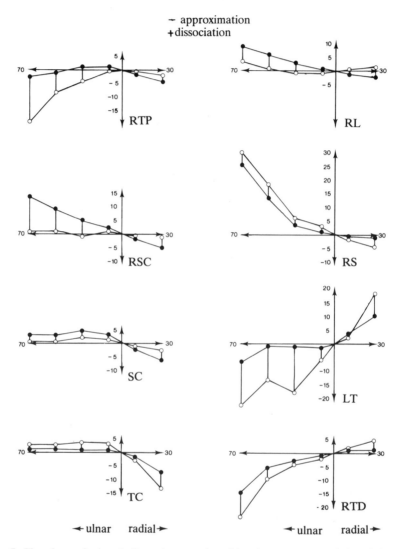

Figure 7. The change in length from the neutral position in percentages of the eigth selected ligaments during simulations of the line-elements from 70 degrees ulnar to 30 degrees radial. Every ligament consists of several line-elements. The data of the most proximal line-element (open circles) and the most distal line-element (closed circles) are given only; the data of the other line-elements are always in between these two extremes.

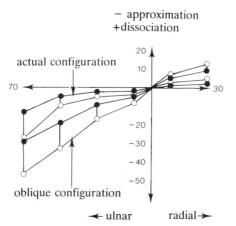

Figure 8. Results of a change in line-element configuration from the configuration resembling the actual situation of the fibre-bundles to a configuration that shows an oblique pattern of line-elements. Axes, open and closed circles are defined in figure 7.

Ligament behaviour is usually derived from simple representations of the ligaments or from simple models of wrist joint movement.[3,17,22] The length changes have experimentally been estimated by de Lange et al.,[6] and by Savelberg et al.[20] Some of the presented simulated elongations are in agreement with their experimental results and some are not. Particularly the results of the RSC and RTD are the same as in the previous cadaver experiments. The pattern of the TC is in agreement with the results of Savelberg et al.,[20] but not with those of de Lange et al.,[6] who calculated a shortening of the TC during ulnar deviation. The results of LT, the large approximation during ulnar deviation and large dissociation during radial deviation, are not in agreement with the experimental results. Both Savelberg et al.[20] and de Lange et al.[6] do not find much elongations or shortening in these ligaments. The RS part and SC part of the RSC have not been measured experimentally before. The RTP does not show its expected behaviour. Instead of dissociation during ulnar deviation, an approach is calculated.

Several sites show approaches during either radial or ulnar deviation and it is assumed that the ligaments shorten during these approximations. Ligaments function only when they are streched during dissociations of insertion sites. This means that the ligaments concerned are not fully shortened during the neutral position of the hand and will shorten more going from this neutral position to one of the extreme positions of the hand. In the cadaver experiments mentioned above these phenomena have been registered also.

The comparison of the experimental and simulated results reveals a short-coming of the CGM method: line-elements remain line-elements during the complete simulations and can not be simulated to become curved. This means that the behaviour of those bones that start pushing or pulling to the side of a ligament can not be mimicked by the CGM. It is discussed by Savelberg et al.[20] that such behaviour of the carpals may add to the length changes of a ligament. The lunate bone is certainly capable of pushing to the larger ligaments, running from the radius to the triquetrum on either dorsal or palmar side of the joint. These phenomena that cannot be accounted for in the CGM have to be taken along in discussions about ligament behaviour.

301

The CGM method is also sensitive for small changes in the configuration of the line-elements (Figure 8). Therefore, a more quantitative connection between line-element configuration and the real fibre-bundle directions in the actual wrist is needed.

The rationale to use this CGM model is that detailed histological information is at hand to make the anatomical model. So, besides bone contour also ligament insertions, joint cartilage and ligaments might be included in the anatomical model and can be studied in kinematic simulations. Therefore this model is a useful tool for further studies about wrist joint behaviour.

REFERENCES

1. R.J. Belsole, D.R. Hilbelink, J.A. Llewellyn, S. Stenzler, T.L. Greene, and M. Dale, Mathematical analysis of computed carpal models, *J. Orthop. Res.* 6:116 (1988).
2. R.A. Berger, R.D. Crowninshield, and A.E. Flatt, The three-dimensional rotational behaviours of the carpal bones, *Clin. Orthop. Rel. Res.* 167:303 (1982).
3. P. Bonjean, J.L. Honton, R. Linarte, and J. Vignes, Anatomical basis for the dynamic exploration of the wrist joint, *Anat. Clin.* 3:73 (1981).
4. A. de Lange, A kinematical study of the human wrist joint, Thesis, University of Nijmegen, The Netherlands (1987).
5. A. de Lange, J.M.G. Kauer, and R. Huiskes, Kinematic behaviour of the human wrist joint. A roentgenstereophotogrammetric analysis, *J. Orthop Res.* 3:56 (1985).
6. A. de Lange, R. Huiskes, and J.M.G. Kauer, Wrist-joint ligament length changes in flexion and deviation of the hand: An experimental study, *J. Orthop Res.* 8:722 (1990).
7. M. Garcia-Elias, K.N. An, W.P. Cooney, R.L. Linscheid, and E.Y.S. Chao, Transverse stability of the carpus: An analytical study, *J. Orthop. Res.* 7:783 (1989).
8. A. Garg, and P.S. Walker, Prediction of total knee motion using a three-dimensional computer-graphics model, *J. Biomech.* 23:45 (1990).
9. E. Horii, M. Garcia-Elias, K.-N. An, A.T. Bishop, W.P. Cooney, R.L. Linscheid, and E.Y.S. Chao, Effect on force transmission across the carpus in procedures used to treat Kienböck's disease, *J. Hand Surg.* 15A:393 (1990).
10. R. Huiskes, A. de Lange, and J.M.G. Kauer, Helical axes of wrist joint motion, in preparation.
11. H.A.C. Jacob, C. Kunz, and G. Sennwald, Zur Biomechanik des Carpus. Funktionelle Anatomie und Bewegungsanalyse der Karpalknochen, *Orthopäde* 21:81 (1992).
12. J.M.G. Kauer, Functional anatomy of the wrist, *Clin. Orthop.* 149:9 (1980).
13. J.M.G. Kauer, and A. de Lange, The carpal joint. Anatomy and function, *Hand Clin.* 3:23 (1987).
14. J.M.G. Kauer, A. de Lange, H.H.C.M. Savelberg, and J.G.M. Kooloos, The wrist joint. Functional analysis and experimental approach, in: The Wrist Joint. Current concepts and challenges. R. Nakamura, R.L. Linscheid, T. Miura (eds.), Springer Verlag, Tokyo: 3 (1992).
15. J.G.M. Kooloos, The insertion sites of selected human carpal ligaments. A quantitative morphological study using three-dimensional computer reconstructions, *Eur. J. Morphol.* 29:77 (1991).
16. J.G.M. Kooloos, and H.H.C.M. Savelberg, Measurements of cross-sectional areas in selected wrist-joint ligaments, *Acta Anat.* 144:235 (1992).
17. J.K. Mayfield, R.P. Johnson, and R.F. Kilcoyne, The ligaments of the wrist and their functional significance, *Anat. Rec.* 186:417 (1976).
18. C.L. Nicodemus, S.F. Viegas, and K. Elder, Kinematic geometry of the wrist, in: "Advances in the Biomechanics of the Hand and Wrist", F. Schuind, K.N. An, W.P. Cooney, M. Garcia-Elias, eds., Plenum, London (1993).
19. L.K. Ruby, W.P. Cooney, K.N. An, R.L. Linscheid, and E.Y.S Chao, Relative motion of selected carpal bones: A kinematic analysis of the normal wrist, *J. Hand Surg.* 13A:1 (1988).
20. H.H.C.M. Savelberg, J.G.M. Kooloos, A. de Lange, R. Huiskes, and J.M.G. Kauer, Human carpal ligament recruitment and three-dimensional carpal motion, *J.Orthop.Res.* 9:693 (1991).

21. G. Selvik, A Roentgenstereophotogrammetric method for the study of the kinematics of the skeletal system, Thesis, University of Lund, Sweden. Reprinted in *Acta Orthop. Scand.* 1989, 90 suppl. 232:1 (1974).

22. J. Taleisnik, The Wrist, Churchill Livingstone, New York (1985).

23. R.G. Volz, M. Lieb, and J. Benjamin, Biomechanics of the wrist, *Clin. Orthop. Rel. Res.* 149:112 (1980).

24. S. Wolfe, L. Katz, and J. Crisco, The use of cine-computed tomography to investigate carpal kinematics, in: "Advances in the Biomechanics of the Hand and Wrist", F. Schuind, K.N. An, W.P. Cooney, M. Garcia-Elias, eds., Plenum, London (1993).

THE STUDY OF THE BIOMECHANICS OF WRIST MOVEMENTS

IN AN OBLIQUE PLANE - A PRELIMINARY REPORT

Philippe Saffar and I. Semaan

Institut Français de la main
15 rue Benjamin-Franklin
F-75116 Paris, France

Almost all the wrist motion studies have concentrated on the mobility of the wrist in flexion-extension (F.E.) and radial-ulnar deviation (R.U.D.).

From a phylogenetical viewpoint, the different species present morphological differences of the wrist anatomy which can be correlated with various locomotive patterns.

Species without prehensile activity use their mouth and this roughly corresponds with quadrupedalism. The development of the prehensile activity of the anterior limb coexists with a flattening of the snout and with alternate bipedalism. But the anterior limb is still specialized in locomotion. Bipedalism, with complete unloading from body support and locomotor constraints has allowed the unique development of the dexterity of the human hand.

SPECIALIZATION OF THE WRIST

In the lower primates

The ulna articulates with the triquetrum and is a weight-bearing structure.

The pisiform is elongate, rod-like and perpendicular to the palm. It is articulated with the ulna and the triquetrum: the powerful flexor carpi ulnaris muscle (F.C.U.) is inserted at its tip and gives a strong impulse in running. It also opposes hyperextension of the wrist (Figure 1).

There is a syndesmosis of the distal radio-ulnar joint (D.R.U.J.) and no pronation-supination. The hamate articulates most often with the lunate in the midcarpal joint. The IVth and Vth carpometacarpal joints (C.M.C.) are stiff.

The thumb is absent or exists as a small digit except when the fourth digit is predominant.

Advances in the Biomechanics of the Hand and Wrist
Edited by F. Schuind *et al.*, Plenum Press, New York, 1994

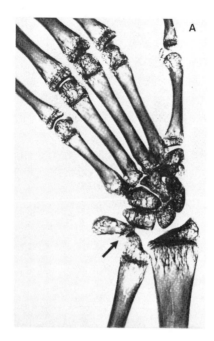

Figure 1. Figure 1-A. Pisiform articulating with triquetrum and ulna. **Figure 1-B**. F.C.U. insertion on the tip of the pisiform (reproduced with permission from Jouffroy[3]).

The consequence is the use of the wrist in F.E. and R.U.D. for locomotion and brachiation (suspension from the trees). Part of the motion and some rotation is located in the midcarpal joint.

In the primates and australopithecus

The ulnar styloid recedes to become an extra-articular structure while pronation-supination and the triangular fibrocartilage complex (T.F.C.C.) develops (as in the gorilla). The pisiform migrates distally but is still rod-like and the disappearance of the ulno-triquetral joint allows pronation-supination and ulnar deviation combined with flexion.[3,4] There is a palmigrade or digitigrade locomotion or knuckle-walking. The thumb is present, and rotation used at times.

Grasping and manipulating objects, sometimes with great skill, is done with a cylindrical grip, without using opposition. The ulnar nerve is predominant.

The elimination of the hand from the locomotor system and its specialization as a prehensile organ coexists with the use of the thumb opposite to the fingers (median nerve).

In homo sapiens sapiens, the pisiform has a pea-shape and is included in the F.C.U. (but it is not a sesamoid). There is usually no articular facet between the lunate and the hamate. On the radial side, the pinch, a precise grip, opposes the thumb to the fingertips. The thumb at rest is in anteposition and rotation relative to the plane of the palm. *These changes allow the wrist to be used in an oblique plane going from extension-radial deviation (E.R.D.) to flexion- ulnar deviation (F.U.D.).*

Little attention has been paid to this oblique plane of motion although Napier,[5] Capener[1] and Fisk[2] have mentioned it as the usual plane of utilization of the wrist (examples: using a hammer, throwing objects, but also writing, pushing or holding a heavy object with one hand).

ANATOMY

Bone and joints

Radiocarpal joint. The radial styloid is volar to the radial articular facet for the scaphoid. The carpus impinges on the radial styloid in radial deviation but not in the movement of E.R.D. (Figure 2).

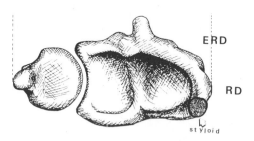

Figure 2. Distal articular surface of the radius.

There is an articular facet of the scaphoid of triangular shape extending dorsally and articulating with the radius in E.R.D. In this position, the proximal pole of the scaphoid is interposed between the radius and capitate, and it is wedge-shaped (like the lunate).

Midcarpal joint. "The force-receiving acetabulum is formed by the distal articular surfaces of the lunate and the proximal two-thirds of the scaphoid".[8] "Inspection of the proximal pole of the capitate suggests that the capitate-scapho-lunate articulation resembles a ball and socket joint capable of rotational motion"[7] (Figure 3).

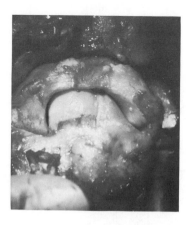

Figure 3. The midcarpal joint.

Fourth and fifth C.M.C. joints. The contra-opposition takes place in these joints.

Muscles

The four groups of wrist movers are not located on the radial and ulnar sides of the wrist, but [1] the extensor carpi radialis brevis and longus are in a dorso-radial location, [2] the flexor carpi ulnaris (F.C.U.) is in a volar ulnar location, [3] the extensor carpi ulnaris (E.C.U.) is in a dorsal-ulnar location, [4] the flexor carpi radialis (F.C.R.) is in a volar radial location.

"No single muscle acts strictly in either radio-ulnar deviation or flexion-extension alone and possesses secondary components"[9] (Figure 4). Contraction of the muscle groups 1 and 2 gives motion to the wrist in the oblique plane.

E.M.G. studies. The closing motion of the hand gives predominantly a contraction of the E.C.R.B. As finger flexion increases, the E.C.U. is recruited, followed by the E.C.R.L.[7] The summation of all muscle forces crossing the carpus is one which tends to place the wrist in a position of flexion and ulnar deviation.

Ligaments

In E.R.D. There is a continuous band of distal carpal ligaments which are taut: the ulnar collateral ligament, the capito-triquetral and the capito-trapezial ligaments; the latter extends across the F.C.R. sheath to insert in the trapezium. This band is attached by oblique fibers to the volar rim of the radius (Figure 5). The radiocapite and the radiolunate ligaments are loose (but a little more tense than in true R.D.). The scaphocapitate ligament is taut.

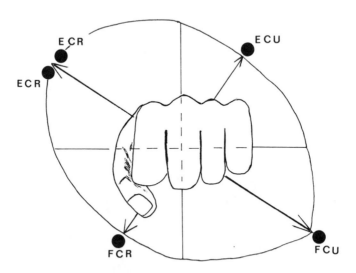

Figure 4. Muscles: the four groups of wrist movers.

In F.U.D. Almost all the volar ligaments (the radio-capite, the radio-lunate and the radiotriquetral ligaments) are loose. The scaphocapitate ligament is loose, allowing 30° of rotational motion of the scaphoid relative to the axis of the capitate.

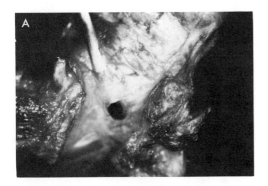
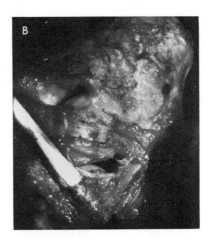

Figure 5. The continuous ligamentous band taut in E.R.D. **Figure 5-A.** Triquetro-capitate. **Figure 5-B.** capito-trapezial.

KINEMATICS

In the plane perpendicular to the axis of the forearm, a line is traced going from the tendon of the F.C.U. to the space between the E.C.R.B. and E.C.R.L. The oblique plane is defined by the plane passing through this line and parallel to the axis of the forearm. From direct visualization and cineradiography, one can see that a greater range of motion of the wrist is achieved in the oblique plane than in the R.U.D. (Figure 4).

When the wrist moves from E.R.D. to F.U.D., the first carpal row slides in a radial direction along the radial slope but stays in flexion; there is no "rotational shift" of the carpus as described by Weber[8] (Figure 6). A significant part of the movement takes place in the midcarpal joint and the movement in the midcarpal joint is oblique. There is a rotational movement of the scaphoid around the capitate that permits the sliding movement of the first row to be in conjunction with the oblique movement of the midcarpal joint. The results of radio-scapho-lunate arthrodesis showed that the residual plane of motion was restricted to an oblique plane extending from dorso-radial to palmar-ulnar.[6] A certain amount of circumduction is probably present.

The movement in the oblique plane is the result of the addition of the translocation of the first row from radial to ulnar in the radio-carpal joint and of the oblique movement in the midcarpal joint (Figure 7).

"Normally, dorsiflexion of the wrist is associated with radial deviation, and volar flexion with ulnar deviation, so that the axis of movement passes obliquely from the dorsum of the lower end of the radius through the capitate to the pisiform".[2] In this plane, the scaphoid has a more significant excursion in extension sliding dorsal to the radial styloid.

DISCUSSION

For heavy grasping, the wrist is in extension and ulnar deviation and the carpal bones are in the "close-packed position", but for movement, it is the oblique plane that is used: [1] there is more mobility and more agility in this oblique plane, [2] this oblique plane utilizes the midcarpal joint to a greater degree, and this can explain the tolerance of some radio-carpal osteoarthritis, [3] this plane is *the plane of opposition*;

when the wrist moves in R.U.D., it is difficult to define the thumb motion because of the position of the thumb relative to the plane of the metacarpals; during motion in the oblique plane, the thumb has a natural consecutive opposition with each finger and its movement is easier to analyze and understand: "the fingers are reviewed by the thumb" during the tenodesis movement of the wrist from F.U.D. to E.R.D. (Figure 8), and [4] there is a definite correlation between thumb opposition, contra-opposition of the IVth and Vth C.M.C. joints and the oblique plane of utilization of the wrist.

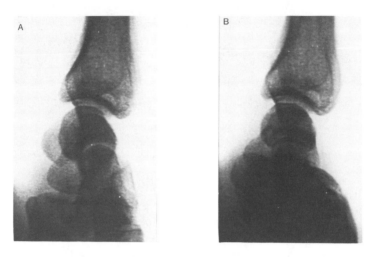

Figure 6A,B. Cineradiography (lateral) from E.R.D. to F.C.U.: there is no "rotational shift".

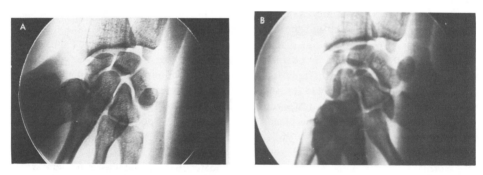

Figure 7A,B. Cineradiography (A.P.) from E.R.D. to F.C.U.: translocation of the first row from radial to ulnar + oblique movement in the midcarpal joint.

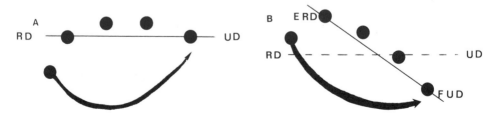

Figure 8. Thumb opposition relative, **Figure 8-A.** to the plane of R.U.D. and **Figure 8-B.** to the oblique plane of motion.

More attention should be paid to this plane of motion because the center of rotation of the carpal bones and of the wrist, the different angles between these bones and their relative motions could be more easily understood if studied in this plane. This preliminary report will be completed by a study with digitizers, electrogoniometers and an "elite" system.

REFERENCES

1. Capener cited by Fisk.[2]
2. G. Fisk, Biomechanics of the wrist joint in: "The Hand" R. Tubiana, W.B. Saunders, Philadelphia (1981).
3. F. Jouffroy, The heelless hand of the human biped, in "Origines de la Bipédie chez les Hominidés" Editions du C.N.R.S., Paris, (1991).
4. J.O. Lewis, Derived morphology of the wrist articulation and theories of hominoid evolution, *J. Anat.* 140:447 (1985).
5. J.R. Napier, The prehensile movement of the human hand, *J. Bone Joint Surg.* 38B:902 (1956).
6. M. Sturzenegger and U. Buchler, Radio-scapho-lunate partial wrist arthrodesis following comminuted fractures of the distal radius, *Ann. Hand Surg.* 10: 207 (1991).
7. R.G. Volz, M. Lieb and J. Benjamin, Biomechanics of the wrist, Clin. Orthop. 149:112 (1980).
8. E.R. Weber, Concepts governing the rotational shift of the intercalated segment of the carpus, *Orthop. Clin. N. Am.* 15:193 (1984).
9. Y. Youm, and A.E. Flatt, Kinematics of the wrist, *Clin. Orthop.* 149:21 (1980).

THE ENVELOPE OF ACTIVE WRIST CIRCUMDUCTION

AN IN VIVO ELECTROGONIOMETRIC STUDY

Patrick Salvia, Paul Klein, José Henri David, and Marcel Rooze

Laboratory for Functional Anatomy
Université Libre de Bruxelles
Route de Lennik 808 C.P. 619, B -1070 Brussels (Belgium)

INTRODUCTION

The knowledge of the functional anatomy of the active wrist is essential for clinical evaluation of upper limb disorders. Range of motion (RoM) is a daily tool in orthopaedic field.[1,29] To allow an evaluation of a functional loss, a total wrist analysis seems necessary.

In this study, the wrist was considered having two degrees of freedom (df), one in flexion-extension (FEM) and the second in radioulnar deviation (RUD). The wrist circumduction seems to permit a more global dynamic analysis of this basic wrist motion.

These two df are fetched from the finite helical axis taken between the first and the actual position. Projecting the rotation vector on appropriate axes, the wrist motion was simulated.

Several methods of wrist kinematics data acquisition are reported in the literature, including sonic digitizing system,[2,8] Roentgen-stereophotogrammetry analysis,[11-13,30,38] uniaxial electrogoniometer,[9] biaxial electrogoniometer,[27] triaxial electrogoniometer,[24] six degree-of-freedom instrumented spatial linkage,[33] biaxial flexible electrogoniometer.[23]

A six degree-of-freedom electrogoniometer has been used to evaluate the in-vivo wrist circumduction quantitatively in terms of common clinical motion (FEM and RUD) and qualitatively in terms of helical axes surface (axode) representation. The device does not require an accurate alignment with the joint centre of rotation.

Kinematic Background

Generally, joint motion has been described using Euler rotations and the translation vector or by helical axes (instantaneous or finite).[5-6,13,19,25,31,36-37]

It is known that the helical (or screw) axis position is undefined in the pure translation case and not precisely determined if the rotation is small.[35] In this approach, joint motion is seen at each instant as a translation along and a rotation about a unique line in space. This axis is completely determined by its unit direction vector **n** and the position **s** of one of its

Advances in the Biomechanics of the Hand and Wrist
Edited by F. Schuind *et al.*, Plenum Press, New York, 1994

points.[3] It is well-known that the helical axis is independent of the position of the chosen coordinate systems. For pure translations, the helical axis is not defined.

To make more communicable joint kinematic results, the reduction of the kinematic parameters in terms of an anatomical or clinical system of joint measurement is a challenge in biomechanics.[16] Moreover, in that case, a precise definition of the coordinate systems (body-fixed and space-fixed systems) is required. Projecting the helical rotation angle onto appropriate axes of the clinical reference frame, a measure of the rotation components in each anatomic plane was performed. A rigourously well defined protocol was followed to ensure an as accurate as possible definition of the coordinate systems.

Figure 1. The coordinate systems. In the reference position (left), the fixed (XYZ) and moving (xyz) axes coincide. After a pure flexion (right), the xyz frame turned with the hand around the Y axis. A radioulnar deviation (RUD) would take place around the x axis. So defined, the RUD commutes with flexion.

MATERIAL AND METHODS

Determination of the Reference Position

Wrist angles were measured with respect to a reference-position defined by orienting the forearm and the hand in the same plane. Before to start our experiments, we have validated a bony structures palpation method. Five anatomic landmarks were taken - four in the frontal plane, the radial styloid, the ulnar styloid, the head of the third metacarpal and the mi-distance between the two styloids - and two in the sagittal plane, the radial styloid and the mi-distance between the dorsal and palmar surfaces of the second metacarpal head. Two lines were drawn on the dorsal face of the hand : one along the third metacarpal and an other between the styloids. These lines and previous markers are covered with metal wires. To obtain the sagittal reference position, two markers were covered with metal to ensure that a proper alignment of the bony structures has been obtained, two X-ray (antero-posterior and lateral) of the wrist in pronation were taken on one subject (Figure 2). It confirms that the bony structures palpation method allows a good alignment of the bones and the reliability of the reference position.

Description of the Coordinate System

In this wrist study, the space-fixed system was linked to the forearm and the body fixed system to the hand. At the reference position, the two positive x axes point dorsally, the two positive y axes point medially, the two positive z axes point proximally. The two origins coincided when the forearm was in pronation (Figure 1).

Subjects

Twenty seven healthy subjects (14 women and 13 men) were tested. They ranged in age from 22 to 40 years (mean 29 years).

Tested movements

We have tested pure motion of flexion-extension, pure motion of radioulnar deviation and the combination of these two motions in circumductions. In order to avoid prosupination of the forearm, the experimental method and device allowed a control of this prosupination. Speed of motion was not imposed.

A

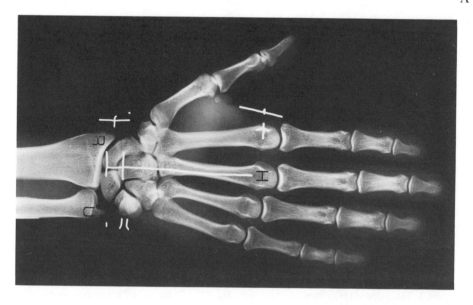

B

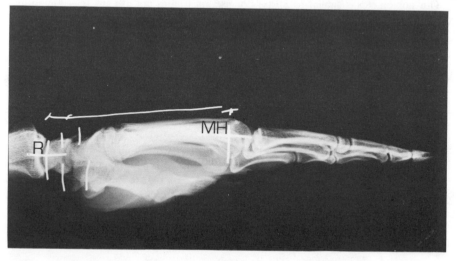

Figure 2. A, Antero-posterior X-ray shows the straight line along the third metacarpal between the head of the third metacarpal H and the mi-distance between the two styloids R and U. The axis of P6 (the hand bound angular potentiometer, Figure 3) has been made parallel to this line. Additionally, but not in use in this study, we determined the lunocapitate interface. **B**, Lateral X-ray shows the two metal cross markers pointing the radial styloid R and the mi-distance between the dorsal and palmar surfaces of the second metacarpal head MH. We aligned these two markers when the start file was recorded. The P6 axis was also made parallel to this virtual line.

315

The six *df* Goniometer

The electogoniometer used is an open chain with seven rigid links and six rotatory potentiometers (FCP22A SAKAE, 10 kΩ, 1% maximum independent linearity, endless). All links are in aluminium except the two translation links that are in stainless steel. The precision potentiometers give the analogical signals. An analog to digital converter (DAP Microstar) was used. No amplification was required. The potentiometers were powered by a 5 volt potential. Each potentiometer was calibrated on a proper angular device. A multiple regression was used to give the best fit.

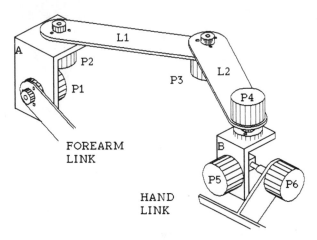

Figure 3. The 6 degrees of freedom electrogoniometer. P1 to P6 are the 6 rotatory potentiometers, L1, L2 the two translation links.

Our goniometer is similar to Shiavi's one,[32] but it does not use any telescoping link. It gives the same results that a Shiavi's goniometer with a translation link from P2 to P4. The angles ϕ_1 and ϕ_2 measured by P2 and P4 in a Shiavi's goniometer are given in function of the angles α_1, at and α_2 measured by P2, P3 and P4 in our goniometer by the formulas:

$$\phi_1 = \alpha_1 + \pi/2 - \alpha\tau/2 ;$$
$$\phi_2 = \alpha_2 + \pi/2 - \alpha\tau/2 .$$

The translation length between P2 and P4 is $2L\sin(\alpha t/2)$ where L is the length of L1 and L2.

The P2, P3, P4 axes are parallel, the axes P1 and P5 are perpendicular to them. P6 is perpendicular to P5. The P1, P2 axes like the P4, P5, P6 axes are meeting at the same point. (See Figure 3). These two points are the cardan crosses. The six potentiometers measure the angles of the basic coordinate transformations. The potentiometers P2, P3 and P4 measure the rotation in the plane orthogonal to them and the translation. The product of the link-to-link matrices gives the overall transformation used in the kinematic calculations.

Fixation Device

To ensure the feasibility of this wrist functional study, we have made a fixation device. It consisted of a hand gauntlet and a hemi bracelet in thermo-plastic material. These two fixations were moulded on a subject's right hand and forearm. On the bracelet, an adjustable fixation component was mounted to permit proper alignment of the goniometer first axis (P1). An other adjustable component allowed to make collinear the axis of P6 and the third metacarpal. A high density PVC foam ensured the adaptation between the bony rigid bodies of the other subjects and the fixation devices (Figure 4).

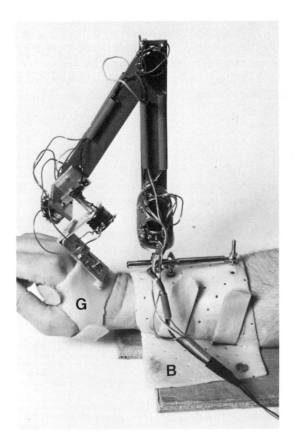

Figure 4. This view shows the electrogoniometer mounted on a subject. On the hand gauntlet G, an adjustment set allows alignment of the P6 axis to the third metacarpal line. On the hemi bracelet B, a cylindric axis and a sliding piece allow to make the P1 axis (the axis of the first potentiometer) perpendicular to the radius.

Experimental Protocol

A strict definition of the coordinate systems and their relative attitudes is absolutely necessary to enable the helical motion analysis.

Before using the goniometer, we zeroed the output voltage of the potentiometers after alignment of the goniometer on a special homemade device. The P1 and P5 axes were set parallel, the two cardan crosses were aligned along the P6 axis. A fixed distance (130mm) between the two cardan crosses was taken. A world (space-fixed) coordinate system was thus defined.

The subsequent steps of the protocol stretched out as follows: [1] the subject was requested to sit on a chair. The shoulder was in 45° abduction, the elbow in 90° flexion, the forearm in pronation. The subject was required to relax his upper arm while the bonds of the goniometer were mounted on him. A reference anatomic position was provided by making collinear the forearm and the third metacarpal (see Determination of the reference position). A start file was recorded and stored. [2] after a short learning of the wrist circumduction, we asked to the subject to rotate his wrist clockwise for the first and the second trials and counter-clockwise for the third with a maximum range of motion. No particular speed was required. 2000 acquisitions were digitized at a 110 Hz frequency. These laps allowed 7-12 full circumductions of the wrist depending on the individual motion speed. [3] additionally, 1000 acquisitions of a pure flexion-extension motion and of a pure radioulnar deviation were also digitized. These motions were active and realised without any plane constraint. [4] finally, a hand goniometer measure of FEM and RUD was performed on the subject in order to establish relationships between the common joint evaluation and our data.

Data Processing

The sequence of the data process was the following: [1] the finite helical axis parameters (see Annexe) were first calculated using programs written in Turbo Pascal 5.5. [2] the wrist circumduction was described like a flexion-extension motion in function of radial-ulnar motion. To define these clinical angles, the helical axis parameters were calculated starting from a reference position (from position 1 to position 2; from position 1 to position 3; and so on...). At this position, the two coordinate systems (fixed and moving) coincide and the helical rotation angle was calculated at each instant relatively to this one. The FEM is the y component of the rotation vector The FEM rotation (flex) was calculated by making the product of the helical rotation angle ψ with the y direction cosinus. The RUD axis is always perpendicular to the FEM axis, it is the x axis of the moving (linked to the hand) referential. The radioulnar angle was obtained by projecting the helical rotation on this axis. Therefore, the RUD angle is given by:

$$RUD = \psi . \text{axinc}$$
where axinc[2] = 0; axinc [1] = cos (flex); axinc[3] = sin (flex).

No smoothing methods were used in this kind of process. An example of these two clinical angles measured during circumduction is shown in Figure 5. [3] these clinical angles gave the locus of the angular changing. A graphical representation of this locus is illustrated on Figure 6A. [4] in order to get the envelope of the active wrist circumduction, we have calculated firstly the polar coordinates r and θ of all FEM and RUD resulting points. Then, we have added 360° times the number of circumductions to the θ value and have eliminated the eventually wrongly increasing values. Next, a cubic spline interpolation was performed to get the r values for the integer θ values. The envelope was calculated by taking the maximal r value for every integer θ modulo 360° value. (Figure 6B). [5] several envelope parameters were calculated: the mean radius r, minimum and maximum radius. The maximum radial deviation, maximum ulnar deviation, maximum dorsal flexion and maximum palmar flexion were also calculated. For all these parameters except the r average, the corresponding normalised (rounded to an integral number of degrees) polar angle θ has been given.

318

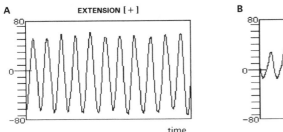
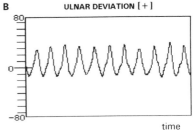

Figure 5. FEM (A) and RUD (B) in function of time during 11 circumduction motions.

[6] The ratio [length /(4π.area)] of this envelope was calculated from:

$$\text{area} = \int_0^{2\pi} \frac{r(\theta)^2}{2} d\theta = \Sigma\ r^2 \cdot \pi/360$$

$$\text{length} = \int_0^{2\pi} \frac{ds}{d\theta} d\theta = \Sigma \quad dr^2 + r^2 \cdot d\theta^2 = \Sigma \quad dr^2 + r^2 \cdot (\pi/180)^2$$

This dimensionless ratio (equal to 1 for a circle, 1.18 for an ellipse whose great axe is twice the small) gives a measure of the shape difference between the envelope and a circle. [7] for one subject the representation of the successive positions of the instantaneous helical axis in one circumduction movement is shown. The instantaneous helical axis parameters were approximated by those of the finite axis of the displacement between two acquisitions (from position 1 to position 2; from position 2 to position 3; and so on...). The angular step between two acquisitions resulting from the 6 signals was approximately 1 degree. It depended on the sector of the motion. When the wrist was in ulnar deviation and made a movement towards the palmar flexion, the motion velocity was higher. When the hand was in flexion and rotated from ulnopalmar to radiopalmar, the velocity was smaller than in the previous sector. A greater speed generally provides a less dense part of the axode (Figure 7 and Figure 8). In this kind of process, we had to use a smoothing technique. A local least squares polynomial fitting[21] has been used. A fourth order polynomial was fitted for n = 10 points. The sampling interval was 0.009 sec.

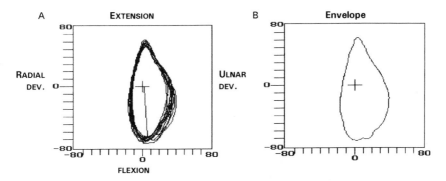

Figure 6. Left:.Locus of 11 circumductions of 1 subject, **Right:** The resulting envelope

Accuracy and Reproducibility

The validity of the goniometer itself was estimated by its accuracy and its precision. The accuracy and precision were tested on a theodolite. The accuracy of the goniometer has been calculated by the 95% confidence interval error estimate

$$\text{Accuracy} = MD^2 + 4 * SD^2$$

where MD is the average of the differences between the imposed angles and the measured angles. SD is the standard deviation of these differences.[17]

The precision has been obtained by comparing a series of repeated measurements from

$$\text{Precision} = (1/n \, \Sigma \, (\text{Diff})^2)$$

where Diff is the difference between the first and the second angle measurement.

The reproducibility has been estimated by comparing two repeated experiments on the envelope parameters. The fixation device was not removed between the repeated circumductions. The results are summarised in Table 3.

An estimation of the error for two repeated circumductions has been given from

$$S.E. = (\Sigma (P_{i1} - P_{i2})^2) / 2n$$

with i varying from 1 to n. n is the number of duplicate values, P_{i1} and P_{i2} are the repeated measures of the envelope polar radius, S.E. is the standard error.[11] The envelope polar radii (360 measures) of two data files from one subject were compared. We have averaged the individual standard errors for the population (n=25).

Statistics

The mean and standard deviation are shown for all parameters of the envelope and for the extreme range of FEM and RUD (Table 2 to 4). Two subgroups (male and female) are also compared (Table 6). The reproducibility is statistically evaluated by a t-test paired between two repeated experiments (Table 5).

RESULTS

Range of Motion

Pure motion. The mean and standard deviation of pure flexion-extension ranges determined by our electrogoniometer and a hand goniometer are given in Table 1. A comparison with other autors studies is presented in the same Table.

Envelope mean parameters (polar radius). The Table 2 graph shows the mean envelope obtained from the 27 individual envelopes. The mean radius was averaged from the population. Its value (35.9 ± 4.5 degrees) is smaller than the 43.6 ± 6.3 degrees presented by Ojima et al..[23] The minimum and maximum radii were also averaged (Table 2). For the polar angle associated with the maximum radius, we had to split the population. 23 peoples reached their maximum radius in flexion while the 4 other one got it in extension.

Circumduction. The mean and standard deviation of FEM and RUD motion components in circumduction were obtained (Table 3). The maximum ulnar deviation and the maximum radial deviation were on the flexion side of the envelope. As the flexion-extension maxima were greater than the radioulnar deviation maxima and as the extension and flexion polar angles were respectively 3 and 4 right of the vertical, the envelope greater axis was nearly vertical. (Table 3 and Figure 5).

Table 1. Comparison between previous studies and our data in pure motion

	n	FEM		RUD		Device used
AAOS [7]		144		52		hand goniometer
Boone [7]	109	148.8	± 12.0	56.4	± 7.8	hand goniometer
Brunfield (male) [9]		137				uniaxial goniometer
Brunfield (female)		147				
Garcia-Elias[15]	6	139.6	± 21.0	45.6	± 15.9	biplanar x-ray
Palmer [24]	10	133		40.5		3 *df* goniometer
Sarrafian [29]	55	121				x-ray
Ryu [27]	40	138.4		58.7		biaxial goniometer
This study						
mixed group	27	132.4	± 12.2	54.9	± 8.4	6 *df* goniometer
male group	13	127.7	± 9.1	52.1	± 7.7	6 *df* goniometer
female group	14	136.8	± 13.1	57.5	± 8.2	6 *df* goniometer
mixed group	27	144	± 22	56	± 13.0	hand goniometer

FEM: flexion-extension motion; RUD: radial-ulnar deviation; n: number of subjects.

Table 2. The polar radius (r) and the ratio k= [length /(4p.area)] parameters. rMEAN is the averaged value of r, rMIN and rMAX, the minimal and maximal values of r. θrMIN and θrMAX are their corresponding polar angles. All these parameters have been averaged over the 27 subjects As the maximal r value occurred 4 times in extension (mark **) and 23 times (mark *) in flexion, we split the population for θrMAX. The curve shown is the mean envelope. Note that the mean envelope parameters are distinct from the averaged parameters of the individual envelopes. r is the polar radius.

r_{Mean}=	35.9 ±	4.2	(28.3 →	46.8)
r_{MIN} =	18.9 ±	5.2	(10.1 →	30.8)
θ_{rMIN} =	166.0 ±	44.0	(19.0 →	208.0)
r_{MAX} =	64.2 ±	5.7	(54.0 →	76.7)
θ_{rMAX} =*	266.0 ±	387	(262.0 →	289.0)
**	89.8 ±	3.3	(86.0 →	94.0)
K =	1.4 ±	0.1	(1.2 →	1.7)

Table 3. Flexion, extension, radial and ulnar deviation (degrees)

EXT_{MAX}	55.1 ± 10.0	(38.6 → 68.9)
$\theta EXTMAX$	87.0 ± 7.0	(59.0 → 98.0)
FL_{MAX}	61.8 ± 5.4	(53.2 → 69.9)
$\theta FLMAX$	274.0 ± 5.0	(264.0 →287.0)
UD_{MAX}	35.4 ± 5.1	(25.7 → 46.0)
$\theta UDMAX$	23.0 ± 18.0	(-44.0 → 29.0)
RD_{MAX}	22.8 ± 5.1	(13.1 → 31.2)
$\theta RDMAX$	194.0 ± 39.0	(116.0→246.0)

Contribution of FEM and RUD in circumduction

Usually, the wrist kinematics has been analysed starting from planar motions. We give a comparison between the pure flexion-extension and radial-ulnar deviation and the same movements when they are measured in circumduction (Table 4).

During the circumduction, the arc of flexion-extension motion (116.8°) was smaller than during the planar flexion-extension motion (88%). The difference was highly significant $p < 0.001$.

Considering radioulnar motion, the arc of motion in circumduction was 58.2°. This value is slightly greater than the 54.9° measured in pure radioulnar deviation (106%). However the difference is not statistically significant.

Table 4. Contribution of pure motion in circumduction.

PURE MOTION			CIRCUMDUCTION		
EXT_{MAX}	64.7°	± 8.6	EXT_{MAX}	55.0°	± 10.0
FL_{MAX}	67.7°	± 7.8	FL_{MAX}	61.8°	± 5.4
FEM_{MAX}	132.4°	± 12.5	FEM_{MAX}	116.8°	± 12.6
% ofpure motion	100%			88% ***	
UD_{MAX}	34.0°	± 5.6	UD_{MAX}	35.4°	± 5.1
RD_{MAX}	20.9°	± 7.1	RD_{MAX}	22.8°	± 5.1
RUD_{MAX}	54.9°	± 8.6	RUD_{MAX}	58.2°	± 6.7
% ofpure motion	100%			106%	

EXT: extension ; FL: flexion ; UD: ulnar deviation ; RD: radial deviation.   *** $p < 0.001$

Reproducibility and effect of rotation sense.

The accuracy of the goniometer was 0.5° in RUD and 0.7° in FEM. Its precision was 0.07°. The mean and standard deviation of the polar radius precision for the population (n=27) were 3.8 ± 1.8 degrees.

The reproducibility is suggested looking at the repetition of the locus (Figure 7). It has been calculated comparing two wrist circumduction trials. No statistically significant differences were found except for r_{MAX} and EXT_{MAX}. A comparison between the sense of the movements is also represented (Table 5). No significant differences have been found except for θ_{FLMAX} and EXT_{MAX}.

Comparison between sexes

A comparaison between sexes are shown in Table 6. The comparison between the male and female subgroup shows a probably significant difference.

Table 5 . Reproducibility in the circumduction between the first and the second trial for the same motion without removing the device (n=23) and comparison with changed motion sense between the first and third trial (n=13).

	1st trial ± s.d.	2nd trial ± s.d.	3rd trial ± s.d.	diff 1-2 ± s.e.m	diff 1-3 ± s.e.m
r_{MEAN}	35.1 ± 4.3	35.7 ± 4.6	36.8 ± 3.9	-0.6 ± 0.1	-0.4 ± 0.2
r_{MIN}	18.8 ± 5.5	19.1 ± 4.9	21.7 ± 3.1	0.4 ± 0.2	-2.1 ± 0.2
r_{MAX}	61.4 ± 6.0	62.7 ± 5.9*	62.4 ± 4.5	-1.4 ± 0.1	0.0 ± 0.4
RD_{MAX}	-24.0 ± 6.0	-23.7 ± 5.8	-28.1 ± 6.2*	-1.3 ± 0.3	3.9 ± 0.5
θ_{RDMAX}	195.0 ± 42.0	194.0 ± 38.0	168.0 ± 33.0	17.0 ± 2.0	22.0 ± 2.0
UD_{MAX}	34.4 ± 5.5	34.4 ± 5.2	35.9 ± 6.9	0.0 ± 0.2	0.5 ± 0.3
θ_{UDMAX}	281.0 ± 120.0	277.0 ± 119.0	329.0 ± 7.0	5.0 ± 5.0	-20.0 ± 5.0
FL_{MAX}	59.1 ± 6.2	60.1 ± 5.9	60.0 ± 5.3	1.0 ± 0.1	-1.1 ± 0.3
θ_{FLMAX}	271.0 ± 6.2	271.0 ± 5.8	279.0 ± 6.5**	-0.7 ± 0.3	-5.0 ± 0.6
EXT_{MAX}	51.6 ± 9.6	54.6 ± 10.3*	53.8 ± 9.3*	-3.0 ± 0.3	1.4 ± 0.6
θ_{EXTMAX}	85.3 ± 8.3	86.6 ± 7.7	93.2 ± 6.7	-1.3 ± 0.3	-5.9 ± 0.5
FEM	110.7 ± 12.4	114.7 ± 13.4*	114.3 ± 9.8	-3.4 ± 0.4	2.4 ± 0.6
RUD	57.1 ± 7.2	58.4 ± 7.8	64.0 ± 11.3*	4.6 ± 0.6	4.6 ± 0.6

t-test paired * p<0.05; ** p<0.01

Representation of the wrist circumduction axode

The successive positions of one circumduction turn of one subject have been subdivided in four motion ranges to attempt to discriminate the participation of FEM and RUD axes in the circumduction. Each range is corresponding to an envelope sector. In the ulnar sector (Figure 7A), the helical axes have the smaller angle with the y reference axis. It demonstrates evolutive axis. In the bottom sector (Figure 7B), the helical axes are located closely along the z reference axis of the fixed coordinate system. In this sector, the hand is in flexion and makes a motion from ulnar to radial deviation. In the radial sector (Figure 7C), the helical axes have the same orientation than in the first sector. In the top sector (Figure 7D), the helical axes are closely along the z reference axis. This piece of axode seems to show a more punctual centre of rotation. We present in Figure 8 the global wrist circumduction axode The length of the cube side is 400mm.

Table 6. Comparison of the mean envelope parameters between sexes for the population (n=27)

	Male (n=13)	Female (n=14)	t-test
r_{MEAN}	34.2 ± 3.6	37.8 ± 4.7	NS
r_{MIN}	17.2 ± 4.9	20.8 ± 5.0	$p < 0.05$
r_{MAX}	61.7 ± 4.8	67.4 ± 5.2	$p < 0.05$
RD_{MAX}	20.9 ± 3.7	25.1 ± 5.9	$p < 0.05$
UD_{MAX}	34.7 ± 5.2	36.1 ± 5.6	NS
RUD_{cir}	55.7 ± 5.5	61.3 ± 6.9	$p < 0.05$
FL_{MAX}	59.3 ± 9.7	64.7 ± 8.8	$p < 0.05$
EXT_{MAX}	53.8 ± 9.3	58.1 ± 10.3	NS
FEM_{cir}	112.6 ±10.8	122.9 ± 11.7	$p < 0.05$
FEM_{pm}	127.7 ± 9.5	136.8 ± 13.6	$p < 0.05$
RUD_{pm}	52.0 ± 8.0	57.5 ± 8.5	$p < 0.05$

DISCUSSION

This study focused on an active wrist circumduction analysis. The six degrees of freedom electrogoniometer has been a reliable tool which gave a good reproducibility. One of the advantages of 6 *df* electrogoniometers is their ability to simulate all other types of goniometers (uni, bi, triaxial goniometers). An other advantage is that with the same data, we have obtained information in terms of clinical angles and in terms of helical parameters.

Usually, the wrist kinematics was investigated during pure motion. "The flexion and deviation about the so called main axes of the wrist, are only a part of the hand displacement".[30] In this population, we found an average full range of pure motion of 132 degrees in FEM and 54 degrees in RUD. These values are in agreement with the other authors data (see Table 1).[24, 27] The comparison between pure motion of the hand and the magnitude of these motions in circumduction presented a difference. In circumduction, 88% of the pure flexion was filled. However, in deviation, 106% of the pure deviation was actuated. The error on the polar radius precision (3.8°) was about 10%. Ryu et al.[27] in a biaxial electrogoniometric study, gave a reproducibility of 3 to 4 degrees for activity of daily living (ADL) motions. Several factors limit the accuracy of the results. First, the relative displacement between soft tissues and the measurement device affect the reproducibility, but the interposition of PVC foam decreased this first factor. Let us note that with a 6 *df* electrogoniometer, the investigated motion was not constrained. Second, the wrist circumduction without prosupination does not look like an usual movement. The wrist firstly stabilises the hand on the forearm.[20] In the active wrist circumduction, the compensation of a joint or coordination loss was less easy.

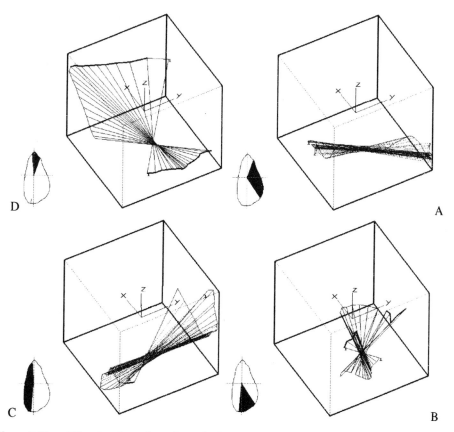

Figure 7. Four different regions of one circumduction axode. Beside each axode piece, the corresponding envelope sector is blackened. The dotted axes intersect the faces orthogonal to Oy. The bolder curves are in the faces perpendicular to Ox.

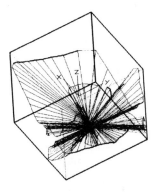

Figure 8. The whole axode of one circumduction.

The value of the mean radius (35.9 ± 4.5 degrees) was smaller than the 43.8 ± 6.3 degrees presented by Ojima.[23] He was the first to present a locus. His locus demonstrated radial deviation values larger than ours. Perhaps, small prosupinations or the limits of a 2 *df* goniometer could have altered his results.

The great axis of the mean envelope did not appear skew. The so called "oblique plane" was not observed here.[28] This physiological plane (from radial and dorsal to ulnar and palmar) as discribed by Fisk[14] appears when the prosupination component is not constrained. The wrist circumduction seems not to be a typical and physiological activity. The oblique plane refers rather to hand circumduction. The summation of all muscle forces crossing the carpus tends to place the hand in a position of flexion and ulnar deviation.[34] In a dynamic wrist motion like circumduction, we have not found this tendency. An analysis of motion velocities will perhaps give more sensible indications.

The kidney shaped envelope of Savelberg[30] page 87 showed a concavity in the radial side of the envelope. This envelope was obtained from the motion of one specimen. A slight depression in the radial side of the envelope was present in our study.

The successive positions of the helical axis presented two major components, one in flexion and the other in deviation. In flexion, the piece of axode showed a more evolutive shape: the axes didn't cross in a very small space. In the both cases, the crossing space was in the ulnar side of the cube.

CONCLUSION

The polar parameters associated with a graphic representation of the wrist circumduction gave a good idea of the wrist pattern in a global motion. The analysis of the wrist full range motion in terms of clinical angles and in the same time in terms of helical axes could give further precious informations for functional loss assessment. A quantitative analysis of helical axis evolution in circumduction remains a challenge.

REFERENCES

1. K. N. An and E. Y. Chao, Kinematic analysis of human movement, *Ann. Biomedical Engineering* 12:585 (1984).
2. J.G. Andrews, and Y. Youm, A biomechanical investigation of wrist kinematics, *J. Biomech.* 12:83 (1979).
3. J. Beggs, Advanced Mechanism, Mac Millan, Ney York (1966).
4. R. A. Berger, R. D. Crowninshield, A. E. Flatt, The three-dimensional rotational behaviour of the carpal bones, *Clin. Orthop.* 167:303 (1982).
5. L. Blankevoort, R. Huiskes, and A. de Lange, The envelope of passive knee joint motion, *J. Biomech.* 21:705 (1988).
6. L. Blankevoort, R. Huiskes, and A. de Lange, Helical axes of passive knee joint motions, *J. Biomech.*. 23:1219 (1990).
7. D. C. Boone, and S. P. Azen, Normal range of motion of joints in male subjects, *J. Bone Joint Surg.* 61A:756 (1979).
8. R.B. Brumbaugh, R. D. Crownshield, W. F. Blair, and J. G. Andrews, An in-vivo study of normal wrist kinematics, *J. Biomech. Eng.* 104:176 (1982).
9. R.H. Brumfield, and J.A. Champoux, A biomechanical study of normal functional wrist motion, *Clin. Orthop.* 187:23 (1984).
10. A. de Lange, R. Huiskes, J.M.G Kauer, and H. J. Woltring, On the application of a smoothing procedure in the kinematical study of the human wrist joint in vitro, In Huiskes Ed: Biomechanics: Principles and Applications. Martinus Nijhoff Pub. pp 303 (1982).
11. A. de Lange, 1987, A kinematical study of the human wrist joint, Thesis Nijmegen (1987).

12. A. de Lange, J.M.G Kauer, and R. Huiskes, Kinematic behaviour of the human wrist joint: a roentgenstereophotogrammetric analysis, *J Orthop. Res.* 3:56 (1985).

13. A. J. Erdman, J. K. Mayfield, F. Dorman, M. Wallrich, and W. Dahlof, Kinematic and kinetic analysis of the human wrist stereoscopic instrumentation, *ASME J. Biomeh. Eng.* 105:136 (1979).

14. G. R. Fisk , La biomécanique de l'articulation du poignet, in Tubiana R. Traité de la chirurgie de la main, vol. 1. Paris, Masson ed. (1980).

15. M. Garcia-Elias , W.P. Cooney , K.N. An , R.L. Linscheid, and E.Y.S. Chao, Wrist kinematics after limited intercarpal arthrodesis, *J. Hand Surg.* 14-A:791 (1989).

16. E. S. Grood, and W. J. Suntay, A joint coordinate system for the clinical description of three-dimensional motions: application to the knee, ASME *J. Biomech. Eng.* 101:124 (1983).

17. R. Huiskes, J. Kremers, A. de Lange, H. J. Woltring, G. Selvik, Th. J. G. van Rens, Analytical stereophotogrammetric determination of three- dimensional knee joint geometry, *J. Biomech.* 18:559 (1985).

18. J.M.G. Kauer, The mechanism of the carpal joint, *Clin. Orthop.* 202:16 (1986).

19. G. L. Kinzel, A. S. Hall, B. M. Hillberry, Measurement of the total motion between two body segments -I. Analytical development, *J. Biomech.* 5:93 (1972).

20. J.N. Kuhlmann, and R. Tubiana, Mécanisme du poignet normal, In: Razemon JP, Fisk GR (eds): Le poignet; Expansion Scientifique Française: pp. 62 (1983).

21. H. Lanshammar, On precision limits for derivates numerically calculated from noisy data, *J. Biomech.* 15:459 (1982).

22. R.L. Linscheid, Kinematic considerations of the wrist, *Clin. Orthop.* 202:27 (1986).

23. H. Ojima, S. Miyake, M. Kumashiro, H. Togami, and K. Suzuki, Dynamic analysis of wrist circumduction: a new application of the biaxial flexible electrogoniometer, *Clinical Biomechanics* 6:221 (1991).

24. A.K. Palmer, F.W. Werner, D. Murphy, and R. Gilsson, Functional wrist motion: a biomechanical study, *J. Hand Surg.* 10A:39 (1985).

25. H. K. Ramakrishnan, and M. P. Kadaba, On the estimation of joint kinematics during gait, *J. Biomech.* 24:969 (1991).

26. L.K. Ruby, W.P. Cooney, K.N. An, R.L. Linscheid, and E.Y.S. Chao, Relative motion of selected carpal bones: a kinematic analysis of the normal wrist, *J. Hand Surg.* 13-A:1 (1988).

27. J. Ryu, W.P. Cooney, L.J. Askew, K.N. An, and E.Y.S Chao, Functional ranges of motion of the wrist joint, *J. Hand Surg.* 16-A: 409 (1991).

28. Ph. Saffar, The study of the biomechanics of wrist movements in a oblique plane. Advances in the biomechanics of the hand and the wrist, NATO Advanced Research Worshop, Brussels 22-23 may p 37 (1992).

29. S. K. Sarrafian, J. L. Melamed, G. M .Goshgarian, Study of wrist motion in flexion and extension, *Clin Orthop* 126:153 (1976).

30. H.H.C.M. Savelberg, Wrist joint kinematics and ligament behaviour, Doctoral Dissertation, University of Nijmegen, Nijmegen,The Nederlands (1992).

31. G. Selvik, Roentgen stereophotogrammetry: a method for the study of the kinematics of the skeletal system, *Acta Orthop. Scand.* 60 (suppl 232) (1989).

32. R. Shiavi, T. Limbird, M. Frazer, K. Stivers, A. Strauss, and J. Abramovitz, Helical motion analysis of the knee -I. Methodology for studying kinematics during locomotion., *J. Biomech.* 20:459 (1987).

33. H. G. Sommer, and N.R. Miller, A technique for kinematic modeling of anatomical joints, *J. Biomech. Eng.* 102:311 (1980).

34. R.G .Volz, M. Lieb, and J. Benjamin, Biomechanics of the wrist, *Clin. Orthop.* 149:112 (1980).

35. H. J. Woltring, R. Huiskes, A. de Lange, and F.E. Veldpaus, Finite centroid and helical axis estimation from noisy landmark measurements in the study of human joint kinematics, *J. Biomech.* 18:379 (1985).

36. H. J. Woltring, A. de Lange, J.M.G. Kauer, and R. Huiskes, Instantaneous helical axis estimation via natural cross-validated splines, In G. Bergman, A. Kölber, and Rohlmann. (eds.) Biomechanics: Basic and applied research. Dordrecht/ Boston/ Lancaster: Martinus Nijhoff Pub. 121 (1987).

37. H. J. Woltring, Representation and calculation of 3-D joint movement, *Human Movement Science* 10:603 (1991).

38. Y. Youm, and A.E. Flatt, Kinematics of the wrist, *Clin Orthop* 149:21 (1980).

APPENDIX

Determination of the helical axis

Let $\mathbf{R}$ be the rotation matrix, explicitly, for a rotation of axis $\mathbf{v}$, angle α,

$$\mathbf{r}'_i = \mathbf{R}_{ij} \cdot \mathbf{r}_j$$

$$\mathbf{R}_{ij} = v_i \cdot v_j \cdot (1 - \cos\alpha) - e_{ijk} \cdot v_k \cdot \sin\alpha + \cos\alpha \cdot \delta_{ij} \qquad \text{where}$$

and e_{ijk} is Levi-Civita's totally antisymmetric density, equal to 1 if ijk is an even permutation of 123, δ_{ij} is Kronecker's symbol (equal to 1 if $i = j$, 0 otherwise). From the definition of $\mathbf{R}$, we get α and the components of $\mathbf{v}$:

$$\cos\alpha = (\text{trace}(R) - 1) / 2$$
$$v_1 = (R_{32} - R_{23}) / (2\sin\alpha) \qquad \text{and circular permutations.}$$

The position of the helical axis is obtained by finding the fixed point in the projection of the movement on a plane perpendicular to the rotation axis.

WRIST RANGE OF MOTION IN ACTIVITIES OF DAILY LIVING

David L. Nelson,[1] Margaret A. Mitchell, [2]
Paul G. Groszewski,[2] Stephen L. Pennick,[2]
and Paul R. Manske[3]

[1]Department of Orthopedic Surgery, Room U471
University of California, San Francisco
San Francisco, CA 94143-0728, USA
[2]Program in Occupational Therapy
Washington University
St. Louis, MO 63110, USA
[3]Division of Orthopedic Surgery
Washington University
St. Louis, MO 63110, USA

INTRODUCTION

The wrist motion used to perform activities of daily living (ADL) is important to know when evaluating wrist function after trauma or surgical procedures that limit wrist motion. Several studies have examined wrist motion in terms of absolute range and, more important, functional range. Palmer et al.,[10] in the most sophisticated study to date, defined the functional range of the wrist to be 30° of extension, 5° of flexion, 10° of radial deviation, and 15° of ulnar deviation. However, these were *mean* values, obtained by averaging the data of *all* subjects in *all* of the activities; the actual motion requirements of *each* individual task was therefore obscured, and range was actually underestimated.

The purpose of the present study was to determine the range of wrist motion used to perform each of the tasks studied by Palmer et al.[10] In contrast with previous investigations, these determinations examined the motion requirements of each task, not the mean of all tasks. In addition, they were made with the aid of a motion tracking device that permits unencumbered motion and thus greater accuracy in measurement.

MATERIALS AND METHODS

Range of Motion

Ten normal volunteers were evaluated in the performance of the same series of activities as that described by Palmer et al.[10] (Table 1). Wrist motion was measured with

Advances in the Biomechanics of the Hand and Wrist
Edited by F. Schuind *et al.*, Plenum Press, New York, 1994

the aid of a Polhemus 3SPACE Tracker (Polhemus Inc., Colchester, VT 05446), which tracks three-dimensional motion with 6 degrees of freedom. The Polhemus was originally invented to control weapons systems in fighter aircraft. The device would track where the pilot was looking and would allow firing without aiming: the pilot merely had to look at the target. We modified the device to track wrist motion. We also used a program that allows real-time data acquisition.[7] The Polhemus device works by means of (1) a sending unit that sets up a magnetic field, and (2) remote sensors that record position and orientation. The output of the Polhemus device is fed into a computer. The sensors were mounted on contoured plastic fittings, which, in turn, were attached to the skin with double-sided adhesive tape that allowed no motion between the skin and the plastic fittings. One sensor was placed over the dorsal distal radius, and a second sensor was placed over the dorsal distal second-third metacarpal interspace. Unlike previous devices,[10,3] the Polhemus 3SPACE Tracker has no mechanical coupling between the two sensors and therefore no internal resistance.

Table 1. Daily Living Activities Tested

---

1. Comb hair.
2. Perineal hygiene.
3. Wring out a washcloth.
4. Button own shirt from top to bottom.
5. Tie and untie own shoes.
6. Cut with a knife.
7. Pierce food with a fork and place in mouth.
8. Turn a pancake-like object over with a spoon.
9. Drink from a cup without a handle, first sipping as if it were hot, then draining it as if it were not hot.
10. Pour from a pitcher, from the side and from the spout.
11. Open a can with a can opener (removing the lid).
12. Stir in a bowl, as if stirring pancake batter.
13. Open a jar, as if the lid were on tight, then remove lid.
14. Turn on a faucet.
15. Turn the pages in a book.
16. Write and print own name.
17. Pick up a telephone, both wall-mounted and desk-mounted, and put up to the ear.
18. Dial "O", both wall-mounted and desk-mounted.
19. Load paper into a typewriter.
20. Turn a steering wheel, both to the left and to the right.
21. Open a doorknob and pull door open.
22. Turn a key in a door lock.
23. Turn a screw with a screwdriver, first as if tight and then as if loose.
24. Throw a ball overhand into a basket ten feet away.

Each activity was performed multiple times over a 10-second period. The sensors were sampled 19 times per second and the data processed by computer using a program previously described.[7] The maximum range of motion (ROM)(flexion, extension, radial deviation, and ulnar deviation) used to complete the task was recorded and averaged for each task for all ten volunteers.

The Polhemus device is based on varying magnetic fields. Large metal objects can affect the fields and this adversely affects the results. Therefore, care was taken to examine the effect of metal objects used in performing the tasks.

The accuracy of the placement of sensors on each volunteer was established by taking over 20 measurements in flexion, extension, radial deviation, and ulnar deviation. Readings were taken simultaneously by the Polhemus device and at least two investigators. This data was the basis for a nonlinear correction that was made to the Polhemus data. This procedure was followed each time the sensors were placed on a volunteer and data collected, and accounted for approximately one-third of the time spent in each testing session.

Table 2. Wrist Range of Motion in Activities of Daily Living.*

Activity	Radial			Ulnar			Flexion			Extension		
	max	min	avg	max	min	avg	max	min	avg	max	min	avg
comb hair	18	-3	10.4	44	8	29.9	74	14	38.4	55	32	42.2
tailbone	24	-3	**11.6**	37	12	24.3	65	28	**49.8**	66	-5	37.8
washcloth	15	-6	6.4	50	17	34.4	74	9	28.3	65	25	49.2
button	14	-5	6.3	40	7	19.1	35	-5	20.3	47	10	29.4
tie shoe	15	2	7.5	31	13	22.7	56	-2	17.9	45	17	29.3
use knife	17	-18	-7.1	37	16	26.4	32	-25	2.9	37	-5	15.0
use fork	16	-9	-0.1	37	12	23.7	22	-20	-4.8	55	28	38.6
spatula	2	-25	-12.1	45	25	37.2	44	8	25.4	53	12	28.9
drink cup	11	-10	0.5	45	18	32.8	6	-27	-11.0	44	19	31.8
pitcher	12	-3	6.1	42	17	31.2	36	-15	5.3	61	29	46.6
can opener	18	-15	4.2	33	16	25.3	10	-22	-5.7	53	22	41.3
stir bowl	18	-30	5.4	43	10	33.8	56	15	32.5	56	-8	21.6
open jar	25	-10	6.9	47	32	39.6	57	24	35.3	52	12	34.8
turn faucet	18	-2	9.0	51	27	**40.0**	50	18	26.3	59	-4	29.6
turn page	17	-19	0	45	6	27.5	52	-17	17.7	40	8	28.3
write name	13	-18	0.9	38	-8	11.6	10	-50	-20.0	54	14	33.9
use phone	17	-3	9.2	32	15	24.7	10	-19	-3.2	63	37	**51.1**
dial "O"	17	-23	0	38	11	24.6	30	-5	10.8	54	7	28.8
typewriter	17	7	11.4	46	18	29.3	42	0	22.0	50	10	36.2
steer.wheel	11	-2	4.7	45	28	34.9	35	-9	9.2	65	17	39.3
door knob	11	-25	-8.3	44	26	36.3	20	-18	-1.1	54	24	41.2
turn key	5	-22	-9.4	47	19	34.6	10	-36	-1.7	45	14	33.5
screwdriver	-12	-25	-18.0	45	18	34.8	25	-9	2.5	45	23	33.4
throw ball	18	-5	6.2	37	13	24.6	31	6	16.4	64	30	48.1

(* Activities with the largest average motion in each of the four directions of motion are in bold letters.)

RESULTS

The results are presented in Table 2. Perineal hygiene (Palmer called this "touch tailbone") used the most flexion (50°), followed by combing one's hair (38°) and opening a jar (35°). The greatest amounts of extension were used for holding a telephone to the ear (51°), wringing out a washcloth (49°), and pouring from a pitcher (46°). Perineal hygiene also used the most radial deviation (12°), followed by loading paper into a typewriter (11°) and combing one's hair (10°). The greatest amounts of ulnar deviation were used for turning a faucet (40°), opening a jar (39°), and turning a doorknob (36°). Thus the limits, or "envelope", of motion, based on the average maximum range of motion in each direction, was 50° of flexion, 51° of extension, 12° of radial deviation, and 40° of ulnar deviation.

Not all of the activities contributed to the limits of the ROM. Eight of the 24 activities used less than 75% of the maximum motion for all four wrist motions, and 14 used less than 90%. Only ten of the activities either helped to define the limits or were within 10% of those limits. These activities included combing hair, perineal hygiene, wringing a washcloth, turning food with a spatula, pouring from a pitcher, opening a jar, turning on a faucet, putting a telephone receiver to the ear, loading paper into a typewriter, and turning a doorknob.

DISCUSSION

Many important decisions concerning the wrist are based on estimates of motion. One such decision is related to the evaluation of various methods of limited intercarpal arthrodesis.[5,8,12] Rozing and Kauer found that 40% of wrist flexion remains after radio-scaphoid arthrodesis, and Meyerdierks et al. reported a 35% loss of flexion with a scaphocapitolunate fusion. However, the ROM needed to perform the activities studied has not yet been defined, so these data cannot be translated into either the per cent loss of wrist function or the per cent loss of activity. Similarly, Trumble, et al. have stated that "the postoperative motion [of limited intercarpal arthrodesis] appears to be sufficient for most daily activities" but provided no data on which to base this conclusion.[14]

The need to know the amount of motion used to perform activities of daily living has prompted several studies. The earliest of these focused on the maximum ROM of the wrist. Sarrafian, et al.[13] examined 55 wrists of normal adults radiographically and found that the maximum flexion was 60° and the maximum extension was 55°. Boone and Azen[2] measured the ROM of 109 normals and found that the maximum flexion was 76°, extension 75°, radial deviation was 22°, and ulnar deviation was 36°. Brumfield, et al.[4] examined ten adults and found maximum flexion to be 82° for women and 73° for men, and maximum extension to be 65° for women and 64° for men. The AAOS has established norms for ROM, based on four studies of joint motion. Their norms are 73° of flexion, 71° of extension, 19° of radial deviation, and 33° of ulnar deviation.[1]

However, maximum ROM alone does not address the amount of motion actually used, so a functional index of wrist motion was developed. Porter and Stockley[11] defined the functional ROM as 45° of extension, 30° of flexion, 15° of radial deviation, and 15° of ulnar deviation but the criteria used to define these values was not indicated. The first study to address wrist ROM in activities was by Brumfield and Champoux,[3] in which a uniaxial electrogoniometer was used to examine the flexion-extension axis. They concluded that the ROM used in ADL was 10° of flexion to 35° of extension. They did not attempt to measure radial/ulnar deviation. A more sophisticated, triaxial study of this problem was that by Palmer, et al.,[10] who concluded that the functional ROM of the wrist was 30° of extension, 5° of flexion, 10° of radial deviation, and 15° of ulnar deviation. This study, however, averaged *all* subjects for *all* activities, rather than averaging *all* subjects for *each* activity. The combining of activities with high degrees of flexion with activities using only small amounts of flexion resulted in lower estimates of motion than were actually used. Stating this another way, the subjects would be able to perform only half of the activities studied if they were limited to the average motion.

Defining Functional Range of Motion

Although we found an increased amount of motion was used in ADL as compared with Palmer, we have been careful not to state that there is a larger "functional" ROM or to state that the patient "needs" the ROM that we found. Palmer's study has been misinterpreted in the literature as defining the minimum acceptable motion for function.[5,9] There are two reasons why this study and previous ones have not defined the functional range of motion of the wrist.

First, there is a conceptual difference between the ROM that is *used* and the ROM that is *needed*. This is an important point that has been overlooked in previous studies of both "functional" wrist motion[10] and "functional" finger motion.[6] Functional motion relates to the amount of motion needed to perform a task, not that usually used to complete a task. All previous studies have examined ROM used, not ROM needed. In order to define the functional ROM of the wrist a study must specifically determine the minimum ROM that is required to complete a set of ADL. Second, previous studies have based their conclusions on experiments that examined somewhat arbitrarily selected tasks.[6,10] In order to define the functional ROM, a study would need to determine which subset of ADL, if able to be performed, would imply that a person could perform the entire set of ADL necessary to be functional.

In addition, the activities that we tested were those used in a prior study[10] which characterized them as "activities of daily living". However, not all the activities would fit such a category, e.g., throwing a ball, loading a typewriter, turning a steering wheel, and turning a screwdriver. While it is recognized that it is important to analyze the ROM used in work and recreational activities, it is likewise important to be clear about what category of activities one is testing and to differentiate between ADL, work, and recreation.

What ROM is required for the wrist to be "functional" has not been determined by this study or by earlier studies.

CONCLUSION

The range of motion of the wrist used to perform 24 activities employed in a previous study was found to be 50° of flexion, 51° of extension, 12° of radial deviation, and 40° of ulnar deviation. This was the range of motion utilized, and should not be interpreted as the range of motion that is required for function.

REFERENCES

1. American Academy of Orthopaedic Surgeons, Joint motion: Method of measuring and recording. Chicago, *American Academy of Orthopedic Surgery* (1965).
2. D.C. Boone and S.P. Azen, Normal range of motion of joints in male subjects, *J. Bone Joint Surg.*, 61A:756 (1979).
3. R.H. Brumfield and J.A. Champoux, A biomechanical study of normal functional wrist motion, *Clin. Orthop.*, 187:23 (1984).
4. R.H. Brumfield, V.L. Nickel, and E. Nickel, Joint motion in wrist flexion and extension, *So.Med. J.*, 59:909 (1966).
5. H. Gellman, D. Kauffman, M. Lenihan, M.J. Botte, and A. Sarmiento, An in vitro analysis of wrist motion: The effect of limited intercarpal arthrodesis and the contributions of the radiocarpal and midcarpal joints, *J. Hand Surg.*, 13A:378 (1988).
6. M.C. Hume, H.G. Gellman, H. McKellop, and R.H. Brumfield, Functional range of motion of the joints of the hand, *J. Hand Surg.*, 15A:240 (1990).

7. S.E. Logan and P. Groszewski, Dynamic wrist motion analysis using six degree of freedom sensors, *Biomed. Sci. Instrumentation* 25: (1989).

8. E.M. Meyerdierks, J.F. Mosher, and F.W. Werner, Limited wrist arthrodesis: A laboratory study. *J. Hand Surg.*, 12A:526 (1987).

9. A. Minami, T. Ogino, and M. Minami, Limited wrist fusions, *J. Hand Surg.*, 13A:660 (1988).

10. A.K. Palmer, F.W. Werner, D. Murphy, and R. Glisson, Functional wrist motion: A biomechanical study, *J. Hand Surg.* 10A:39 (1985).

11. M.L. Porter and I. Stockley, Functional index: A numerical expression of post-traumatic wrist function, *Injury*, 16:188 (1984).

12. P.M. Rozing and J.M.G. Kauer, Partial arthrodesis of the wrist: An investigation in cadavers, *Acta Orthop. Scand.*, 55:66 (1984).

13. S.H. Sarrafian, J.L. Melamed, and G.M. Goshgarian, Study of wrist motion in flexion and extension, *Clin. Orthop.*, 126:153 (1977).

14. T. Trumble, C.J. Bour, R.J. Smith, and G.S. Edwards, Intercarpal arthrodesis for static and dynamic volar intercalated segment instability, *J. Hand Surg.*, 13A:384 (1988).

CHANGES IN THE TFCC ARTICULAR DISK DURING FOREARM

ROTATION: A STUDY OF CONFIGURATION AND SURFACE STRAINS

Brian D. Adams and Kathy A. Holley

Department of Orthopaedics and Rehabilitation
University of Vermont College of Medicine
Given Building
Burlington, VT 05405, USA

INTRODUCTION

The triangular fibrocartilage complex (TFCC) of the wrist is a multifunctional structure composed of several anatomical components.[10] The horizontal portion of the TFCC, often referred to as the triangular fibrocartilage proper or TFC, is triangular in shape and composed of the articular disk and the dorsal and palmar radioulnar ligaments. The TFC forms a continuation of the distal radial articular surface from its radial attachment at the sigmoid notch to its apical attachment in the eccentric concavity of the ulnar head (fovea) and the projecting ulnar styloid. The disk provides an interface between the ulnar head and ulnar carpus, with the biconcave shape serving to reduce the geometric incongruencies between the bony surfaces. The peripheral margins of the TFC (radioulnar ligaments) are thicker and composed of longitudinally oriented collagen fibers, structurally adapted to bear tensile loading. The central portion (articular disk) is thinner and the collagen fiber pattern has multiple obliquities to the surface, implying that variable loading conditions occur.[3,4]

Several mechanical functions have been attributed to the disk, including load bearing, load distribution, shock absorption, and stabilization.[3,10] The multiplicity of functions is related to the complexity of motions and loading in the forearm and wrist. During forearm motion, the distal radius undergoes both rotation and translation relative to the ulna, with the translational component occurring primarily at the extremes of pronation and supination.[5,6,12] There is minimal resistance to forearm movement during most of the motion arc because the TFC attachment to the ulnar head is near the axis of motion.[1] However, as the TFC and other soft tissues become taut at the extremes of motion, the discrepancy in the radii of curvatures between the radial and ulnar articular surfaces of the distal radioulnar joint allows translation.[2,12] This translational motion produces asymmetrical loading in the TFC resulting in nonuniform strain distribution.[2,11] During injury loading conditions, areas of higher tensile strain would be at greater risk for initiation and propagation of tears. Although the radioulnar ligaments have been the

Advances in the Biomechanics of the Hand and Wrist
Edited by F. Schuind *et al.*, Plenum Press, New York, 1994

focus of several biomechanical studies, the articular disk has received little attention regarding distribution of tensile strain.

Despite advancements in the diagnosis and treatment of articular disk injuries, the mechanics of traumatic injury are poorly understood. Increasing our knowledge of injury mechanisms has important implications for prevention, treatment, and rehabilitation of TFC tears. Some injury theories propose that susceptibility of injury and site of tear are related to forearm position at the time of loading.[3,8] Based on clinical histories of patients presenting with tears, and observations of TFC motion made during arthroscopy, a relationship seems probable. However, to test this hypothesis requires a better understanding of the effects of forearm motion on the TFC. The purpose of this experiment was to study the relationship between surface strains in the articular disk and forearm position.

METHODS

A video imaging system integrated with a microcomputer was used to study the articular disk. Six unmatched, fresh-frozen upper extremities from human cadavers were used in this laboratory experiment. Specimens were disarticulated through the distal radiocarpal and ulnocarpal joints with care to preserve the articular disk, radioulnar ligaments, and the proximal attachments of the ulnocarpal ligaments of the TFCC. All specimens were from young adults and absence of preexisting injury to the TFCC was required. Specimens were held in a custom jig with the elbow flexed at 90 degrees. Rigid fixation to the jig was obtained with bone screws into the ulna and humerus. The ulna was mounted horizontal and the humerus was mounted vertical to the platform. A Steinmann pin inserted into the radial styloid was used to passively position the radius at desired degrees of forearm rotation. A stationary protractor was used to measure the angle of forearm pronation and supination. This experimental setup provided an unrestricted arc of forearm motion and provided a direct view of the distal articular surfaces of the radius and TFC.

In order to track surface changes over the entire disk, black circular markers measuring 1.1mm in diameter were adhered to the surface with droplets of cyanoacrylate. Four markers were applied to the distal radial articular surface adjacent to the TFC insertion at the edge of the sigmoid notch (radial margin of lunate fossa). An array of three rows of four markers each were placed in a grid-like fashion across the surface of the disk, spanning the area between the palmar and dorsal radioulnar ligaments and the radial and ulnar TFC attachment sites. Video images of the marker array were taken with a high-resolution CCD video camera (Ultrachip, Javelin Electronics). The images were captured to a frame-grabber board (PC Vision Plus, Imaging Technologies, Inc.) mounted in a Dell 320LX, 386-based microcomputer. Image acquisition and analysis was controlled by an imaging software package (Optimas, BioScan, Inc.). Calibration of the system was performed with a thin, high-precision ruler placed at the level of the markers. Camera field size was set to visualize only the TFC during its arc of motion. The black markers were selected as screen objects based on a gray scale threshold set by the operator. A user-defined object class recognition scheme was used to further evaluate the screen objects for size and circularity. Edges of the resulting objects were then outlined by the system. Centroids of each object were found by the supporting software and written to a data file. The sixteen centroids were used to delineate nine discrete regional areas (A) of the surface: three across the radial margin of the disk, three across the central portion, and three along the ulnar margin (Figure 1). In addition, 12 longitudinal lengths (L) and 9 transverse lengths (T) were defined by the intercentroid distances (Figures 2 and 3).

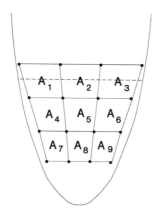

Figure 1. Schematic of regional areas (A) evaluated in this study.

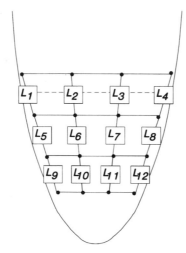

Figure 2. Schematic of longitudinal lengths (L) evaluated in this study.

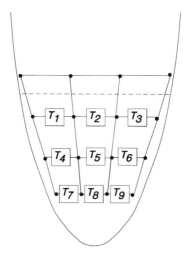

Figure 3. Schematic of transverse lengths (T) evaluated in this study.

A close-up video image of the markers was acquired at every ten degrees of forearm motion, beginning with full pronation and moving to full supination. Neutral position (zero degrees) was defined as parallel alignment of the longitudinal axis of the distal radial surface with the humeral shaft. The experiment was then repeated in the reverse direction of forearm motion. The entire sequence was repeated for each specimen.

Areas of the regions and magnitudes of the lengths were measured with the software for each image and stored to a data file. Strains were calculated from changes in centroid locations according to the equations:

$$L = (l_i - l_g)/ l_g \qquad \text{and} \qquad T = (t_i - t_g)/ t_g$$

where l_i and t_i are the respective distances between two centroids at i degrees of forearm motion, and l_g and t_g are the distances between the same centroids with the forearm in neutral position (gage length). Twelve longitudinal strains and nine transverse strains were calculated for each image. Strains were plotted versus degree of forearm rotation. Changes in areas, "area strains", were calculated for each of the nine regions, with the neutral position as reference, by the equation:

$$A = (a_i - a_g)/ a_g$$

Area strains were also plotted versus degree of forearm rotation. The first derivative of total disk area strain was then calculated according to the equation:

$$A' = (A'_i - A'_{i+1})/\Delta i$$

where A'_i is the total area strain at i degrees of forearm motion, and A'_{i+1} is the strain at the next forearm position. The plot of the results demonstrates rate of change in total area during forearm motion, i.e., rate of change in different forearm positions.

System error is calculated according to the pixel density of the video frame buffer and the field of view. Given a 640 by 480 pixel density in the frame buffer and a 20mm by 15mm field of view, the resolution is 0.03 mm/pixel. An object centroid is calculated by taking the average x and y values of points lying on the object perimeter. The minimum number of points needed to define a 1.1mm object is 64. Thus, error in calculating centroids was:

$$\frac{\sigma}{\sqrt{N}} = 0.03 \frac{mm}{\sqrt{64}} = 0.00375 mm$$

RESULTS

The sum of all nine regional areas marked on the disk decreased as forearm motion proceeded from neutral to either pronation or supination (Figure 4). However, the distribution of strain was not uniform and depended upon forearm position. Areas decreased across the entire disk in all specimens during supination (Figure 5). Greater decreases occurred in the central and radial regions of the disk than in the ulnar regions. During pronation, the opposite strain distribution was found. The central and ulnar regional areas decreased while the radial regions remained nearly isometric when compared to neutral. Rate of change in total disk area increased as the full pronation and supination were approached (Figure 4).

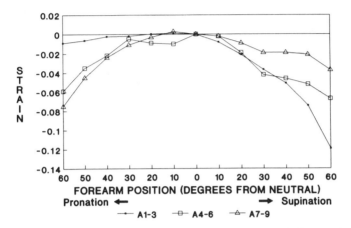

Figure 4. Average regional area strains (A) in all specimens are shown divided into three groups.

Area strains consistently paralleled longitudinal strains (Figure 6). Transverse strains were negligible (< 0.02) in all regions for all forearm positions. Thus, area strains and configuration changes were caused by changes in the longitudinal axis, radial-ulnar, of the disk and not the transverse axis, dorsal-palmar (Figure 7). Similar to area strains, longitudinal strains in the radial portion of the disk, L1 through L4, had positive strains or remained isometric during pronation. Thus, the radial portion of the disk behaved differently than the central and ulnar portions during pronation. Although the radioulnar ligaments were not evaluated, strains along the dorsal and palmar margins of the disk were calculated. Longitudinal strains along the dorsal margin decreased more during supination, while strain along the palmar margin decreased more during pronation.

DISCUSSION

Contemporary advances in diagnostic imaging techniques and wrist arthroscopy have made it possible to regularly identify and treat lesions in the articular disk of the TFCC.[8,9] Although the care of TFCC injuries has greatly improved with these technologies, our knowledge of injury mechanics is sparse. Current injury theories are

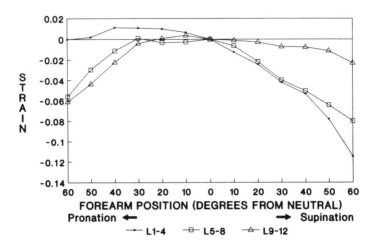

Figure 5. Average longitudinal strains (L) in all specimens are shown divided into three groups.

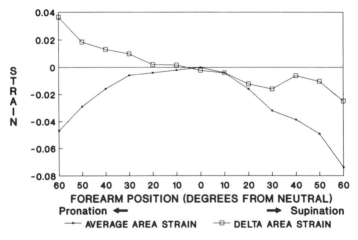

Figure 6. Average total area strain in all specimens. Rate of change in total area strain is also shown.

derived primarily from interpretations of clinical observations, gross anatomy dissections, and reviews of post mortem findings. Laboratory studies that support these theories are not available, however, some conclusions regarding distribution of stresses in the TFC have been made based on anatomical studies, force versus joint displacement analyses, and forearm axial load transmission experiments.[2,3,9,10] The results provide strong evidence that both compressive and tensile loads are regularly borne by the TFC. Bowers has suggested that a load conversion mechanism may occur in the TFC during axial loading of the ulnocarpal joint.[3] According to this mechanism, distal radioulnar joint distraction caused by axial loading is resisted by the TFC, thus converting some compressive stress in the disk to tensile stress in the radioulnar ligaments.

Most investigators agree that the TFCC is the primary stabilizer of the distal radioulnar joint, however, controversy exists regarding the relative tension in the radioulnar ligaments of the TFC during forearm motion. In a biomechanical study, Schuind, et al, demonstrated greater tension in the palmar ligament in full supination, and greater tension in the dorsal ligament in full pronation.[11] Based on an anatomical study, Ekenstam, et al, claimed the opposite occurred in the ligaments.[2] Both investigators, however, found reciprocal laxity in the other radioulnar ligament. In the present study, greater longitudinal shortening occurred along the dorsal margin of the disk in supination, and along the palmar margin in pronation. Thus, the results were consistent with the findings of Schuind, et al. Despite this controversy, the translational component of distal radioulnar joint motion is considered the cause of asymmetrical distribution of tension in the TFC. Using computed tomography (CT), translation is seen to occur primarily at the extremes of pronation and supination.[5,12] Anatomical studies have demonstrated that the difference in radii of curvatures between the radial and ulnar articular surfaces allows translation to occur.[2,3]

In kinematic studies, the translational component is demonstrated by a slight, but definite, shift in the axis of rotation during forearm motion.[6] The axis passes through the fovea of the ulnar head, a site of insertion for the TFC.[1] Because the TFC attachment is near the axis location, the TFC can guide and restrain distal radioulnar joint motion during simple rotation with minimal distortion of its anatomy and minimal impedance to joint motion. However, as soft tissues become taught at the extremes of pronation and supination, gliding along the incongruous articular surfaces occurs and results in asymmetrical changes in TFC configuration.[2,11] The purpose of this study was to quantify, by region and by direction, surface strains in the disk due to these configuration changes.

Although strains cannot be directly related to injury mechanics, the present results are consistent with previously reported clinical and histological findings. According to most authors, traumatic injuries to the TFCC usually result from an acute rotational injury to the forearm, an axial load and distraction injury to the ulnar border of the forearm, or a fall on the pronated outstretched upper extremity.[3,8] The most common site of a traumatic tear is located 2 mm ulnar to the radial attachment of the TFCC and oriented palmar to dorsal.[7,8] This site corresponds to the junction of short thick and radially oriented collagen fibers emanating from the radius with the remaining interweaving and obliquely oriented fibers in the central disk.[4] As collagen fibers tend to align with principle stresses, the transition in fiber arrangement suggests that a distinct change in functional requirements and material properties occurs in this region. The results in this study demonstrate that strains occur primarily in the longitudinal axis of the disk, with transverse strains being negligible. Progressively negative strains were consistently found across the disk during supination when compared to neutral. During pronation, regions along the radial attachment of the disk remained unchanged.

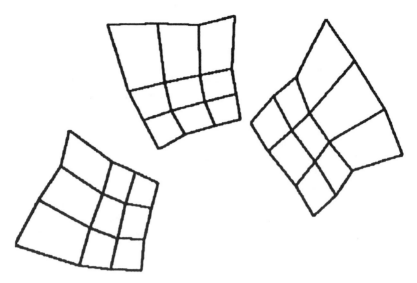

Figure 7. Intercentroid connecting grid for one specimen in pronation, neutral and supination is shown.

Therefore, tension is reduced across the entire disk in supination, while pronation leaves the radial portion of the disk under tension. As disparity in tissue architecture combined with asymmetrical loading can result in locally high stress gradients, these findings are consistent with a tensile loading mechanism of injury for disk tears near the radial attachment when the forearm is in the pronated position.

CONCLUSION

This study confirms our clinical impressions that changes in articular disk configuration consistently occur during forearm pronation and supination. These changes result in a nonuniform distribution of surface strains which is dependent upon forearm position. Although the findings are consistent with our current knowledge of disk injury, the full implications to injury mechanics will depend upon additional studies that evaluate the effects of tensile loading in different forearm positions. This experimental model and the current results will provide a basis for future study.

REFERENCES

1. B.D. Adams, Detrimental effects of resections in the articular disk of the TFCC, Proceedings of the 37th Annual Meeting, *Orthopaedic Research Society*, 212 (1991).
2. F. af Ekenstam, C.G. Hagert, Anatomical studies on the geometry and stability of the distal radio ulnar joint, *Scand. J Plast. Reconstr. Surg.* 19:17 (1985).
3. W.H. Bowers, The distal radioulnar joint, *in*: "Operative Hand Surgery," D. Green, ed., Churchill Livingstone, Inc., New York (1989).
4. L. K. Chidgey, P.C. Dell, E.S. Bittar, and S.S. Spanier, Histologic anatomy of the triangular fibrocartilage, *J. Hand Surg.* 16A:1084 (1991).

5. R.O. Cone, R. Szabo, D. Resnick, R. Gelberman, J. Talesnik, and L.A. Gilula, Computed tomography of the normal radioulnar joints, *Investigative Radiology* 18:541(1983).

6. G.J. King, R.Y. McMurty, J.D. Rubenstein, S.D. Gertzbein, Kinematics of the distal radioulnar joint, *J. Hand Surg.* 11A:798 (1986).

7. A.L. Osterman, R.G. Terrill, Arthroscopic treatment of TFCC lesions, *Hand Clinics* 7:277 (1991).

8. A.K. Palmer, Triangular fibrocartilage complex lesions: A classification, *J. Hand Surg.* 14(A):594 (1989).

9. A.K. Palmer, F.W. Werner, Biomechanics of the distal radioulnar joint, *Clin. Orthop.* 187:26 (1984).

10. A.K. Palmer and F.W. Werner, The triangular fibrocartilage complex of the wrist: Anatomy and function, *J. Hand Surg.* 6(2):153 (1981).

11. F. Schuind, K.N. An, L. Berglund, et al., The distal radioulnar ligaments: A biomechanical study, *J. Hand Surg.* 16A:1106 (1991).

12. R. Wechsler R et al., Computed tomography diagnosis of distal radioulnar subluxation, *Skeletal Radiol.* 16:1(1987).

THE ROLE OF THE SCAPHO - TRAPEZIAL - TRAPEZOIDAL LIGAMENT

COMPLEX ON SCAPHOID KINEMATICS

Christian L. Jantea,[1] Kai-Nan An,[2] Ronald L. Linscheid,[2]
and W.P. Cooney III[2]

[1]Orthopedic Department, Heinrich-Heine-University Düsseldorf, Germany
[2]Orthopedic Biomechanics Laboratory, Mayo Clinic/Mayo Foundation,
Rochester, MN, USA

INTRODUCTION

Instability of the scaphoid is a common factor in instabilities of the wrist.[17,26] Most scientific interest has focused on the scapholunate ligament with regard to the instability pattern of the scaphoid.[16,18,20]

There are few clinical reports where a complete or partial disruption of the scapho-trapezial-trapezoidal (STT) ligamentous complex has occured. This may be the case in a subluxation[11,15] or a dislocation of the trapezium.[22,32] The consequence of this condition generally is the development of osteoarthritis of the STT joint which may require operative treatment.[5,28]

In an experimental descriptive paper on the STT ligaments, a gap between the trapezium and the distal pole of the scaphoid was observed in ulnar deviation when the STT ligaments were sectioned.[7] However a more precise biomechanical and kinematic analysis of the STT ligament complex is desirable in order to understand the function of the STT ligaments.

PURPOSE OF THE STUDY

The purpose of this study is to assess the role of the STT ligament complex in stabilization of the distal scaphoid. The central questions are: [1] do the STT ligaments contribute to the stability of the scaphoid during flexion-extension motion (FEM) and radial-ulnar deviation (RUD) of the wrist and how is the motion pattern of the scaphoid altered when the STT ligaments are transsected ? [2] how does loading of the hand influence the kinematic behavior of the scaphoid with intact or transsected STT ligaments ?

Advances in the Biomechanics of the Hand and Wrist
Edited by F. Schuind *et al.*, Plenum Press, New York, 1994

MATERIAL AND METHOD

Material

A kinematic analysis was done in 10 fresh human cadaver wrists. The premortem histories were studied to exclude diseases deletorius to ligamentous integrity. Plain X-ray and CT scans (General Electrics model 9800) were obtained to exclude other bony and soft tissue pathology. One specimen with a scapholunate dissociation had to be excluded leaving 9 wrists of the 6 donors, 3 women and 3 men, while 3 (2 females, 1 male) contributed both wrists. The age ranged from 29 to 62 years with a meall of 42 years. These extensive pre-experimental studies were done to exclude carpal instabilities as the experiments were done without opening the carpal joints.

The skin and subcutaneous fat was removed, fiberglass rods were inserted in the scaphoid, trapezium, 1st and 3rd metacarpals so as not to interfere with the tendon excursion during the kinematic experiments.

Experimental setup

The radius was cemented in a custom device and rigidly fixed together with the source, which generated, the magnetic field to the same platform (Figure 3). The orientation of the source's coordinate system corresponds to the axis defined for the wrist motion (Figure 1).

Before implantation of the fiberglass rods a drill hole with a diameter of 2.2 mm was made in each bone without opening any of the carpal joints. Fiberglass rods do not interfere with the magnetic field as this would be the case with metal implants. Superglue (cyanoacrylate ester) was used to fix the rods in the bones to prevent axial rotation. Direct contact between the superglue and soft tissue was avoided in order not to alter the tissue properties. Small plastic platforms were fixed to the rods and the sensors of the Isotrack system were fixed to the latter by plastic screws (Figure 3).

The rod was inserted from the dorsum into the scaphoid between the tendons of the extensor pollicis longus and extensor carpi radialis muscles. The insertion area corresponds to the dorsal rim of the bone to which the dorsal transverse carpal ligament (ligament radio-carpeum arcuatum dorsum) running from the triquetrum over the midcarpal joint to the scaphoid is attached.[10,14,31] This dorsal rim also is the insertion area of the radio-scaphoid joint capsule. The dorsal STT joint capsule and ligaments insert on the distal part of this dorsal rim.[10,27]

A second rod was inserted in the trapezium from a palmar direction through the origin of the tendon of the abductor pollicis longus muscle. This pin was drilled through both the trapezium and trapezoid in order to avoid motion between these bones. The trapezial-trapezoidal (TT) joint motion can be clinically neglected as there is only about 3 degrees of relative motion between these bones.

A third rod was inserted into the first metacarpal adjacent to the carpo-metacarpal (CMC) joint of the thumb. The hand was moistened through the two hours average duration of the experiments with vaporized physiologic saline solution.

The sensors in the first metacarpal and the trapezium were used to calculate the relative motion between them. This has to be zero by definition in order not to interfere with recording the scaphoid motion at the STT joint level.

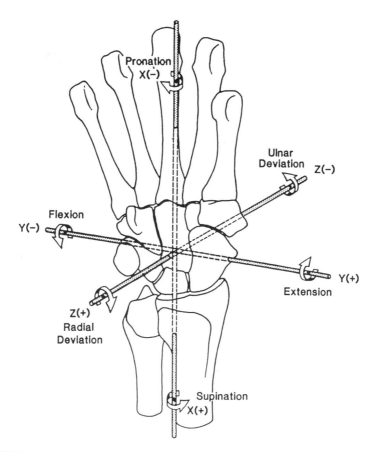

Figure 1. XYZ coordinate system used to describe the 3 motion components of the scaphoid during flexion-extension motion (FEM) and radial-ulnar deviation (RUD) of the wrist. Rotation around the X-axis describes the pronation-supination of the scaphoid while pronation is defined when the value of the angle becomes negative. The rotation around the Y-axis describes the extension-flexion from positive to negative angle values. Rotation around the Z-axis describes the ulnar to radial deviation while a positive angle defines the ulnar deviation.

Kinematic data acquisition was obtained from the sensors while the wrist was moved by another rod inserted in the 3rd metacarpal. The latter glided in a special device along a frame which was aligned with the flexion-extension motion (FEM) and radial-ulnar deviation (RUD) plane of the wrist (Figure 3).

The kinematic data of the scaphoid in different experimental conditions is presented.

Measurement method for kinematic data acquisition and analysis

Magnetic field technology provides a unique method to register the data required for analysis of the position and motion of a coordinate system attached to a moveable sensor with a fixed spherical coordinate system. The relative position and motion of the sensor to a source has to be described in terms of translation and rotation of the coordinate system located in the sensor. These variables (translation and rotation matrix) are necessary to describe the 6 degrees of freedom (DOF). These 6 variables were registered during the guided motion of the wrist in the different experimental conditions.

The analytical method for the description of 6 DOF spatial motion used in this study is the screw displacement axis (SDA).[1,25,29] The relative displacement of a moving segment from one position to another can be desribed in terms of a rotation around and a translation along a unique axis called the SDA which is fixed in the segment (Figure 2). The advantage of using a screw axis is that the orientation of the SDA remains invariant, regardless of the reference coordinate axes used. The SDA is a true vector quantity; its magnitude can be decomposed along any coordinate axes used for analysis. So the description of the motion pattern of the scaphoid around the 3 defined X-, Y- and Z-axes of the cartesian coordinate system is possible (Figure 1).

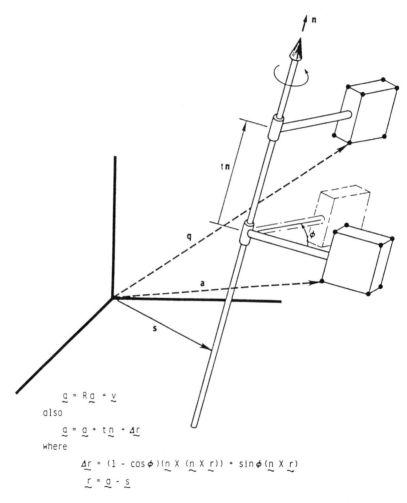

$$\underset{\sim}{q} = R\underset{\sim}{a} + \underset{\sim}{v}$$

also

$$\underset{\sim}{q} = \underset{\sim}{a} + t\underset{\sim}{n} + \Delta\underset{\sim}{r}$$

where

$$\Delta\underset{\sim}{r} = (1 - \cos\phi)(\underset{\sim}{n} \times (\underset{\sim}{n} \times \underset{\sim}{r})) + \sin\phi(\underset{\sim}{n} \times \underset{\sim}{r})$$

$$\underset{\sim}{r} = \underset{\sim}{a} - \underset{\sim}{s}$$

Figure 2. Description of the generalized 6 DOF joint motion by using the concept of the SDA. The relative displacement of a moving segment from the position to another can be defined in terms of rotation around and translation along a unique SDA. The SDA is a true vector quantity; its magnitude can be decomposed along any coordinate axis used for analysis. So the description of the motion pattern of the scaphoid is expressed in terms of rotation around the three defined X-, Y- and Z-axes of a cartesian coordinate system as defined in figure 1.

Figure 3. Device for kinematic data acquisition using the 3-Space Isotrack magnetic field technology. The source and radius are rigidly fixed on the same platform. Fiberglass rods are inserted in the scaphoid, trapezium and first metacarpal. These fiberglasrods are placed without opening the joint capsule or sectioning neither the retinaculae nor the carpal ligaments. To these rods the sensors of the Isotrack system were rigidly fixed. The rod in the 3rd metacarpal is used to guide the motion during FEM and RUD of the wrist. The thumb was rigidly fixed to the 3rd metacarpal to avoid relative motion at the STT joint level. The tendons proximal to the wrist are prepared. All tendon sheaths and the transverse carpal ligament remain intact. All tendons of the finger-, thumb- and wrist-muscles are loaded using two different loading conditions ("LL" = 1.4 kg and "HL" = 23.5 kg).

Magnetic field technology allows a continuous data acquisition from the sensors used through the whole range of motion. It has proved its accuracy (0.016" and 0.1 degree at 60 measurements per second) in other experimental kinesiologic studies.[2] The main advantage of this technology is that the performed motion does not require interruption for data acquisition. In this study a frequency of 15 Hz for data acquisition was chosen during which FEM and RUD of the wrist were performed over 10 seconds interval for one motion cycle.

Kinematic experiments for analysis of the motion pattern of the scaphoid

The following experiments were performed in order to assess the role of the STT ligaments in scaphoid motion.

Experiment "N-LL": normal/intact STT ligaments, hand low loaded. Normal muscle tension was simulated by loading all tendons of the hand in a low loading condition ("LL") whith intact STT ligaments ("N-LL"). Force distribution to each tendon was 100g, resulting in a total of 1.4 kg for all tendons of the hand (EPL 100g, EPB 100g, APB 100g, FPL 100g, EXT DIG COM 200g, FLEX DIG SUP 100g, FLEX DIG PROF 100g, ECRB 100g, ECRL 100g, ECU 200g, FCR 100g, FCU 100g). The kinematic data obtained from this experiment is the reference for the normal scaphoid motion, which may simulate the physiologic situation with intact STT ligaments.

Experiment "N-HL": normal/intact STT ligaments, hand high loaded. This experiment was to determinate how a high loading condition changes the normal motion pattern of the scaphoid with the STT ligaments intact. A total weight of 23.5 kg was distributed to the tendons of the hand (EPL 1.0 kg, EPB 0.5 kg, APB 0.5 kg, FPL 1.5 kg, EXT DIG COM 4.0 kg, FLEX DIG SUP 2.0 kg, FLEX DIG PROF 2.0 kg, ECRB 0.75 kg, ECRL 0.75 kg, ECU 3.5 kg, FCR 3.5 kg, FCU 3.5 kg).

Experiment "2-LL": STT ligaments dissected, hand low loaded. In this experiment an isolated STT dissociation was simulated by circumferentially cutting of the capsule and ligaments of the STT joint. The specimen was left attached to the fixation device. A needle was passed through the STT joint from dorsal to palmar in order to identify the STT ligaments accurately. The floor of the flexor carpi radialis (FCR) tendon sheath covers the palmar aspect of the STT ligaments and it is not possible to separate the two structures from each another. The flexor retinaculum covering the FCR sheath was left intact while sectioning the underlying STT ligaments. The tensile load on the tendons of the hand crossing the wrist was the same as in the experiment "N-LL", so a direct comparison between both experiments is possible.

Experiment "2-HL": STT ligaments dissected, hand high loaded. In this experiment the load was increased to 23.5 kg while the STT ligaments were sectioned to simulate an instability pattern of the scaphoid under a high axial load of the hand.

Analysis of the results. Differences in kinematics between the four experimental conditions ("N-LL", "N-HL", "2-LL", and "2–HL") were registred and statistical analysis of variance (ANOVA) using commercialy available software was performed.

RESULTS

Normal kinematics of the scaphoid during flexion-extension motion of the wrist

The 3 motion-components of the scaphoid are illustrated as rotation around the 3 axes of the cartesian coordinate system (Figure l). The curves for all specimens were interpolated to the same increment (of l degree).

For FEM of the wrist an interval from -20 degrees of extension to +40 degrees of flexion was chosen. For RUD the interval ranged from -20 degrees for ulnar deviation to +15 degrees for radial deviation.

The abscisse of each graph represents the motion of the wrist while the ordinate represents the scaphoid motion in respect to each axis of the cartesian coordinate system.

Figure 4 represents the average motion of the scaphoid for all specimens when FEM of the wrist was performed: [1] when the wrist is moved from extension to flexion the main motion component of the scaphoid is extension-flexion; in 20 degrees of wrist extension the scaphoid is extended in 22 degrees; at the neutral position of the wrist (point 0/0) the scaphoid is in a slightly extended position of 6 degrees; with the wrist flexed in 40 degrees the scaphoid flexes to 28 degrees (SD= 4.2 degrees); [2] the second motion component of the scaphoid during the FEM of the wrist is pronation and supination (rotation around the X-axis); the position of the scaphoid is 6 degrees of supination with the wrist extended, 2.3 degrees of supination in the neutral position of the wrist and 2 degrees of pronation at 40 degrees of wrist flexion (SD= 2.2 degrees); [3] rotation around the Z-axis which describes the RUD of the scaphoid is less during FEM of the wrist. In extension the scaphoid is in 2 degrees of ulnar deviation and rotates to the neutral position of 0 degree of rotation (SD= 2.3 degrees) during flexion of the wrist.

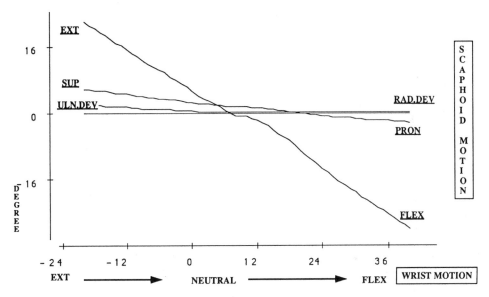

Figure 4. The 3 motion components of the scaphoid during the extension-flexion motion of the wrist with intact STT ligaments in an low-load condition with 1.4 kg (experiment "N-LL"). In the extended wrist position the scaphoid is extended, ulnar deviated and supinated (see on the left part of the graph). When the hand is flexed the scaphoid moves to a flexed, pronated and radial-deviated position.

Normal scaphoid motion during radial-ulnar deviation of the wrist

Figure 5 describes the motion pattern of the scaphoid around the 3 defined axes (Figure 1) during RUD of the wrist: [1] with the wrist in ulnar deviation of 20 degrees, the scaphoid is extended to 14 degrees; when the wrist is radially deviated to 15 degrees, the scaphoid is flexed to nearly 7 degrees (SD= 3.8 degrees); [2] the scaphoid rotates around the X-axis from 8 degrees of supination to 6 degrees of pronation (SD= 2.1 degrees) during RUD of the wrist; [3] the smallest amount of rotation of the scaphoid occurs around the Z-axis during RUD of the wrist; during an ulnar to radial deviation of the wrist, the scaphoid is ulnar deviated to 1.2 degrees and rotates to a radially deviated position of 0.8 degree.

Effect of STT dissociation and loading on the scaphoid motion during flexion extension motion of the wrist

The effects of sectioning of the STT ligaments and loading are presented separately for all 3 motion components of the scaphoid (Figures 6,7,8).

The FEM motion component of the scaphoid around the Y-axis when FEM of the wrist is performed shows no statistically significant difference (p= 0.098) between experiments. All curves show similar slope and configuration; however, loading sligthly decreases the FEM component of the scaphoid (Figure 6).

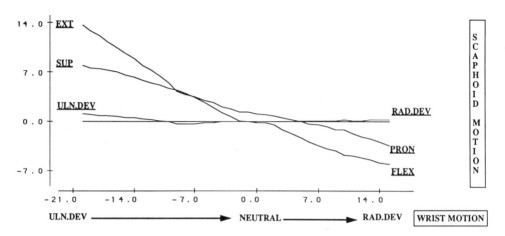

Figure 5. The 3 motion components of the scaphoid during RUD of the wrist with intact STT ligaments in an low-load condition with 1.4 kg (experiment "N-LL"). In ulnar deviation of the wrist the scaphoid is extended, supinated and ulnar-deviated (left part of the graph). At the end of radial deviation of the wrist the scaphoid is flexed, pronated and radial-deviated (right part of the graph).

Figure 7 illustrates the RUD scaphoid rotation around the Z-axis during FEM of the wrist. Sectioning of the STT ligaments (experiment "2-LL") decreased that range of motion. In a high loading condition (experiment "2-HL") the motion pattern is similar to the low loading condition (experiment "2-LL"), but displaced upwards, showing that the scaphoid remains ulnar-deviated during the FEM motion of the wrist. In summary, loading and sectioning of the STT ligaments significantly decrease the range for RUD of the scaphoid during FEM of the wrist.

Figure 8 illustrates the changes in the pronation supination motion component of the scaphoid during FEM of the wrist. In the normal situation (experiment "N-LL") and after sectioning of the STT ligaments in a low loading condition (experiment "2-LL") the curves have the same configuration and slope. However in the latter the amount of rotation of the scaphoid is considerably reduced. When the wrist is loaded (experiment "N-HL" and "2-HL") there is basically no pronation and supination of the scaphoid: both curves parallel the abscissis. In summary, when a qualitative comparison of the 3 motion components for the scaphoid is done during FEM of the wrist (Figures 6,7,8) the pronation supination motion component of the scaphoid is influenced most by loading the wrist or sectioning the STT ligaments.

Effect of STT dissociation and loading on the scaphoid motion during the radial-ulnar deviation motion

Figure 5 illustrates the 3 motion components of the scaphoid during RUD of the wrist with the STT ligaments intact and the muscles of the hand loaded with a total weight of 1.4 kg (experiment "N-LL"). The main motion component of the scaphoid is the FEM, followed by the pronation and supination during RUD of the wrist. This motion pattern is also observed in the other experimental conditions (Figures 9,10,11).

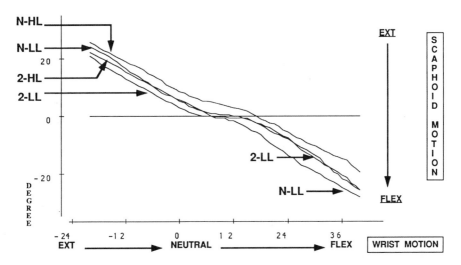

Figure 6. The motion component for extension-flexion motion of the scaphoid during extension-flexion motion of the wrist in different experimental conditions with the STT ligaments intact/sectioned under low/high load condition (experiments "N-LL", "N-HL", "2-LL", "2-HL"). Loading decreases the FEM of the scaphoid (compare experiment "-HL" to "-LL"), while sectioning of the STT ligaments decreases the range of FEM of the scaphoid (compare experiment 2- to N-).

Figure 9 demonstrates that sectioning the ligaments (experiment "2-LL") has the most dramatic effect on the FEM component of scaphoid motion, as the motion arc is reduced to 70 % (11.5 degrees) compared to the normal (experiment "N-LL" > 100%, or 16 degrees). Loading reduces the FEM component of the scaphoid regardless if the STT ligaments are intact or cut, though this not statistically significant.

Figure 10 illustrates scaphoid rotation around the Z-axis which describes the RUD of the scaphoid during the RUD of the wrist. With the STT ligaments intact (experiment "N-LL") the scaphoid motion occurs close to the neutral position, as this curve crosses the abscissa. When the STT ligaments are sectioned (experiment "2-LL") or the wrist is loaded (experiment "N-HL" or "2-HL") the scaphoid is deviated ulnarly, illustrated by the parallel shift of the curves to a positive value on the ordinate.

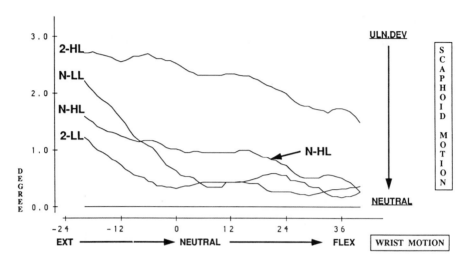

Figure 7. The motion component for radial to ulnar deviation of the scaphoid during extension-flexion motion of the wrist in the different experimental conditions with the STT ligaments intact/sectioned under low/high load condition (experiments "N-LL", "N-HL", "2-LL", "2-HL"). Loading and sectioning of the STT ligaments decrease the RUD motion component of the scaphoid.

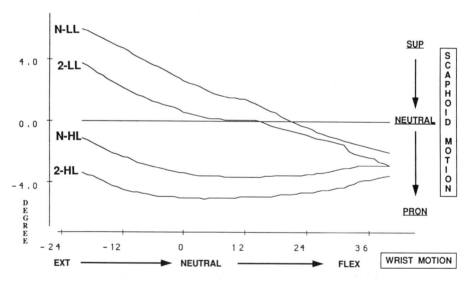

Figure 8. The motion component for supination-pronation of the scaphoid during the extension-flexion motion of the wrist in the different experimental conditions with the STT ligaments intact/sectioned under low/high load condition (experiments "N-LL", "N-HL", "2-LL", "2-HL"). Loading and dissection of the STT ligaments decrease the pronation-supination motion of the scaphoid.

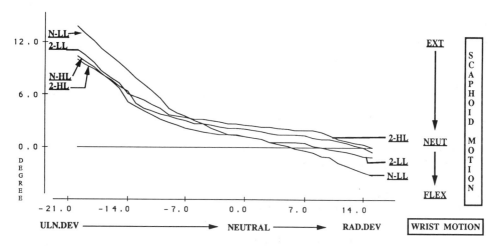

Figure 9. The motion component for extension-flexion of the scaphoid during ulnar-radial deviation motion of the wrist with the STT ligaments intact/sectioned under low/high load condition (experiments "N-LL", "N-HL", "2-LL", "2-HL"). Loading and sectioning of the STT ligaments decrease the extension-flexion motion component of the scaphoid.

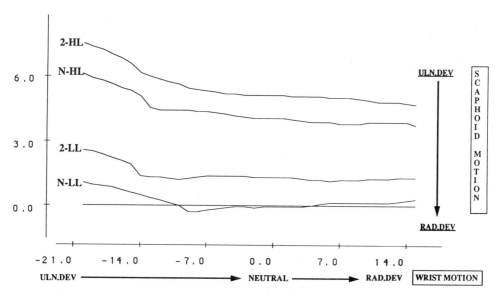

Figure 10. The motion component for ulnar-radial deviation of the scaphoid during ulnar-radial deviation motion of the wrist in the different experimental conditions with the STT ligaments intact/sectioned under low/high load condition (experiments "N-LL", "N-HL", "2-LL", "2-HL"). Under loading and after sectioning of the STT ligaments the scaphoid is translated in ulnar deviation (as the curves become displaced to the top of the graph).

Using the ANOVA method for a statistical comparison, there is a significant difference (p < 0.001) between the kinematic behavior of the scaphoid RUD motion component, during the RUD of the wrist, with a high loading (experiment "N-HL", "2-HL") compared to the experimental setup with a low load (experiment "N-HL", "2-HL"). However, no statistic significant difference could be found between the motion patterns of the scaphoid in the low load conditions with intact or sectioned STT ligaments (experiment "N-LL" compared to experiment "2-LL") .

Figure 11 shows the rotation of the scaphoid around the X-axis which represents the pronation and supination motion component of the scaphoid during RUD of the wrist. With intact STT ligaments (experiment "N-LL") the range of rotation is 14 degrees, which is considerably reduced (p < 0.001) after sectioning the STT ligaments in experiment "2-LL". This difference is significant only for the motion arc from ulnar deviation to the neutral position of the wrist. From neutral position to maximal radial deviation, the scaphoid pronates 6 to 7 respectively degrees for the experimental condition "N-LL" and "2-LL". For this second part of the motion no statistical difference was found.

Another interesting pattern is shown in figure 11: when the wrist is loaded pronation and supination of the scaphoid is reduced to 50% in the experiments "N-HL" and "2-HL", as compared to experiment "N-LL" and "2-LL". This difference is also statistically significant (p < 0.001). There is no statistical significant difference concerning the motion pattern of the scaphoid in the high load conditions whether the STT ligaments are intact or not (experiment "N-HL", "2-HL").

In summary the pronation and supination motion component of the scaphoid is the most affected motion component when sectioning the STT ligaments or loading is performed during the RUD motion of the wrist.

DISCUSSION

The role of the scaphoid in understanding carpal instabilities has been challenging. Interest has focused on the scapholunate ligament as being responsible for the rotatory instability of the scaphoid (RIS).[17,18] However a lesion of the STT ligaments may occur in traumatic conditions which are clinically seen as a luxation of the trapeziun or of the scaphoid itself. Concurrent attenuations of the ligaments at either end of the scaphoid may also occur.

The aim of this study was to assess the importance of the STT ligaments for the kinematics of the scaphoid during standardized wrist motion. To our knowledge there are no previous studies of the kinematic behavior of the scaphoid with special regard to the STT ligaments using a magnetic field tracking system and measurernent method (3-Space Isotrack).

Our previous experience on kinematic data analysis was based on the roentgen-stereophotogrammetric method applied in different experimental protocols related to carpal kinematics.[8,12,18,23] The disadvantages of this method are that an incision of the joint capsule has to be made for implantation of 4 metal markers in each carpal bone and a continuous registration of sensors motion cannot be done. X-ray exposure has to be performed in a static condition. The analysis of kinematic measurements of bones in biomechanical and kinematic studies[13] has reached a high technological standard.[21,30] The most accurate results are obtained with the roentgen-stereophotogrammetric system,[19,30] the sonic digitizer method,[3,33,34] and the six-degree-of-freedom spatial linkage.[24] Photographical-stereometric methods using optical signals from a light emitting diode (L.E.D.) such as in gait analysis are not suitable for an analysis of carpal

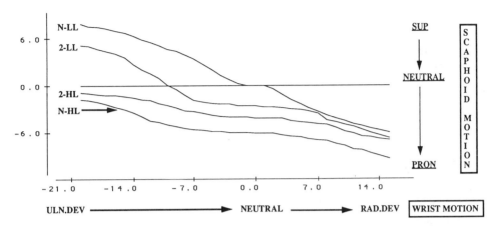

Figure 11. The motion component for supination-pronation motion of the scaphoid during ulnar-radial deviation motion of the wrist in the different experimental conditions with the STT ligaments intact/sectioned under low/high load conditions (experiments "N-LL", "N-HL", "2-LL", "2-HL"). Sectioning of the STT ligaments and loading decrease the amount of supination-pronation motion of the scaphoid (compare "N-LL" to "2-LL" and "N-HL" to "2-HL").

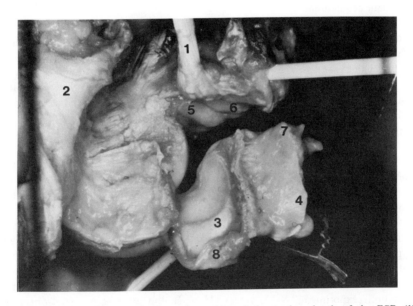

Figure 12. Palmar view from proximal on the wrist and the tendon sheath of the FCR (1) after exarticulation of the wrist by sectioning the palmar radiocarpal ligaments. The flexor retinaculum (2) is cut from its insertion at the scaphoid (3). The tendon sheat of the FCR (4) is opened and the course of the tendon along the palmar side of the scaphoid (4), trapezoideum (5) and trapezium (6) is visible. The STT ligaments and the palmar capsule of the STTjoint (7) are dissected. The scapholunate ligament (8) is also cut to visualize the scaphocapitate joint. The FCR tendon and the palmar STT ligaments have to be considered as a "functional unit" for stabilizing the distal pole of the scaphoid during the wrist motion.

kinematics.[29] All these methods require a fixed position of one of the rigid bodies when measurement is performed. With the sonic digitizer, data collection is performed in 25 intervals of the wrist motion.[3] Using the roentgen-stereogrammetric method, measurements of wrist motion in increments of 10 degrees were chosen.[30] The linkage method has not been applied for measurements of carpal kinematics. The motion between the different positions of the bone has to be extrapolated using different algorithms.

Most kinematic analyses performed in human cadaver hands were obtained from a population aged between 60-80 years, as this corresponds to natural mortality in an average population. Some authors report on carpal kinematics in hands of younger donors: 57 and 44 years in the study of Woltring et al.,[30] however this particular study was done in only 2 hands. In most papers specification of sex and age of the hands used for kinematic data analysis is not found. As kinematic data analysis is very time consuming, most papers only report on experiments performed with 2 to maximal 6 specimens.[8,12,23,30] In the present study the data of nine wrists with no pathologic findings detected in preexperimental examination is reported.

This study introduces for the first time the magnetic field technology (Isotrack 3-Space) for an experimental analysis of the kinematic behavior of the scaphoid. Advantages of this technology are the capability for continuous data acquisition and incremental motion analysis. This allows a graphic display of the continuous motion, rather than the static defined end positions of the scaphoid, such as the roentgen-stereophotogrammetric method. This technology allows comparison of the scaphoid motion in intermediate positions during motion under different conditions, with continous data acquisition.

Although described in more technological biomechanical papers, the concept of motion analysis presented in this paper is based on the kinematics of rigid bodies in 6 DOF. The methodologic approach is briefly summarized: a rigid body is defined in classical mechanics as "a system of mass points subject to the holonomic constraints that the distances between all pairs of points remain constant throughout the motion".[9] This can be assumed for the scaphoid as no change of the shape of the bone is to be expected, even when the hand is loaded. For a rigid body 3 reference points are required to determine the position and orientation of the object in space. A kinematic analysis with a complete and accurate quantitative description of motion requires 15 data variables.[1,29] These include the position vectors, linear velocity and acceleration of the rigid bodies segment's center of mass, angular orientation, angular velocity and angular acceleration. All these criteria are available with magnetic field technology (Isotrack 3-Space System).

In the experimental setup the physiologic motion pattern of the scaphoid was simulated with 1.4 kg crossing the wrist. In the second experimental condition 23.5 kg were applied. Under clinical circumstances this is a realistic force transmission through the wrist.

The different motion patterns of the scaphoid with intact STT ligaments it a low load condition (experiment "N-LL") are: [1] when the wrist is moved from extension to flexion in a 60 degrees arc, the scaphoid performs a flexion arc of 50 degrees; pronation and supination amount to 8 degrees and the RUD of the scaphoid has a range of 2 degrees (Figure 4); [2] during the RUD of the wrist in an arc of 35 degrees the main motion component for the scaphoid is FEM of 18 degrees. The amount of pronation and supination motion is 14 degrees. The scaphoid rotates around the Z-axis only 2.5 degrees during RUD of the wrist (Figure 5).

After sectioning the STT ligaments all three components of scaphoid motion are reduced. This could be found for FEM as well as for RUD of the wrist when a low load condition was simulated (experiments "N-LL" and "2-LL"). These experiments underline the importance of the STT ligamentous complex for the stability of the scaphoid. Our data for the normal motion of the scaphoid with the STT ligaments intact can be compared with other kinematic studies where the scaphoid motion was analyzed using the screw axis concept.[30] No description of a similar experimental protocol on the effect of loading was found in the literature. Loading led the scaphoid from its normal position to a flexed, ulnar deviated and pronated position when RUD motion of the wrist was performed (Figures 9,10,11). When FEM of the wrist was done the scaphoid moved under loading in a flexed, radial-deviated and pronated position (Figures 6,7,8).

Sectioning of the STT ligaments leads to a slight decrease of the 3 motion components (rotation of the scaphoid around the X-, Y-, and Z-axes) of the scaphoid during FEM of the wrist. During FEM of the wrist, the scaphoid is "stabilized" against rotation by loading which reduces the pronation supination considerably ($p < 0.001$). This happens whethler the STT ligaments are intact or not. In the high load condition there is no difference in the motion pattern of the scaphoid between the experiments "N-HL" and "2-LL".

The pronation supination motion component of the scaphoid is altered most significantly when the STT ligaments are sectioned. The other two motion components of the scaphoid (FEM and RUD) are not changed in their characteristics after sectioning of the STT ligaments was performed. Loading reduces the range of motion for all three motion components of the scaphoid during RUD of the wrist. Under the high load condition (experiment "N-HL" and "2-HL") there is no difference in the kinematic behavior of the scaphoid whether the STT ligaments are intact or cut.

Scaphoid motion decreases after sectioning the STT ligaments. The motion characteristics for the 3 motion components of the scaphoid remain however the same without regard to the STT ligaments intact or sectioned. Loading of the hand results in an overall decrease of the spatial motion of the scaphoid during FEM and RUD of the wrist. This phenomenon is due to the close attachment of the tendon sheath of the FCR to the palmar aspect of the scaphoid. Anatomic dissection of the specimens used in this study showed that the tendon sheath of the flexor carpi radialis is in direct contact to the palmar side of the trapezium and scaphoid. The FCR tendon inserts at the base of the second metacarpal. The palmar STT ligaments cannot be isolated from the tendon sheath of the FCR as shown in figure 12.

This finding and the kinematic data suggest that the FCR has to be considered as a "dynamic" stabilizer of the scaphoid during the wrist motion. The effect of the FCR tendon on the kinernatics of the scaphoid is explained by the "bow stringing" effect which causes an extension moment on the distal pole of the scaphoid when this tendon is loaded (experiment "N-HL" and "2-HL" - figure 13).

The present concept of classification of carpal instabilities describes the position of the lunate with regard to its angular inclination to the radius as flexed or extended, as an intercalated segrnent (VISI and DISI deformity). A similar description of the position of the scaphoid concerning the different instability pattern is not available. Based on these experiments, this study suggests a new classification for the RIS which may be described in a similar fashion as for the lunate: [1] for the rotatory instability of the scaphoid due to the scapholunate dissociation alone, the abbreviation RIS-A for this type can be used; [2] if dissociation between the scaphoid, trapezium and trapezoid (STT-D) is responsible for the collapse deformity of the scaphoid, the abbreviation RIS-B can be used; [3] if both ligamentous structures of the scaphoid, the distal intrinsic STT ligaments and the proximal intrinsic scapholunate and/or extrinsic scaphoradial ligaments are ruptured or attenuated, this should be described by the abbreviation RIS-C.

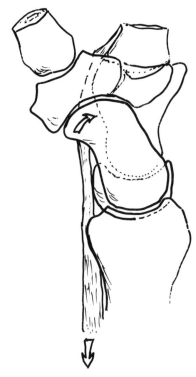

Figure 13. Lateral view of the wrist with the tendon of the FCR. When a tensile load is applied on the FCR tendon a force vector is directed on the distal pole of the scaphoid. This anatomic finding, the "bow stringing effect" of the FCR is consistent with the kinematic data, therefore this study suggests to consider the FCR muscle as a "dynamic" stabilizer of the scaphoid during the wrist motion.

Furthermore our clinical experience may suggest that a secondary attenuation of the STT ligaments occurs as a result of chronic scapholunate dissociation. Under clinical circumstances the evaluation of the instability pattern of the scaphoid should include all assessment of the STT ligaments according to the previous suggested classification for the rotatory instability of the scaphoid (RIS).

This may be helpful for a further classification of carpal instabilities as treatment options for the rotatory instability of the RIS will vary arnong the different conditions in the 3 different RIS-A, RIS-B and RIS-C types of scaphoid collapse due to ligamentous carpal instabilities.

ACKNOWLEDGEMENTS

This study was supported in part by the Deutsche Forschungsgemeinschaft D..F.G. (German Research Society). This experimental study was done by the first author at the Orthopedic Biomechanics Laboratory of the Mayo Clinic, Rochester, Minnesota, USA.

REFERENCES

1. K.N. An, and E.Y.S. Chao, Kinematic analysis of human movement, *Annals of Biomedical Engineering*, 12:585 (1984).

2. K.N. An, M.C. Jacobsen, L.J. Rerglund, and E.Y.S. Chao, Application of magnetic tracking device to kinesiologic studies, *J. Biomech.* 21:613 (1988).

3. J.G. Andrews, and Y. Youm, A biomechanical investigation of wrist kinematics, *J. Biomech.* 12: 83 (1979).

4. D. Buck-Gramcko, Carpal instabilities, *Orthopäde*, 15:88 (1986).

5. E.B. Crosby, R.L. Linscheid, and J.H. Dobyns, Scaphotrapezial trapezoidal arthrosis, *J. Hand Surg.* 223 (1978).

6. A. de Lange, A kinematic study of the human wrist joint, thesis accepted by the university of Nijmegen, Konoklijke Bibliotheek, Den Haag, The Netherlands (1987).

7. J.J. Drewniany, A.K. Palmer, and E. Flatt, The scaphotrapezial ligament complex: an anatomic and biomechanical study, *J. Hand Surg.* 10A:492 (1985).

8. M. Garcia-Elias, W.P. Cooney, K.N. An, R.L. Linscheid, and E.Y.S. Chao, Wrist kinematics after limited intercarpal arthrodesis, *J. Hand Surg.* 14A , 5:791 (1990).

9. H. Goldstein, The kinematics of rigid body motion, Chapter 4 in Classical Mechanics, 2nd edition Reading, Mass. Addison-Wesley, 1970.

10. L. Guyot, Anatomy of the limbs, Springer Verlag Berlin (1980).

11. F.M. Hankin, P.C. Amadio, E.M. Wojtys, and E.M. Braunstein, Carpal instability with volar flexion of the proximal row associated with injury to the scaphotrapezial ligament: report of two cases, *J. Hand Surg.* 13B:298 (1988).

12. E. Horii, M. Garcia-Elias, K.N. An, A.T. Bishop, W.P. Cooney, R.L. Linscheid, and E.Y.S. Chao, A kinematic study of luno-triquetral dissociations, *J. Hand Surg.* 16 A:355 (1991).

13. R. Huiskes, A. Kremers, A. de Lange, H.J. Woltring, G. Selvik, and T.J.G. van Rens, Analytical stereophotogrammetric determination of three-dimensional knee-joint geometry, *J. Biomech.* 18:559 (1985).

14. A. Kapandji, Biomechanics of the carpus and the wrist, *Orthopäde*, 12:60 (1986).

15. E. Kuur, and A.M. Boe, Scaphoid trapezium trapezoid subluxation, *J. Hand Surg.* 11B:434 (1986).

16. R.L. Linscheid, Traumatic instability of the wrist, diagnosis, classification and pathomechanics, *J. Bone Joint Surg.*, 54A:1612 (1972).

17. R.L. Linscheid, Kinematic considerations of the wrist, *Clin. Orthop.* 202:27 (1986).

18. L.K. Ruby, K.N. An, R.L. Linscheid, W.P. Cooney, and E.Y.S. Chao, The effect of scapholunate ligament section on scapholunate motion, *J. Hand Surg.* 12A:767 (1987).

19. G. Selvik, A roentgenenstereophotogrammetric method for the study of kinematics of the skeletal system, Dissertation, University of Lund, Sweden (1974).

20. G. Sennwald, The Wrist, Springer Verlag Berlin (1990).

21. R. Shapiro, Direct linear transformation method for three dimensional cinematography, *Research Quarterly*, 49:197 (1978).

22. D.A. Sherlock, Traumatic dorsoradial dislocation of the trapezium, *J. Hand Surg.* 12A:262 (1987).

23. D.K. Smith, K.N. An, W.P. Cooney, R.L. Linscheid, and E.Y.S. Chao, Effects of a scaphoid waist osteotomy on carpal kinematics, *J. Orthop. Res.*, 4 :590 (1989).

24. H.J. Sommer, and N.R. Miller, A technique for kinematic modeling of anatomical joints, *J. Biomech. Engin.* 102:311 (1980).

25. C.W. Spoor, and F.E. Veldpaus, Rigid body motion calculated from spatial coordinates of markers, *J. Biomech.* 13:391(1980).

26. J. Taleisnik, The Wrist, Churchill Livingstone Inc. (1985).

27. T. von Lanz, W. Wachsmuth, Praktische Anatomie, Arm, Springer Verlag, Berlin (1959).

28. K. Wilhelm, A. Role, and A. Hild, Die Skaphoid Trapezium Trapezoid Arthrose (STT), *Unfallchirurg.* 92:59 (1988).

29. D.A. Winter, Biomechanics of Human Movement, New York: John Wiley and Sons, Inc., 10 (1979).

30. H.J. Woltring, R. Huiskes, A. de Lange, and F.E. Veldpaus, Finite centroid and helical axis estimation from noisy landmark measurements in the study of human joint kinematics, *J. Biomech.* 18:379 (1985).

31. G.E. Woyasek, and H. Laske, Der Bandapparat des Kahnbeines, *Handchir. Mikrochir. Plast. Chir.* 23 :18 (1991).

32. L. Yao, and J.K. Lee, Palmar dislocation of the trapezoid: case report, *J. Trauma* 29:405 (1987).

33. Y. Youm, and Y.S. Yoon, Analytical development in investigation of wrist kinematics, *J. Biomech.* 12:613 (1979).

34. Y. Youm, and A.E. Flatt, Kinematics of the wrist, *Clin. Orthop.* 149:21 (1986).

THREE-DIMENSIONAL IN-VIVO KINEMATIC ANALYSIS OF FINGER

MOVEMENT

Sandro Fioretti

Dipartimento di Elettronica ed Automatica
Universita' di Ancona
Ancona, Italy

INTRODUCTION

A large part of biomechanics literature concerning the hand has focused its interest on the long fingers and in particular on the index. In fact, this latter is characterized by the highest mobility with respect to the other long fingers and has an important functional role in allowing the movement of opposition with the thumb. The joint that mostly provides the index finger with the mobility and the stability necessary to perform useful work is the metacarpophalangeal (MCP) joint. Many investigators have studied different aspects of this joint such as its anatomy or its kinematic and dynamic behaviour. But most of these studies have been performed in-vitro and, as far as kinematics is concerned, in-vivo studies were limited only to the determination of range of motion or to the assessment of finger orientation in static conditions.

In this paper the methods and the results of a three-dimensional and in-vivo kinematic analysis of the MCP joint of the index finger will be presented. Beside classical kinematic description in terms of angular parameterisation, emphasis will be given to the description of the relative movement of contiguous body segments by means of the Instantaneous Helical Axis (IHA) concept. Under the assumption of rigidity of body segments, at each moment in time the relative movement of one body segment with respect to the adjacent one can be described as the composition of a translation velocity along and a rotation velocity about an axis in space (i.e. the IHA).

In general, if referred to anatomical coordinate systems, the instantaneous axis of rotation can give a significant perception of movement evolution. Moreover, lever arms of ligaments and muscles, as well as other dynamic quantities may be computed with respect to this axis. Consequently, IHA determination can be very useful in fields like articular physiology, prosthetic design, replacement and reconstructing surgery.

The instantaneous axis of rotation has been often approximated by its finite counterpart,[5,13,19,24,25] i.e. by the so called Finite Helical Axis (FHA). In fact one usually disposes of a sampled version of a movement. In this case, every finite displacement can be characterized by a finite translation along, and a finite rotation about the FHA. Unfortunately, FHA parameters are highly sensitive to measurement noise affecting landmark coordinates.[24,25] Moreover, in the practical case of noisy observations, the approximation of the instantaneous axis of rotation by means of the FHA degrades as the sampling interval decreases.

It has been shown[25] that the determination of the instantaneous axis of rotation can be improved if a continuous time rigid-body model is adopted. In this case, with respect to its finite counterpart, IHA parameters are characterized by less incertitude but their estimates require the knowledge of landmarks velocity.

As it is well known, derivative estimates obtained by numerical differentiation of noisy data are highly sensitive to measurement noise (see for instance[14,18]). This implies that reliable estimates of velocity data can be obtained both reducing as much as possible the noise superimposed on measured kinematic data and employing very accurate stereophotogrammetric and numerical differentiation algorithms.

The accuracy in the IHA determination depends not only on the goodness of kinematic data but also on the way it is calculated and on the experimental protocol adopted.

Consequently, the following sections will be devoted to describe, in sequence, the measurement system employed, the stereophotogrammetric and smoothing/differentiating methods suitably developed, the kinematic equations used to describe the relative movement at a joint and the experimental protocol followed. Finally, results will be presented and discussed.

MEASUREMENT SYSTEM

The measurement system employed is CoSTEL.[15,16] It is an optoelectronic system constituted by three cameras and a main unit connected to a personal computer. Each camera is based on a linear CCD array sensor and an anamorphic objective with nodal axis perpendicular to the linear CCD array. Every sensor can locate the plane passing through the lens nodal axis and the tracked landmark. By the intersection of three independent planes (one plane per each sensor) it is possible to uniquely identify the landmark in the 3-D space.

Figure 1. Stereophotogrammetric apparatus: CoSTEL system and calibration object.

Active markers are required. One marker at a time can be tracked so that time multiplexing of IR-LED firing is required.

Up to 20 active markers can be tracked by the system with a sampling frequency ranging from 12.5 to 200 Hz. The resolution is 1/4096 the largest dimension of the observed scene.

Figure 1 shows the actual geometrical disposition of the cameras as results from a compromise between the specifications relative to the visibility of markers by each sensor and those related to the best reconstruction of the depth coordinate. The two lateral sensors are 52 cm far from the central one, and all three sensors are 120 cm far from the centre of the experimental volume.

STEREOPHOTOGRAMMETRY

The stereophotogrammetric algorithm adopted is described in detail in Fioretti et al.,[7] and in Fioretti et al.[9] It is based on a black box approach to describe the functional relationship between object and image space coordinates.

If $\mathbf{X}=[X_1,X_2,X_3]^T$ are the object space coordinates of a point and $\mathbf{x}=[x_1,x_2,x_3]^T$ are the corresponding image coordinates of the same point seen by the three cameras of CoSTEL system, the projective mapping between object and image spaces can be described by a set of nonlinear equations of the following form:

$$\mathbf{g}(\mathbf{x},\mathbf{X})=\mathbf{0} \tag{1}$$

It can be shown[11] that the vector function $\mathbf{g}(.,.)$ in (1) can be explicitly solved for $\mathbf{X}$, giving:

$$\mathbf{X}=\mathbf{f}(\mathbf{x}) \tag{2}$$

where $\mathbf{f}(.)$ can be expanded in Taylor's series.

In case of CoSTEL system in the actual configuration, this expansion can be truncated at a relatively low order (four). Equation (2) can thus be approximated by:

$$X_i = \mathbf{C}(\mathbf{x})^T\mathbf{p}_i \quad i=1,2,3 \tag{3}$$

where $\mathbf{p}_i$ is the vector of parameters relative to the i-th object coordinate and $\mathbf{C}(\mathbf{x})$ is a vector of coefficients.

The parameters $\mathbf{p}_i$ are not directly related to any internal characteristics of the optical sensors or to any geometrical quantity describing the position of the cameras in the object space. This makes this approach suitable also for other kinds of sensors such as, for example, classical photographic cameras.

In order to estimate the values of the photogrammetric parameters $\mathbf{p}_i$, a calibration procedure must be run. Parameters $\mathbf{p}_i$ can be estimated when both image and object coordinates of a suitable set of control points are known. Moreover a suitable stochastic model of the experimental incertitude must be defined (see for instance[17]).

The assumed stochastic model is:

$$X_i(j)=\mathbf{C}(\mathbf{x}(j))^T\mathbf{p}_i + v_i(j) \qquad i=1,2,3 \qquad j=1,2,...,N \tag{4}$$

where j is the index of the j-th control point and $v_i(.)$ is the random residual supposed zero-mean and normally distributed.

In order to determine the $\mathbf{p}_i$ parameters, equation (4) must be solved in a context of overdeterminacy by means of classical least squares techniques.

The number of parameters that must be estimated is 35. Consequently a calibration object with at least 35 control points is needed. The quality of the estimate depends also on the regularity of control-point distribution within the object-space. The calibration object shown in Figure 1 distributes with quadrantal symmetry a maximum of 504 control points in a volume of 0.42x0.56x0.56 m^3. The control points are spaced 70 mm in the three space directions.

The overall accuracy of the stereophotogrammetric algorithm for a measurement field of 0.28x0.42x0.42 m^3 resulted in residuals root mean square values of 0.43, 0.31 and 0.25 mm for the depth, width and height directions respectively. These values have been calculated with reference to 126 points of known position in the 3-D space and not coincident with the control points used in the calibration procedure. Moreover the residuals resulted uncorrelated and gaussian distributed at a 5% confidence limit.

SMOOTHING AND DIFFERENTIATION

A fixed-lag Kalman smoother for smoothing/differentiating noisy data has been developed. Mathematical details are given in Fioretti and Jetto,[8] and in Fioretti et al.[9] This technique is based on a minimum variance, state space approach, according to the well-known advantages of this approach with respect to the precision of the estimates. In fact the maximum precision theoretically obtainable is more sensitive to the model noise than to the measurement noise. Hence, the inferior limit of the estimate error covariance matrix can be made small at will provided that an adequate signal model is available. The signal model is based on the assumption of a band-limited signal and it is defined as follows:

$$\mathbf{X}((k+1)\Delta) = \mathbf{A}\,\mathbf{X}(k\Delta) + \mathbf{W}(k\Delta)$$

$$y(k\Delta) = \mathbf{C}\,\mathbf{X}(k\Delta) + v(k\Delta) \qquad\qquad k=0,1,2,...,N$$

where:

$$\mathbf{A} = \begin{bmatrix} 1 & \Delta & \Delta^2/2! & ... & \Delta^M/M! \\ 0 & 1 & \Delta & ... & \Delta^{M-1}/(M-1)! \\ ... & ... & ... & ... & ... \\ 0 & 0 & 0 & ... & 1 \end{bmatrix}$$

$\mathbf{X}(.)$ is the state vector that must be estimated. It is composed of the signal and its time derivatives up to a predefined order (M);
$\mathbf{W}(.)$ is the state noise vector;
Δ is the sampling interval;
$y(.)$ is the noisy discrete sequence of the N observations of the signal;
$\mathbf{C}$ is the (M+1) row vector [1, 0, ... , 0];
$v(.)$ is the measurement noise sequence supposed to be zero mean, white and normally distributed.

The variance of the measurement noise must be given to the filter; the inaccuracy of the model with respect to the true signal is quantified by a state-noise term supposed uncorrelated with the measurement noise. The state-noise variance is automatically determined by a variance-matching algorithm. A third order (M=3) Kalman smoother has been adopted because this choice guarantees very reliable estimates of the first order derivative.

KINEMATICS

To locate at any instant in time the position of any point in a rigid-body segment relative to an arbitrarily chosen reference system it is necessary to define a local coordinate system (l.c.s.) fixed to the segment and a global coordinate system (g.c.s.) fixed to the observer. Given the position vector x_{lo} of any point in the l.c.s., the relevant position vector x_{gl} in the g.c.s. is given by:

$$x_{gl} = R \; x_{lo} + p$$

where R is a 3x3 orthonormal attitude matrix and p is the position vector of the origin of the l.c.s. in the g.c.s.

The R matrix and the p vector have been calculated by the algorithm described in Spoor and Veldpaus.[20]

Given two contiguous rigid-body segments, joint movement is usually defined as the movement of the distal body segment with respect to the proximal one. If the proximal and distal segments are described in the g.c.s. by the R_p and R_d matrices and p_p and p_d vectors respectively, then the corresponding relative joint kinematics is given by:

$$x_p = R_j \; x_d + p_j$$

where R_j and p_j are given by:

$$R_j = R^T_p R_d$$

$$p_j = R^T_p (p_d - p_p)$$

Once R_j matrix is calculated, it is then possible to calculate cardan angles according to any possible attitude parameterisation. In particular the same convention adopted in Chao et al.,[3] has been used. The finite angular displacement of the distal body with respect to the proximal one is decomposed into the following sequence of elementary rotations: the first rotation occurs about the flexion/extension axis of the proximal coordinate system, the third rotation is along the longitudinal axis (internal/external rotation) of the distal coordinate system, and the second rotation is along a floating axis (lateral-lateral deviation) which is perpendicular to the two other ones and embedded in neither coordinate system.

An alternative description of joint kinematics can be given in terms of the IHA. The relative parameters can be derived when one disposes of the angular velocity vector w. In particular, an unbiased estimate of w can be given as follows:[22]

$$\mathbf{w} = \left[\sum_{i=1}^{L} \left[\mathbf{S}\{\mathbf{y}_i - \mathbf{p}\} \right]^T \left[\mathbf{S}\{\mathbf{y}_i - \mathbf{p}\} \right] \right]^{-1} \sum_{i=1}^{L} \left[\mathbf{S}\{\mathbf{y}_i - \mathbf{p}\} \right] (\dot{\mathbf{y}}_i - \dot{\mathbf{p}})$$

where:

$\mathbf{S}\{\mathbf{a}\}$ is an anti-symmetric matrix defined from its axial vector $\mathbf{a} = [a_1, a_2, a_3]^T$ as follows:

$$\mathbf{S}\{\mathbf{a}\} = \begin{bmatrix} 0 & -a_3 & a_2 \\ a_3 & 0 & -a_1 \\ -a_2 & a_1 & 0 \end{bmatrix}$$

$\mathbf{y}_i$ is the position vector of the i-th landmark in the reference coordinate system;
$\dot{\mathbf{y}}_i$ is the velocity vector of the i-th landmark in the reference coordinate system;
$\mathbf{p}$ is the position of the landmarks geometric centre in the reference coordinate system;
$\dot{\mathbf{p}}$ is the velocity vector of the landmarks geometric centre in the reference coordinate system;
L is the number of landmarks rigid to the moving segment.

The parameters that characterize uniquely the IHA can be calculated as follows:

$w = \sqrt{\mathbf{w}^T \mathbf{w}}$ rotation speed about the IHA

$\mathbf{n} = \mathbf{w}/w$ unit direction vector

$v = \dot{\mathbf{p}}^T \mathbf{n}$ translation speed along the IHA

$\mathbf{s} = \mathbf{p} + \mathbf{n}_* \dot{\mathbf{p}}/w$ position of the IHA

The IHA is thus that axis about which, and along which, the moving segment or joint is moving at a given time, with rotation and translation speeds w and v respectively. The position of the IHA is defined by the projection $\mathbf{s}$ of $\mathbf{p}$ onto it, i.e. by the projection of the landmark geometric centre onto the rotation axis.

As well known, the determination of the IHA parameters is highly sensitive to noise. In order to test the efficacy of the smoothing/differentiating algorithm on the accuracy of the IHA parameters, a simulation has been performed. It consisted in estimating the IHA's for a cylinder rolling with a constant angular velocity over a plane surface (Figure 2).

The IHA at each moment in time is the generatrix of the cylinder lying on the XZ-plane. The radius R of the cylinder has been chosen equal to 100 mm. Three points have been considered at the vertices of an equilateral triangle inscribed on the rotating base of the cylinder. Their trajectories have been sampled in such a way to have available 64 samples per each revolution of the cylinder. The sampling period has been assumed equal to unity. The sampled data have been corrupted with additive, zero mean, white noise (standard deviation equal to 0.5 mm). Table 1 shows the results in comparison with those obtained smoothing data by means of quintic splines.[21]

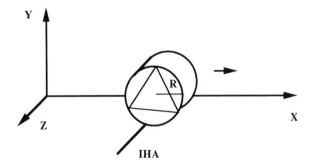

Figure 2. Rolling cylinder along X direction, over XZ-plane.

Table 1. Mean values (μ) and standard deviations (σ) of residuals between estimated and simulated values of IHA parameters: w=rotation speed about IHA; v = translation speed along IHA; I_x, I_y, I_z = Direction cosines of IHA; d_x, d_y = intersection point of IHA with XY-plane. Simulation performed with additive, white gaussian noise standard deviation = 0.5 mm.

	Quintic Splines		Kalman Smoother	
	μ (x10^{-3})	σ (x10^{-3})	μ (x10^{-3})	σ (x10^{-3})
w (rad/s)	-0.16	0.92	0.10	0.28
v (mm/s)	-2.59	126.0	0.35	22.0
I_x	0.15	10.8	-0.36	1.65
I_y	-2.18	12.5	-1.12	2.82
I_z	0.13	0.16	0.006	0.008
d_x (m)	0.19	0.53	0.09	0.32
d_y (m)	-0.16	1.27	0.18	0.30

Simulation results shown in table 1 put into evidence that Kalman smoother, with respect to quintic splines smoother, introduces less bias in the estimation of IHA parameters and gives a better identification of the IHA direction with reduced scattering around the true one. Moreover, an acceptable accuracy has been obtained in the estimation of the translation velocity and of the coordinates of the IHA intersection with the XY-plane. In particular the standard deviation of the error in the intersection point coordinates is comparable with the standard deviation of the noise superimposed to simulated data.

EXPERIMENTAL PROTOCOL

The attention has been focused on the flexion of the index finger starting from hyperextended position.

Subjects were asked to flex their finger as fast and as much as possible. In order to reduce skin motion artifacts, the following experimental protocol has been adopted:

distal phalanges are kept in line with the first one by a light rigid bar placed on the anterior aspect of the index finger. Soldered with the bar there is a cluster of active markers. Four IR-LEDs (OD100) are placed at the vertices of a rectangle with sides of 68 and 45 mm as shown in figure 3. Though a cluster of three markers would be sufficient to uniquely locate one rigid body in the 3-D space, a fourth marker is employed in order to obtain better estimates of the angular velocity vector and, consequently, of the IHA parameters. In particular the variances associated with the incertitude in the estimate of the rotation and velocity speeds are inversely proportional to the number of landmarks employed.[23]

The cluster of markers is placed on the finger so that the plane passing through the LEDs resulted coincident with the anatomical sagittal plane of the finger. Care must be given in order to minimise the distance between the geometric centre of the marker distribution and the axis of rotation at the MCP-joint. In fact the accuracy of IHA parameters depends inversely on this distance. On average, the geometric centre of the cluster resulted about 60 mm far from the centre of the MCP joint.

A fifth marker is placed on the skin of the MCP joint, approximately in correspondence of its axis of rotation.

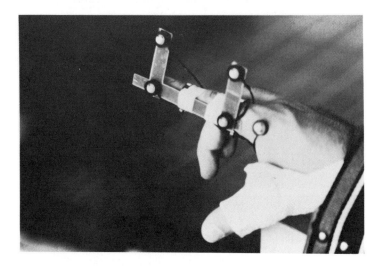

Figure 3. Cluster of markers used for experiments on index finger.

The cluster of markers identifies the technical coordinate system relative to the finger.

An analogous technical coordinate system placed in correspondence of the first metacarpus resulted too much sensitive to skin movements. Consequently, in order to associate a coordinate system to the proximal segment, the finger is kept in line with the first metacarpus and it is supposed that in this position the orientation of the coordinate system, fixed to the metacarpus, is coincident with the orientation of the coordinate system placed on the finger. In order to prevent changes in the orientation of the metacarpal system during flexion, hand and forearm are firmly attached on a suitably shaped holder.

Figure 4 shows a skematic representation of the technical coordinate systems adopted. An anatomical calibration of finger and metacarpus is carried out[2] in order to relate the technical coordinate systems with the anatomical ones. With the finger kept in line with the first metacarpus, a linear stick carrying two markers at known distance is used to point the position of the following points: the tip of the index finger, the superior and lateral prominences of the metacarpal head and a point placed on the proximal end of the first metacarpal bone. The dimensions of the finger and of the metacarpal head are also measured using a calliper.

The origin of the metacarpal and phalangeal technical coordinate system are placed, respectively, coincident with the centre of the metacarpal head and the geometric centre of the cluster of markers.

The maximum sampling frequency (200 Hz) allowed by CoSTEL system is adopted.

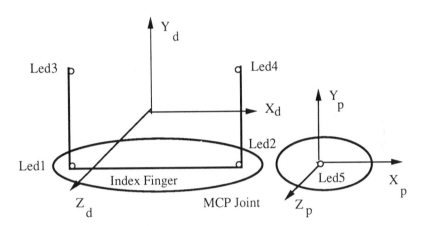

Figure 4. Technical coordinate systems and marker placement.

RESULTS

Sixteen subjects with no pathological condition of the hand joints have been analysed. Table 2 shows the average ranges of motion calculated on the basis of the cardan angles described in the Kinematics section.

These results put into evidence that the voluntary movement, i.e. flexion, is accompanied by limited radial-ulnar-deviation and pronation-supination rotations. This latter movement is surely a passive one, mainly due to the shapes of the articular surfaces and to the constraints constituted by capsulo-ligamentous tissues surrounding the joint and in particular by the asymmetrical collateral ligaments.

Slight differences in the shape of the angular trajectories exist both in the same subject and among different subjects when experiments are repeated. However, in most cases, the set of results evidences the pattern shown in figure 5. When finger has not yet reached the neutral position, the flexion movement is accompanied by radial deviation and pronation; then, from the neutral position to the last 20-30 degrees of flexion, supination and ulnar deviation are more evident, while the last 20-30 degrees of flexion are characterised again by more evident pronation and radial deviation.

As far as the IHA's are concerned, they form a bundle of axes that remain almost parallel and perpendicular to the metacarpal sagittal plane. In particular table 3 shows the direction cosines of the Mean Helical Axis (MHA) direction and the dispersion angle of the IHA's with respect to the MHA.

Table 2. Average ranges of motion (degree) for index flexion starting from hyperextended position

	Average value	Standard deviation
Flexion	77.9	14.9
Extension	17.0	9.4
Radial-Deviation	3.8	2.4
Ulnar-Deviation	3.9	1.4
Pronation	0.9	2.5
Supination	5.8	3.0

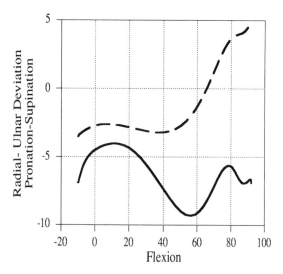

Figure 5. Radial-ulnar-deviation (dashed line) and pronation-supination (continuous line) vs flexion angle. Pronation and radial-deviation if shapes are increasing. All values in degrees.

Table 3. Direction cosines (c_x, c_y and c_z) of the Mean Helical Axis and dispersion angle χ (deg) of the IHA's with respect to the MHA. Mean (standard deviation) values.

c_x	-0.056 (0.057)
c_y	-0.046 (0.084)
c_z	0.993 (0.09)
χ	4.8 (0.06)

It is evident from data reported in table 3, that in all experiments the MHA resulted almost perpendicular to the $X_p Y_p$-plane ($c_z \geq 0.99$ with a very low standard deviation) and that the dispersion χ of the IHA's was lower than 5 degrees.

In the majority of the tested subjects, the intersections of the IHA's with the metacarpal sagittal plane show a common behaviour: [1] they tend to become more and more palmar as flexion increases, [2] with respect to the "calibrated" centre of the metacarpal head, IHA's tend to be distally displaced.

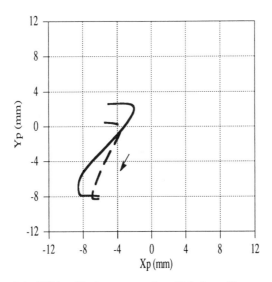

Figure 6. Intersections of the IHA's with the metacarpal sagittal plane. They are relative to one subject tested twice in different days. The origin of the graph is coincident with the calibrated centre of the metacarpal head. The arrows indicates the direction of flexion.

This is evident in figure 6 which shows the typical trend as resulted on one subject that has been analysed twice in different days. Consequently, the same figure gives an idea of the repeatability of the results obtainable with the adopted experimental protocol.

CONCLUSIONS

In this paper particular attention has been given to the methods and techniques necessary to obtain the IHA as a descriptor of the relative motion at the MCP joint.

Both the reduced dimensions of the body district and the sensitivity of IHA to noisy data make the reliable estimation of IHA parameters a critical problem. However, both simulation and experimental results have shown that it is possible to obtain quite satisfactory results in IHA determination. Moreover, the experimental protocol adopted, allowed a rather high level of repeatability of the results as put into evidence by experiments performed in different days on a same subject.

Results put into evidence that the MCP joint, in flexion, cannot be considered to have a fixed axis of rotation but sliding and rolling occur during flexion. In fact IHA's tend to dispose distally and volarly as flexion increases.

Looking at the angle trajectories, the experimental findings are in accordance with most relevant functional anatomy findings.[6,12] Slight variations exist among results obtained with a same subject, even in a same experimental session. These seem mainly due to different initial conditions in the relative position of the finger with respect to the metacarpus.

All procedures run on a PC. In particular the experimental protocol and the related acquisition and data processing procedures constitute a computer aided movement analysis system[10] developed with the aim to bring into clinics the set of movement analysis methods and techniques thus contributing to transfer the laboratory-based research findings to clinical practice.

REFERENCES

1. B.D. Anderson, and J.B. Moore, "Optimal filtering", Prentice-Hall, Englewood Cliffs (1979).
2. A. Cappozzo, Gait analysis methodology, *Human Movement Science,* 3:27 (1984).
3. E.Y.S. Chao, K.N. An, W.P. Cooney, and R.D. Linsheid, "Biomechanics of the hand. A basic research study", World Scientific Pub., Singapore (1989).
4. A. de Lange, J.M.G. Kauer, R. Huiskes, and H.J. Woltring, Carpal bone motion axes and pivots in flexion and extension of the hand, *in:* "Proc. 32nd Ann. Orthopaedic Research Society Meeting", New Orleans (1986).
5. J. Dimnet, and M. Guinguand, The finite displacements vector's method: an application to the scoliotic spine, *J.Biomech.* 17:397 (1984).
6. J. Dubousset, Anatomie fonctionnelle de l'appareil capsulo-ligamentaire des articulations des doigts (sauf le pouce), *in:* Traumatismes ostéoarticulaires de la main, Conférence du G.E.M., Masson, Paris (1974).
7. S. Fioretti, A. Germani, and T. Leo, Stereometry in very close-range stereophotogrammetry with non-metric cameras for human movement analysis,. *J.Biomech.* 18:11,831 (1985).
8. S. Fioretti, and L. Jetto, Accurate derivative estimation from noisy data: a state-space approach, *Int. J. Systems Sci.* 20:1,33 (1989).
9. S. Fioretti, L. Jetto, and T. Leo, Reliable in-vivo estimation of the instantaneous helical axis in human segmental movements, *IEEE Trans. on BME,* 37:4,398 (1990a).
10. S. Fioretti, T. Leo, E. Pisani, and M.L. Corradini, A computer-aided movement analysis system, *IEEE Trans. on BME,* 37:8,812 (1990b).
11. S.K. Ghosh, "Analytical Photogrammetry", Pergamon Press, New York (1979).
12. R.W. Hakstian, and R. Tubiana, Ulnar deviation of the fingers. The role of joint structure and function, *J. Bone Joint Surg.* 49A:2,299 (1967).
13. G.L. Kinzel, B.M. Hillberry, A.S.Jr. Hall, D.C. Van Sickle, and W.M. Harvey, Measurement of the total motion between two body segments, II. Description of application, *J.Biomech.* 5:283 (1972).
14. H. Lanshammar, On precision limits for derivatives numerically calculated from noisy data, *J. Biomech.* 15:459 (1982).
15. T. Leo, and V. Macellari, An optoelectronic device-microcomputer system for automatized gait analysis", *in:* "Changes in Health Care Instrumentation due to Microprocessor Technology", F.Pinciroli and J.Anderson Ed., North-Holland, Amsterdam (1981).
16. V. Macellari, CoSTEL: a computer peripheral remote sensing device for 3-dimensional monitoring of human motion, *Med. and Biol. Engng. and Comput.* 21:311 (1983).
17. E.M. Mikhail, "Observations and Least Squares", IEP Dun-Donnelley, New York (1976).
18. V.A. Morozov, "Methods for Solving Incorrectly Posed Problems", Springer-Verlag, New York (1984).
19. M.M. Panjabi, M.H. Krag, and V.K. Goel, A technique for measurement and description of three-dimensional six degree-of-freedom motion of a body joint with an application to the human spine, *J.Biomech.* 14:447 (1981).
20. C.W. Spoor, and F.E. Veldpaus, Rigid-body motion calculated from spatial coordinares of markers, *J.Biomech.* 13:4,391 (1980).
21. H.J. Woltring, A Fortran package for generalized, cross-validatory spline smoothing and differentiation, *Advances in Engineering Software,* 8:2,104 (1986).
22. H.J. Woltring, Representation and calculation of 3-D joint movement, in: Proceedings of the Workshop on: Assessment of Clinical Protocols, Ancona, October 16-17, 1989, Deliverable 6 of CAMARC project, CEC AIM Programme, Public Report (1989).

23. H.J. Woltring, Model and measurement error influences in data processing, *in* "Biomechanics of Human Movement: Applications in Rehabilitation, Sports and Ergonomics", N.Berme and A. Cappozzo, ed., Bertec Corporation, Pub., Worthington (1990).
24. H.J. Woltring, and R. Huiskes, A statistically motivated approach to instantaneous helical axis estimation from noisy, sampled landmark coordinates, *in:* "Biomechanics IX-B", D.A. Winter et al., eds, Human Kinetics Pub., Champaign (1985).
25. H.J. Woltring, R. Huiskes, and A. de Lange, Finite centroid and helical axis estimation from noisy landmark measurements in the study of human joint kinematics, *J.Biomech.* 18: 379 (1985).
26. H.J. Woltring, A. de Lange, J.M.G. Kauer, and R. Huiskes, Instantaneous helical axis estimation via natural, cross-validated splines, *in:* "Biomechanics: Basic and Applied Research", G.Bergmann, R. Kolbel, and A. Rohlmann, eds., Martinus Nijhoff, Dordrecht (1987).

THREE-DIMENSIONAL FOUR VARIABLES PLOT FOR THE STUDY OF

THE METACARPO-PHALANGEAL JOINT KINEMATICS

Antonio Merolli,[1] Paolo Tranquilli Leali,[1] Sandro Fioretti,[2]
and Tommaso Leo[2]

[1]Clinica Ortopedica dell'Universita' Cattolica
Largo Gemelli 8, Roma, I-00168, Italia
[2]Dipartimento di Elettronica ed Automatica
Facolta' di Ingegneria, Universita' di Ancona, Italia

INTRODUCTION

A 2-D plot describes the relationship between two variables (X,Y); the way to produce and read this kind of plot is long well established. A 3-D plot describes the relationship among three variables (X,Y,Z); the 2-D plot can be defined as a 3-D plot in the particular case of a constant value for one of the three variables. Three-dimensional plots have been pretty arduous to produce before the advent and widespread distribution of microcomputers: this limited their use and consequent knowledge on their practical reading.

The kinematics of the metacarpo-phalangeal joint of the index finger in the human has been studied by a computerized opto-electronic system and the three Eulerian angles of flexion-extension, abduction-adduction and axial rotation have been measured as a function of time. Plotting these three angular parameters versus time gave us three 2-D plots which have been compared together to investigate a possible standard pattern of motion. Anyway the comparison of the three different 2-D plots did not performed as a confortable procedure to provide an intuitive description of the pattern of motion.

We noticed that the imposed experimental condition of having an ever increasing value in one parameter (namely flexion) gave us the opportunity to use a 3-D plot to describe effectively the relationship among the 4 variables measured (1-flexion-extension, 2-abduction-adduction, 3-axial rotation and 4-time). A graphical software devoted to this task has been produced and the analysis of 3-D/4-variable plots of a series of experimental measurements has been enterprised.

MATERIALS AND METHODS

The movement analysed is the flexion of the index finger starting from an hyperextended position with respect to the metacarpus kept fixed by a suitably shaped

Advances in the Biomechanics of the Hand and Wrist
Edited by F. Schuind *et al.*, Plenum Press, New York, 1994

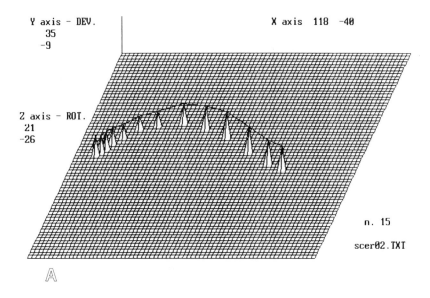

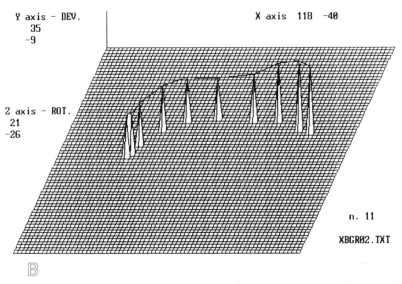

Figure 1. In the canonical form of a rapid movement (A) rotation values are high at the beginning, decrease in the middle and, then, rise up again; a concentration of peaks is present at the start; there is a bell-shaped profile of flexion angular velocity without any agglomeration of peaks at the end of the movement; values of deviation have a trend to increase. A degree of coupling can be described between deviation and rotation: at the inversion in sign of rotational values there is a slow-down in the rate of increase of deviation velocity (B).

holder. Markers have been fixed to the finger. The subjects have been asked to perform two kinds of movement: fast or slow.

Thirty-seven measurements in 13 normal subjects have been done, each subject contributing 1 to 4 measurements (1 meas.= 1 subject; 2 meas.= 2 subjects; 3 meas.= 8 subjects; 4 meas.= 2 subjects). The three Eulerian angles of flexion extension, abduction-adduction and axial rotation (pronation supination) have been measured as a function of time; positive progression of numerical values was adopted from extension to flexion, from adduction to abduction and for pronation (from now on it will be referred to flexion extension as "flexion"; for abduction-adduction as "deviation" and for axial rotation as "rotation"). Great care has been given to the accuracy of the experimental procedure: a highly accurate optoelectronic stereo-photogrammetric system (CoSTEL) has been adopted, and dedicated stereo-photogrammetric and smoothing procedures have been developed.[2] A frequency of sampling of 100 Hz has been used.

Since the values of flexion always increase, according to the experimental protocol, it was thought to link the reading of time to that of flexion: in detail, the constant time interval between two values (0.01 sec) is visualized by two 3-D peaks and the magnitude of time is given by the density of peaks along the direction of flexion. A dedicated software has been developed under MS-DOS environment in Microsoft Quickbasic.[3,6] X and Z axis lie in the transverse plane: flexion is read on the X-axis which is frontal and positive rightward; rotation is read on the Z axis which is sagittal and positive proximalward. Y-axis is the vertical axis and deviation is read on it, positive upward. A second option is given to exchange rotation and deviation axis and produce a second plot with rotation on the Y-axis and deviation on the Z-axis. The peaks of each tetrad of values are connected by a line and a simple animation routine produces their sequential blinking and gives an highly effective perception of variation in flexion angular velocity during movement. Scale limits for each parameter have been set to the minimum and maximum values considering all the thirty-seven measurements.

RESULTS

The maximum ranges of motion were: 94 degrees for flexion and 10 degrees for deviation and rotation.

The subjects were asked to perform fast or slow movements but analysis of the plots revealed the existence of three classes of movement that have been named: [1] rapid movements, [2] intermediate movements, [3] slow movements.

Duration of rapid movements is up to 0.15 sec and they are present in 8 subjects (11 measurements).

Intermediate movements last between 0.16 and 0.41 sec and are present in 9 subjects (12 measurements).

Slow movements have been recorded from 0.48 to 1.76 sec and are present in 10 subjects (14 measurements).

DISCUSSION

From the analysis of all the 3-D/4-variable plots a clean and simple representation of each class of movement can be derived (a canonical form).

In the canonical form of a rapid movement (Figure 1A) rotation values are high at the beginning, then decrease in the middle and, last, rise up again. A concentration of peaks[3,4] is present at the start (slow start) then there is a bell-shaped profile of flexion angular velocity without any agglomeration of peaks at the end of the movement (no significant "brake" before the end). Values of deviation have a trend to increase,

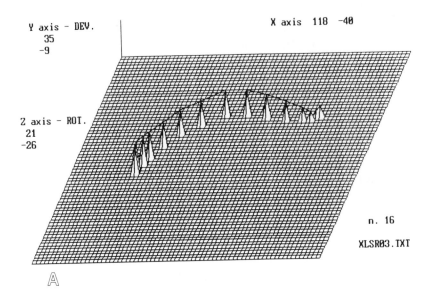

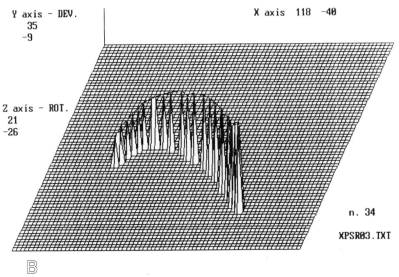

Figure 2. Intermediate movements include ones which are considerable still as fast (A) or already as slow (B). The canonical form of an intermediate movement mantains, basically, the same profile of rotation values of rapid movements which are high at the beginning, decrease in the middle and, then, rise up again. Anyway, it is difficult to discern the presence of the other characteristics ascribed to rapid movements.

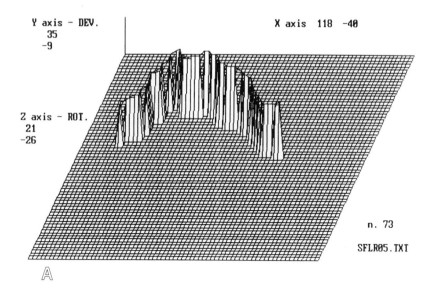

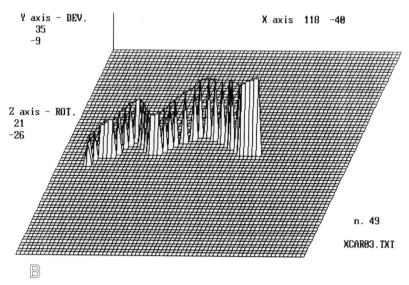

Figure 3. The canonical form of a slow movement is typically characterized by gross variations in the rate of growth of angular velocity: flexion angular velocity "starts and stops" more times during the movement (A): it can be proposed that slow movements are generated by a superimposition of rapid movements (B).

anyway in six subjects a sharp fall in the last value has been noted and turned out to be not an artifact. A degree of coupling can be described between deviation and rotation: in detail, at the inversion in sign of rotational values there is a slow-down (flexure point) in the rate of increase of deviation velocity (Figure 1B).

The canonical form of a intermediate movement (Figure 2) mantains, basically, the same profile of rotation values which are high at the beginning, decrease in the middle and, then, rise up again. Anyway, it is difficult to discern the presence of the other characteristics ascribed to rapid movements .

The canonical form of a slow movement is typically characterized by gross variations in the rate of growth of angular velocity; this peculiarity is pretty well visualized by the animation utility which shows how flexion angular velocity "starts and stops" two, three or more times during the movement. It can be proposed that slow movements are generated by a superimposition of rapid movements (Figure 3).

From these observations it can be inferred that rapid movement plot describes the pattern of motion basically due to the mechanical constraints of the joint (for studies on constraints: see references [1,4,5]). In intermediate and, more properly, in slow movements the pattern of motion reveals the influence of control mechanisms presumably exerted by the neuro-motor control system.

CONCLUSIONS

The three-dimensional four-variable plot of angular velocity values gives a single descriptive picture of the complex three-dimensional motion of the MCP joint. Its application to the analysis of our experimental measurements led to the identification of three classes of movements: rapid, intermediate and slow. A rapid movement probably derives from the purely mechanical constraints of the joint; it can be, maybe, produced also as a component of a slow movement. In the slow movement there is a fluctuation in the rate of growth of angular velocity and an higher level of neuro-motor control should be a responsible for this.

ACKNOWLEDGEMENTS

The first author would like to acknowledge SEIPI SpA of Milano for the contribution given.

REFERENCES

1. M.C. Hume, H. Gellman, H. McKellop, R.H. Brumfield Jr., Functional range of motion of the joints of the hand, *J. Hand Surg.* 15:240 (1990).
2. S. Fioretti, L. Jetto, and T. Leo, Reliable in vivo estimation of the instantaneus helical axis in human, *Trans. Biomed. Eng.* 37:398 (1990).
3. S. Fioretti, T. Leo, A. Merolli, and P. Tranquilli Leali, An in-vivo 3-D kinematic study of the metacarpo-phalangeal joint (abstract), *J. Biomech.* 24:476 (1991).
4. R.J. Schultz, A. Storace, and S. Krishnamurthy, Metacarpophalangeal joint motion and the role of the collateral ligaments, *Int. Orthop.* 11:149 (1987).
5. K. Tamai, J. Ryu, K.N. An, R.L. Linscheid, W.P. Cooney, and E.Y. Chao, Three dimensional geometric analysis of the metacarpophalangeal joint, *J. Hand Surg.* 13:521(1988).
6. P. Tranquilli Leali, A. Merolli, T. Leo, and S. Fioretti, Studio biomeccanico dell'articolazione metacarpo-falangea e considerazioni sulle sue attuali sostituzioni protesiche, *Minerva Ortopedica*, (in press).

BIOMECHANICAL EXAMINATION OF POSTOPERATIVE FLEXOR TENDON

ADHESIONS IN RABBIT DIGITS

Lars Hagberg

Department of Hand Surgery
University of Lund
Malmö Allmänna Sjukhus
S-214 01 Malmö, Sweden

INTRODUCTION

Restoration of gliding function following injury in the tendon sheath area of flexor tendons is a major issue in hand surgery.[8,12] The main problem is the formation of restrictive adhesions between the tendon and the tendon sheath or other tissues. Despite increased knowledge about tendon healing and the introduction of gentle surgical techniques and sophisticated postoperative treatments, the individual results of tendon repair in zone II still remain highly unpredictable.[2,3,7,9,11,13,14] There is a need for futher research and experimental work in the field of postoperative prevention of adhesions. In most experimental work the evaluation is based on macroscopic and histologic investigations. Determination of biomechanical characteristics of restrictive adhesions is very important and has been approached in some experimental studies.

Previously described test instruments have quantified adhesions by either measuring force versus tendon excursion (or angulation) to a certain endpoint or by measuring total excursion and angulation for a given force. The first approach has been used by Lane et al.[6] and Peterson et al.[10] The second approach was used by Woo et al.[17] and Younger et al. [19]

The purpose of tendon excursion is joint motion. This is therefore the most important parameter to which other parameters should be related. The first method can not continuously relate the registered "work of flexion" to the resulting joint motion. The test has to be brought to a defined endpoint. The load necessary for rupture of adhesions cannot be estimated by the second method. Ability to infinitely increase the tensile load until adhesion rupture occurs is necessary if the adhesion tensile strength is to be determined. The development of a test instrument for measurements of joint motion, tendon excursion and gradually increased tensile load is necessary for detailed biomechanical analysis of experimental adhesions. In the present study such a test instrument was developed for determination of tensile load required to obtain defined DIP joint motion and simultaneous recording of tendon excursion in rabbit hindpaw digits. An experimental model was developed for biomechanical characterization of postoperative adhesions and determination of tensile strength of flexor tendon repairs. The purpose was to keep the experiment as similar to the clinical situation as possible.

MATERIALS AND METHODS

Animals

Loop rabbits of both sexes (weight 2.3 - 4.5 kg) were used for the experimental studies and conditioned in the laboratory for at least 14 days before experimentation. All procedures were performed under fentanyl-fluanisone anesthesia (Hypnorm®; Janssen, Belgium; 0.7 ml/kg b.w. was given i.m., supplemented by doses of 0.25-0.5 ml i.m. when required, approximately every 30 min). Cefuroxime 30 mg/kg b.w. was given intramuscularly for prophylaxis against infection. The hindpaws were shaved, prepared, and covered in sterile drapes.

Basic anatomy of the flexor tendon digital unit

Great similarities exist between man and several animals in the anatomy of the flexor tendon-finger unit. Experimental studies on tendon healing have consequently been performed in monkeys, rabbits and dogs. The rabbit hindpaw has four digits, conventionally numbered 2-5 starting medially.[1] At rest the distal interphalangeal (DIP) joints are in about 80 degrees of hyperextension,[19] the proximal interphalangeal (PIP) joint in about 60 degrees of flexion and the metatarsophalangeal (MTP) joint in neutral position. The digital flexor tendon sheath consists of two main structures, a continuous synovial tendon sheath lining and fibrous anchorages of the tendon sheath to the volar periosteum, the pulleys. The rabbit flexor tendon sheath has three pulleys. The most proximal pulley is located at the MTP joint and called the plantar annular ligament. Each digit has two long flexor tendons, the deep (flexor digitorum fibularis - FDF - or flexor hallucis longus) and the superficial (flexor digitorum superficialis, FDS).[1]

Treatment protocol

Digits subjected to three different treatments were investigated: [1] normal digits, [2] digits operated with tenorrhaphy immediately prior to the study, and [3] two weeks after tenorrhaphy.

The third digits on both hindpaws were used for tendon repair. The FDF tendon was completely transected at the level of the proximal phalangeal bone between the plantar annular ligament (i.e. the pulley over the MTP joint) and the first digital annular ligament distally on the proximal phalanx. The surgical technique of the repair of the FDF tendon was identical to the "state of the art" clinical procedure, i.e. direct tendon repair by a modified Kessler suture followed by a continuous adaptation suture of the tendon ends and closure of the tendon sheath. However, two deviations from the clinical situation were allowed. First, the FDS tendon was resected along the proximal phalangeal bone to make the digit a one flexor tendon unit. Second, the FDF and FDS tendons were transsected proximal to the paw to unload the tendon repair. A firm dressing was applied which kept the ankle in flexion. All operations were performed under the microscope using microsurgical instruments.

Test instrument and tendon testing

The skin was carefully removed from the digits and all structures divided at the proximal level of the metatarsal bone. The digits were mounted in a test instrument capable of continuous and simultaneous registration of angular rotation in the DIP joint, tendon excursion and gradually increasing tensile load of the FDF tendon (Figure 1). These parameters were registered by precision potentiometers and graphically recorded. The test was run until the tendon or the tendon repair ruptured.

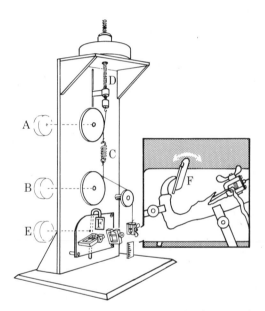

Figure 1. Test instrument. Precision potentiometer B measured tendon excursion and was used together with precision potentiometer A to measure tensile load. A coil spring (C) was mounted between the two potentiometer wheels. One end of the wire was connected to a motor-controlled rod (D). A third precision potentiometer (E) was connected to a plate (F) which could be moved by the claw nail of the digit and measured angular rotation in the DIP joint. Reprinted with permission from Hagberg et al.[5]

The recorded curves were digitized and analyzed in a computer. The tensile load at different values of angular rotation, the tensile load required for ruptures of adhesions, and the maximal tensile load recorded to reach different degrees of DIP joint rotation were determined. Energy consumed to reach different values of angular rotation (work of flexion) were calculated from the curves of tendon excursion and tensile load according to the following expression for energy at clamp motion x:

$$W_x = \int_0^x F(x).dx$$

where F is the tensile load, and x is the tendon clamp motion in the direction of the tensile load.

RESULTS

Load-strain analysis of tendon and wire gave a total strain of less than 10% at 6 N and about 20% at 30N (Figure 2a-b).

Typical recordings from a normal digit and a digit with tendon adhesions are shown in figure 3 a-b. Tenorrhaphy did not by itself disturb the tendon gliding ability within the tendon sheath. Mean values of tensile load at different DIP joint motions are shown for each category studied (Figures 4,5). The test instrument could identify partial rupture (development of gap, Figure 6) of the tendon repair as well as the tensile strength of the tenorrhaphy.

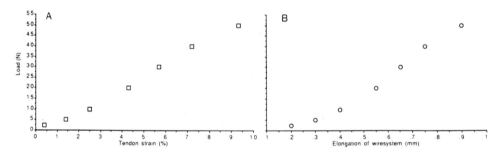

Figure 2. Load/strain curve of **Figure 2-A**) FDF tendons, **Figure 2-B**) the wire system.

Immediately after tenorrhaphy the tendon repair ruptured at 18 ± 1 N. Restrictive adhesions and ruptures of adhesions were easily detected by the different shapes of the recorded curves. In normal digits the tensile load curve formed a plateau up to 50° of rotation and thereafter started to increase more rapidly. An explanation for this may be that the physiological range of motion in the DIP joint is about 50° under the experimental circumstances when the other joints were locked in a standardized position in the test instrument. It was therefore suggested that the tensile strength of postoperative adhesions is best described by the maximum tensile load recorded to achieve 50° of DIP joint motion (MTL50), starting from resting position in hyperextension. A method for computer assisted calculation of energy consumed to reach a certain DIP joint motion was developed as well.

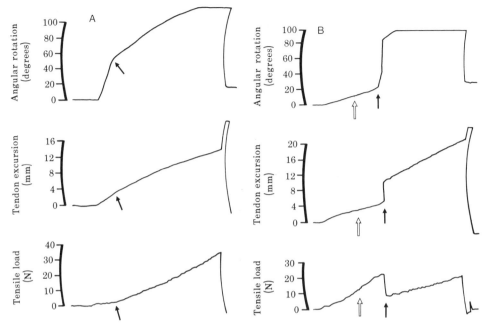

Figure 3. Typical recording of **Figure 3-A)** an unoperated digit. Top curve: angular rotation; Middle curve: tendon excursion; and lower curve: tensile load. There is an increased resistance to flexion at the arrow; **Figure 3-B)** a 15-day-old tendon suture. The open arrow indicates the zone where adhesions prevent angular rotation. At the black arrow there is rupture of peritendineal adhesions. Reprinted with permission from Hagberg et al.[5]

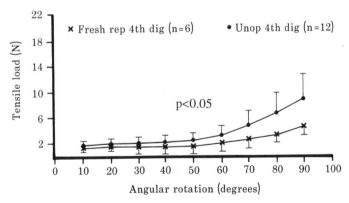

Figure 4. Tensile load plotted against angular rotation in unoperated 4th digits and digits subjected to tenorrhaphy immediately before examination in the same animals. Reprinted with permission from Hagberg et al.[5]

387

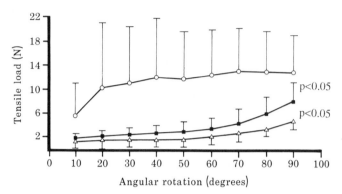

Figure 5. Tensile load plotted against angular rotation immediately after (▲) and 15 days after tenorrhaphy. The latter group of digits was separated into two groups: digits with curve forms indicative of restrictive adhesions (o) and those without (■). Reprinted with permission from Hagberg et al.[5]

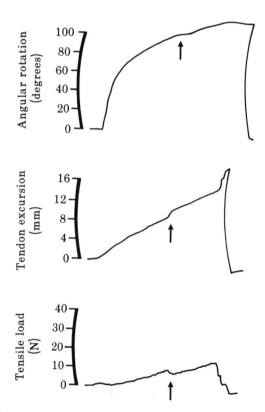

Figure 6. Recording from an operated digit. Partial rupture of the tendon suture occurs at the arrow. No angular rotation occurs (top), although there is a sudden increase in tendon excursion (middle) and a drop in tensile load (bottom). Reprinted with permission from Hagberg et al.[5]

DISCUSSION

The purpose of tendon excursion is joint motion. Continuous recording of the DIP joint motion is therefore suggested as being the most important variable for use as a measure of the ability of active flexion. Proximal tendon excursion is a more indirect parameter. Ability to infinitely increase the tensile load to achieve adhesion rupture is necessary to determine the adhesion tensile strength and must be continuously recorded. Our aim was also to determine the work of flexion, and tendon excursion therefore had to be recorded. As it was possible to perform direct tendon suture and tendon sheath closure in rabbit hindpaw digits using microsurgical techniques, this animal was chosen for practical reasons. Our instrument is designed to comply with these criteria.

The equipment developed by Woo et al.[17] was designed to measure the angular motion of the DIP joint in dogs simultaneously with the tendon excursion when a set force was applied to the flexor tendon. This was a development of the method used by Wray et al.[18] who placed chicken digits in an Instron tensile tester. The result was expressed as joint flexion and tendon excursion from a single load. Younger et al.[19] applied this approach in a model for rabbits, and argued that their method did not destroy the adhesions but allowed dissection and histology studies after the test. In other words, they did not measure the absolute tensile strength of the adhesions.

Lane et al.[6] used the Instron tensile testing machine on rat digits. This machine records tensile load and tendon excursion continuously, and these variables were followed by the authors until full digital flexion was reached. Peterson et al.[10] used the same technique in chickens. In this method the work of flexion can be determined, and this represents the work of forces that resist tendon gliding and the rotation of the interphalangeal joints. As this approach does not include measurement of DIP joint motion, the test always has to be run to a defined endpoint, in this case called "full flexion". The tensile load needed for full flexion normally increases considerably in the last part of the range of motion, and it may not be quite clear where the limit is for physiologic joint rotation. Detailed information about the course of events during the test procedure, such as identifying at what tensile load and DIP angle adhesion ruptures occur and distinguishing them from partial ruptures of tendon repairs, are not recorded by these methods.

Examination immediately after tenorrhaphy did not result in impaired tendon gliding, which implies a good specificity in diagnosing functional impairment from later postoperative adhesions.

Digits examined 15 days after tenorrhaphy which did not have curve irregularities indicating adhesions, required significantly less tensile load to reach a certain joint motion compared to digits with the characteristic curve features of adhesions. But the former digits did not show as low mean values as normal digits, which indicated the existence of minor adhesions not revealed by curve irregularities. The definition of adhesions evidently had to be based on something other than the shape of the recorded curves.

A model for calculation of work of flexion was developed and easily applied in normal digits. However, in operated digits the events during sudden adhesion ruptures were often very rapid, with profound changes of tensile load and tendon excursion. It was difficult to precisely relate the actual load to the tendon position during this fast course of events. Furthermore, the recorded tendon excursion includes a number of possible errors such as varying tendon deformation by elongation, elongation in the tendon repair, elongation of the test instrument's wire system and possible clamp slippage. The tendon excursion necessary to achieve a certain DIP motion should be identical in digits of the same size, irrespective of existing adhesions. As the force is the only parameter that should vary due to adhesions, it does not seem rationale to include the parameter of tendon excursion. Work of flexion is a product of force and distance and would be influenced to a considerable degree by the irrelevant variation of recorded "tendon excursion" which gives indirect information about the tensile load required to obtain joint motion. It was decided that the most reliable, simple and still adequate expression for adhesion strength was the maximum load recorded to achieve a physiologic range of DIP joint motion. The suggested expression for tensile strength of adhesions: "Maximum tensile load recorded to 50° of DIP joint motion" (MTL50 - Figure 7a), correlated well with the recorded tensile strength of individual adhesions, especially for the stronger adhesions (Figure 7b). However, the strength of a population of weak, or functionally less severe, adhesions that ruptured at an angle of more than 50° of angular rotation was underestimated by MTL50.

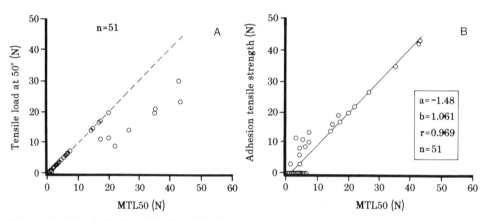

Figure 7-A. Tensile load at 50° plotted against maximal tensile load recorded between an angular rotation of 0 and 50° (MTL50) in animals operated with tenorrhaphy 15 days previously. The line y=x is also indicated. **Figure 7-B.** Adhesion tensile strength plotted against MTL50 in the same animals. The regression curve and equation are also indicated. Reprinted with permission from Hagberg et al.[5]

In normal digits the mean tensile load at 50° of DIP motion with the addition of two standard deviations was 4.5 N. It was suggested that when MTL50 was used, the definition of restrictive adhesions would be values exceeding 4.5 N. Welsh et al.[15] found that the "toe region" of the load/strain curve of rabbit tendons was within 0 - 1.5 % of elongation, the transitional zone was 1.5 - 3 % and the linear region was beyond 3 %. FDF tendons tested in the present instrument required a mean load of 5 N to reach 1.5 % of strain, the upper limit of the toe region. According to Viidik[16] functional activity occurs within the toe region of the load/strain curve, and 50° therefore seems to be within a physiological range of motion at the mean tensile load (+ 2 SD) of 4.5 N. Furthermore, in the experimental model described by Younger et al.[19] 4 N was reported to be the approximate force required to flex the DIP joint to what they meant was 80% of the normal range of motion.

Because adhesion preventive methods may impair tendon healing, the capability to determine the tensile strength of a tendon repair simultaneously with the examination of adhesion characteristics is a very important feature. The absolute strength of the tendon repair during healing also determines how early active flexion excercises can be introduced during rehabilitation.

REFERENCES

1. R. Barone, C. Pavaux, P.C. Blin, and P. Cuq, "Atlas of Rabbit Anatomy," Masson, Paris (1973).
2. J.H. Boyes, and H.H. Stark, Flexor-tendon grafts in the fingers and thumb: a study of factors influencing results in 1000 cases, *J. Bone Joint Surg.* 53A:1332 (1971).
3. O. Eiken, J. Holmberg, L. Ekerot, and S. Salgeback, Restoration of the digital tendon sheath. A new concept of tendon grafting, *Scand. J. Plast. Reconstr. Surg.* 14:89 (1980).
4. O. Eiken, and G. Lundborg, Experimental tendon grafting within intact tendon sheath, *Scand. J. Plast. Reconstr. Surg.* 17:127 (1983).
5. L. Hagberg, O. Wik, and B. Gerdin, Determination of biomechanical characteristics of restrictive adhesions and of functional impairment after flexor tendon surgery: a methodological study of rabbits, *J. Biomech.*, Vol. 24, No. 10, Pergamon Press Ltd., Copyright 1991.
6. J.M. Lane, J. Black, and F.W. Bora, Gliding function following flexor-tendon injury; a biomechanical study of rat tendon function, *J. Bone Joint Surg.* 58A:985 (1976).
7. G.D. Lister, H.E. Kleinert, J.E. Kutz, et al., Primary flexor tendon repair followed by immediate controlled mobilization, *J. Hand Surg.* 2:441 (1977).
8. P.R. Manske, Review Article: Flexor tendon healing, *J. Hand Surg.* 13B:237 (1988).
9. M.A. McClinton, R.M. Curtis, and E.F.S. Wilgis, One hundred tendon grafts for isolated flexor digitorum profundus injuries, *J. Hand Surg.* 7:224 (1982).
10. W.W. Peterson, P.R. Manske, C.C. Kain, and P.A. Lesker, Effect of flexor sheath integrity on tendon gliding: A biomechanical and histologic study, *J. Orthop. Res.* 4:458 (1986).
11. J.W. Strickland, and S.V. Glogovac, Digital function following flexor tendon repair in Zone II: A comparison of immobilization and controlled passive motion techniques, *J. Hand Surg.* 5:537 (1980).
12. J.W. Strickland, Flexor tendon repair, *Hand Clinics* 1:55 (1985).
13. J.W. Strickland, Flexor tendon surgery. Part 1: Primary flexor tendon repair, *J. Hand Surg.* 14B:261 (1989).
14. J.W. Strickland, Flexor tendon surgery. Part 2: Free tendon grafts and tenolysis, *J. Hand Surg.* 14B:368 (1989).
15. R.P. Welsh, I. Macnab, and V. Riley, Biomechanical studies of rabbit tendon, *Clin. Orthop.* 81:171 (1971).
16. A. Viidik, Functional properties of collagenous tissues, *Int. Rev. Connect Tissue Res.* 6:127 (1973).
17. S. L-Y. Woo, R.H. Gelberman, N.G. Cobb, D. Amiel, K. Lothringer, and W.H. Akeson, The importance of controlled passive mobilization on flexor tendon healing. A biomechanical study, *Acta Orthop. Scand.* 52:615 (1981).
18. R.C. Wray, and P.M. Weeks, Experimental comparison of techniques of tendon repair, *J. Hand Surg.* 5:144 (1980).
19. E.W. Younger, N.A. Sharkey, and R.M. Szabo, An apparatus to measure flexor tendon excursion and angular motion of the distal interphalangeal joint in a rabbit model, *J. Orthop. Res.* 6:462 (1988).

Part IV

CLINICAL APPLICATIONS

ASPECTS OF THE BIOMECHANICS OF FRACTURE FIXATION

Franz Burny, Yves Andrianne, Monique Donkerwolcke,
Maurice Hinsenkamp, Jean Quintin, and Frédéric A. Schuind

Department of Orthopaedics and Traumatology
Cliniques Universitaires de Bruxelles, Hôpital Erasme
808 Route de Lennik, B-1070 Brussels, Belgium

INTRODUCTION

The word *osteosynthesis* was first used by Albin Lambotte in 1907: "Mon but est surtout d'étudier la suture osseuse, ou, pour parler plus exactement, l'ostéo-synthèse" ("my aim is mainly to study bony suture, or, to be more precise, osteosynthesis"). That definition of the word "osteosynthesis" was expressed for the first time by Lambotte in 1902.[4] Osteosynthesis is defined in the French dictionary "Larousse" as a "surgical procedure to mechanically stabilize the bone fragments of a fracture with a metallic device to allow the bone union with callus formation". We define osteosynthesis as a means of fixation of bone fragments with direct anchorage of the implant in the bone.[2,7] The definition includes external fixation.

Most of the mechanical failures in fracture fixation result from a faulty technique or from an inappropriate selection of the treatment, regardless of the theoretical merit of the method used. It is important to understand the causes of the failures and to adapt the treatment. The relationship host-implant is a difficult problem which should be considered in the framework of a single general theory, formulated for the first time by Burny et al. in 1983 and called "the implanted system theory".[1,2,3,7]

THE IMPLANTED SYSTEM

An implanted system is a temporary or definitive association of a host tissue with a foreign material. The system is formed by a set of single units, with mutual connections. It may not be reduced to individual entities, as the whole is more than the addition of each individual unit. The degree of complexity of an implanted system depends on the number and types of connections between these units. The individual units include: [1] the biological environment, [2] the bone, which remodels following a biofeedback adaptation, [3] the principal and the connecting implants, and [4] the interfaces (Table 1, Figure 1).

Table 1. Elements of a general implanted system in Orthopaedics and Traumatology.

B	(1--->n)	Bone Fragments
C	(1--->n)	Connecting Implants
P	(1--->n)	Principal Implants
IBB	(1--->n)	Interfaces Bone Bone (callus, time-related)
IBS	(1--->n)	Interfaces Bone Soft-tissue
IBC	(1--->n)	Interfaces Bone Connecting implants
IBP	(1--->n)	Interfaces Bone Principal implant
IPP	(1--->n)	Interfaces between Principal implants
ICP	(1--->n)	Interfaces Connecting Principal implants
ISC	(1--->n)	Interfaces Soft-tissues Connecting implants
ISP	(1--->n)	Interfaces Soft-tissues Principal implants

(B represents a bone fragment with its mechanical and biological characteristics, C represents a connecting implant and its mechanical characteristics, in a biological and biomechanical environment, and P represents the principal implant-reproduced with permission from Burny et al.[3]).

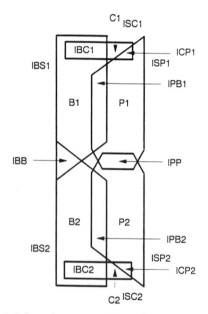

Figure 1. Schematic representation of the implanted system.

CHARACTERISTICS OF THE IMPLANTED SYSTEM

We insist on two characteristics of the implanted system.

The principle of symmetry

An osteosynthesis or a total joint replacement represents a succession of elements symmetrically distributed proximally and distally to a plane of symmetry represented by the fracture line or by the prosthetic joint. An implanted system is a succession of elements, the strength of the whole being that of the weakest element.[1,3,7] This "symmetry" has no geometrical representation but means a symmetry of the mechanical and of the biological properties at both sites of a very specific interface of the system (Figure 2).

Characteristics of the time-related interfaces

The implanted system is submitted to biological and mechanical influences from the time of implantation throughout life. The interfaces responsible for callus formation determine the future of any osteosynthesis: at the beginning of fracture healing most of the stress flows through the implants; callus formation decreases this stress as demonstrated theoretically and clinically (Figure 3).[2]

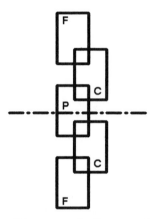

Figure 2. The principle of symmetry. We represent the fragments, the connecting implants, the principal implant and the plane of symmetry.

CLINICAL APPLICATION AT THE HAND AND WRIST

We present some applications of the theory for the rational treatment of hand and wrist fractures.

Wrist external fixation

In comminuted epiphyseal fractures affecting the distal radius, the reduction is often obtained by transarticular distraction, as the capsular and ligamentous structures are usually preserved. Prolonged transarticular traction maintained by a radius-second metacarpal external fixator is the treatment of choice (Figure 4).[5-7]

We assume that during the healing of a wrist fracture, no significant modification occurs in the mechanical characteristics of the bones. The invariants are [1] the radius, carpal and second metacarpal bones, [2] the clamps and the connecting rod (principal implants), [3] the pins (connecting implants). The bone-pin, fracture and joints interfaces are modified with time. The bone-pin interface, an important factor of the symmetry of the osteosynthesis, may vary with time, for example in case of loosening of the fixation device. Table 2 indicates the quality of pin anchorage at the time of pin retrieval.

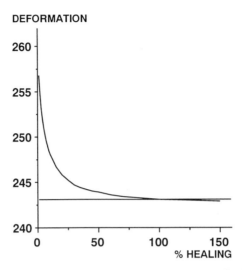

DEFORMATION

Figure 3. Hyperbolic relationship between the deformation of an external fixation rod (ordinate) and the percentage of healing (abscissa), theoretical study.

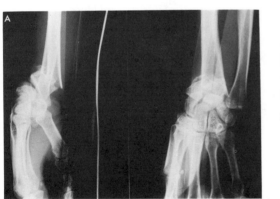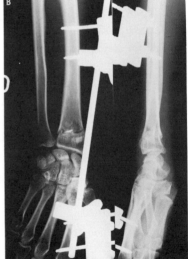

Figure 4. Fixation of a wrist fracture by Hoffmann external fixation in half-frame configuration. **Figure 4-A**: preoperative X-ray. **Figure 4-B**: external fixation.

Table 2. Manual evaluation of the clinical anchorage of external fixation pins at the time of retrieval of the fixation.

	Metacarpal bone	Radius bone
Perfect anchorage	91%	83%
Slight motion	4%	4%
Important motion	2%	11%
Manual extraction	2%	1%
Spontaneous pull-out	1%	1%

At the level of the wrist itself, two interfaces are described, both of them varying with time: [1] the bone fragments interface, depending on the initial quality of the contact between the bone fragments and the progressive formation of callus, and [2] the wrist joints. Little is known about the biomechanics of ligamentous distraction. Research should be undertaken to assess which ligaments are under tension and what the mechanical consequences of prolonged distraction on these structures may be. There could also be a role of the atmospheric pressure: we have measured a substantial decrease in the intraarticular pressure during wrist distraction at the time of the reduction of the fracture (Figure 5). This negative pressure could be responsible for a suction effect which could play a significant role in the reduction of intraarticular fragments.

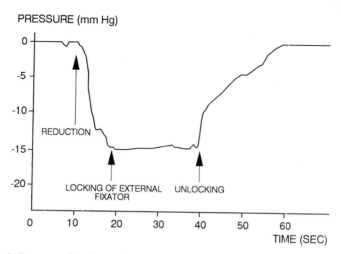

Figure 5. Decrease of the intraarticular pressure during the reduction of a wrist fracture.

In vivo biomechanical study of hand fractures

External fixation is an extrafocal means of fixation of a fracture, with direct anchorage of percutaneous pins into bone. It can represent a mechanical connection between the bone fragments and recording devices, allowing measurements of the mechanical characteristics of fracture healing. Strain gauges may be attached to the connecting rod of an external minifixator. We have used this principle to monitor the healing of metacarpal and phalangeal fractures, and to assess indirectly the stresses applied to the hand bones during passive or active motion.[7,8]

CONCLUSIONS

A single general theory, the implanted system theory, takes into account most of the mechanical and biological problems related to osteosynthesis. The theory is useful, as [1] it permits simple definitions, [2] it permits comparisons between various implants, and [3] equilibrium factors may be found.

REFERENCES

1. Y. Andrianne, F. Burny, J. Quintin and M. Donkerwolcke, Aspect of the failure of the implanted systems in Orthopaedics and Traumatology, In "Biocompatibility of Co-Cr-Ni alloys", Hildebrand H.P., Champy M. Eds., NATO-ASI Series, Plenum Press, New York, 158:249, (1988).
2. F. Burny, R. Bourgois and M. Donkewolcke, Pitfalls in osteosynthesis, In "Current concepts of internal fixation of fractures", ed. Uhthoff H.K., Springer-Verlag, Berlin, Heidelberg, New-York, 37 (1980).
3. F. Burny, Y. Andrianne, M. Donkerwolcke, and J. Quintin, Clinical manifestations of biomaterials degradation in orthopaedics and traumatology, In "European Materials Research Society Monographs, Biomaterials Degradation. Fundamentals Aspects and Related Clinical Phenomena", Volume 1; Barbosa M., Muster D., Hastings G., Burny F., Dörre E., Cordey J., Tranquilli-Leali P. Edts; Elsevier Pub., Amsterdam; 291 (1991).
4. A. Lambotte, L'intervention opératoire dans les fractures récentes et anciennes envisagée particulièrement au point-de-vue de l'ostéosynthèse avec la description de plusieurs techniques nouvelles, Bruxelles, Lamertin (1907).
5. F. Schuind, M. Donkerwolcke, and F. Burny, External fixation of wrist fractures, Orthopedics, 7, 841 (1984).
6. F. Schuind, M. Donkerwolcke, C. Rasquin, and F. Burny, External fixation of fractures of the distal radius: a study of 225 cases, *J. Hand Surg.* 14-A (Part 2), 404 (1989).
7. F. Schuind, and F. Burny, New techniques of osteosynthesis of the hand, principles, clinical applications and biomechanics with special reference to external minifixation, Karger, Basel, 1 (1990).
8. F. Schuind, M. Donkerwolcke, and F. Burny, Hand fractures: an "in vivo" biomechanical study, Proceedings of the 7th meeting of the European Society of Biomechanics, Aarhus, Denmark, 28 (1990).

DISLOCATION AND INSTABILITY OF THE DISTAL RADIO-ULNAR JOINT

Adalbert I. Kapandji

Department of Orthopædics and Hand Surgery
Clinique de l'Yvette
Longjumeau, F-91160 France

INTRODUCTION

Lately a subject of interest, the Distal Radio Ulnar Joint (D.R.U.J.) is now a matter of discussion between surgeons and physiologists, thanks to anatomical and radiological studies, especially Computerized Tomography (C.T.), Magnetic Resonnance Imaging (M.R.I.), arthrography and arthroscopy. Therefore, the "Wrist Sprain Syndrome" has been discarded.

DEFINITION

D.R.U.J. instability may occur either after trauma to a previously normal joint - an *acquired instability*, or during a degenerative process in a joint already affected by post-traumatic changes or a congenital malformation - *secondary instability*.

The definition is essentially *clinical:* the pronation-supination range of motion (ROM) is compromised by pain, so that the hand is unable to work with its normal function and strength.

This study clarifies the fact that D.R.U.J. lesions are usually associated with traumatic changes of the forearm,[27] and that misdiagnosis leads to severe pronation-supination problems with consequences for the overall wrist function.

D.R.U.J. BIOMECHANICS

Definition and Range of Motion

In order to be evaluated separately, the pronation and supination ROM must be measured from a *zero position* (Figure 1), with the plane of the hand vertical, the thumb directed upward and the elbow flexed at a right angle.[11] The supination range is 90° and the pronation is 85-90°. Limitation in pronation may be better compensated, owing to shoulder abduction. A good illustration of total supination is the "Waiter's Test" (Figure 2) when the hand is able to hold a tray with a water-filled glass without spilling it.[9] Measuring from complete supination is a poor method because it does not allow evaluation *separately* of pronation and supination.

With two joints, the D.R.U.J and the Proximal Radio-Ulnar Joint (P.R.U.J.), pronation-supination gives to the wrist a third degree of freedom, the longitudinal rotation of the forearm, that could not be obtained so perfectly with an *impossible enarthrosis joint.*[8]

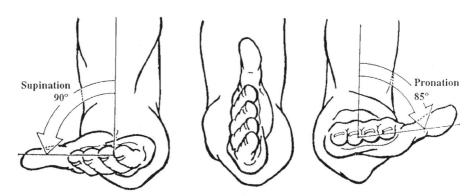

Figure 1. The measurement of supination and pronation (front view). Zero position (mid): the thumb is pointed upward. Supination (left): the palm is oriented upward. Pronation (right): the palm is oriented downward.

Figure 2. The "Waiter's test". In the *bearing position* (I), the tray is held on the right shoulder, the elbow flexed, the wrist in *complete pronation and extension*; in the *serving position* (II), the waiter is straightening his elbow as his wrist is supination; then (III) the *supination is total* so that the water in the glass cannot be spilled.

Joint surfaces and means of union

The D.R.U.J. is a *trochoid joint*: the ulnar head surface is convex/convex, born by the ulnar neck and prominent *anteriorly and laterally*. It corresponds to the radial sigmoid notch which is concave/concave. The shapes of these two surfaces are not exactly congruent, allowing a certain *looseness,* necessary during motion.[14] The joint *congruity* is maximum (Figure 3) in the zero position (midpronation or midsupination...).[9,10,12,13,14,15]

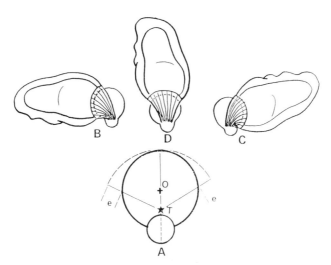

Figure 3. The triangular ligament motion on the ulnar head (inferior view). Bottom: schematic inferior view of the ulnar head showing the center (O) and the insertion point of the triangular ligament (T). The distance **e** shows (A) the relative ligament lengthening in mid-position (D) when the ligament is the tightest by reason of its eccentric insertion. The ligament "sweeps" the inferior facet of the ulnar head like a windshield wiper, in supination (B) and pronation (D).

Means of Coaptation of the D.R.U.J.

We must assume continued contact and concordance of the joint surfaces during motion, in *transverse* and *longitudinal coaptations.*

There are ligaments and one *muscle,* the pronator quadratus, stretched transversely between the extremities of the two bones, which are maintained in close proximity by the muscle, whose strength is proportional to its contraction, as in pronation (*active transversal coaptation).* Longitudinal muscles, for example the flexor digitori, are important factors in the *longitudinal coaptation,* preventing the downward slipping of the radius that is not attached to the humerus, particularly when the hand is bearing a load. Anterior and posterior ligaments are *thin*, and play a secondary role in coaptation and limitation of supination (*anterior ligament)* and pronation (*posterior ligament).*

The *triangular fibro-cartilage complex* (T.F.C.C.) is the principal linkage of the D.R.U.J. (Figure 4). Simultaneously it is a joint surface: [1] as a *joint surface*, also called *meniscus* (Figure 5), its superior aspect is articulated with the ulnar head and its inferior one is in contact with the lunate and the triquetrum; it is *biconcave*, with two reinforcements - anterior and posterior - and a *thin central part*, often open in the elderly: 80% after age 50 years;[21] a communication through this septum between the D.R.U.J. and the radio-carpal joint (R.C.J.) may be *normal* or *abnormal,* after trauma. On the other hand, there is often a normal split at the inferior edge of the radial sigmoid notch; [2] as a linkage between the distal extremities of ulna and radius, stretched between the inferior edge of the radial sigmoid notch and the ulnar styloid process, continuing in the *medial collateral ligament* of the R.C.J. The coaptation effect of the T.F.C.C. is maximum in the mid position (Figure 3), when it is tightened.

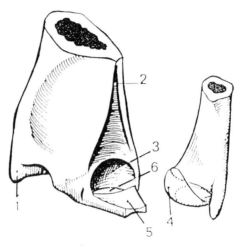

Figure 4. The triangular ligament (supero-medial view). 1- Radial styloid process 2- Bifurcation of the medial edge 3- Radial sigmoid notch 4- Peripheral aspect of the ulnar head 5- Superior aspect of the triangular ligament, thickened on its two borders 6- Dystrophic crack.

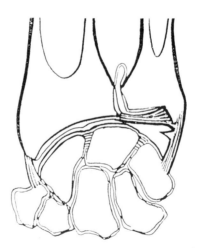

Figure 5. The triangular ligament (frontal wrist cut). The triangular ligament is both a linkage between the radius and the ulna and an articular surface for the ulnar head and the carpus. Just above it, the meniscus ligament of Taleisnik is prominent in the joint cavity.

The T.F.C.C. is constrained along the three directions of the space (Figure 6) and participates in the *medial fibrous node of the wrist,*[15] thanks to its numerous fibrous connections (Figure 7) with the ulnar styloid process, the triquetrum and the base of the fifth metacarpal through the extensor carpi ulnaris sheath. On its anterior and posterior edges are attached ulnocarpal ligaments which are connected with the *extensor carpi ulnaris sheath* under the *expansion of the dorsal retinaculum extensorum*, wrapping around the medial wrist border.

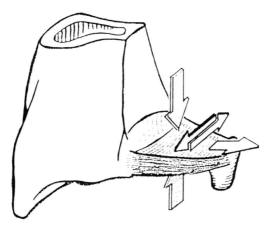

Figure 6. The stresses on the triangular ligament. Three types: Tearing (white arrow) compression/crushing (hatched arrows) - hacking (parallel white arrows).

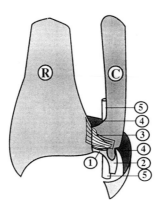

Figure 7. The triangular fibrocartilage complex (T.F.C.C.). Several fibrous sheets surrounding the ulnar head. 1- the triangular ligament, attached on the inferior aspect of the ulnar head, on the ulnar styloid process, on the deep aspect of the medial collateral wrist ligament (2); some of its fibers continue to the fifth metacarpal. 3- The anterior fascicle of the inferior radio-ulnar ligament. 4- The medial expansion of the dorsal carpal ligament. 5- The extensor carpi ulnaris tendon in its sheath, tightened against the posterior aspect of the ulnar head by the deep septa of the dorsal carpal ligament.

Therefore, the *medial fibrous node of the wrist* is an important element of ulnar head stability and also the *extensor carpi ulnaris sheath*. It prevents the occurrence of potential instability of the ulnar head to during pronation.[30] Consequently, a T.F.C.C. lesion is the main cause of D.R.U.J. instability, due to a loss of coaptation, whereas the motion of supination is diminished. On the contrary, the ulnar head, embedded in this fibrous node, contributes to the stability of the medial wrist, because of the efficacity of the extensor carpi ulnaris tendon: [1] the *dorsal carpal ligament* contributes to the constitution of the *medial fibrous node of the wrist*; [2] the *interosseous membrane* with interwoven oblique fibers (Figure 8), prevents the widening of the radioulnar space and also downward slippage of the radius. The load-bearing hand is attached to the radius and the radius to the ulna, not to the humerus.

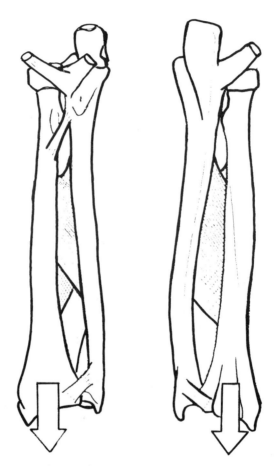

Figure 8. The interosseous membranes. There is a strong linkage between radius and ulna: oblique fibers are tightened when the radius slips down beside the ulna (white arrow). The Weitbrecht ligament and the two radio-ulnar ligaments play the same role.

Radio-Ulnar motion during Pronation-Supination

During pronation, the radius turns around the ulna, which is the hinge of the motion. Therefore, the two radio-ulnar joints are *linked* and cannot work separately. The slightest problem with one may affect the other. Moreover, the axis of each must to be aligned with the axis of the other, or there will be malfunction: they are *co-axial* and this fact may be compromised by any modification in shape or curvature. The problem is the same with a luxation or a sub-luxation of one of these joints. A two-hinged door[11] illustrates this important point (Figure 9).

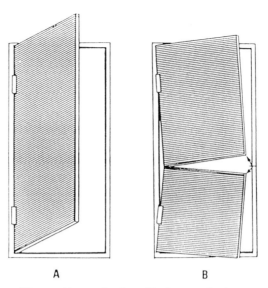

A   B

Figure 9. The two hinges. The two hinges of a door (A) play exactly the same mechanical role as the two radio-ulnar joints: one of them cannot work without the other one. They are mechanically linked and, moreover, their axes must correspond to each other: this coaxiality is the necessary condition for good function. If the two hinges of the door no longer correspond, it cannot open, except if it is separated into two parts.

However, the *effective axis of pronation-supination* is not this immutable hinge, passing through the centers of each radio-ulnar joint; it is the axis of the *real movement of the radius around the ulna,* as in a ballet.

As for every axis in biomechanics, *the true pronation supination axis is constantly evolving idealized.*[14] It is located "somewhere" in the radial epiphysis, hence *the radius rotates around itself,* combined with a *circular motion of the lower ulna* (Figure 10) on a segment of a circle without any rotation around itself. This motion needs a slight elbow extension combined with a lateral shift, because of the longitudinal humerus rotation.[6]

Furthermore, when the elbow flexes or extends[31] there occurs some automatic radial rotation on its longitudinal axis.

Tilting and lengthening motion of the ulnar head in the radial sigmoid notch

Longitudinal motion. From being parallel to the ulna in supination, the radius becomes crossed above it in pronation. This "crossing over" is made possible by the slight anteriorly concave curvature of the radial shaft, which faces the inverse curvature of the ulna. Becoming *diagonal,* the radius is slightly shortened in comparison with the ulna. Moreover, the ulnar head tilts in the D.R.U.J. because of the changing of the angle between radius and ulna. One may also consider the radius as stable and the ulna as mobile: the relative vertical movements of the radius in relation to the ulna (Figure 11) generate physiologic variations of the ulnar variance (U.V.): as the radial head is close to the humeral condyle, the ulnar head moves slightly upward in the radial sigmoid notch during supination, and, inversely, downward during pronation.

Tilting Motion. Meanwhile, the ulna, from parallel, becomes slightly oblique: the slight sphericity of the ulnar head surface depends on this tilting.

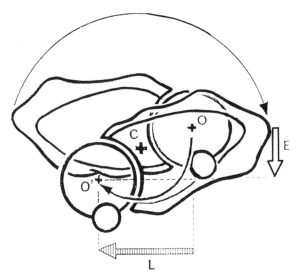

Figure 10. The "ballet" of the radius around the ulna. From pronation to supination the radius moves in rotation (large circular arrow); but, simultaneously, the ulna moves without any rotation, along a *circular path*, from the O to O' position thanks to a combination of an extension component (E) with a lateral component (L). So, the axis of motion (C) is *evolutive* and is located "somewhere" in the radial epiphysis.

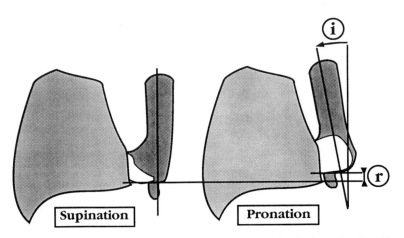

Figure 11. The accessory motion of the ulnar head. Besides its rotation in the radial sigmoid notch, the ulnar head has two accessory motions. If the radius is supposed stable, the ulnar head motions may be understood as follows: in supination, the two bones are parallel and the ulna seems to be almost equal to the radius: it is the *normal U.V.* In pronation, the radius crosses over the ulna and their axes are no longer parallel; therefore, the ulna is inclined (i) with regard to the radius; moreover, the radius seems to be longer (r) than the ulna with a decrease or even an *U.V. inversion*. In conclusion, during pronation, there is a *tilting* and a *shortening* of the ulna with regard to the radius.

Muscles

Two groups (Figure 12) are linked to the *radial crank*:

Supinators. Their strength is 1.00 kg in men and 0.60 kg in women: [1] *the supinator brevis* (Figure 12-A), a mono-articular muscle, acts by unrolling on the radial neck; simultaneously, it is *a coaptor* of the proximal radio ulnar joint; [2] *the biceps* (Figure 12-B) is *the most powerful supinator*; at the same time, it is an elbow flexor; on the flexed elbow, it dislocates the radial head anteriorly and acts as a coaptor when the elbow is extended; [3] *the supinator longus* is also a mono-articular muscle; it is mainly an elbow flexor, but also a supinator from a complete pronation position; it is a coaptator muscle of the P.R.U.J.in every wrist position.

Pronators. Their strength is 0.8 kg in men and 0.5 kg in women: [1] *the pronator teres* (Figure 12-C) is a mono-articular muscle; it flexes the elbow; it is a pronator muscle in full supination; it is a P.R.U.J. coaptor in every elbow position; [2] *the pronator quadratus* (Figure 12-C) is also a mono-articular muscle; it works by unwinding around the ulnar neck, and it is an *active element* of the D.R.U.J. coaptation.

Stresses. D.R.U.J. dislocation is stressed in three directions: [1] *forward in supination*: there is no abutment, and if motion surpasses physiologic limits, the ulnar head will shift forward, tearing the anterior ligament of the D.R.U.J. and the T.F.C.C; [2] *backward in pronation*: the abutment of the radius on the ulna is softened by the "mattress" of the anterior muscles and the stretching of the interosseous membrane. During a regular pronation motion, a dorsal subluxation of the ulnar head is seen, which can be measured with C.T. Further pronation will cause a fracture of the radial shaft combined with a posterior D.R.U.J. dislocation: this is Galeazzi's fracture; [3] *downward*: D.R.U.J. dislocation is prevented by the longitudinal muscle action.

CLINICAL EXAMINATION

Pain is located on the medial side of the wrist, just under the ulnar head bump, increased by ulnar deviation combined with flexion and also pronation that lengthens the ulna with regard to the radius. It is also located on the dorsal aspect on the articular line of the D.R.U.J., exaggerated by maximum pronation. Pain may be either intermittent, appearing only during the use of the wrist or some movements, or permanent, occurring even if the hand is not used.

Pronation-supination limitation is the second important sign of D.R.U.J. instability, mostly in supination, and easily compensated by shoulder abduction. This limitation of supination is made obvious by the "Waiter's Test": a glass filled with water cannot be borne by the hand on a plate without spilling it (Figure 2).

Clicks are detected by the patient himself and sometimes by the examiner, during pronation or supination motion. They indicate a T.F.C.C. disorder; they are often associated with violent pain or simply problems with a brief block.

Hand strength loss, especially during pronation or supination motion, with important consequences for professionnal or daily use, such as holding a saucepan.

Tenderness when palpating the articular D.R.U.J. posterior line during pronation or supination.

Pain provoked by transverse compression of the wrist is observed in more than half of the cases.

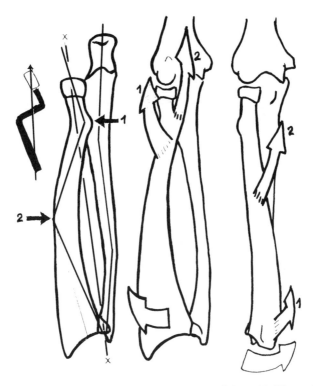

Figure 12. The pronation supination muscles inserted on the "radial crank". **Figure 12-A**. In the front view in supination, *the radius must be considered as a crank* with two bends: the supinating (1) and the pronating (2). **Figure 12-B**. The supinators: the supinator (1) acting by unwinding the radial neck - the biceps brachialis (2) acting on the superior bend of the radial crank, *eccentrically*. **Figure 12-C**. The pronators: the pronator quadratus (1) acting by unwinding on the ulnar neck - the pronator teres (2) acting on the inferior bend of the radial crank, *eccentrically*.

Ulnar paresthesias may be reported by some patients.

Abnormal mobility of the ulnar head, so-called "keyboard mobility", may be seen on the prominent bump of the ulnar head; it is painful to reduce, and it indicates *posterior instability* in pronation.

Screwing-unscrewing test consists in pronation or supination against resistance. The screwing-test is painfully impeded pronation *in a right-handed lesion*, which indicates posterior instability and even lesions of posterior osteoarthritis. *In a left-handed lesion*, the unscrewing-test has the same significance.

EXTRA-CLINICAL INVESTIGATIONS

Plain Radiographs

Important clues may be seen on front and side view radiographs, especially when compared with the other side.

On the side view, normally the ulnar head is totally hidden by the radial epiphysis; if it juts out posteriorly over the radius by comparison with the normal side, a posterior subluxation is possible.

On the front view, a gap in the D.R.U.J. articular line should draw attention: normally, the distance between the ulnar head and the lateral limit of the radial sigmoid notch is 2-3 mm, and, generally, the ulnar head is slightly superimposed over the posterior edge of the sigmoid notch. These changes may also be appreciated on front views taken in pronation and supination.

The ulnar variance (U.V.) is an important point for diagnosis: on the front views *in supination*, it may be measured by the difference of level between the inferior limit of the ulnar head and that of the inferior edge of the sigmoid notch. Normally, the ulnar head level is *slightly above* that of the inferior edge of the sigmoid notch: the U.V. is then conventionally considered as *negative*. Its normal value is **-2 mm,** which represents the thickness of the T.F.C.C.

The U.V. may be affected by several factors: [1] *position of the forearm*: during *pronation*, when the radius shortens relative to the ulna, the ulnar head lengthens downward and the *absolute value* of the U.V. decreases from supination (Figure 13) to pronation (Figure 14) with the difference d; [2] *sex*: in women, the absolute value is *less than in men* from -2 to 0; [3] *age*: in aging people, the ulna seems to be longer relative to the radius and the U.V. becomes zero or even *positive*. This fact is attributed to the thinning of the condylo-radial joint in the P.R.U.J. This progressive downward pushing of the ulnar head compresses the T.F.C.C. on the superior aspect of the lunate and the triquetrum, causing wear, until it articulates *directly with the lunate*. This ulno-lunate impingement is seen on plain radiographs (Figure 15) as a positive U.V., with a decrease in the ulno-lunate distance and increased density of the cortex of the lunate at the point of contact of the ulnar head before destruction of its superior pole.

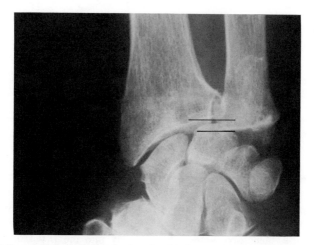

Figure 13. The U.V. variations during supination. On this front wrist radiograph, the U.V. is the shortest in supination. Here, it is not normal, because it is already *positive*, in aged people.

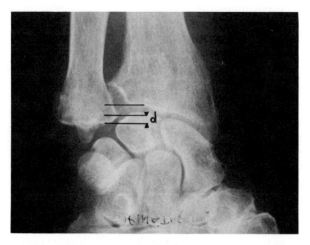

Figure 14. The U.V. variations during pronation. On this front wrist radiograph, the U.V. is the longest in supination. Here, the distance **d** shows tthe increase in the U.V., which is quite abnormal because it is an aged wrist with an impingement syndrome.

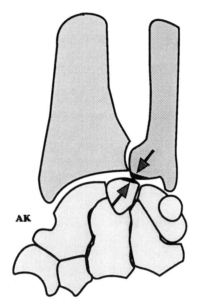

Figure 15. The ulno-lunate impingement syndrome. It is now known that in aged people, the radius lifts up because of the thinning of the condylo-radial joint. This apparent shortening makes the ulna a little longer and its head goes down, laminating the triangular ligament and pressing directly and forcefully on the supero-medial pole of the lunate. Therefore cartilages are worn out, and the cortex of the two bones becomes denser.

Wrist Arthrography

When tearing of the T.F.C.C. is suspected, wrist arthrography is used first.[3] It must always be done with *three successive injections into the three compartments:*[28] first in the radio-carpal line, then, *two hours later,* dorsally into the D.R.U.J., and finally into the medio-carpal joint. A tear of the T.F.C.C may be seen directly, or identified by a leak in the adjacent compartment. Other lesions, such as rupture of the scapho-lunate ligament, may be obvious at the same time, but there are sometimes *valve effects*. Therefore, it is important to use a direct injection into the D.R.U.J., the only way to show a *medial avulsion of the T.F.C.C.* However, wrist arthrography does not show *cartilaginous lesions* visible by arthroscopy or M.R.I.

Wrist Arthroscopy

Never used primarily, wrist arthroscopy is indicated after radiographs and perhaps arthrography, if doubt remains about a T.F.C.C lesion. Among the many ports of entry, only radio-carpal introduction is useful in D.R.U.J. instabilities: it is the only way to show the inferior aspect of the T.F.C.C, which normally appears white, smooth, even, concave and depressed under palpation, continuing into the radial glena surface.[25] The examiner must localize by palpation the slightest fissure, with its localization, appearance and area. According to Osterman,[24] lesions may be classified into three types: [1] type I: simple fissure (34%), that, when isolated, does not necessarily produce instability;[3] [2] type II: central perforation (46%), easily mistaken for a degenerative one, whose frequency increases in people over 45 years old, possibly because of the positivity of the U.V.; [3] type III: peripheral or medial perforation (20%), at the the T.F.C.C. radial insertion, which causes instability. In this category, a wide and mobile flap was observed in only three cases. Chondromalacia of the ulnar head is common when there is a central tear or perforation, and wear of the lunate cartilage also occurs with positive U.V. It is also very important to appreciate capsular ruptures, which cause major instability when associated with tearing of the T.F.C.C.

Wrist arthroscopy also permits therapeutic maneuvers, discussed later.

Computerized Tomography (C.T.)

A simple C.T. of the D.R.U.J. provides important information[23] about the distance between joint surfaces and their status (Figure 16), a possible fracture of the posterior edge of the sigmoid notch, and mainly the stage of a possible luxation or subluxation of the ulnar head in relation to the sigmoid notch. The difficulty is to get comparable slices of the opposite side *at the same level*. However, the C.T. slices do not show the status of the ligaments. The threee-dimensional reconstruction is of no interest in this matter.

A new C.T. procedure has been recently proposed:[17] C.T. slices are made in *stressed pronation and in supination against resistance* with relaxed flexor muscles and, in a second stage, *with simultaneous contraction of the supinator or pronator muscles* and also *flexor muscles*, a so-called constrained C.T. In pronation *against resistance without muscular contraction*, the ulnar head is strongly pressed close to the posterior edge of the sigmoid notch, but when there is D.R.U.J. instability a posterior subluxation becomes obvious (Figure 17). Flexor muscle contraction diminishes the posterior subluxation, hence the coaptor role of the pronator quadratus.

Magnetic Resonnance Imaging (M.R.I.)

M.R.I., in the T2 "spin echo" mode, shows perfectly soft tissues, particularly the ligaments[4] and T.F.C.C, in slices in three directions. The cartilage status can be well appreciated. However, it is still a difficult and very expensive examination, and in most of the cases, it only gives a confirmation of elements already obtained by other investigations, especially by plain radiographs; consequently, M.R.I. should be used only when diagnosis has not been made by other methods. In this case, it provides excellent images of partial or total T.F.C.C. rupture: in total rupture there is a large communication between the two compartments, whereas in partial rupture, it exists only toward one side. In some cases, the ligaments appear intact, but there are "abnormal intra-ligamentous signals" indicating "degenerative lesions" or "interstitial rupture".

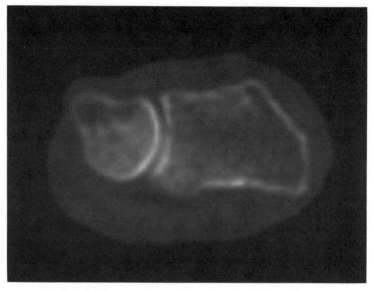

Figure 16. C.T. of the D.R.U.J. in pronation. This slice displays a normal D.R.U.J. cut, with perfect congruity.

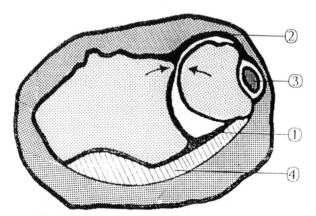

Figure 17. C.T. of the D.R.U.J. in pronation under stress. This slice displays the anterior ligament (1), the posterior ligament (2), the extensor carpi ulnaris tendon in its sheath (3), the pronator quadratus (4) which is a coaptor of the joint. Under stress, the D.R.U.J. has lost its congruity and the ulnar head, in subluxation position, presses forcefully on the posterior edge of the radial sigmoid notch.

ETIOLOGY

Trauma

Sprains. The "wrist sprain syndrome" has now been discarded; after wrist trauma, the first question is whether there are intra-carpal or D.R.U.J. lesions, or an association of the two types.[26] D.R.U.J. sprains are observed after trauma in extreme pronation or supination, most of the time combined with other movements of the radio-carpal or intra-carpal joint. A precise examination is necessary to recognize signs associated with the wrist joints or to the D.R.U.J.: so, a slight "keyboard" mobility, some clicks or jumps, must be analyzed with regard to their location and trigger movements. Point tenderness at the D.R.U.J level is also a good element for the diagnosis. But a "wrist sprain" diagnosis must always be completed with further investigations to define the etiology: arthrography is indicated first, and may show a simple fissure in the T.F.C.C. considered as a *first stage* lesion. Arthroscopy allows a much more precise diagnosis.

Dislocations. D.R.U.J dislocations require more significant trauma than that of simple wrist sprains. Lesions are defined as *second stage* lesions: *hyper-pronation* tears the posterior radio-ulnar ligament and the T.F.C.C. or the ulnar styloid process, and the extensor carpi ulnaris tendon is displaced medially. *Hyper-supination* tears the anterior radio-carpal ligaments and produces an anterior D.R.U.J. subluxation. Subluxations are often unnoticed, because they reduce spontaneously; they are important causes of D.R.U.J. instability. Generally the diagnosis is confirmed thanks to comparative plain radiographs, side and front views in pronation and supination. Comparative C.T. of the D.R.U.J. is useful to appreciate the importance of any displacement.

In the true luxations, there are *third degree* lesions, with significant displacement of the whole radio-carpal block around the ulna made possible by the rupture of all inferior radio-ulnar ligamentous connections. There is sometimes a fracture of the ulnar styloid process associated with Colles fractures. In this case, a leak of the arthrographic liquid is seen in the extensor carpi ulnaris sheath.

These dislocations are rarely isolated, and H. Judet showed in 1913 the functional solidarity between the two forearm bones and their own joints. Since then, many authors[19] have stressed this point. D.R.U.J. dislocations are usually associated with lower radius fractures, called Galeazzi's fractures.

Galeazzi's Fractures

Eighteen cases were described in 1934 by Galeazzi, but this type of fracture was already noted by Asthley-Cooper in 1822 and confirmed by Darrach in 1912 and Milch in 1926. The Maurel's work[22] was a precise review. A study of Mansat in 1977,[20] based on 50 cases of lower radius fractures explored with D.R.U.J. arthrography, shows 22 cases (44%) of obvious luxations, 6 associated fractures of the ulnar styloid process (14%) and 9 antero-posterior instabilities (18%). The 13 remaining cases, of which 10 were explored by arthrography, included 5 T.F.C.C. lesions.

Therefore, with an apparently isolated fracture of the radial shaft, it is of the utmost importance to rule out an associated D.R.U.J. lesion, i.e. a Galeazzi fracture. Pain on the medial side of the wrist, aggravated by pronation-supination, is a good indicator. Plain and comparative radiographs in pronation and supination give an early diagnosis of the D.R.U.J. instability or dislocation. Arthrography must also be done. The primary treatment of this type of fracture combines an *anatomic* reduction of the fracture, sometimes fixed with an osteosynthesis, with the suture of the ligamentous lesions, for example fixation of the avulsed T.F.C.C., with a temporary stabilizing wire, screwing of the ulnar styloid process or simple immobilization with a cast in *a position of stability*. D.R.U.J. instability remains an element of *poor prognosis* in Galeazzi's fractures, because when the ligamentous lesion has gone unnoticed or when it has been impossible to repair, there is secondary chronic D.R.U.J. instability, sometimes associated with radial malunion. Only after treating the radial malunion can the D.R.U.J. instability be corrected with one of the numerous procedures.

Monteggia's Fractures

One might think that Monteggia's fracture, associating a fracture of the ulnar shaft with an anterior radial head luxation, has no place in the D.R.U.J. instability syndrome. In fact, as the two radio-ulnar joints are *functionally linked*, the D.R.U.J. may be involved with radial head lesions after an initial dislocation, especially if they have not been diagnosed and treated from the beginning.

Quadruple dislocation of the two forearm bones

The simultaneous dislocation of the wrist and the elbow, and of the two radio-ulnar joints, described by Vesely, is exceptional and does not remain undiagnosed.

Double D.R.U.J. and P.R.U.J. dislocation

This type of double dislocation has been observed once by Mansat, and was already old - the diagnosis had been missed. The radial head was displaced forward in supination and the ulnar head was dislocated backward in pronation; consequently, it was not a stable position, with treatment being difficult and disappointing.

Radial head fractures

Radial head fractures are associated, in more than the half of the cases, with D.R.U.J. lesions, that correspond to an Essex-Lopresti syndrome, described in 1950, although Boehler, Curr and Coe had mentioned it before. These fractures are classified into three types: type I, nondisplaced (37.8%); type II, displaced (41.6%) with categories according to the importance of the fracture: one-third, one-half or two-thirds of the head; type III, comminution (6.7%). The D.R.U.J. problems are caused by *P.R.U.J. compression*, which causes a relative radial shortening and a vertical dislocation of the D.R.U.J. with a *positive U.V.* This radial shortening may also be caused by an avulsion of the radial head that interrupts the harmonious relationship between the two radio-ulnar joints. Secondarily, the T.F.C.C., stuck between the lunate and the ulnar head, is progressively worn down, and a lunate-ulna impingement syndrome develops.

Radiographs (Figure 15) show a depression of the ulnar head, a thinning of the lunate ulnar space with a bony condensation of the ulnar head and of the lunate at the point of contact. Arthrography shows the fissure and the rupture of the T.F.C.C. Arthroscopy confirms the deep ulceration of the superior pole of the lunate cartilage.

In recent radial head fractures, it seems, to the contrary of what has been said by Thomine et al.,[7] that primary radial head avulsion should be abandoned: the radial head must absolutely be saved, *with its normal height,* to prevent secondary D.R.U.J. problems. In emergency treatment, D.R.U.J. instability requires temporary pinning. If the radial head must be removed, it must be replaced by an implant *with a normal height.* Long-term tolerance is uncertain, so a secondary ablation of the implant is catastrophic.

Radius malunions in children

Some radius malunions in children may cause very important D.R.U.J. dislocations in relation to growth: the ulna becomes too long *(cubitus longus)* and the U.V. is positive: 20 - 25 millimeters, which causes painful pronation-supination and unesthetic club-hand.

As the D.R.U.J. linkage is completely worn out by ulnar lengthening, isolated ulnar shortening is not sufficient and a D.R.U.J. arthrodesis combined with an ulnar pseudarthrosis (M. Kapandji - L. Sauvé procedure) is necessary.

Lower radius fractures can be associated in children with a lower ulnar epiphyseal fracture which may cause a spontaneous epiphysiodesis, generating a *cubitus brevis* with a secondary D.R.U.J. dislocation. Such an unfavorable evolution must be detected by radiographic surveillance until the termination of growth. Ulnar lengthening indicates a radial epiphysiodesis, which equalizes the two bones.

Colle's fractures

In 1934, Taylor and Parson stressed the importance of the D.R.U.J. and the T.F.C.C. in Colles fractures. Many authors, among them Mansat,[18] pointed out the severe consequences of radius malunion on the D.R.U.J. and on pronation-supination. In fact, lower radius fractures almost always produce a complex D.R.U.J. dislocation (Figure 18). The three components, associated to a variable degree (Figure 19) are very difficult to correct. The radial epiphysis lifts up, with a positive U.V. causing a secondary lunate/ulnar impingement, as the radial glena is tilted posteriorly (Figure 20). A radio-ulnar diastasis due to a T.F.C.C. tear or medial avulsion from the ulnar styloid process occurs in 50% of cases. The radial sigmoid notch tilt, frequently posterior, results in the axes of the two surfaces no longer remaining coincident.

Rupture of the posterior radial glenoid margin (69%), generating a "third postero-medial fragment", destroys the posterior border of the sigmoid notch and leaves the ulnar head in permanent posterior subluxation. D.R.U.J. instability follows the distal radius fracture like a *shadow.*

The prognosis is quite different depending on whether the D.R.U.J. is intact or damaged. The study of several series shows poor results in 20%, an optimistic view. Clinical experience shows that in lower radius malunion, loss or limitation of pronation-supination R.O.M. is more hampering than flexion-extension limitation, especially when the loss of supination cannot be compensated by the shoulder. Secondary procedures that allow an increase in pronation-supination R.O.M. are definitively more useful than those that correct radius malunion itself. The improved treatment of this secondary D.R.U.J. instability and of pronation-supination problems is a perfect primary treatment, which is unobtainable in spite of the recent technical improvements.

Isolated T.F.C.C. lesions with D.R.U.J. instability

In this case, no trauma has occurred evocating a "wrist sprain syndrome". It occurs frequently in an elderly manual worker, who has pain in wrist extension, pronation, and adduction, with grasp weakness. Sometimes, it manifests itself as a bump during some movements, but clicks are not necessarily pathologic. Diagnosis is difficult and plain radiographs appear to be normal if an U.V. inversion is not sought, along with lunate-ulnar impingement, that increases with age. Arthrography, then arthroscopy, shows more or less complete T.F.C.C. destruction, explaining the D.R.U.J. instability. Arthroscopic treatment is sometimes possible[2] associating a partial ligament resection with shaving of the part of the ulnar head damaging the lunate.

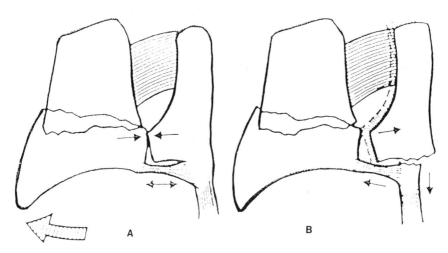

Figure 18. The D.R.U.J. problems in Colles' fractures. A- In the first stage, the triangular ligament does resist and there is an impaction in the joint. B- When the triangular ligament is not ruptured, it generates a fracture of the basis of the ulnar styloid process.

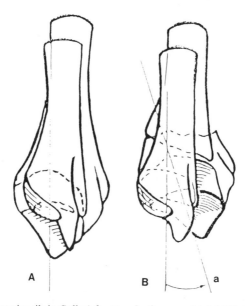

Figure 19. The radial posterior tilt in Colles' fracture. In the normal joint (A), the axis of the ulnar head and the one of the radial sigmoid notch are coincident. In Colles' fractures, the posterior tilt dislocates the joint and the axes are no longer coincident (a): the congruity is lost.

Rhumatoid Diseases

In rhumatoid arthritis of the wrist, the D.R.U.J. is destabilized, in two ways, first by destruction of the ulnar head, secondly by destruction of the *medial fibrous node*, particularly the extensor carpi ulnaris sheath, whose anteriorly-luxated tendon becomes a wrist flexor. This *painful* ulnar head instability must be treated early, before much more severe ulnar head destruction occurs. For many reasons *it is very important to save the ulnar head*; moreover, this procedure must be associated with an *extensor synovectomy*, so as to prevent tendon destruction, and with realignment of the radial axis. This is accomplished by repositioning the extensor carpi ulnaris tendon into its proper place and transposing the extensor carpi radialis brevis. Saving the ulnar head is important for stability of the remaining *medial fibrous node*. It avoids the medial secondary shift of the carpus that very frequently occurs after ulnar head avulsions.

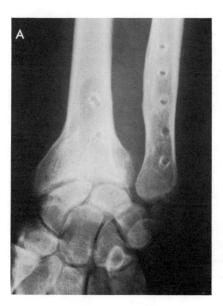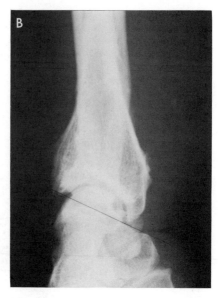

Figure 20. Radial malunion with a D.R.U.J. dislocation after a Colles' fracture. **Figure 20-A.** The radius shortening generates an inferior dislocation of the D.R.U.J. with a very positive U.V., a widening of the radio-ulnar space and an ulno-lunate impingement. Note the ulnar styloid process pseudarthrosis. **Figure 20-B.** The radius glenoid line is inverted because of the posterior tilt and the D.R.U.J. is dislocated.

Congenital Malformations

Madelung's disease. This well-tolerated congenital malformation includes bending of the radius, which is too long relative to the ulna. The special conformation of the two forearm bones generates conditions conducive to D.R.U.J. instability: a transverse space between the radius and ulna, lifting of the ulnar head with a very negative U.V., changing the orientation of the articular surfaces, and, particularly, insertion of the superior pole of the lunate between the radius and ulna (Figure 21). In the mild forms, a simple "wrist sprain syndrome" may indicate Madelung disease. The D.R.U.J. dislocation problem is complicated by the lunate insertion and by radial sigmoid notch atrophy.

Treatment of D.R.U.J. instability associated with congenital malformations is very difficult. The solution to this problem may be a D.R.U.J. arthrodesis combined with an ulnar pseudarthrosis (Kapandji-Sauvé procedure) with an ulnar radial graft and a resection of the superior pole of the lunate.

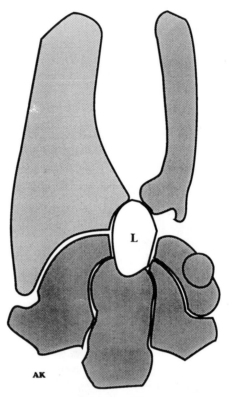

Figure 21. The D.R.U.J. in Madelung disease. This schema shows the abnormal curvature and length of the radius, with an atrophic radial sigmoid notch. Note also the shortness of the ulna whose head is distorted and especially the *lunate interposition between radius and ulna*, which modify significantly the structure and the shape of the triangular ligament.

TREATMENT

Nonoperative treatment

Sometimes, in recent luxations, nonoperative treatment is possible with a one-month cast involving the elbow and blocking the wrist in a position of reduction: [1] pronation is used for anterior luxations generated by hypersupination, with flattening of the ulnar head bump; [2] supination is used for posterior luxations generated by hyperpronation, the most frequent form (60%), with an exaggerated ulnar head bump.

Arthroscopic treatment

This treatment is used more and more frequently.[2]

Arthroscopic resection of the T.F.C.C., more or less completely damaged parts, resolves the clicks and bumps but does not stabilize the wrist. Arthroscopy also permits, when there is "ulnar-radial impingement" with chondromalacia or destruction of the cartilage of the superior aspect of the lunate, performance of a *customized resection of the parts of the ulnar head that are causing the damage.*

Operative Treatment

Osteosynthesis. When the anatomical relations between the two forearm bones are compromised, an osteotomy of the radius or ulna is essential before any action on the D.R.U.J. The simple screwing of a pseudarthrosis of the ulnar styloid process associated with a repositioning of the extensor carpi ulnaris is sometimes sufficient to get a good result.

Ligamentoplasties. Many D.R.U.J. ligamentoplasty procedures have been described; they are divided into two groups: [1] *passive ligamentoplasties* are difficult to adjust and may cause limitation of pronation-supination; recurrence is frequent; the most recent, the Sheckler procedure, is very good: it is done through an anterior approach (Figure 22), and uses a free graft of the palmaris longus passed first obliquely downward and medially in a tunnel from the lateral side of the radius to the inferior edge of the sigmoid notch, then through another tunnel, obliquely upward and medially, from the point of attachment of the triangular ligament to the antero-medial aspect of the ulna; the graft is fixed to the bones with two stitches, after its tension in pronation-supination is tested; it seems to be possible to do this procedure by arthroscopy; [2] *dynamic or active ligamentoplasties* seem to be more reliable; both are described. The *Hamlin procedure* uses the extensor carpi ulnaris tendon looped obliquely around the neck and the head of the ulna. It is fixed on the radius after passing over the posterior aspect of the D.R.U.J. It achieves simultaneously a static and dynamic fixation. The *Mansat procedure*[26] is executed through a dorso-medial approach; after the dorsal retinaculum is opened, the fibrous node is dissected, the extensor carpi ulnaris tendon is isolated in its sheath, as is the ulnar head, and the triangular ligament is resected. The ulnar head with the extensor carpi ulnaris tendon is repositioned in its proper place and fixed by the fibrous tissues with trans-osseous stitches. When the extensor carpi ulnaris sheath is weak, a plasty of the triangular ligament must be done with a palmaris longus tendon graft passed through a radial tunnel parallel to the inferior edge of the sigmoid notch, then in a triangular shape through a tunnel drilled at the base of the ulnar styloid process. A temporary pinning during for two weeks in zero position is followed by 4 to 6 weeks of immobilization with a cast.

Ulnar shortening. The procedure, described by Milch, may be used with an inverse U.V. *without severe D.R.U.J. dislocation* (Figure 23). The shaft resection may be transverse and segmentary, with a screw plate (Figure 23-A), or made with a simple oblique section without any resection by sliding the fragments on each other, fixed with an anterior screw plate after customized shortening obtained with ulnar wrist deviation (Figure 23-B). It is a very interesting procedure which does not cause a definitive damage and gives good results *when the D.R.U.J. is not severely dislocated.*

Ulnar head avulsions. There are three types (Figure 24): [1] *Total avulsions*: the Moore-Darrach procedure was for a long time the usual procedure; now, we know that it is imperative to do a short resection of no more than 25 mm; some save the ulnar styloid process, but it is important to reposition in its proper place the extensor carpi ulnaris tendon; however, it is now well known that the ulnar head avulsion makes the wrist medially unstable, and it does not appear that the Swanson ulnar head implant modifies the results; [2] *partial avulsion* (Bowers) or hemi-resection of the ulnar head with fibrous or tendinous interposition, without saving part of the ulnar head, provides better wrist stability; [3] *"matched ulna" resection* (Watson): the subtotal ulnar head resection is made obliquely from the ulnar styloid process, for a length of 4 to 5 centimeters, so that the lateral aspect of the ulna becomes a convex and fusiform surface parallel to the radial medial aspect, with an inferior spike extremity. This "custom made" resection allows better rotation of the ulna close to the radius.

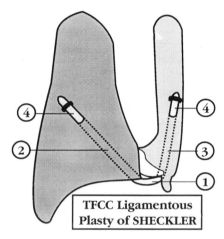

Figure 22. The D.R.U.J. ligamentoplasty Shekler procedure. The palmaris longus is used as a free graft and passed through two tunnels to be fixed with two staples, after adjustment of the tension. 1- The triangular ligamentoplasty. 2- The radial tunnel. 3- The ulnar tunnel. 4- The two graft extremities fixed with a staple.

D.R.U.J. arthrodesis associated with an above-ulnar pseudarthrosis

The so-called Kapandji-Sauvé procedure[29] (Figure 25) was wrongly named the Lauenstein procedure by English-speaking authors. Buck-Gramcko recently defined this historical controversy.[5] This procedure[13] gives good results,[1] and is now used more frequently than the Darrach-Moore technique, even in rhumatoid indications. The condition for success is *not to transform D.R.U.J. instability in to a painful ulnar stump instability*. So, a precise procedure is indispensable, particularly with respect to the location and the width of the ulnar shaft resection: it must be *as distal and as narrow as possible*.

D.R.U.J. PROSTHESIS

We are just beginning to use the D.R.U.J. prosthesis (Figure 26). Initially, it was conceived[16] for managing painful and unstable ulnar stumps, either after total or extensive ulnar head avulsions, or after Kapandji-Sauvé's procedure with a too proximal and too wide ulnar shaft resection. There is not enough experience with this prosthesis to give a valid assessment, but it shows much promise for solving the most difficult problems.

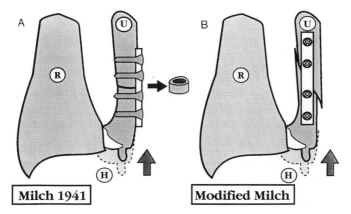

Figure 23. The ulnar shortening procedures (Milch and others). The shortening may be done with a segmentary transverse resection fixed with a screw plate, according to the original Milch procedure (**Figure 23-A**), or by sliding an oblique osteotomy fixed with an anterior screw plate (**Figure 23-B**).

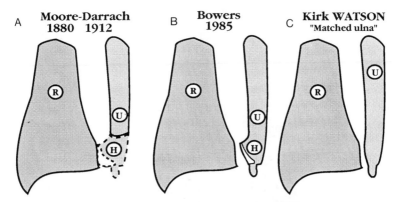

Figure 24. The ulnar head avulsions are of three types. A- Total avulsion of the ulnar head (Moore-Darrach, 1912). B- Partial resection of the ulnar head, with or without a fibrous interposition (Bowers, 1985). C- Customized ulnar head resection or "matched ulna" (Kirk Watson, 1989).

Kapandji-Sauvé Procedure1936

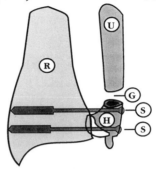

Figure 25. The Kapandji-Sauvé procedure. Invented in 1936, this procedure consists in a D.R.U.J. arthrodesis associated with a planned superior pseudarthrosis of the ulnar shaft. To obtain a good result it is necessary to do the cut as low as possible and the ulnar resection as narrow as possible (7-8 mm). There are two risks: the painful instability of the ulnar stump if the two preceding conditions have not been observed and the ulnar shaft reconstitution if the resection has been too short.

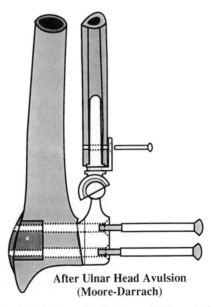

After Ulnar Head Avulsion
(Moore-Darrach)

Figure 26. The D.R.U.J. prosthesis (A.I. Kapandji). This is model II, used after total avulsion of the ulnar head (Moore-Darrach). It is a *cementless* prosthesis, fixed with screws. Its surfaces are *spherical*, with an inferior metallic head fitted with a polyethylene coated cup.

424

CONCLUSION

Chronic D.R.U.J. dislocations and instability are very difficult to treat. Treatment possibilities are extensive, but the procedure must be selected knowing its limits and potential complications. A precise primary diagnosis is necessary to make the correct decision. In the choice of procedure, it is indispensable to consider the hand dominance, the possible bilaterality, the age, the sex, the profession and the patient's preferences.

REFERENCES

1. C. Baciu, L'opération de Kapandji-Sauvé dans le traitement des cals vicieux de l'extrémité inférieure du radius, *Ann. Chir.* 31:323 (1976).
2. F.W. Bora Jr, A.L. Osterman, E. Maitin, and J. Bednar, The role of the arthroscopy in the treatment of disorders of the wrist, *Contemp. Orthop.* 12:28 (1986).
3. C. Bour, L'imagerie du poignet, L'Expansion Scientifique, Paris, *Monographie des Annales de Chirurgie de la Main*, 51 (1989).
4. G. Brunelli, and Ph. Saffar, L'imagerie du poignet, in Monographies du Poignet, *Springer-Verlag*, Paris, 214 (1991).
5. D. Buck-Gramcko, The priorities of the publication of some operative procedures on the distal end of the ulna, *J. Hand Surg.* 15B, 416 (1990).
6. H.C. Dbjay, L'humérus dans la prono-supination, *Rev. Méd. de Limoges,* 3,147 (1972).
7. L. Elayoubi, P.Y. Milliez, J.M. Thomine, and B. Tardif, Modification de la prono-supination entraînée par la résection de la tête radiale et par son remplacement prothétique; étude anatomique de 20 cas, *Rev. Chir. Orthop.* 77:77 (1991).
8. A.I. Kapandji, Pourquoi l'avant-bras comporte-t-il deux os? *Ann.Chir.* 29:463 (1975).
9. A.I. Kapandji, Le membre supérieur, soutien logistique de la main, *Ann. Chir.* 31: 1021 (1977).
10. A.I. Kapandji, La radio-cubitale inférieure vue sous l'angle de la prono-supination, *Ann. Chir.* 31:1031 (1977).
11. A.I. Kapandji, Physiologie Articulaire, Schémas Commentés de Mécanique Humaine, Tome I: le Membre Supérieur. 5ème Edition Paris Maloine Ed. (1980).
12. A.I. Kapandji, L'articulation radio-cubitale inférieure, Anatomie fonctionnelle, in *Le poignet*, J.P. Razemon, G.R. Fisk Ed. L'Expansion Scientifique Française, Paris (1983).
13. A.I. Kapandji, Opération de Kapandji-Sauvé, techniques et indications dans les affections non rhumatismales, *Ann. Chir. Main*, 5:181 (1986).
14. A.I. Kapandji, Biomécanique du poignet, *Ann. Chir. Main*, 6:147 (1987).
15. A.I. Kapandji, La biomécanique "Patate", *Ann. Chir. Main*, 6:260 (1987).
16. A.I. Kapandji, La prothèse de l'Articulation Radio-Cubitale inférieure, G.E.M. (in press, *Ann. Chir. Main)*.
17. A.I. Kapandji, Y. Martin-Bouyer, and S. Verdeille, Etude du carpe au scanner à trois dimensions sous contrainte de prono-supination, *Ann. Chir. Main* 10:36 (1991).
18. M. Mansat, R. Gay, C. Mansat, and C. Martinez, Cals vicieux de l'extrémité inférieure du radius et dérangements de l'articulation radio-cubitale inférieure, *Ann. Chir.* 31:297 (1977).
19. C. Mansat, and M. Mansat, Lésions de l'articulation radio-cubitale inférieure au cours des traumatismes de l'avant-bras et du poignet, in: *Poignet et Médecine de Rééducation*, Masson, Paris, 38 (1981).
20. M. Mansat, C. Martinez, and R. Gay, La luxation fracture de Galéazzi, *Rev. Chir Orthop.* 46:50 (1977).
21. M.J. Maraval-Bonnet, Etude anatomo-radiologique du ligament triangulaire du poignet, Thesis, Toulouse, (1969).
22. E. Maurel, Fractures du radius avec luxation radio-cubitale inférieure, Thesis, Toulouse (1970).
23. D.E. Mino, A.K. Palmer, and E.M. Levinsohn, Radiography and computerized tomography in the diagnosis of incongruity of the distal radio ulnar joint, *J. Bone Joint Surg.*, 67A:247 (1985).
24. A.L. Osterman, Arthroscopic debridement of triangular fibrocartilage complex tears, *Arthroscopy* 6:120 (1990).
25. A.K. Palmer, Triangular fibrocartilage disorders: injury patterns and treatment, *Arthroscopy*, 2:125 (1990).
26. J.P. Razemon, and G.R. Fisk, Le poignet, *L'Expansion Scientifique Française*, Paris (1983).

27. G. Rieunau, R. Gay, C. Martinez, C. Mansat, and M. Mansat, Lésions de l'articulation radio-cubitale inférieure dans les traumatismes de l'avant-bras et du poignet, *Rev. Chir. Orthop.* 57:253 (1971).
28. J.H. Roth, and R.G. Haddad, Radiocarpal arthroscopy and arthrography in the diagnostic of ulnar wrist pain, *Arthroscopy* 2:234 (1986).
29. L. Sauvé, and M. Kapandji, Une nouvelle technique des luxations récidivantes isolées de l'extrémité cubitale inférieure, *J. Chir.* 47:4 (1936).
30. C. Sirvente, Contribution à l'étude de l'anatomie fonctionnelle du cubital postérieur, Thesis, Toulouse, (1978).
31. E. Viel, and Cabanal, Oral presentation, Course on Upper Extremity Biomechanics, Lyon, France, (1990).

KINEMATIC DYSFUNCTION OF THE DISTAL RADIOULNAR
JOINT AFTER DISTAL RADIAL FRACTURES

Ronald L. Linscheid

Consultant, Section of Surgery of the Hand
Mayo Clinic and Mayo Foundation
Professor of Orthopedics
Mayo Medical School
Rochester, MN 55905, USA

NORMAL RELATIONSHIPS OF THE DISTAL RADIOULNAR JOINT

The distal radioulnar joint is part of an extended joint system from the elbow to the wrist. The axial motion that this joint system produces is responsible for the torque that may be generated for manipulation of materials. Proximally, the radius spins in an elongated conical fashion about the capitellum, and distally, it describes a cam-like motion around the ulnar head. This provides an arc of approximately 160° of pronosupination. The axis of rotation describes an irregular course that passes near the center of the roughly circular radial head to an area around the foveal attachment of the triangular fibrocartilage to the ulnar head.[8] Duchenne[5] pointed out that the distal radius and ulna described arcs about each other during pronosupination, with the former being larger than the latter.

The complex curvature of the forearm bones allows pronosupination with only minimal displacement in the planes parallel to the plane of rotation. These bones are constrained actively by the surrounding musculature and, at the extremes of excursion, by the ligamentous system. At the distal radioulnar joint, this constraint system consists of primarily the dorsal and volar radioulnar ligaments, which are fibrous inclusions within the triangular fibrocartilage. These ligaments resist lateral displacement well, resist proximodistal displacement poorly, and resist dorsovolar displacement primarily at the terminus of either pronation or supination.[6,8,12]

The ulnar head presents a roughly cylindric surface to the radial sigmoid notch in supination and a conical surface in pronation. The latter morphology permits the cam-like motion in pronation. The ulnar head describes a rolling, sliding translation in the sigmoid notch from volar to dorsal during supination to pronation. This adjustment can be accomplished without significant shear stress on the articular surfaces because of the differing radii of curvature of the ulnar head and sigmoid notch and the eccentric attachment of the radioulnar ligaments medial to the axes of rotation. The dorsal ligament reaches its greatest length and, thus, its tensile function at full pronation and the volar ligament at supination.[12]

The alignment of the distal radioulnar joint in the frontal plane is usually at approximately 20° to the longitudinal axis of the radius but roughly parallel to the rotation axis in patients with ulnar-neutral wrists. The minimal congruency and small contact area of the radioulnar articulation suggest that transverse compressive loads are small. In patients with significant ulnar minus variance, this angle is usually greater, and in those with ulnar plus variance, the angle is often reversed.

These same morphologic factors determine the distribution of joint compressive forces to the radius and the ulna.[10]

Radiocarpal motion is also somewhat dependent on ulnar variance. An ulnar plus variance tends to block ulnar deviation and, conversely, the lunotriquetral complex may block the dorsovolar translation of the ulnar head.

KINEMATIC ALTERATIONS WITH DISTAL RADIAL FRACTURES

The most common distal radial fractures in adults are the Pouteau-Colles' type, in which fracture occurs under tensile stress volarly with extensile loading. Dorsal cortical comminution occurs in shear stress, as in a cantilever beam loading model. The distal fragment displaces and angulates dorsoradially. This usually produces sufficient tensile stress on the triangular fibrocartilage to avulse its attachments or the styloid process of the ulna. Maintenance of reduction in the postfracture period is seldom sufficient to prevent some deformity from recurring.[1,2,4,7,9,13]

Late deformity, thus, produces dorsal angulation, radial angulation, dorsal displacement, radial displacement, and proximal displacement. Because the restraining qualities of the radioulnar ligaments are attenuated, the ligaments have less effect on stability.

Congruency of the distal radioulnar joint is diminished by both the proximal migration and the radial angulation. The proximal rim of the ulnar head has virtually point contact. The dorsal displacement and angulation results in malalignment of the articular surface. Full supination may be blocked as a result of the ulnar head already resting on the volar rim of the sigmoid notch, preventing further translation. Conversely, pronation is also partially blocked because dorsal translation is limited. Forceful rotation induces increased shear stress on the cartilage surfaces.

Radial shortening effectively produces an ulnar plus variance, increasing the component of joint compressive force on the ulnar pole. The ulnar head, already in a volar position on the sigmoid notch, may be limited further by encroachment of the lunate and triquetrum. If these two bones become displaced dorsally because of the dorsally angulated radial articular surface, they lie dorsal to the ulnar head and further restrict dorsal translation with attempted pronation.[11,14]

Pressure from the carpus also tends to distract the ulnar head from the radius during forceful grasp. Excision of the ulnar head has been a popular treatment for the problem of distal radioulnar dysfunction, but residual rotatory dysfunction due to malunion may persist.[3]

DISCUSSION

Despite the alterations produced by displacement of the distal fragment of the radius dorsoradially, some patients adjust with minimal difficulty. In part, this may be because most injuries occur in older patients, in whom physical demands are less. On the other hand, some patients are markedly inhibited by the pain, limited motion, and crepitus that may accompany the altered kinematics associated with distal radial fractures. For the most part, restoration of alignment and length seems to be the most important feature of treatment. Even if the ulnar head is still unstable, it is rarely troublesome if the joint is returned to an anatomic position.

REFERENCES

1. R.W. Bacorn and J.F. Kurtzke, Colles' fractures: a study of two thousand cases from the New York State Workmen's Compensation Board, *J. Bone Joint Surg. [Am]* 35:643 (1953).
2. D.R. Bickerstaff and M.J. Bell, Carpal malalignment in Colles' fractures, *J. Hand Surg. [Br]* 14:155 (1989).
3. E.J. Bieber, R.L. Linscheid, J.H. Dobyns, and R.D. Beckenbaugh, Failed distal ulna resections, *J. Hand Surg. [Am]* 13:193 (1988).
4. W.P. Cooney III, J.H. Dobyns, and R.L. Linscheid, Complications of Colles' fractures, *J. Bone Joint Surg. [Am]* 62:613 (1980).
5. G.B.A. Duchenne. "Physiology of motion: demonstrated by means of electrical stimulation and clinical observation and applied to the study of paralysis and deformities," (Translated and edited by E.B. Kaplan.) W.B. Saunders Company, Philadelphia (1959).
6. F.W. Ekenstam af, The distal radio ulnar joint: an anatomical, experimental and clinical study with special reference to malunited fractures of the distal radius. Thesis, Uppsala Universitet, Uppsala, Sweden, 1 (1984).
7. J.J. Gartland Jr. and C.W. Werley, Evaluation of healed Colles' fractures, *J. Bone Joint Surg. [Am]* 33:895 (1951).
8. R.L. Linscheid, Biomechanics of the distal radioulnar joint, *Clin. Orthop.* 275:46 (1992).
9. M. McQueen and J. Caspers, Colles fracture: does the anatomical result affect the final function? *J. Bone Joint Surg. [Br]* 70:649 (1988).
10. A.K. Palmer and F.W. Werner, Biomechanics of the distal radioulnar joint, *Clin. Orthop.* 187:26 (1984).
11. D.J. Pogue, S.F. Viegas, R.M. Patterson, P.D. Peterson, D.K. Jenkins, T.D. Sweo, and J.A. Hokanson, Effects of distal radius fracture malunion on wrist joint mechanics, *J.Hand Surg. [Am]* 15:721 (1990).
12. F.A. Schuind, R.L. Linscheid, K.N. An, and E.Y.S. Chao, A normal data base of posterior anterior radiographic measurements of the wrist, *J. Bone Joint Surg. [Am]* 74:1418, 1992.
13. G. Sennwald, "The wrist: anatomical and pathophysiological approach to diagnosis and treatment," Springer-Verlag, Berlin (1987).
14. J. Taleisnik and H.K. Watson, Midcarpal instability caused by malunited fractures of the distal radius, *J. Hand Surg. [Am]* 9:350 (1984).

DISTAL RADIUS FRACTURES: EVALUATION OF INSTABILITY AND THERAPEUTIC APPROACH

Thierry Authom, Michel Lafontaine, Dominique C.R. Hardy, and Philippe Delincé

Orthopedic Department
Hôpital Universitaire Saint-Pierre
Rue Haute 322, 1000 Brussels, Belgium

INTRODUCTION

For many years, various factors have been considered to be associated with secondary displacement of distal radius fractures. However, the assessment of the instability and the limits of conservative treatment often remain based on the clinical experience of the surgeon. In a previous retrospective study, Lafontaine et al.[4] considered 5 supposed severity factors: age over 60 years, dorsal comminution, dorsal angle greater than 20°, intra-articular radio-carpal fracture and fracture of the ulnar styloïd process. They found a strong correlation between the sum of these five factors and the risk of secondary displacement after orthopedic reduction and cast immobilization.

This prospective study is a continuation of this previous series and has for its purpose : [1] to compare surgical and orthopedic treatments in subgroups with similar characterisctics, [2] to elucidate more specific indications for surgical treatment, and [3] to specify the relative importance of each supposed severity factor.

MATERIAL AND METHODS

From January 1988 to November 1991, one hundred consecutive patients with a dorsally displaced fracture of the distal radius were randomly separated into two therapeutic groups. Exclusion criteria were : undisplaced fractures, highly comminuted fractures and the presence of a growth plate. The mean age was 60 years in the conservatively treated group versus 65 years in the surgically treated group. The distribution of fractures in the Older[6] (Figure 1) and Frykman[2] (Figure 2) classification systems showed them to be slightly more severe in the surgically treated group. Fourteen percent of the fractures were due to high energy trauma and this proportion was similar in each group.

Advances in the Biomechanics of the Hand and Wrist
Edited by F. Schuind *et al.*, Plenum Press, New York, 1994

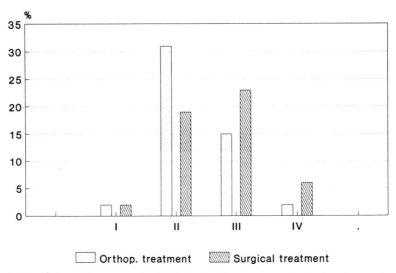

Figure 1. Distribution of fractures in the OLDER classification system for the two therapeutic groups.

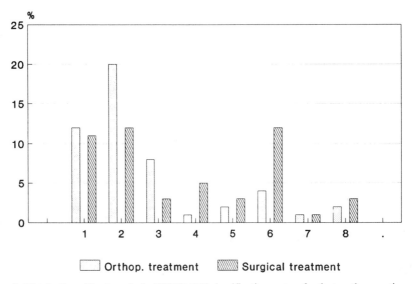

Figure 2. Distribution of fractures in the FRYKMAN classification system for the two therapeutic groups.

Fifty patients underwent orthopedic reduction under general (64%) or locoregional (36%) anesthesia and were immobilized with a forearm cast for 5 to 6 weeks. Ten patients with secondary displacement accepted repeat reduction. Five were carried out orthopedically and five surgically. The final results in these patients were considered to be the results obtained before this second reduction.

In the 50 surgically treated patients, 21 (42%) underwent intrafocal double pinning as described by Kapandji,[3] 11 (22%) underwent other types of pinning and 18 (36%) external fixation. No repeat reductions were necessary.

Five severity factors were taken into account for each patient before treatment : over 60 years of age, the presence of dorsal comminution on the radiograph before reduction, dorsal angle greater than 20°, intra-articular radio-carpal fracture, and fracture of the ulnar styloïd process. Additionally, radiological assessment using Stewart's method[9] (Table 1) was performed before reduction, after reduction and at union of the fracture. This scoring method uses the measurements of dorsal angle, loss of radial angle and loss of radial length. Each criteria is scored from 0 to 4. The scores are added and their sum provides a grade. This grade is an evaluation of the anatomical position of the distal radius.

Statistical analysis was performed using the chi-square test of the GB Stat[TM] program and with $p < 0.05$ considered as significant.

Table 1. Method of scoring and grading of anatomical position.[6]

	Score for each measurement			
	0	1	2	4
Dorsal angle (degrees)	Neutral	1-10	11-14	15+
Loss of radial angle (degrees)	0-4	5-9	10-14	15+
Loss of radial length (mm)	0-3	4-6	7-11	12+
Total grade (sum of the above 3)	Excellent 0	Good 1-3	Fair 4-6	Poor 7-12

RESULTS

The mean Stewart's score before treatment is more severe in the surgical group, indicating slightly more important displacement of the fracture. However, the comparison of the radiological results of orthopedic and surgical treatments shows a lower mean score of the latter representing a better anatomical result (Figure 3).

The relationship between the mean final Stewart's score and the number of severity factors (Figure 4) shows that, in the orthopedic group, in patients with more than three severity factors, the mean final Stewart's score is higher, indicating unsatisfactory results. On the contrary, the flattened curve in the surgical group indicates better results whatever the number of severity factors.

The two therapeutic groups can be divided into 5 subgroups defined by the number of severity factors (Figures 5,6).

The mean Stewart's score at fracture time increases with the number of severity factors. The score at reduction is relatively good. However, there is a greater tendancy towards secondary displacement in the orthopedic group than in the surgical group. The comparison of the radiological results with Stewart's grading system in the two therapeutic groups (Table 2) shows only 6 fair and no poor results in the surgical group versus 17 and 3, respectively, in the orthopedic group. In the latter, the proportion of fair and poor results regularly increases with the number of severity factors. There is a significant correlation between the sum of the severity factors and the final Stewart's grade (p = 0.047) (Table 3). This correlation is not observed in the surgical group.

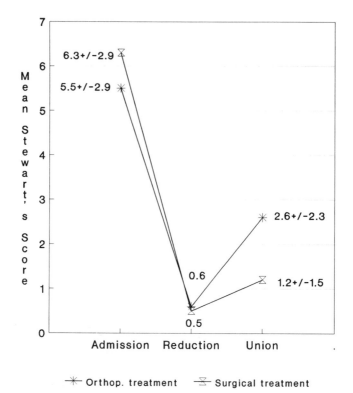

Figure 3. Evolution of the anatomical position as defined by the Stewart's score at admission, reduction and union in the orthopedic and surgical treatment groups.

Table 2. Grades of radiological results according to Stewart at union for both therapeutic groups.

	Orthopedic	Surgical	TOTAL
Excellent	11	18	29
Good	19	26	45
Fair	17	6	23
Poor	3	0	3
	50	50	100

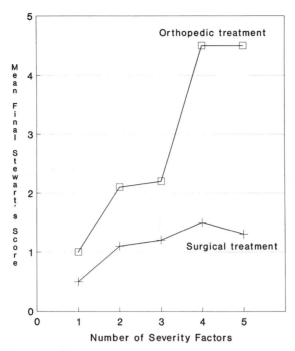

Figure 4. Relationship between the number of severity factors and the mean final Stewart's score in the two therapeutic groups.

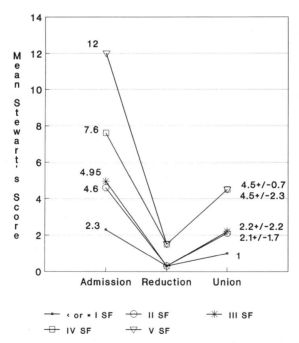

Figure 5. Evolution of the mean Stewart's score at admission, reduction and union in the orthopedic treatment group related to the number of severity factors.

Table 3. Grades of radiological results according to Stewart at union for both therapeutic groups in function of the number of severity factors.

Number of severity factors	Orthopedic treatment		Surgical treatment	
	Excellent or good	Poor or fair	Excellent or good	Poor or fair
0	2	0	1	0
1	1	0	1	0
2	12	3	10	1
3	12	8	16	3
4	3	7	13	2
5	0	2	3	0

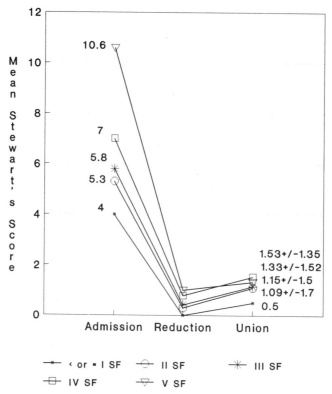

Figure 6. Evolution of the mean Stewart's score at admission, reduction and union in the surgical treatment group related to the number of severity factors.

In the orthopedic group, the relationship between the presence or the absence of each severity factor and the Stewart's grade at union of the fracture is presented on table 4. Three of the 5 severity factors are strongly correlated with the final radiological grade. These are age greater than 60 years ($p = 0.038$), presence of dorsal comminution ($p = 0.033$) and, especially, dorsal angle greater than 20° ($p = 0.006$). The influence of the fracture of the ulnar styloïd process is not significant. The proportion of favourable or unfavourable results is the same whether or not there is an intra-articular fracture.

Table 4. Relationship between the presence or the absence of each severity factor and the radiological grade at union according to Stewart in the orthopedic group.

Criteria		Results according to Stewart	
		Excellent + Good	Fair + Poor
Age > 60	+	12	14
years*	-	18	8
Dorsal	+	24	20
comminution*	-	6	0
Dorsal angle	+	14	17
> 20°*	-	16	3
Intra articular	+	7	4
radio-carpal	-	23	16
fracture			
Ulnar styloid	+	14	13
fracture	-	16	7

* p < 0.05

DISCUSSION

The radiological score according to Stewart can be easily calculated before and after reduction and at time of union. Its value is a reflection of the displacement of the distal end of the radius. The slope of the lines joining the scores at reduction and consolidation is a measure of the tendancy to secondary displacement. The final Stewart's score is a measure of residual displacement in relation to the physiological situation and should not be considered as having a prognostic value in terms of ulterior functional result. Two particularities limit the use of Stewart's score for prognostic evaluation: [1] the value which is attributed to loss of radial length in calculating Stewart's score is not sufficiently large when compared to that of loss of radial angle and of dorsal angle; a shortening of 6 mm is attributed only 1 point; a dorsal angle of 1° recieves the same score of 1; however, such shortening has much more unfavourable functional repercussions[1,5,7,8,10] than a 1° dorsal angle, [2] the presence of an intra-articular fracture or radio-carpal and radio-ulnar incongruity are not taken into consideration.

Interpretation of the results based on mean Stewart's scores is complicated by the high standard deviation. In the orthopedically treated patients, subgroups with 2 and 3 severity factors had mean final Stewart's scores of 2.1 and 2.2. Better differentiation between these 2 groups is obtained by interpreting the results with Stewart's grading system. Indeed, the percentage of fair and poor results is greater in patients with 3 than in those with 2 severity factors (40% versus 20%).

In our series, orthopedic treatment gives better results than those obtained in a previous study in wich fracture reduction was performed under local anesthesia in the emergency room.[4] This previous study included 112 cases with the same general characteristics as those in our series except that the fractures seemed to be slightly less severe (lower mean Stewart's score at fracture time = 3.0 versus 5.5 in this study). This difference can be explained by the fact that the previous study was retrospective, selecting more favourable cases for orthopedic treatment than for surgical treatment, while the second was prospective using random selection.

We cannot conclude that the use of general or locoregional anesthesia is the only factor responsible for this improvement due to the contribution of other factors. These are related to the differences between emergency treatment and programmed treatment: specialized surgical team, adequate assistance and optimal technical material.

Unlike the orthopedic treatment group, showing an increase of unfavourable results in relation to the sum of the severity factors, surgical treatment gives better results whatever the number of severity factors. The superiority of surgical treatment is seen in each subgroup of severity factors except the first two where results are similar and good or excellent (Table 3). Patients with no or one severity factor may be treated conservatively without risk of secondary displacement. With two or more severity factors, the advantages of surgery increase with the number of severity factors. In patients with 4 or 5 severity factors, 75% show fair and poor results with orthopedic treatment versus 11% with surgical treatment. These patients should be systematically operated. Subgroups with 2 and 3 severity factors (35 patients) represent an intermediate situation. In our experience, 68.5% of these fractures show no important secondary displacement when orthopedically treated. Therefore, in patients with 2 or 3 severity factors, closed reduction should be attempted first under general or locoregional anesthesia. At this time, all criteria for adequate reduction (restitution of volar angle, loss of radial length less than 3 mm, apposition of the volar cortex) were clearly present in 16 cases. All of these cases led to excellent or good radiological results. Of the 7 cases where at least one criteria was absent, none showed a good final result.

In the orthopedic group, the study of the influence of each severity factor on the final Stewart's grade (Table 4) shows a significant correlation for the criteria age greater than 60, presence of dorsal comminution and, especially, dorsal angle greater than 20° This latter may be due in part to the fact that dorsal angle is used to calculate the Stewart's score.

In the absence of ulnar styloïd process fracture, there are twice as many excellent or good radiological results, but this is not satistically significant. It may be considered as a minor severity factor. On the contrary, radio-carpal involvement does not influence the final radiological result. Whether present or absent, the results are similar to the overall mean (highly comminuted fractures being excluded).

CONCLUSION

When treating the fracture of the distal radius orthopedically, the strong correlation between the number of severity factors and the final radiological grade according to Stewart confirms the prognostic value of these factors already demonstrated by Lafontaine et al.[4] Indeed, this study confirms the unfavourable influence of age over 60, dorsal comminution and the dorsal angle greater than 20° on fracture stability after orthopedic reduction. Contrarily, the presence of a radiocarpal intra-articular fracture has no influence on the anatomical position of the distal radius at union, highly comminuted fractures being excluded.

Furthermore, the correlation between the sum of the severity factors and the final radiological results permits us classify the patients and their fractures according to the numer of severity factors, to asses the risk of secondary displacement and evaluate the indication for immediate surgical treatment in patients with 4 or 5 severity factors.

REFERENCES

1. A.F. De Palma, Comminuted fractures of the distal end of the radius treated by ulnar pinning, *J. Bone Joint Surg.* 34-A:651 (1952).
2. G. Frykman, Fracture of the distal radius including sequelae, *Acta Orthop. Scand.* 108:30 (1967).
3. A. Kapandji, L'ostéosynthèse par double embrochage intra-focal, *Ann. Chir.* 30:903 (1976).
4. M.A. Lafontaine, Ph.E.A. Delincé, D.C.R. Hardy, M. Simons, L'instabilité des fractures de l'extrémité inférieure du radius: à propos d'une série de 167 cas, *Acta Orthop. Belg.* 55:203 (1989).
5. A. Lidström, Fractures of the distal end of the radius, a clinical and statistical study of end results, *Acta Orthop. Scand.* 41:1 (1959).
6. T.L. Older, E.V. Stabler, W.H. Cassebaum, Colles' fracture: evaluation and selection of therapy, *J. Trauma.* 5:469 (1965).
7. S. Solgaard, Classification of distal radius fractures, *Acta Orthop. Scand.* 56:249 (1984).
8. S. Solgaard, Early displacement of distal radius fracture, *Acta Orthop. Scand.* 57:229 (1986).
9. H.D. Stewart, A.R. Innes, F.D. Burke, Factors affecting the outcome of Colles' fracture: an anatomical and functional study, *Injury* 16:289 (1985).
10. R.N. Villar, D. Marsh, N. Rushton, R.A. Greatorex, Three years after Colles' fracture, a prospective review, *J. Bone Joint Surg.* 69-B:365 (1987).

GEOMETRIC ANALYSIS OF OPENING-CLOSING WEDGE CORRECTIVE

OSTEOTOMY OF THE DISTAL RADIUS AFTER MALUNITED FRACTURE

Ranko Bilic[1], Vilijam Zdravkovic[1], and Vasilije Nikolic[2]

[1]Department of Orthopaedic Surgery
[2]Anatomy Department
School of Medicine
University of Zagreb
Zagreb 41000, Croatia

INTRODUCTION

It is well known that after conservative treatment fractures of the distal radius often do not heal in a good position. Consequently, function of the wrist is reduced, as well as rotation of the forearm because of dislocation and instability of the distal radioulnar joint. Relations in the distal radioulnar joint are usually the main reason for altered function of the wrist after malunion,[1,2,6,9,11] with special accent on radius shortening.[7,12]

In some cases it is necessary to perform an opening-wedge corrective osteotomy to correct the position and shortening of the malunited radius. In the literature there are several techniques described of corrective osteotomy differing with regard to the place of the bone graft removal, and the principles of osteosynthesis, but all include correction of the ulnar and volar tilt with simultaneous lengthening of the radius.

Occasionally it is necessary to correct ulnar and volar tilt, by lengthening the radius on one side of the distal radioulnar joint surface, while shortening it on the other. Because lengthening is achieved with an opening wedge and shortening with a closing wedge corrective osteotomy, fulfilling the above-mentioned task is not easy.

The aim of this article is to define and classify the types of rotations of the distal end of the radius in the frontal plane in cases of Colles and Smith's extraarticular fractures and malunion. We have therefore performed geometric analyses of lateral X-rays of patients treated with a corrective osteotomy of the distal end of the radius, planned using the computer-assisted BIZCAD method. Also, the incidence of different types of corrective osteotomies was established for all cases treated.

Advances in the Biomechanics of the Hand and Wrist
Edited by F. Schuind *et al.*, Plenum Press, New York, 1994

441

MATERIALS AND METHODS

From 1987 to 1992, 42 corrective osteotomies were performed in the Department of Orthopaedic Surgery, School of Medicine, University of Zagreb. Preoperative planning was performed by the BIZCAD method, with which, by using a computer, it was possible to simulate and analyze corrections in 3 dimensions.[4,13]

In 38 patients corrective osteotomy was performed because of a malunited Colles fracture, 3 cases because of a Smith's fracture, and in one because of a pseudo-Madellung deformation. In 40 cases the fractures were extraarticular, while in 2 cases the fractures were also intraarticular. Of the 2 cases one osteotomy was extraarticular, while in the other the osteotomy was double.

Of the total number of patients, 25 were male, and 17 were female. The average age of our male patients was 34.4 years ranging from 15 to 59 years, while in female patients the average age was 43.1 years ranging from 19 to 64 years.

Geometric analysis of all cases was performed on the lateral X-rays of the healthy and the malunited radius. First, we drew on transparent paper the contours of the distal 5 cm of both healthy and malunited radiuses (Figure 1).

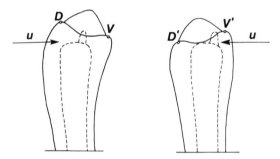

Figure 1. Drawing of the lateral projection of the healthy (left) and affected radius (right). D = dorsal edge of the ulnar notch. V = volar edge of the ulnar notch. U = distal end of the ulna.

Drawings were superimposed in such a manner that the X-ray of the healthy radius was reversed and placed exactly over the corresponding X-ray of the fractured radius. Axes of the radiuses matched exactly, as well as the distal ends of the ulnas ("U" on figure 1). Thus, the relative displacement of some landmark points of the fractured radius could be studied. We determined the following landmark points: volar edge of the sigmoid notch on the healthy (V) and malunited (V') radius, and dorsal edge of the sigmoid notch on the healthy (D) and malunited (D') radius.

Connecting points V and V' and D and D' (Figure 2 a) we defined two circles with the same center (Figure 2 b). This was taken to be the center of rotation of the fractured segment of the radius in the frontal plane for a certain angle (Figure 2 c).

GEOMETRIC ANALYSIS OF ROTATION IN THE FRONTAL PLANE

The analyzed model was the distal part of the radius as seen in the lateral projection. The proximal segment is considered to be a stable part, while the fractured (or osteotomied) part of the radius is considered a mobile segment.

At the moment of fracture we can assume that the mobile segment:

a) Translates proximally.

b) Translates in the volar-dorsal direction.

c) Rotates around some hypothetical axis in the frontal plane.

In cases where one or both translations occurred at the moment of fracture, and healing in malposition followed, the need for an opening corrective osteotomy is indicated. However, as rotation of the mobile segment sometimes occurs around different axes, some cases may need an opening-closing wedge corrective osteotomy.

Three characteristic positions of the center of rotation can be defined (Figure 3).

Position A is the case when the center of rotation is outside the distal part of the radius. Position B is when the center is close to the volar or dorsal rim of the distal radius. Finally, position C is the case when the center of rotation is inside the distal part of the radius.

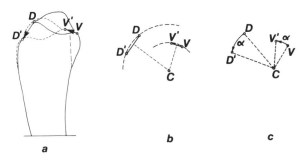

Figure 2. a. Overdrawing of healthy (reversed) and malunited radius shows displacement of joint surface. D, D' = dorsal edge of the ulnar notch of a healthy and affected radius respectively. V, V' = volar edge of the ulnar notch of a healthy and affected radius respectively. b. Displacement of the volar and dorsal edges define two circles with the same center. C = center of rotation. c. Displacement of the mobile segment had occurred around the center of rotation C for an angle alpha. The same angle should be corrected with osteotomy.

The first case (Figure 3 a), where the center of rotation may be outside the distal radius, can be corrected with opening corrective osteotomy of the distal radius. In this case the graft should be of a certain thickness on both the volar and dorsal sides.

In the second case (Figure 3 b) the center of rotation is on, or close to, the volar rim of the distal radius in a Colles fracture, while for a Smith's fracture it is on, or close to, the dorsal rim of the distal radius. These fractures can be treated with opening corrective osteotomy with a graft in the form of a triangle, where the volar or dorsal edge of the graft are close to zero thickness.

The third case (Figure 3 c), with the center of rotation inside the distal radius, can be corrected with opening-closing wedge corrective osteotomy.

It is therefore possible to define the type of osteotomy needed only by defining the center of rotation (Figure 4). It also roughly defines the type of graft needed for a particular patient.

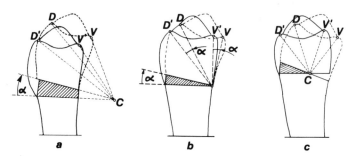

Figure 3. a. A case where the center of rotation is outside the radius. b. Center of rotation close to the volar or dorsal cortex of the distal radius. c. Center of rotation inside the distal end of the radius.

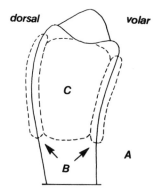

Figure 4. Classification of the rotation of the mobile segment based on the position of the center of rotation.

Among our patients there were 23 cases (54.8%) with the center of rotation in zone A, 4 cases (9.5%) with the center of rotation in zone B, and 15 cases (35.7%) where rotation was found in zone C.

In 15 cases opening-closing wedge corrective osteotomy was needed. Among these cases, 14 were Colles type and only one Smith's type.

In three cases shortening of the ulna was considered because a graft of more than 12 mm was needed according to the BIZCAD plan. Two of these patients had a Colles fracture, and one had a Smith's fracture; and with regard to the center of rotation, two had the center in zone A, and one in zone C.

POSSIBILITIES FOR PLANNING OPENING-CLOSING WEDGE CORRECTIVE OSTEOTOMY WITH THE BIZCAD METHOD

In the case when the center of rotation is in zone C, it is necessary to perform an opening-closing wedge corrective osteotomy. If the plan is performed as if the center of rotation was in zone A, then overcorrection of radius length occurs (Figure 5). This overcorrection is not significant if the height of the closing wedge is less than 2mm.

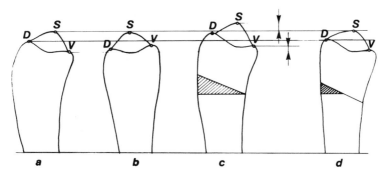

Figure 5. a. Lateral shape of the healthy radius. b. Lateral shape of the malunited radius. c. Overcorrection of the radius length. Note volar tit is perfectly corrected. d. With opening-closing wedge corrective osteotomy anatomic correction is achieved.

To perform an opening-closing wedge corrective osteotomy in practice, it is necessary to carry out a precise preoperative plan in 3 dimensions. This is possible by using the computer assisted BIZCAD method for preoperative planning. For performing an opening-closing wedge corrective osteotomy more information is needed than for a simple osteotomy (Figure 6). It is necessary to know not only the dimensions of the bone graft (lengthening), but also the dimensions and shape of the segment of the radius that is to be resected (shortening).

The opening and closing sides of the osteotomy are not necessarily equal. Among our 15 patients who needed an opening-closing wedge corrective osteotomy, 14 had unequal opening and closing sides.

Because the BIZCAD planning and technique for surgical procedure are based on measurement of dimensions in millimeters, rather than angles, the whole procedure is simplified (Figure 7).

The patient treated with corrective osteotomy for a fracture with the center of rotation in zone C is shown in Figure 8.

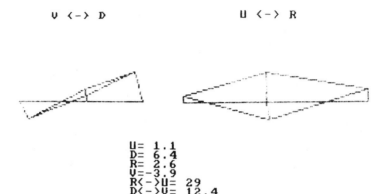

ひ <-> D   U <-> R

```
U=  1.1
D=  6.4
R=  2.6
V=-3.9
R<->U=  29
D<->V=  12.4
```

Figure 6. Printout of the bone graft needed for the opening-closing wedge corrective osteotomy. Parameters are: D - dorsal height of the graft, V - volar edge to be resected, R - radial height of the graft, U - ulnar height of the graft, R-U - width of the graft in the radio-ulnar direction, V-D - width of the graft in the dorso-volar direction.

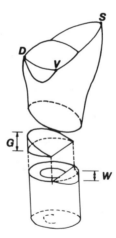

Figure 7. Technique of opening-closing corrective osteotomy. G - height of the graft on the dorsal side. W - height of the part of the radius edge to be resected. G is not necessarily equal to W.

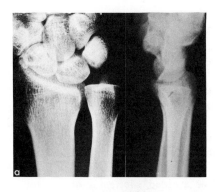

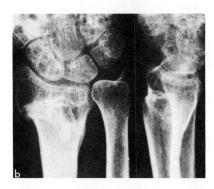

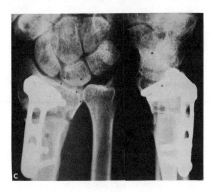

Figure 8. X-rays of the patient treated with an opening-closing edge corrective osteotomy. a. Healthy hand (reversed for better comparison). b. Affected hand. c. Postoperative control.

DISCUSSION

The connection between shortening of the radius after fracture and reduced function is well known. Cooney and al.[5] found that 20 of their patients with various complications after treatment of Colles fractures had painful movement of the wrist, and/or mechanical obstruction of movements. This state was called osteoarthritis, and was detected in 37 patients. Radioulnar osteoarthritis more often found (27 cases) than radiocarpal osteoarthritis (10 cases). This complication was most often found in fractures of the type VI, VII and VIII according to Frykman's classification.[8] When treatment of the patients was performed with external fixation, resulting in better correction of the radius length, a lower incidence of oteoarthritis (4 patients) was found despite the fact that such treatment was usually applied in patients with more comminuted fractures.

Shortening of the radius, as well as changes of volar and ulnar angle, increase the stress in the distal radiocarpal joint and change the loading pattern.[14]

All these facts stress the importance of restoration of the length of the radius immediately, while the fracture is fresh, or later, with corrective osteotomy.

Opening-closing wedge corrective osteotomy of the malunited radius has already been reported in literature,[10] however, only in combination with resection of the ulna. This additional procedure may not be necessary in some cases, but detection of such cases is sometimes possible only with 3 dimensional preoperative planning.

Shortening of the ulna dictates how the bone graft should be designed. In the case that bone grafting should be avoided, shortening of the ulna can turn an opening wedge into an opening-closing wedge, or even a closing wedge corrective osteotomy. We had a particular case where, after shortening of the ulna by 10mm, only an opening-closing wedge corrective osteotomy could restore the exact radial-ulnar relation.

Planning of corrective osteotomy with the aim of correcting volar and ulnar tilt, together with radius length, is rather complicated. However, it is even more complicated if an opening-closing wedge osteotomy is to be performed. Therefore, besides the drawbacks already noted,[3] the overlay drafting method for planning is completely unsuitable for an opening-closing corrective osteotomy. Furthermore, precise planning (the key to good anatomic and functional results) is impossible with the standard overlay drafting method because of the complexity of possible translations and rotations of the mobile segment.

In this article the center of rotation of the mobile segment in the frontal plane has been studied. The question is however: is there, and in which cases, a common center of rotation in 3 dimensions? If not, how are the different rotational axes, types of fractures, and extent of deformation related? Special mention should be made of rotation in the horizontal plane (axial rotation). Further investigation is needed to determine the relations of different rotation centers.

CONCLUSION

The information obtained on the basis of the simple geometric analysis of X-rays after malunion have direct implications for the type of corrective osteotomy and the shape of the needed bone graft.

We conclude that it is occasionally necessary to perform an opening-closing wedge corrective osteotomy for precise reconstruction of the radius.

REFERENCES

1. H. Abbaszadegan, U. Jonsson and K. von Sivers, Prediction of instability of Colles fractures, *Acta Orthop. Scand.* 60:646 (1989).
2. F. af Ekenstam, C.G. Hagert, O. Engkvist, A.H. Tornvall and H. Wilbrand, Corrective osteotomy of malunited fractures of the distal end of the radius, *Scand. J. Plast. Reconstr. Surg.* 19:175 (1985).
3. R. Bilic and V. Zdravkovic, Planning corrective osteotomy of the distal end of the radius. 1. Improved method, *Unfallchirurg.* 91:571 (1988).
4. R. Bilic and V. Zdravkovic, Planning corrective osteotomy of the distal end of the radius. 2. Computer-aided planning and postoperative follow up, *Unfallchirurg.* 91:575 (1988).
5. W. Cooney, J.H. Dobyns and R.L. Linscheid, Complications of Colles fractures, *J. Bone Joint Surg.* 62-A:613 (1980).

6. D.L. Fernandez, Radial osteotomy and Bowers arthroplasty for malunited fractures of the distal end of the radius, *J. Bone Joint Surg.* 70-A:1538 (1988).

7. H. Förstner, The distal radioulnar joint. Morphologic aspects and implications for orthopedic surgery, *Unfallchirurg.* 90:512 (1987).

8. G. Frykman, Fracture of the distal radius including sequelae - Shoulder-hand-finger syndrome, disturbance of the distal radio-ulnar joint and impairment of nerve function. A clinical and experimental study, *Acta Orthop. Scand.*, Suppl. 108 (1987).

9. P. Massart and Ph. Merloz, Segmental shortening of the ulna in some malunions of the distal radius. *Ann. Chir. Main* 1:65 (1982).

10. G. Sennwald, The Wrist, Springer-Verlag, Berlin, (1987).

11. S. Solgaard, Function after distal radius fracture, *Acta Orthop. Scand.* 59:39 (1988).

12. R.N. Villar, D. Marsh, N. Rushton and R.A. Greatorex, Three years after Colles fractures, *J. Bone Joint Surg.* 69-B:635 (1987).

13. V. Zdravkovic and R. Bilic, Computer-assisted preoperative planning (CAPP) in orthopaedic surgery, *Comput. Methods and Programs in Biomedicine* 32:141 (1990).

14. T. Ziger, V. Nikolic, R. Bilic and V. Zdravkovic, Distribution of pressure in the radiocarpal joint after malunited distal radius fracture, "Proceedings of the 7th meeting of the European Society of Biomechanics", Aarhus, 1990 (1990).

SCAPHOID AND PERISCAPHOID INJURIES:

MATHEMATICAL PATTERN

Pascal Ledoux

Centre S.O.S. Main Bruxelles
Clinique du Parc Léopold
Bruxelles, Belgium

INTRODUCTION

Classical studies of the carpus in columns or in rows do not explain carpal injuries. In order to propose logical treatements it is important to understand the mechanisms of these injuries.

Vectorial analysis of transmission of loads through the wrist may be a good way to determine the conditions of several types of scaphoid and periscaphoid injuries.

BASIC CONCEPT

Equilibrium is a condition which exists when all the forces and moments applied to a body are balanced. Some loads are known, and by deduction, through vectorial analysis, it is possible to establish the theoretical transmission through the different carpal bones.

A fracture or a ligamentous injury exists when the capacity of kinetic energy absorption of the system is exceeded.

METHOD

The comparison between transmission of loads when the wrist is in 45° or 100° extension can explain, in theory, the circumstances under which a fracture or a ligamentous injury will occur.

Advances in the Biomechanics of the Hand and Wrist
Edited by F. Schuind *et al.*, Plenum Press, New York, 1994

VECTORIAL STUDY

The transmission must be studied separately at the lunate and the scaphoid levels, after which the resultants are added together.

The transmission of loads through the wrist in 45° and 100° extension were studied concurrently in order to understand step by step the difference between the two situations.

LUNATE LEVEL

During a fall on the wrist in extension the load (F1) is transmitted from the radius to the lunate (Figure 1).

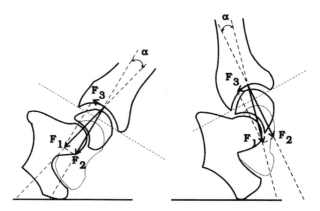

Figure 1. See text.

Because of a 15° angle between the glenoid and the axis of the radius we must consider the change in direction of the force. F1 is separated into two components acting at right angles to each other: F2 and F3.

F3 is a stress which tends to cause slipping of the radius on the lunate. This force is canceled by the elasticity of the anterior ligaments of the wrist. F2 is the load transmitted from the radius to the lunate.

The force F2 is applied from the lunate to the capitate with a new change of direction according to the axis of the lunate, and the vector is resolved into two new vectors : F4 and F5 (Figure 2).

When the wrist is in 100° extension, the disposition of the different bones and the importance of F5 can cause a peri-lunate dislocation. This vectorial approach confirms the classical concept according to which injury needs forced hyperextension.

F4 can be broken down into two new components: F6 in the axis of the capitate and F7 toward the ground. F7 will produce a moment (M = F7.d) with a twisting effect on the capitate and the hand (the axis of rotation is the line joining the fulcra of the trapezium and the hook of the hamate) (Figure 3).

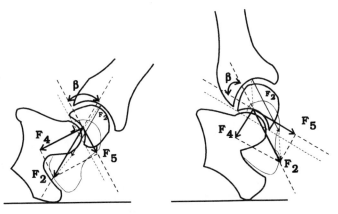

Figure 2. See text.

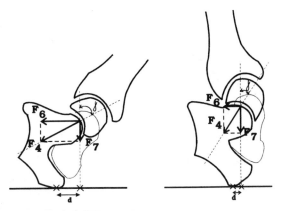

Figure 3. See text.

SCAPHOID LEVEL

The force is transmitted from the radius to the scaphoid with a change of direction because of the angle between the radius and the scaphoid (Figure 4). The decomposition at right angles gives a component of translation. It is interesting to note that the direction of this component is dorsal when the wrist is in 45° extension, whereas it is palmar when the wrist is in 100° extension (Figure 4).

The vector F10 is translated to the distal pole of the scaphoid and is transmitted to the trapezium perpendicular to the joint space. This new change of direction produces a vectorial decomposition at right angles: F11 is a force which tends to make the scaphoid slip off the trapezium, and F12 is the load transmitted to the trapezium (Figure 5).

This vector F12 is translated to the fulcrum of the trapezium and there is broken down at right angles into a vector perpendicular to the ground (F14) and a vector (F13) which tends to produce slippage of the hand on the ground (Figure 6).

Of course this force is greater when the wrist is in 45° extension. This force is partly counterbalanced by the force F11.

453

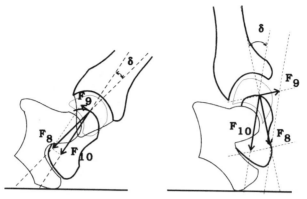

Figure 4. See text.

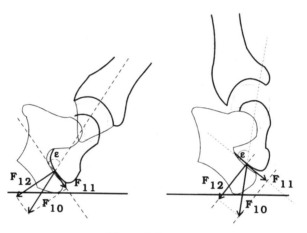

Figure 5. See text.

The scaphoid is, in this analysis, subject to two forces on the two poles. When the wrist is in 45° extension, the two forces act in opposite directions, whereas for the wrist in 100° extension the forces are roughly additive. This is a very important point for the understanding of scaphoid and periscaphoid injuries (Figure 7). The position of the radio-scapho-capitate ligament is of first importance. Figure 8 represents the components of shearing.

When the wrist is in 45° extension these loads act in opposite directions between the scaphoid and the lunate, and when these loads exceed the capacity of absorption of kinetic energy, the scapho-lunate ligament will break (Figure 9).

Of course the force F11, which contributes along with the force F9 to rotate the scaphoid around the radio-scapho-capitate ligament, increases the shearing load on the scapho-lunate ligament.

This rotation of the scaphoid occurs in the opposite direction to the rotation of the distal carpus owing to the force of F7.

This opposite rotation shears the scapho-trapezial ligament, which can of course break. The scapho-trapezial ligament is stronger than the scapho-lunate ligament. In spite of this, at this stage of our study, it is difficult to say which ligament will break first.

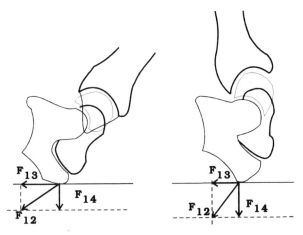

Figure 6. See Text.

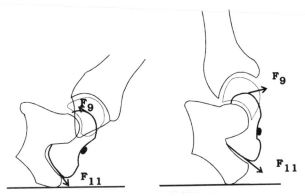

Figure 7. See Text.

When the wrist is in 100° extension, the translation loads on the upper pole of the scaphoid are superimposed and therefore additive. The resultant force (F15) is the vectorial sum of F5 and F9. The two forces F11 and F15 applied on the two poles of the scaphoid with the radio-scapho-capital ligament as fulcrum may cause a fracture of the scaphoid by tensile stress. In this situation the stress on the scapho-trapezial ligament is less than when the wrist is in 45° extension, and it is likely that ruptures of this ligament will occur in the latter situation.

CONCLUSION

The following conclusions may be drawn from this vectorial analysis: [1] the angle of extension of the wrist determines the type of injury: fracture or ligamentous rupture, [2] fracture of the upper and lower poles of the scaphoid are equivalent to ligamentous ruptures, [3] the disposition of the bones in the carpus permits good absorption of the loads.

455

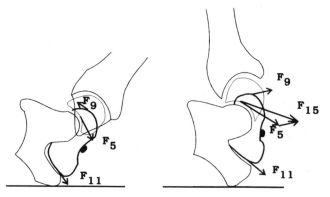

Figure 8. See Text.

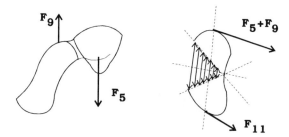

Figure 9. See Text.

REFERENCES

1. F. Bonnel, and Y. Allieu, Les articulations radio-cubito-carpienne et médio-carpienne, Organisation anatomique et bases biomécaniques, *Ann. Chir. Main*, 3:287 (1984).
2. G. Sennwald, "L'entité radius carpe", Springer-Verlag (1987).
3. E.R. Weber, and E.Y. Chao, An experimental approach to the mechanism of scaphoid waist fracture, *J. Hand Surg.* 3:142 (1978).

IN VIVO LOAD-DISPLACEMENT BEHAVIOR OF THE CARPAL SCAPHOID LIGAMENT COMPLEX: INITIAL MEASUREMENTS, TIME-DEPENDENCE, AND REPEATABILITY

Joseph J. Crisco III, and Scott W. Wolfe

Department of Orthopaedics and Rehabilitation
Yale University School of Medicine
333 Cedar Street, New Haven, CT, 06510, USA

INTRODUCTION

Hyperextension injuries of the wrist account for a wide variety of wrist pathology, from what may be perceived as a mild "sprain" to the extreme of fracture-dislocation. Ligamentous injuries, which are notoriously difficult to diagnosis with imaging techniques, may progress to overt instability and carpal collapse. Without radiographic carpal malalignment, wrist instability may be assessed by isolating and manually displacing portions of the carpus. This clinical "scaphoid shift test" is an attempt to dorsally subluxate the scaphoid with a manual load applied at its tubercle.[8] An observation of increased carpal displacement, especially if symptomatic, may indicate ligamentous injury or disruption. However, a clinical observation such as this is subjective.

The goal of this work was to develop an objective *in vivo* technique that quantitates and documents scaphoid shift. We designed and built an apparatus that applies a dorsally directed load to the scaphoid tubercle and records the resulting displacement. In this paper we present the repeatability of the technique, the effect of preconditioning and rest on the load-displacement behavior, and the initial measurements on 16 healthy wrists.

MATERIALS AND METHODS

Sixteen wrists in male volunteers (mean 29 years; range 19-35 years), with no prior history of wrist injury or disease, participated in this study. The load-displacement characteristics of the carpal scaphoid ligament complex were measured with a specially designed apparatus. The supinated forearm and wrist were strapped to a horizontal support platform (Figure 1). An upper platform, situated over the wrist, supported a pair of linear bearings that guided the free vertical travel of a steel plunger. The plastic tip of the plunger was dimpled to accept the scaphoid tubercle. A handle at the top of the plunger allowed the examiner to apply a dorsally directed load to the scaphoid. A strain gage cell, constructed and calibrated in our laboratory, recorded the load applied by the plunger. A linear transducer (1000 HR-DC, Schaevitz, Pennsauken, NJ) recorded the displacement of the plunger.

The elasticity of the testing apparatus was studied by loading the plunger against an aluminum block. The total deformation recorded at 40 N was approximately 0.8 mm, most of which was assumed to be in the tip of the plunger. The deformation of the plunger during an actual test on a subject may be less since the soft tissues of the wrist distribute the stresses over the surface of the plastic tip.

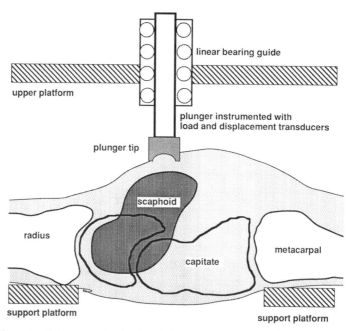

Figure 1. Schematic of the apparatus developed for measuring the load-displacement behavior of the carpal scaphoid ligament complex *in vivo*.

During testing, a personal computer sampled the analog signals from the load cell and displacement transducer simultaneously. An examiner manually applied the load. The computer audibly alerted the examiner as he reached the maximum load of 40N, determined previously to be the maximum load without producing discomfort. Each test consisted of five loading cycles. The maximum displacement (mm) was the displacement at 40N. The stiffness (N/mm) was the slope of the linear least squares fit (MATHPAK 87, Precision Plus Software, Ontario, Canada) of the load-displacement curve. A zero load in the first cycle defined the zero position. After the first cycle, the position of the plunger at a zero load defined the residual displacement (mm) (Figure 2).

To investigate the time dependent response of the scaphoid complex, we tested 1 subject at 9 specific times with rest intervals ranging from 2 min to 4 hr. Each test consisted of 5 cycles. The significance of the differences in the maximum displacements with time was determined using ANOVA with a post-hoc Fisher LSD test at a confidence level of 95%.

Repeatability was addressed by two approaches. The first examined the repeatability of three tests performed at approximately 4 and 1 hour rest intervals. A standard test consisting of 5 cycles was performed (Test 1). The subject's hand was removed from the apparatus and then retested at 4 hours (Tests 2) and again after 1 hour (Test 3). The repeatability of the measurements was evaluated by plotting the maximum and residual displacement at each cycle for all tests.

The second approach examined the repeatability of preconditioning on 16 wrists. After preconditioning the wrist with a single test of 5 cycles, several tests were performed. Thirteen subjects were tested 3 times, 1 was tested 4 times, and two were tested 2 times. Each test consisted of 5 cycles; only the data of cycle 5 was analyzed. This data was grouped and analyzed to describe the variation between subjects (Table 1). The variance in the repeated tests was analyzed by determining the standard deviation within each subject (Table 2).

RESULTS

Typical results for a single test of 5 cycles are presented as load-displacement curves in Figure 2. Each curve demonstrated an approximately linear behavior, with a trend toward increased stiffness with increasing load. With each successive loading cycle the curves tended to shift along the displacement axis as residual and maximum displacement increased, while stiffness was unchanged.

The means and standard deviations of the maximum displacement of 1 test (5 cycles) are plotted as a function of time for the 9 sequential tests in Figure 3. In the first test, at t = 0, the average maximum displacement was 2.5 mm. This increased

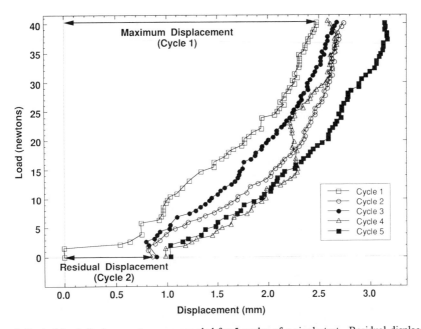

Figure 2. Typical load-displacement curves recorded for 5 cycles of a single test. Residual displacement was defined as the difference in the displacement from Cycle 1 at 0 N. The maximum displacement was defined at 40 N. The stiffness of each curve was a linear least-squares fit.

significantly (p = 0.01) after 2 min rest to 3.1 mm (t = 2 min) and then significantly again after another 2 min rest to 3.9 mm (t = 4 min). After a 15 min rest it decreased significantly to 3.2 mm. The two following tests performed with 2 min rest returned to values of 3.8 mm and 3.7 mm, respectively. Neither of these tests were different from the test at t = 4 min. The next test was performed after a 4 hour rest and we observed almost a complete recovery to 2.7 mm that was not statistically different from t = 0. Another test after 2 min again significantly increased the maximum displacement, which recovered completely with a 60 min rest to 2.7 mm.

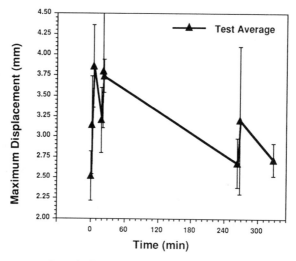

Figure 3. The average and standard deviation (5 cycles) of the maximum displacement recorded from 1 subject at 9 points in time.

Plotted in Figure 4 are the residual displacements and the maximum displacements recorded in tests at t = 0, 4, and 5 hr. At each cycle, the maximum displacements were almost identical. The maximum displacements also tended to increase with each successive cycle. The residual displacements were constant from cycle 2 to cycle 5, but there was a significant difference with each test. Note that we defined the residual displacement at cycle 1 to be 0 mm.

Listed in Table 1 are the parameters describing cycle 5 for 47 tests on 16 wrists. As expected, we recorded an apparent normal distribution of values. The relatively small magnitudes of the standard deviations (1.8 mm for maximum displacement and 4.6 N/mm for stiffness) suggest a homogeneous population. The best measure of repeatability is the variance in the measurements, as quantified by the standard deviation. For the 3 tests on each subject the standard deviations of the cycle 5 parameters were calculated. The average standard deviation was 0.4 mm in residual displacement, 0.5 mm in the maximum displacement, and 1.8 N/m in the stiffness (Table 2). Also listed in Table 2 are the maximum values.

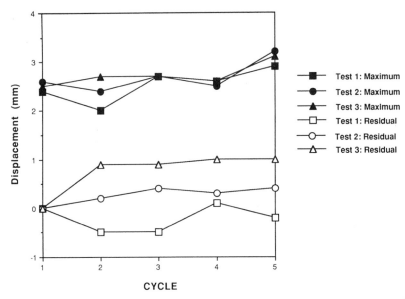

Figure 4. The residual and maximum displacement for cycles conducted at t = 0 (Test 1), 4 hours (Test 2), and 5 hours (Test 3).

Summarizing the results, the significant viscoelastic nature of the carpal scaphoid complex is illustrated in Figure 3. With sufficient rests between tests the results were highly repeatable (Figure 4). Furthermore, in the homogeneous population of our subjects (Table 1), preconditioning with 5 cycles also generated highly repeatable results (Table 2).

Table 1. The range, average, and standard deviations of the cycle 5 mechanical parameters from all tests.

	Residual Displacement (mm)	Maximum Displacement (mm)	Stiffness (N/mm)
Average	0.8	4.7	11.1
Maximum	2.1	9.1	24.8
Minimum	0.0	2.4	4.3
s.d.	0.5	1.8	4.6
n	47	47	47

DISCUSSION

Markolf et al.[4] have reported a technique that measures the load-displacement behavior of the knee *in vivo*. They quantified the mechanical parameters of the knee by laxity (defined herein as residual displacement) and stiffness in 49 patients, and found the technique accurate and highly repeatable. These *in vivo* measurements correlated well with previous *in vitro* measurements on 35 cadaveric knees.[4] In studies involving patients with unilateral tears of the anterior cruciate ligament, significant differences in displacements have been measured as compared to the uninjured side.[1,2,7] These studies document a relationship between mechanical parameters and instability of the knee.

Table 2. Within each subject, repeatability was described by the standard deviation of the mechanical parameters (cycle 5) from 3 tests of 13 wrists, 4 tests of 1 wrist, and 2 tests of 2 wrists. The average and maximum standard deviations are listed.

	Residual Displacement (mm)	Maximum Displacement (mm)	Stiffness (N/mm)
Average	0.4	0.5	1.8
Maximum	0.9	1.7	4.2
n	16	16	16

A number of devices are now available to complement clinical assessment of knee instability (e.g., KT200, MEDmetric Co., San Diego, CA; Knee Laxity Tester, Orthopedic Systems Inc, Hayward, CA), and are accepted as reliable techniques for quantification of ligamentous integrity.[1,2,6] We have applied this biomechanical approach to the quantification of the carpal scaphoid ligamentous complex. We described the mechanical behavior of the scaphoid ligamentous complex by residual displacement, maximum displacement, and stiffness. To our knowledge such a study has not been previously performed.

Savelberg et al.[5] tested isolated (*in vitro*) bone-ligament-bone preparations of several carpal ligaments. They reported average stiffness values that ranged from 10 N/mm (s.d. 8 N/mm, n= 7) at 6.9 N for the radioscaphoid collateral ligament to the largest value of 47 N/mm (s.d 21 N/mm, n= 6) at 44N for the radiotriquetrum ligament. Although a direct comparison is not possible, it is interesting that our average stiffness value of 11 N/mm is in the neighborhood of the those values. Our value includes unknown contributions from the skin, underlying fascia, and numerous ligamentous structures loaded indirectly in tension.

Under a dorsally directed load the carpal scaphoid complex exhibits noteworthy time dependent responses. Tests demonstrated that this viscoelastic behavior was repeatable with sufficient preconditioning (5 cycles). Preconditioning has also been shown to provide repeatable results *in vitro*.[5] Herein, the effect of preconditioning was reversible. After a rest period of 1 hour the ligamentous complex completely recovered. Direct extrapolation to the clinical setting may not be possible due to the non physiological loading direction in this study. These viscoelastic properties of the scaphoid ligament complex are being studied further.

The subjects of this study had no prior history of wrist injuries. The range of mechanical parameters recorded for these subjects is believed to be associated with natural variations in ligamentous laxity and in gross anatomy. Quantifying scaphoid displacement may provide valuable information to complement the clinical examination of the injured wrist. *In vitro* tests in our laboratory are currently determining changes in the biomechanical properties of the scaphoid ligamentous complex following ligament injury.

CONCLUSIONS

A technique to quantify the *in vivo* mechanical behavior of the carpal scaphoid ligament complex was developed. The load-displacement behavior was repeatable with either sufficient preconditioning or with sufficient rest between tests. In 47 tests of 16 healthy wrists the average maximum displacement and stiffness at 40 N was 4.7 mm (s.d. 1.8 mm) and 11.1 N/mm (s.d. 4.6 mm), respectively.

ACKNOWLEDGEMENTS

We would like to thank Eric Vajda, B.S. for his assistance in constructing the load-displacement apparatus and Robert Brown, M.D. for his assistance in analyzing the data.

REFERENCES

1. B.B. Bach, R.F. Warren, W.M. Flynn, M. Kroll, and T.Z. Wichiewicz, Arthrometric evaluation of knees that have a torn anterior cruciate ligament, *J. Bone Joint Surg.* 72A:1299 (1990).
2. D.M. Daniel, M.L. Stone, R. Sachs, and L. Malcom, Instrumented measurement of anterior knee laxity in patients with acute anterior cruciate ligament disruption, *Am. J. Sports Med.* 13(6):401 (1984).
3. K.L. Markolf, A. Graff-Radford, and H.C. Amstutz, In vivo knee stability: a quantitative assessment using an instrumented clinical testing apparatus, *J. Bone Joint Surg.* 60-A:5:664 (1978).
4. K.L. Markolf, J.S. Mensch, and H.C. Amstutz, Stiffness and laxity of the knee-The contributions of the supporting structures, A quantitative in vitro study, *J. Bone Joint Surg.* 58-A:583 (1976).
5. H.H.C.M. Savelberg, J.G.M. Kooloos, R. Huiskes, and J.M.G. Kauer, Stiffness of the ligaments of the human wrist joint, *J. Biomech.*25(4):369 (1992).
6. K. Shino, M. Inoue, S. Horibe, M. Nakamara, and K. Ono, Measurement of anterior instability of the knee, A new apparatus foe clinical testing, *J. Bone Joint Surg.* 69-B:608 (1987).
7. M.E. Steiner, C. Brown, B. Zarins, B. Brownstein, P.S. Koval, and P. Stone, Measurement of anterior-posterior displacement of the knee, *J. Bone Joint Surg.* 72A:1307 (1990).
8. H.K. Watson, D. Ashmead, and M.V. Makhlouf, Examination of the scaphoid, *J. Hand Surg.* 13A: 657 (1988).

THE USE OF CINE-COMPUTED TOMOGRAPHY TO INVESTIGATE CARPAL KINEMATICS

Scott W. Wolfe,[1] Joseph J. Crisco III,[1] and Lee D. Katz[2]

[1]Department of Orthopaedics and Rehabilitation
[2]Department of Diagnostic Imaging
Yale University School of Medicine
333 Cedar Street
New Haven, CT 06510, USA

INTRODUCTION

Hyperextension wrist injuries have been associated with a high degree of wrist morbidity.[10,21,23] The inexorable progression of scaphoid rotary subluxation or nonunion to carpal collapse and subsequent degenerative arthritis has been well documented in the literature.[15,23] Examination of the acutely injured wrist is limited by pain and swelling, and initial radiographs are frequently normal. Subtle instabilities of the wrist are notoriously difficult to study with plain films alone. Carpal overlap, normal variations, and inconsistencies in wrist positioning make interpretation and comparison of plain films inaccurate.[6,14] Cineradiography is useful, but is still limited by osseous overlap and can not be quantitated. More exacting measurements used in the laboratory are not applicable to use in the clinical situation, as these often require dissection of the wrist and placement of radiographic markers or light-emitting diodes.[1,14,27] In fact, few kinematic studies of the normal wrist have been performed *in vivo*, due to these limitations.

It has been postulated that scapho-lunate dissociation will progress over time to dorsal intercalated instability.[10,18,25] There is indirect evidence that scapho-lunate dissociation will progress to intercalated segment instability, as 21/23 patients in Linscheid's series with dissociation had lunate rotational instability.[10] Attempts at producing this collapse pattern in the laboratory by ligament sectioning have been unsuccessful.[10,11] It is likely that the ligamentous damage produced by a radial sided wrist injury alters the ability of the scaphoid to synchronize the motion of the proximal and distal carpal rows. Subsequent time-dependent attenuation of the stout lunate ligamentous anchors may allow the lunate to assume a dorsiflexed attitude under the axial load of the capitate, and the unopposed forces on the medial side of the carpus.

We hypothesize that abnormal kinematics of the carpus precedes static radiographic evidence of carpal instability. We hypothesize that "abnormal patterns" of motion occur within the capitate-lunate-radius complex prior to the presentation of dorsal intercalated instability. Thus our **goal** is to develop a radiographic technique which may accurately measure normal and abnormal carpal kinematics, and one that may be used *in vivo*.

MATERIALS AND METHODS

Ten volunteers with no prior history of wrist disease or injury, were selected for kinematic analysis. Clinical examination consisted of range of motion measurements, and results of the Watson "scaphoid shift" test.[21,22] Radiographic analysis consisted of antero-posterior, and true lateral views of the index wrist. A true lateral radiograph was defined as one in which the axis of the third metacarpal and long axis of the radius were colinear. Radiographs were analyzed by measuring the radiolunate angle, the scapholunate angle, and the volar tilt of the distal radius, using the tangential techniques of Larsen et al.[8] Radiolunate angle was corrected for deviations from the true lateral by subtracting fifty percent of the radiometacarpal angle from the radiolunate angle (corrected radiolunate angle RL_c). This assumes that global wrist motion in this range is equally split between radiolunate and lunocapitate motion.[14] This assumption is borne out by the kinematic measurements of normal wrists in this study.

Volunteers were grouped based on measurements of the corrected radiolunate angle. Three volunteers with an RL_c of negative fifteen degrees or greater were grouped and analyzed separately from the six with normal radiolunate angles. All three patients in this group had a positive scaphoid shift test. It was assumed that this group, who had no history of prior wrist injury or discomfort, represented "physiologic VISI."

Three patients with hyperextension wrist injuries were also studied. Two patients had sustained the injury six weeks prior to the examination. One patient had normal radiographs at the time of injury, but had developed early signs of scapholunate dissociation at the time of exam. The lateral radiolunate angle measured fifteen degrees, the scapholunate angle measured sixty-seven degrees, and scapholunate gap measured 1.5 mm. Radiographs of the second patient showed abnormally extended lunates bilaterally, measuring thirteen degrees and scapholunate angles of sixty-four degrees. Only her injured side was symptomatic. Additional studies included a three compartmental arthrogram for the first patient, which documented a tear of the scapholunate ligament; and a magnetic resonance imaging scan of the second patient's symptomatic wrist, which documented abnormally increased signal in the scapholunate ligament, consistent with an acute ligament tear. The third had chronic scapholunate dissociation with widening of the scapholunate interosseous space and dorsal intercalated segment instability evident on radiographs (radiolunate angle = thirty degrees, scapholunate angle = eighty-five degrees).

Cine-computed Axial Tomography (C-CAT)

Each subject underwent computed tomographic imaging of the wrist using a General Electric 9800 CT scanner. The subject was positioned standing aside the gantry, with the elbow flexed, such that a sagittal computed tomographic image could be obtained through the distal radius, lunate, capitate, and the third metacarpal. The hand and forearm was positioned in a custom plexiglass device, which fixed the wrist at ten degree increments from flexion to extension (Figure 1A-B). All sequential scans were obtained at the same sagital location, using 1.5 mm thick sections and a 16 cm field of view. The data was copied on to magnetic tape and transferred to an ISG Technologies independent work station (Toronto, Canada). Using Camera S-200 Allegro software (ISG Technologies, Inc., Toronto, Canada), the data was reformated into a cine loop of carpal motion, for the entire flexion/extension arc. The investigators reviewed the dynamic display of each subject for abnormalities of carpal motion (Figure 2).

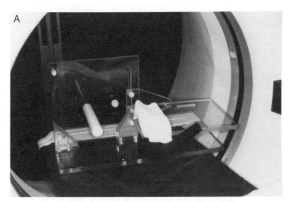
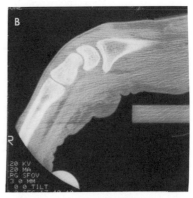

Figures 1. Figure 1-A. Computed tomography data acquisition. The custom jig in position within the CT gantry. **Figure 1-B.** A mid-sagital image through the radius, lunate, capitate and third metacarpal.

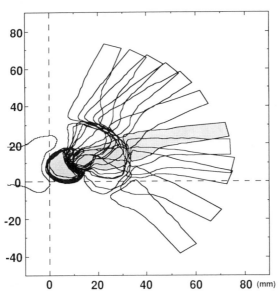

Figure 2. A composite image of carpal contours following digitization of the computed tomographic images.

Kinematic Analysis

To determine the relative motion of the carpal bones, tomographic images were projected on to a digitizer (Altek, 63x63 cm2, accuracy $\pm$ 0.075 mm, Altek Corp, Silver Spring, MD), which was controlled by specifically written software on a personal computer. To provide a consistent coordinate system from image to image, we defined a coordinate system fixed to the radius. Two points on each bone were also chosen to define the lunate, capitate and third metacarpal. Carpal motion between each successive tomographic image was calculated using the two bony points and under the assumption of planar rigid body kinematics. To improve the consistency of identifying the same two points on each carpal image, a transparent template of each bone was created, which identified the two reference points. The template contour was positioned to a provide "best fit" with the tomographic image, and the points digitized. To reduce variability, these points were digitized three times and the data averaged. The data generated included incremental rotation values for radiometacarpal, radiocapitate, and radiolunate motion between successive images.

Incremental capitolunate rotation was defined as the difference between radiocapitate rotation and radiolunate rotation. Incremental rotations between images were then added to provide summed capitolunate and radiolunate rotations at each radiometacarpal position.

Total capitolunate and radiolunate rotation was expressed as a percentage of global wrist motion for each group. Global wrist motion was defined as the total of all incremental rotations of the third metacarpal with respect to the radius coordinate system.

Global wrist motion was further broken down for analysis into an "extension arc" (neutral to full extension), and a "flexion arc" (neutral to full palmar flexion). The neutral tomographic image was defined as that in which the third metacarpal and the longitudinal axis of the radius were colinear. Rotation in extension was assigned a positive value, and rotation in flexion was assigned a negative value. Capitolunate rotation was plotted as a function of radiolunate rotation for each subject, in both the flexion arc and the extension arc. A linear least-squares fit to each curve was derived, the slope of which was termed the LC/LR ratio. The LC/LR ratio represents the number of degrees of capitolunate motion for each degree of radiolunate motion. (Figure 3). Significant differences between groups was determined using ANOVA with a Fischer LSD post-hoc test (Statview SE+, Abacus Concepts Inc., Berkeley, CA.), to compare normal and abnormal kinematics.

RESULTS

Average corrected radiolunate angle measured -7 degrees in the normal group (range -3 to -13). Scapholunate angle measured 54 degrees (range 48-62). The three subjects in the VISI group had an average radiolunate angle of -22 degrees (range -18 to -25), with an average scapholunate angle of 39 degrees. The three patients with scapholunate dissociation had an average radiolunate angle of 19 degrees (range 13-30) and scapholunate angle of 72 degrees.

In normal wrists, global wrist motion in the flexion/extension plane was nearly equally divided between capitolunate and radiolunate rotation. Radiolunate rotation among the seven normal subjects accounted for an average of 48.4 percent of total wrist flexion/extension (S.D. $\pm$ 8.5). Average radiolunate contribution for the three subjects in the VISI group was similar (46.9 $\pm$ 4.5 percent). Among the injured wrists however,

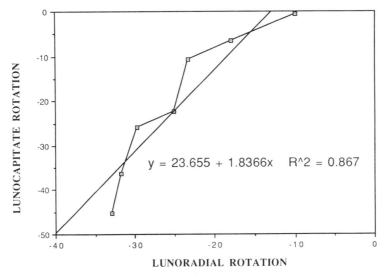

Figure 3. Representative graph of normal wrist flexion, which plots lunocapitate rotation as a function of radiolunate rotation. The slope of the curve is the LC/LR ratio.

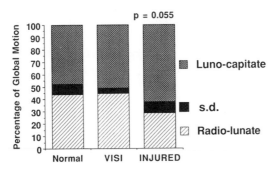

Figure 4. Total lunate rotation as a percentage of global wrist motion. Note the marked decrease in lunate rotation in patients with scapholunate dissociation.

radiolunate motion was markedly restricted, contributing only 31.5 ± 8.3 percent to the flexion/extension arc (Figure 4). This value was different from the normal and VISI groups at $p = 0.055$.

Data for LC/LR ratios in flexion and extension for all groups is displayed in table 1. In the normal wrists, extension was equally divided between lunocapitate and radiolunate joint motion, while palmar flexion occured predominantly at the midcarpal joint. In the uninjured wrists with a VISI pattern, palmar flexion occured almost exclusively at the lunocapitate joint, while extension occured predominantly at the radiolunate joint. This pattern is different from both the radiographically normal wrists and the injured wrists in flexion, with a significance level of $p = 0.055$. In injured wrists, lunocapitate motion increased dramatically in extension, with a large increase in the LC/LR ratio. This finding is significantly different from both the normal wrists and the VISI group in extension ($p < .01$). Interestingly, LC/LR ratio in flexion for the injured group was not significantly different from the normal wrists (Figure 5).

Table 1. Lunocapitate/lunoradius (LC/LR) ratio for extension and flexion arcs.

Extension	LC/LR	St. Dev.	Flexion	LC/LR	St. Dev.
Normal	1.0	0.6	Normal	1.9	1.3
VISI	0.1	0.44	VISI	**3.5**	**1.7**
Injured	**2.9**	**0.8**	Injured	1.0	1.0

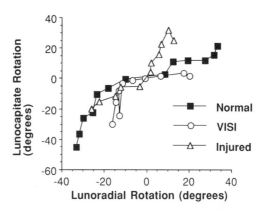

Figure 5. Representative graphs of carpal rotation for each group. Note the significant increase in lunocapitate extension in the injured group, and the increase in lunocapitate flexion in the VISI group.

DISCUSSION

Previous studies have helped to define the kinematics of the normal wrist in the in vitro situation.[1,5,14,27] There is considerable controversy, however, concerning the degree of contribution of the proximal and distal carpal rows to wrist flexion and extension. Ruby et al. reported three dimensional relative carpal motion in cadaveric specimens using orthoradiography and sonic digitization. They concluded that flexion and extension was nearly equally divided between the motion at the midcarpal and radiocarpal joints, with a greater contribution of the midcarpal joint to flexion than extension.[14] Andrews and Youm recorded uniplanar motion of the wrist on one normal and one rheumatoid patient, using a combination of cineradiography and sequential Schonander Xray films.[1] These authors maintain that the normal wrist functions essentially as a hinge joint, flexing about a stationary center of rotation located in the proximal pole of the capitate. They report "coaxial" motion of the capitate and lunate throughout the entire arc of flexion and extension.[1,26,27] In designing a wrist replacement prosthesis, Volz studied normal carpal kinematics and concluded that approximately two-thirds of the arc of extension occurs at the radiocarpal joint, while most of carpal flexion (60%) arises at the midcarpal joint.[20] Saraffian's work demonstrated similar conclusions.[16]

Our analysis of cine-computed tomographic data confirms the consensus opinion that normal wrist flexion from neutral occurs to a larger extent from the midcarpal joint (57%) than radiocarpal joint (43%). Neutral to wrist extension involved an equal contribution of radiocarpal and midcarpal motion.

A theoretical limitation of C-CAT to studies of carpal motion is its two dimensional analysis of complex three dimensional motion, and this introduces a source of error. Previous studies in *in vitro* models, have calculated individual carpal screw displacement axes, and concluded that wrist motion is mulitplanar, and that significant degrees of rotation and deviation of both rows occurs during wrist motion.[2,14,17] All studies agree, however, that the highest degree of "out of plane" motions occur during ulnar/radial deviation. Savelberg et al. concluded that flexion and extension is almost uniplanar, by the fact that the ligament structures are almost perpendicular to the plane of movement. Furthermore, these authors found *exclusively* planar motion of the lunate and capitate in the arc from neutral to extension.[17] Each of these studies is limited in its application to an *iv vivo* analysis, by the need for placement of interosseous markers, light-emitting diodes or sonic spark-gap emitters. Because of the small "out-of-plane" motions which occur in the flexion/extension arc, C-CAT is able to provide a two dimensional kinematic analysis *in vivo*, which is likely highly accurate. Work is underway at this institution to correlate wrist kinematics as measured by C-CAT with a three dimensional stereoradiographic analysis.

Few kinematic studies of the abnormal wrist have been performed. Youm stated, "Quantitative measurements of wrists with pathologic processes have been useful but fraught with problems".[26] The application of laboratory studies to the *in vivo* situation is again limited by the need for intraosseous placement of radiographic markers. Arkless' cineradiographic study of wrists affected by rheumatoid arthritis, Kienbock's disease, and trauma documented dynamic carpal "shifts" during motion, but was not quantitative.[2]

In one of the only kinematic studies of the injured wrist, Smith et al. simulated a mid-waist scaphoid nonunion, and documented changes in carpal motion using implanted carpal markers and stereoradiography.[19] Total lunate motion increased slightly following the osteotomy. The distribution of lunate motion following osteotomy (50% extension, 50% flexion) was markedly different from the normal state (68% extension, 32% flexion). This corresponded to a dorsiflexed position of the lunate in static radiographs.

Our analysis of abnormal wrists, following scapholunate disruption, differed markedly from *in vitro* kinematic studies.[13,19] We demonstrated a significant <u>decrease</u> in radiolunate motion with a proportionate increase in midcarpal (capitolunate) motion. The largest decrease in lunate motion was seen in the arc from extension to neutral. It may be postulated that the discrepancy with the *in vitro* data may be related to post-traumatic capsular contractures about the abnormally extended lunate. Whether the altered degree of motion documented in this study relates to known patterns of degenerative disease, in which the radiolunate joint is spared at the expense of accelerated capitolunate degeneration,[30] is speculative.

Our analysis of uninjured wrists with a radiographic VISI pattern demonstrates a marked alteration of intercarpal motion in the sagital plane. Total lunate and capitate rotation is unchanged from the radiographically normal wrists, however, lunate rotation occurs predominantly in the extension arc, while capitate rotation occurs almost exclusively in the flexion arc. This phenomenon has not been described previously, to our knowledge.

Cine-computed tomography allows a dynamic display of carpal motion in both normal and abnormal wrists, and is able to generate data at multiple sampling sites throughout the arc of motion. While de Lange demonstrated that carpal motion is smooth and synchronous throughout motion in normal wrists,[4] we have presented data which indicates the reverse is true for those wrists with an abnormally rotated lunate. Dramatic shifts of carpal motion have been described in certain instability states.[2,9] We believe that it is necessary to study the entire arc of motion to detect subtle abnormalities in carpal kinematics.

SUMMARY

Cine-computed axial tomography (C-CAT) provides an accurate demonstration of the complex motion of the proximal and distal carpal rows without the osseous overlap inherent in plain radiography and fluoroscopy. Flexion of the wrist in normal volunteers occured predominantly at the midcarpal joint, while wrist extension occurred equally at the radiolunate and lunocapitate joints. Three patients with scapholunate ligament disruption demonstrated a significant decrease in radiolunate motion, which was most pronounced during extension. Three uninjured subjects with a physiologic VISI pattern showed a marked alteration in the distribution of intercarpal motion during flexion and extension.

Cine-computed tomography is a simple and non-invasive technique to study wrist kinematics, which can be successfully applied in the clinical situation.

REFERENCES

1. J.G. Andrews, and Y. Youm, A biomechanical investigation of wrist kinematics, *J. Biomech.* 12:83 (1978).
2. R. Arkless, Cineradiography in normal and abnormal wrists, *Amer. J. Roentgenology*, 96:837 (1966).
3. R.A. Berger, W.F. Blair, R.D. Crowninshield, and A.E. Flatt, The scapholunate ligament, *J. Hand Surg.* 1:87 (1982).
4. A. deLange, J.M.G. Kauer, and R. Huiskes, Kinematic behavior of the human wrist joint: a roentgen-stereophotogrammetric analysis, *J. Orthop. Res.* 3:56 (1985).
5. A.G. Erdman, J.K. Mayfield, F. Dorman, M. Wallrich, and W. Dahlof, Kinematic and kinetic analysis of the human wrist by stereoscopic instrumentation, *J. Biomech. Eng.* 101:124 (1979).
6. M. Garcia-Elias, K. An, P.C. Amadio, W.P. Cooney, and R.L. Linscheid, Reliability of carpal angle determinations, *J. Hand Surg.* 14A:1017 (1991).
7. E. Horii, M. Garcia-Elias, K.N. An, A.T. Bishop, W.P. Cooney, R. L. Linscheid, and E.Y.S. Chao, A kinematic study of luno-triquetral dissociations, *J. Hand Surg.* 16A(2):355 (1991).
8. C.F. Larsen, F.K. Mathiesen, and S. Lindequist, Measurements of carpal bone angles on lateral wrist radiographs, *J. Hand Surg.* 16A:888 (1991).
9. D. M. Lichtman, J.R. Schneider, A.R. Swafford, and G.R. Mack, Ulnar midcarpal instability; clinical and laboratory analysis, *J. Amer Hand Surg.* 6:515 (1981).
10. R.L. Linscheid, J.H. Dobyns, J.W. Beabout, and R.S. Bryan: Traumatic instability of the wrist; diagnosis, classification and pathomechanics, *J. Bone Joint Surg.* 54(8):1612 (1972).
11. T.D. Meade, L.H. Schneider, and K. Cherry, Radiographic analysis of selective ligament sectioning at the carpal scaphoid: a cadaver study, *J. Hand Surg.* 15A:855 (1990).
12. L.K. Ruby, Wrist biomechanics, In Instructional Course Lectures, v. XLI, ed. R.E. Eilert, American Academy Orthopaedic Surgeons, Park Ridge, IL, 25 (1992).
13. L. K. Ruby, K.N. An, R.L. Linscheid, W.P. Cooney, and E.Y.S. Chao, The effect of scapholunate ligament section on scapholunate motion, *J. Hand Surg.* 12A(5):767 (1987).
14. L.K, Ruby, W.P. Cooney, K.N. An, R.L. Linscheid, and E.Y.S. Chao, Relative motion of selected carpal bones: a kinematic analysis of the normal wrist, *J. Hand Surg.* 13A:1 (1988).
15. L.K. Ruby, J. Stinson, and M.R. Belsky, The natural history of scaphoid non-union, a review of fifty-five cases, *J. Bone Joint Surg.* 67A:428 (1985).
16. S.K. Sarrafian, J.L. Melamed, and G.M. Goshgarian, Study of wrist motion in flexion and extension, *Clin. Orthop. Rel. Res.* 126:153 (1977).
17. H.H.C.M. Savelberg, J.G.M. Kooloos, A. deLange, R. Huiskes, and J.M.G. Kauer, human carpal ligament recruitment and three-dimensional carpal motion, *J. Orthop. Research*, 9:693 (1991).
18. J.R. Sebald, J.H. Dobyns, and R.L Linscheid, The natural history of collapse deformities of the wrist, *Clin. Orthop.* 104:140 (1974).

19. D.K. Smith, K.N. An, W.P. Cooney, R.L. Linscheid, and E.Y. S. Chao, Effects of a scaphoid waist osteotomy on carpal kinematics, *J. Orthop. Res.* 7:590 (1989).

20. R.G. Volz, Biomechanics of the wrist, *Clin. Orthop.* 149:112 (1980).

21. H.K. Watson, Intercarpal arthrodesis, in: "Operative Hand Surgery", 2nd ed., D.P. Green, ed., Churchill Livingston, New York, (1988).

22. H.K. Watson, D. Ashmead, and M.V. Makhlouf, Examination of the scaphoid, *J. Hand Surg.* 13A:657 (1988).

23. H.K. Watson, and F.L. Ballet, SLAC wrist: scapholunate advanced collapse of degenerative arthritis, *J. Hand Surg.* 9A:358 (1984).

24. H.K. Watson, J. Ryu, and A. DiBella, An approach to Kienbock's disease: triscaphe arthrodesis, *J. Hand Surg.* 10A(2):179 (1985).

25. E.R. Weber, Concepts governing the rotational shift of the intercalated segment of the carpus, *Orthop. Clin. North Am.* 15(2):193 (1984).

26. Y. Youm, and A.E. Flatt, Kinematics of the wrist, 149:21 (1980).

27. Y. Youm, R.Y. McMurtry, A.E. Flatt, and T.E. Gillespie, Kinematics of the wrist, *J. Bone Joint Surg.* 60A(4):423 (1978).

SURGICAL TREATMENT WITH EXTERNAL FIXATOR FOR

POSTTRAUMATIC CARPAL INSTABILITY

Kozo Shimada, Minoru Nishino, Tohru Kadowaki, and Tomio Yamamoto

Department of Orthopaedic Surgery
Osaka Kosei-nenkin Hospital
4-2-78 Fukushima Fukushima-ku
Osaka, 553 Japan

INTRODUCTION

In case of posttraumatic carpal instability, especially a chronic case with malunion or nonunion after a fracture in the wrist region, we have reconstructed the bony structure at the originally injured site with a bone graft, repaired the ligaments and maintained it under the effect of ligamentotaxis by the external fixator "Orthofix Pennig Model" (Figure 1). The purpose of this paper is to describe our surgical procedure and clinical experience, and also clarify the effectiveness and problems of this procedure. We also present a cadaver study to demonstrate the effect of this procedure from a biomechanical point of view.

MATERIALS AND METHODS

Materials

Six cases of chronic posttraumatic carpal instability were treated with this procedure. All cases had had a bony lesion of the wrist (one case of transscaphoid-perilunate dislocation, three cases of dorsal intercalated segmental instability - DISI - deformity after nonunion of the scaphoid and two cases of DISI deformity after malunion of the distal radius), and they were all treated using the same concepts: [1] reconstruction of the normal bony anatomy, [2] repair of the injured ligament, [3] maintenance of the anatomical structure. We have used the combined technique of anatomical reconstruction with a wedge-shaped bone graft and maintenance by an external fixator. Age at operation ranged from 19 to 56 years old and the mean

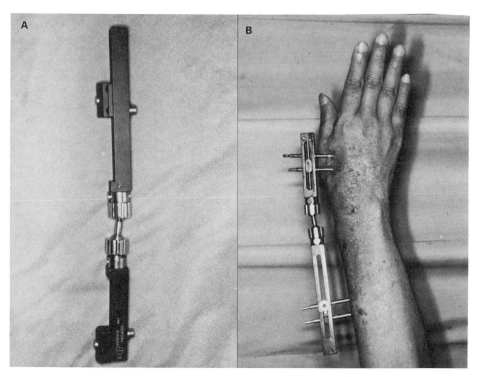

Figure 1. External fixator "Orthofix Pennig model" (**Figure 1-a**) and its application on the wrist (**Figure 1-b**).

age was 42 years old. The follow-up period was six to twelve months, with a mean of 8 months.

Surgical Procedure

We set the external fixator on the patient's radial forearm over the wrist joint. Fixation pins are set on the diaphysis of the radius and on the proximal third of the second metacarpal bone. Distraction force is applied until the carpal bones are well aligned (ligamentotaxis). Release or osteotomy at the fracture site and reconstruction of bony anatomy with a wedge-shaped bone graft is performed.

Nonunion of the scaphoid. We have chosen the conventional volar approach of Russe because of its direct visualization of the site of the scaphoid defect and the malposition of the lunate. Release or resection of the nonunion site is performed. Volarly-based bicortical wedge-shaped iliac bone is grafted at the waist of the scaphoid to correct the volarly flexed and shortened deformity.

Malunion of the radius. Through a dorsal approach the radius is osteotomized at the site of deformation and dorsally-based wedge-shaped iliac bone is grafted in the space to reconstruct the normal anatomy of the wrist.

Ligaments. Injured carpal ligaments are repaired at the same time, if necessary. Four to 6 weeks later, the ball-joint of the fixator is loosened and flexion-extension at the wrist joint is allowed under the condition that the fixator still distracts the joint. In ten to twelve weeks, the external fixator is removed, and exercise is encouraged.

Evaluation

Clinical results (pain, range of motion, grip strength) at the final follow-up were examined and scored with a modified Green and Cooney's rating system (Table 1). Roentgenographical assessment (union of grafted bone, radio-lunate angle, scapho-lunate angle, carpal height ratio, ulnar variance and volar tilting angle of the end of the radius in case of radial malunion) were also evaluated.

Table 1. Scoring system for clinical evaluation.

Pain (50 points)	No pain	50 points
	Cold weather symptoms	40
	Mild, no effect on activity	25
	Moderate after activity	10
	Severe	0
ROM (35 points)	140 degrees or more	35
	100 to 140 degrees	25
	70 to 100 degrees	15
	40 to 70 degrees	5
	Less than 40 degrees	0
Grip strength (15 points)	Normal	15
	Greater than 70% of normal	10
	Greater than 50% of normal	5
	Less than 50 % of normal	0

(Modification of Green and Cooney's scoring system) .

RESULTS

Carpal Instability Due To Nonunion of the Scaphoid (Table 2)

Reduction of the scaphoid was easy under distraction by this device. In 3 cases of nonunion after fracture of the scaphoid, good correction was achieved at surgery and it was maintained. In the operative field, ligamentotaxis realigned the carpal bones well. As a result, pathological rotation of the distal part of the scaphoid was improved and the space for the bone graft was well visualized. The external fixator also helped.

In a case of chronic transscaphoid-perilunate dislocation, we planned a two-stage operation. Only application of the fixator over the wrist was done at the first operation. At the second operation, reduction was easy after gradual lengthening for ten days (0.75 millimeter a day with the threaded rod of this device) because of improvement in the soft tissue contracture.

In all 4 cases, wedge-shaped iliac bone was grafted on the defect at the fracture site of the scaphoid with internal fixation by a polylactate screw or Kirchner wires. The carpal bones were realigned. After that, the carpal ligaments (usually the radio-scapho-capitate ligament) were repaired with nonabsorbable sutures. In those 4 cases of unstable wrist, the postoperative radiolunate, radioscaphoid and capitolunate angles were 0, 34, and 0 degrees, respectively. This procedure was quite useful to improve the carpal instability due to a bony lesion of the scaphoid.

Table 2. Cases of carpal instability due to nonunion of the scaphoid.

Preoperative Clinical/Radiological Status

Case	Age	Sex	Side	Follow-up Period (months)	Pain	ROM F/E	ROM R/U	P/S	Grip Strength (%)	Clinical Score	RLA	SLA	CHR
YK	56	F	R	12	moderate	85	45	170	43	25	-40	102	0.41
TJ	29	M	L	10	moderate	90	35	180	56	30	-28	73	0.55
IM	19	M	L	8	moderate	105	40	180	76	45	-32	89	0.49
KY*	21	M	R	8	severe	30	15	90	72	10	-8	65	0.39

Clinical/Radiological Status at Follow-up

Case	Age	Sex	Side	Follow-up Period (months)	Pain	ROM F/E	ROM R/U	P/S	Grip Strength (%)	Clinical Score	RLA	SLA	CHR
YK	56	F	R	12	none	125	50	180	65	80	-8	56	0.48
TJ	29	M	L	10	none	110	35	180	85	85	-4	46	0.57
IM	19	M	L	8	none	105	40	170	75	85	2	45	0.51
KY*	21	M	R	8	mild	80	30	120	70	50	-6	52	0.55

* KY is a patient with transcaphoid perilunate dislocation. The others are the patients with DISI deformity with nonunion of the scaphoid. Grip strength is indicated as a percentage to the intact wrist. RLA: radiolunate angle SLA: scapholunate angle CHR: carpal height ratio.

Case 1 Y.K. 56 y.o. female (Figure 2). The patient fell down on the road and injured her right wrist in an extended position in 1985. She did not receive any medical care. In 1991, she visited our clinic suffering from mild wrist pain with a carpal tunnel syndrome and tenosynovitis of her thumb flexor. Roentgenographic examination revealed volar subluxation and a dorsiflexion deformity of the lunate with nonunion of scaphoid.

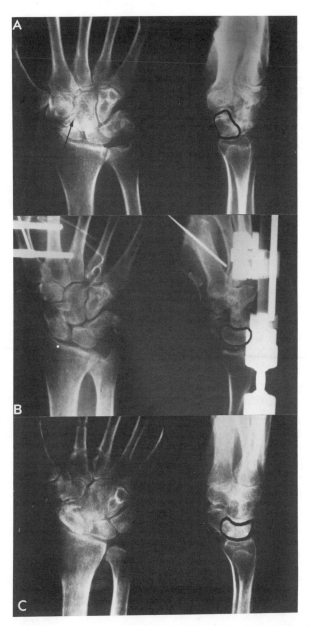

Figure 2. Case 1. Before operation (**Figure 2-a**), the lunate was volarly subluxed and dorsiflexed (outlined). Sclerotic nonunion site of the scaphoid was shown (arrowhead). X-ray immediately after the operation (**Figure 2-b**), and at follow-up (**Figure 2-c**) showed union of the scaphoid and reduction of the lunate.

At surgery, we observed a rupture of the volar radio-scapho-capitate ligament. The fractured segment of the scaphoid protruded and impinged on the flexor pollicis longus tendon. Distracting force by the device was applied on the wrist, the lunate was reduced and the scaphoid was realigned. As a result, the bone defect was seen on the volar aspect of the waist of the scaphoid. Wedge-shaped iliac bone was grafted in the space and fixed with a polylactate screw.

The radio-scapho-capitate ligament was repaired at the same time. The carpal bones were properly realigned and maintained with an external fixator.

Twelve weeks after surgery, the fixator was removed. Eight months after surgery, her right wrist pain had disappeared and the range of motion was improved. Total motion of the arc of the right wrist was 80% of the intact left wrist. Grip strength was 50% compared to the intact wrist, but it increased after the surgery. An X-ray shows normal alignment of the carpal bones. The carpal height ratio was improved to 0.48 from 0.42 preoperatively. The radio-lunate angle and the scapho-lunate angle were improved to -8 degrees and 55 degrees from -40 degrees and 102 degrees, respectively.

Carpal Instability after Malunion of the Distal Radius (Table 3)

In 2 cases of malunion after fracture of the radius, we tried to reconstruct the normal anatomical shape of the radius with a wedge-shaped iliac bone graft after the osteotomy at the malunited site. In the operative field, distraction was applied by the external fixator but the force was not strong enough to distract sufficient length to correct the shape of radius. Better alignment of the carpus was achieved but positive variance and dorsal tilting of the distal end of the radius still existed.

Postoperative ulnar variance, radiolunate angle and volar tilting of the end of the radius were 2 mm positive, -18 degrees, and -15 degrees, respectively. In those two cases, short-term clinical results were not as poor, and preoperative wrist pain disappeared but radiological assessment revealed insufficient anatomical correction.

Case 2   N.Y. 52 y.o. male (Figure 3). The patient suffered from chronic wrist pain after malunion of the radius since 1989 and visited our clinic in 1991. An X-ray revealed a 4 mm plus variant and 24 degrees dorsal tilting of the distal end of the radius. This deformity resulted in secondary dorsally flexed malposition of the lunate.

We tried to reconstruct a normal anatomy of the wrist with a wedge-shaped bone graft at the site of malunion through a dorsal approach. However, the distracting force by the device was not strong enough to correct the deformity of the radius. As a result, the correction was not long enough and resulted in tilting of the distal end of the radius.

After 8 months follow-up, the wrist pain has resolved, and the range of motion has increased, but the grip strength is the same as before surgery. The X-ray still shows a 2 mm plus variance and 15 degrees dorsal tilting of the distal end of the radius. The radio-lunate angle is -12 degrees, and the DISI pattern still exists.

CADAVER STUDY

Examination

We examined the distraction effect for reduction of carpal malalignment with a fresh frozen cadaver.

Table 3. Cases of carpal instability due to malunion of the radius.

Preoperative Clinical/Radiological Status

Case	Age	Sex	Side	Follow-up Period (months)	Pain	ROM F/E	ROM R/U	ROM P/S	Grip Strength (%)	Clinical Score	UV	PT	RI	RLA
AY	33	M	R	13	severe	55	30	80	25	10	2.5	-18	16	2
NY	52	M	R	11	severe	110	40	160	45	25	4.0	-24	13	-17

Clinical/Radiological Status at Follow-up

Case	Age	Sex	Side	Follow-up Period (months)	Pain	ROM F/E	ROM R/U	ROM P/S	Grip Strength (%)	Clinical Score	UV	PT	RI	RLA
AY	33	M	R	13	none	75	40	100	55	70	0.0	-14	14	-9
NY	52	M	R	11	none	105	40	170	60	80	2.0	-15	18	2

Grip strength is indicated as a percentage compared to the intact wrist.
UV: ulnar variance   PT: palmar tilt   RI: radial inclination   RLA: radiolunate angle.

The PENNIG model was set on a normal fresh frozen cadaver wrist by the usual technique. Gradual distracting force was applied to the wrist by lengthening the threaded rod. The lengthening effect was checked roentgenographically at 0 mm, 2.5 mm, 5.0 mm, 7.5 mm and 10.0 mm lengthening of the rod. An X-ray was taken in the same way and in the same position (one meter distance from the X-ray tube to the film, A-P view and lateral view at the neutral wrist position and the neutral forearm rotation).

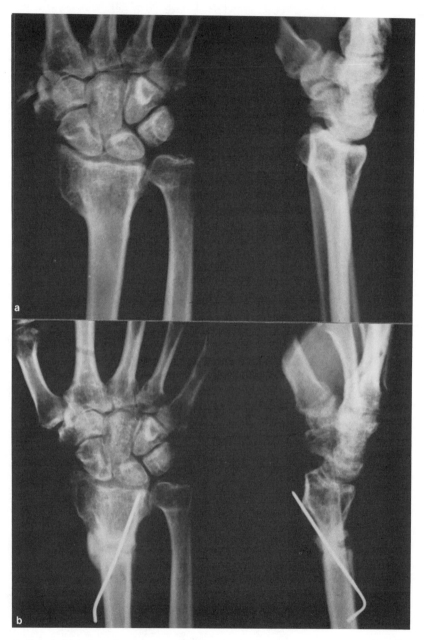

Figure 3. Case 2. Malunion after fracture of the distal end of the radius (**Figure 3-a**). Correction of the anatomical shape of the radius was not achieved and malalignment of the scaphoid still existed (**Figure 3-b**).

We also examined a model of a scaphoid fracture in a fresh frozen cadaver. Russe's volar approach was chosen, and the radio-scapho-capitate ligament was cut at the waist of the scaphoid. After that, osteotomy at the waist of the scaphoid was done and a volar wedge-shaped fragment was removed to reproduce an unstable wrist (the distal fragment of the scaphoid was volarly flexed and the total length of scaphoid was shortened). The distraction effect was checked as in the same way as a normal wrist.

Normal Wrist

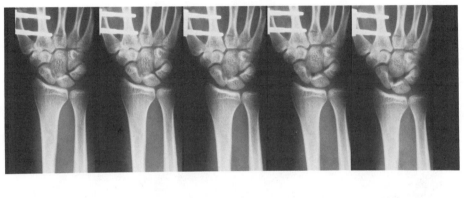

| 0.0mm | 2.5mm | 5.0mm | 7.5mm | 10.0mm |

Figure 4. Cadaver study (normal wrist). The widening of the radiocarpal joint and the rotation of the scaphoid were observed in an early stage of distraction. The midcarpal joint was more stable.

Results

In a normal wrist, distracting forces rotated the scaphoid and made the scaphoid stand up. They also widened the radiocarpal joint at the same time. This is well recognized as parallel curves of scaphoid length and of distance between the radius and the proximal carpal row on the A-P view. This means the proximal carpal row is located in a straight position with respect to the radius by ligamentotaxis. Next, the distracting force widened the midcarpal joint, and finally the wrist was bent ulnarly (Figure 4). In the scaphoid fracture model (unstable wrist), the same tendency was observed not only in the radiocarpal joint but also in the midcarpal joint because of insufficient scaphocapitate ligamentous stability. In this condition, the proximal carpal row is more unstable and its position is easily influenced by its surroundings. As a result, the wedge-shaped bone graft on the scaphoid easily and effectively maintains this normal alignment. Too much distraction bends the wrist ulnarly (Figure 5).

DISCUSSION

The conception of treatment is "Good functional results follow good anatomical reduction". According to this conception, the principle of treatment for posttraumatic carpal instability after fracture in the wrist region is reconstruction of the normal bony anatomy.

The association of nonunion of the scaphoid and dorsiflexed malalignment of the lunate (DISI) has been well recognized and analyzed by various authors.[2,5,6,8,10,13] In these cases the scaphoid becomes shorter because of its volarly flexed position or displacement at the fracture site which results in an increased scapholunate angle and dorsiflexed position of the lunate. Fisk[7] has stressed the importance of correction of the flexion deformity and restoration of normal scaphoid length which reestablishes normal tension in the palmar radiocarpal ligament and corrects the pathological rotation of the lunate. For this purpose, he recommended the radial wedge graft technique. Fernandez[3] showed a modification of the original Fisk procedure. It consists of preoperative planning to achieve the normal length of the scaphoid, the use of the palmar approach, the insertion of a bicortical iliac bone graft at the site of nonunion and internal fixation.

Scaphoid Fx. Model

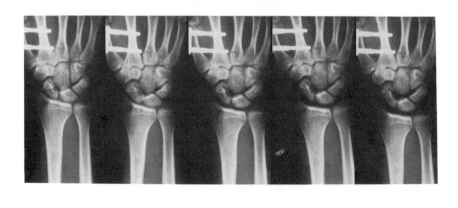

| 0.0mm | 2.5mm | 5.0mm | 7.5mm | 10.0mm |

Figure 5. Cadaver study (scaphoid fracture model). The widening of the radiocarpal joint and midcarpal joint was observed at the same stage. In this condition, the proximal carpal row is very unstable and easily reducible. A bone graft on the scaphoid, to reconstruct the normal length, is effective for stabilization.

We also often recognize the association of malunion of the radius and dorsiflexion of the lunate after Colles' fracture. The initial change in this deformity is the change in length and the pathological tilt of the distal radius. The relationship of the lunate and distal end of the radius is normal. But shortening and dorsal tilting of the radius result in ulnocarpal impingement and malalignment of the carpal bones especially in the relationship between the lunate and the capitate. It causes carpal instability as the secondary change. Reconstruction of the normal anatomy is essential in the treatment of this kind of deformity.[11]

For this purpose, our choice of treatment is a bone graft at the fracture site after the resection of the nonunion or osteotomy at the deformed site and maintenance by the use of an external fixator.

The fixed-pin traction technique in the wrist region was introduced for the treatment of comminuted fracture of the distal radius in 1929 (published in 1956) by Böhler.[1] The external fixator provides good stability for the fixed-pin traction techniques, and now this method has been widely recommended. The concept of ligamentotaxis was shown by Vidal et al.[12] in 1983, and this provided a biomechanical backbone for this method. Leung[9] reviewed the very good results for comminuted fractures of the distal end of radius with the use of ligamentotaxis and bone graft. Fernandez[4] used this method for complex carpal dislocations. Our conception is similar to Fernandez's and for this, the Pennig model is more useful because of its gradual lengthening system. In our experience, for the purpose of correction of the radius, this device seems to be too weak, but for the correction of carpal alignment, especially with nonunion of the scaphoid, our procedure, the combination of bone graft and external fixation, is quite effective and this device is quite useful for the purpose.

CONCLUSION

Our procedure, the combination of bone graft and external fixation, is useful for posttraumatic carpal instability, especially in the case of carpal malalignment after fracture of the scaphoid. The characteristics of this procedure were clarified by clinical and radiological assessment. Cadaver studies were also undertaken to demonstrate the effect of this procedure from a biomechanical point of view.

REFERENCES

1. L. Böhler, Translated from the 13th German ed. by H. Tretter, H.B. Luchini, K. Kreuz, O.A. Russe, R.G.B. Bjornson, The treatment of fractures, Grune & Stratton, New York (1956).
2. W.P. Cooney, J.H. Dobyns, and R.L. Linscheid, Non-union of the scaphoid; analysis of the results from bone grafting, *J. Hand Surg.* 5:343 (1980).
3. D.L. Fernandez, A technique for anterior wedge-shaped grafts for scaphoid nonunions with carpal instability, *J. Hand Surg.* 9-A:733 (1984).
4. D.L. Fernandez, and R. Ghillani, External fixation of complex carpal dislocation: a clinical report, *J. Hand Surg.* 12-A:335 (1987).
5. G.R. Fisk, Carpal instability and the fractured scaphoid, *Ann. R. Coll. Surg. Engl.* 46:63 (1970).
6. G.R. Fisk, Overview of wrist injuries, *Clin. Orthop.* 149:137 (1980).
7. G.R. Fisk, Volar wedge grafting of the carpal scaphoid in non-union associated with dorsal instability patterns (discussion), *J. Bone Joint Surg.* 64-B:632 (1982).
8. J.M.G. Kauer, The mechanism of the carpal joint, *Clin. Orthop.* 202:16 (1986).
9. K.S. Leung, W.Y. Shen, P.C. Leung, A.W. Kinninmonth, J.C. Chang, and G.P. Chan, Ligamentotaxis and bone grafting for comminuted fractures of the distal radius, *J. Bone Joint Surg.* 71-B:838 (1989).
10. R.L. Linscheid, J.H. Dobyns, J.W. Beabout, and R.S. Bryan, Traumatic instability of the wrist; diagnosis, classification and pathomechanics, *J. Bone Joint Surg.* 54-A: 1612 (1972).
11. K. Shimada, K. Tada, T. Yoshida, and T. Shibata, Treatment for ulnar wrist disorders (in Japanese), *J. Jpn Soc. Surg. Hand* 7:626 (1990).
12. J. Vidal, Ch. Buscayret, M. Paran, and J. Melka, Ligamentotaxis, in: D.C. Mears ed. External skeletal fixation, Williams & Wilkins, Baltimore (1983).
13. E.R. Weber, Biomechanical implications of scaphoid waist fractures, *Clin. Orthop.* 149:83 (1980).

X-RAY STEREOPHOTOGRAMMETRIC EXAMINATIONS OF FLEXOR TENDON EXCURSION IN THE TENDON SHEATH AREA AFTER TENDON REPAIR

Lars Hagberg

Department of Hand Surgery
University of Lund
Malmö Allmänna Sjukhus
S-214 01 Malmö, Sweden

INTRODUCTION

A number of early controlled mobilization techniques following direct repair of flexor tendons have been established and described to improve the final results.[2,5,6,7,8,9,11,13,18] The efficiency of the different methods are, however, not well documented, controlled clinical studies are rare and there are also negative reports.[4,19] Lane et al.[12] found in an experimental study a rapid decrease in tendon gliding occurring within hours of the surgical procedure. They concluded that postoperative hematoma and edema restricted gliding long before collagenous adhesions were synthesized. This may be an explanation to unpredictable results from passive mobilization methods in cases where peroperative tendon gliding were satisfactory. Would then the addition of limited active flexion be necessary in these cases and could that be performed without excessive dehiscence? An investigational method is needed to study the effects of different controlled mobilization techniques on tendon excursion at the level of injury and the gap formation in the tendon repair during the treatment. In this study, Roentgen stereophotogrammetric analysis (RSA) has been applied on tendon tissues for clinical studies on movements of flexor profundus tendons during different postoperative treatments.

MATERIALS AND METHODS

Marker Stability in Tendon Tissue

Nine rabbits had two pair of markers implanted with a special istrument into the flexor digitorum fibularis tendon[1] above the ankle in each hind leg (Figure 1). The animals were allowed immediate mobilization. The intemal distance between the two markers in each pair was followed by repeated RSA during 62 days (Figure 2). Two x-ray tubes with 40 degrees between their central rays were used to make simultaneous

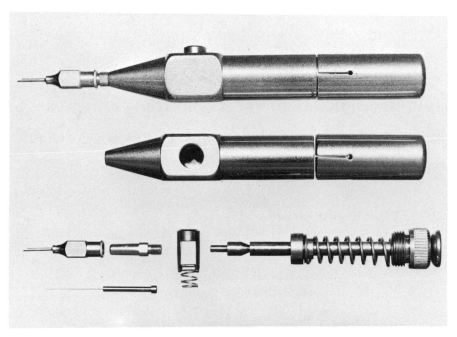

Figure 1. Instrument used for implantation of tantalum markers.

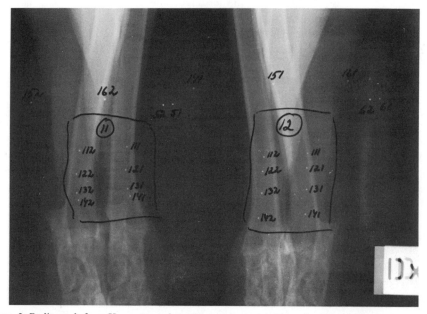

Figure 2. Radiograph from X-ray stereophotogrammetric examination of rabbit hindlegs. There are four tantalum markers in each flexor digitorum fibularis tendon.

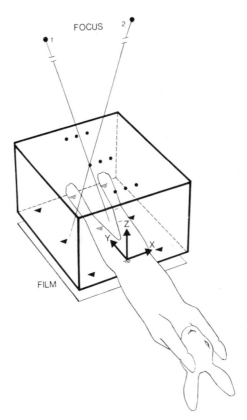

FOCUS   2

Z

Y   X

FILM

Figure 3. The rabbit to be examined by RSA is held in a calibration cage of plexiglass at roentgen exposure. The markers in the walls of the cage make it possible to reconstruct the bundles of rays from the roentgen foci and thus also the bodymarkers' 3-D coordinates.

exposures on one film (Figure 3). After final examination, the tendons were excised and in vitro x-ray done to show the presence of all markers within the tendon tissue. The tendons were taken to histologic examination.

Clinical Investigation

With a special applicator a tantalum ball with a diameter of 0.5 mm was implanted in tendon tissue at each side of the flexor profundus (FDP) lesions within the tendon sheath in 19 digits in 17 patients at the time of tenorrhaphy (Figure 4). Four tantalum balls were implanted into either the proximal or the middle phalangeal bone to be used as reference markers at the level of injury.

Primary tendon repair was performed using a Kessler-Tajima suture with 4-0 monofilament or multistrand stainless steel and a continuous adaptation suture with 6-0 Polydioxanon. Tendon sheaths were closed if possible. The study was approved by the local ethics committee and all patients were informed and had signed consent prior to surgery.

Methods of Early Controlled Mobilization

Each patient had one of the four alternative postoperative treatments described below started on the third postoperative day: [1] the injured digit mobilized by passive flexion (rubber band) and active extension in a dorsal splint as described by Kleinert et al.[11] and Lister et al.[13] (K-1, Figure 5-A); [2] all four digits mobilized by passive flexion (rubber band) and active extension in a dorsal splint (K-4, Figure 5-B); [3] passive flexion-active hold as described by Cannon and Strickland[2] who have been using the method as a "frayed tendon program" following tenolysis; in a dorsal splint passive finger flexion including all fingers is followed by active maintenance of the flexed position; finally the fingers are actively extended (A-h, Figures 6-A,B,C); [4] limited active flexion following full range passive movements in a dorsal splint (A-f).

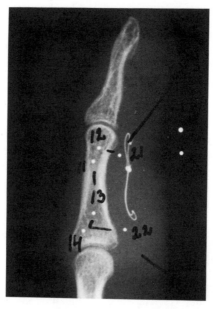

Figure 4. Radiograph illustrating a repair of the profundus tendon. There are two tantalum markers in the tendon, one on each side of the tendon suture. Four tantalum markers have been implanted in the middle phalanx to be used as reference points.

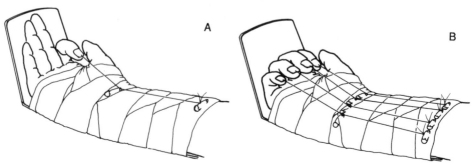

Figure 5. Figure 5-A. Kleinert-mobilization including 1 digit (K-1). **Figure 5-B.** Kleinert-mobilization including all 4 digits (K-4).

Radiographic Stereophotogrammetric Analysis

RSA was performed as described by Selvik.[15,16] For the clinical investigation two X-ray tubes with 90 degrees between their central rays were used to make simultaneous exposures on two films (Figure 6-D). Exposures were made at the beginning of early controlled mobilization with the hands in their regular radiolucent splints (Baycast®). RSA was repeated if possible once a week as long as controlled mobilization was continued.

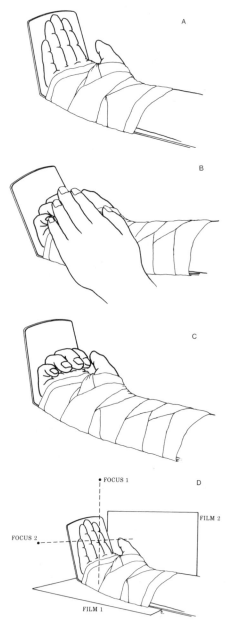

Figure 6. Figure 6-A,B,C. Passive flexion followed by active hold (A-h). **Figure 6-D.** The hands were examined by RSA in their regular radiolucent splints.

For analysis of tendon excursion one true lateral view with RSA was done for each operated digit. Data from all other examinations from the same digit could then be reoriented and evaluated by computer at exactly the same lateral view and thereby compared to each other. Only tendon markers located at the level of the marked phalangeal bone were used for calculations. In the lateral view motion along the digital phalanx in proximal or distal direction was registered as positive or negative. Motion perpendicular to the phalangeal bone was registered as zero.

In most patients different methods of controlled mobilization were tried in the same digit at each time of examination regardless of the treatment given between the examinations. The individual digit could thus be used as its own control and paired data for comparison of treatments were received.

Dehiscence of tendon repairs were calculated as increased three-dimensional absolute distance between tendon markers when interphalangeal joints were kept in extension.

The methodological error of RSA is about 0,02 mm.

RESULTS

Marker Stability

Minor instabilities were measured but the changes in distances between the 2 markers of each pair did never exceed 1 mm over the period of 62 days (Figure 7). The median value was 0.22 mm. All markers proved to be within tendon tissue at the end of the study. Histological examination showed granulation tissue limited to the triangular space developed in the tendon tissue by the metal marker separating the parallel collagen bundles. There was no evidence of any tissue reaction along the surface of the collagen in direct contact with the metal marker.

Clinical study

Most patients were examined once a week with RSA during the period of controlled mobilization. Two patients had ruptures of the tendon repair easily verified by X-ray. Both ruptures occurred in cases with injury level at A3. Six patients were lost for late RSA follow up at four months.

Most notable results from measurements of tendon excursion during controlled mobilization were the poor mean values produced by the Kleinert methods at A3 and A4 levels of injury (Figure 8). In A4 injuries the Kleinert methods even produced backward (distally directed) tendon motion during flexion.

Another significant finding was a general decrease in tendon excursion in A4 injuries during the third postoperative week (Figure 9). This was not seen in digits with more proximal injuries.

Statistical analyses of paired observations of tendon excursion by different mobilization methods irrespective of injury level proved the A-h procedure to be superior to the K-1 and K-4 procedures during the first three postoperative weeks. The differences were highly significant.

The internal distance between markers (IDM) correlated significantly with TAM at late follow up surprisingly in a positive direction (Figure 10). Thus dehiscence was not produced to the extent causing severe impairment of the final TAM.

DISCUSSION

The amount of marker instability in tendon tissue was well below clinically relevant dehiscence of tendon repairs. Tendon excursion measurements are independent of the long term marker instability. It was thus concluded that marker stability was sufficient for the application of RSA on tendon tissues with an expected error of less than 1 mm.

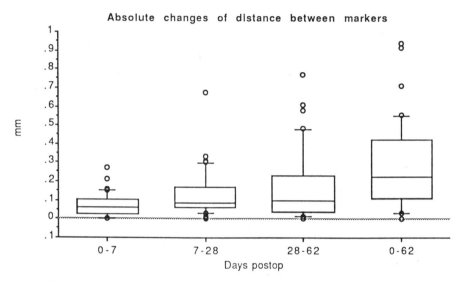

Figure 7. Absolute changes in distance between markers in each pair during different time intervals. The box plot show 10, 25, 50, 75 and 90th percentiles. Reprinted with permission from Hagberg et Selvik.[10]

As RSA only determines the 3-D marker positions and calculates the straight distance between different markers difficulties appear when a marker moves along a curved line. This will happen when a tendon marker is not positioned over the marked phalangeal bone (Figure 11). During flexion the tendon marker will continuously change its direction of motion in relation to the marked bone due to the fact that the adjacent bone and the corresponding part of the tendon (with the tendon marker) is moving in relation to the marked bone. Therefore, only tendon markers positioned and moving along the marked bone can be used for examination of tendon excursion. All motions have to be compared in the same lateral view and the main directions of motions decided. Dehiscence was measured using a standardized position with the interphalangeal joints in full extension.

Tendon excursion during treatment showed great differences depending on the level of tendon injury with very small values registered at A3 and A4 levels (Figure 8). At these levels the K-1 and K-4 mobilization techniques did not attain the desired tendon excursion and the A-h technique was significantly better. By introducing a component of active flexion to the early controlled mobilization the tendon excursion was improved without simultaneous development of deleterious dehiscence.

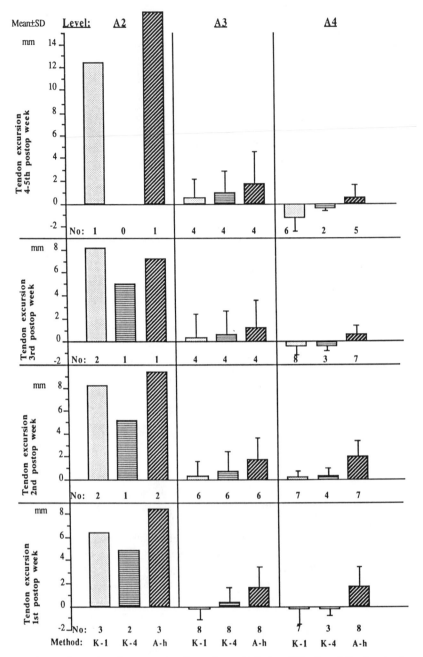

Figure 8. Mean values of tendon excursion by early controlled mobilization following tendon repair. Results are presented separately for each level of tendon injury and for each postoperative week. Tendon motion in distal direction during finger flexion is presented as negative. In most fingers 3 different mobilization methods were examined each time by RSA. Reprinted with permission from Hagberg et Selvik.[10]

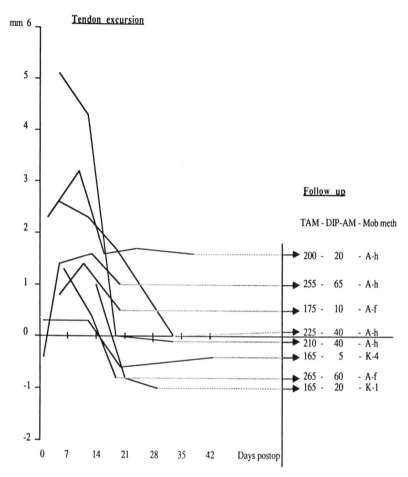

Figure 9. Tendon excursion followed in individual digits with A4 injury level during the period of early controlled mobilization. Reprinted with permission from Hagberg et Selvik.[10]

The difference between K-1 and K-4 indicated the importance of mobilizing all digits to avoid the quadriga phenomenon.

A general impairment of tendon excursion was specifically seen at the A4 level during the third postoperative week irrespective of mobilization method (Figure 9). This probably reflects the small amount of normal tendon excursion at this level[14] and the tight conditions under the A4 pulley. Early controlled active motion therefore seems to be important at this type of injury and is in concordance with recent results.[3,17]

Dehiscence was not a significant clinical problem. We found a positive correlation between IDM and TAM at follow-up probably indicating that an efficient mobilization will produce more dehiscence (Figure 10).

RSA applied on tendon tissue is a unique method that provides knowledge not achievable in any other way.

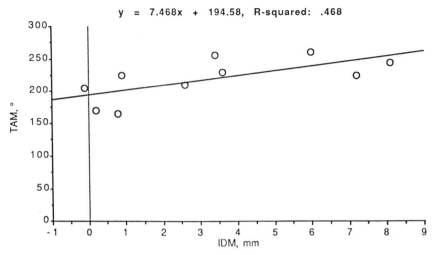

Figure 10. Correlation between tendon dehiscence (IDM) and TAM at follow up (r=0,684; p=0,029). Reprinted with permission from Hagberg et Selvik.[10]

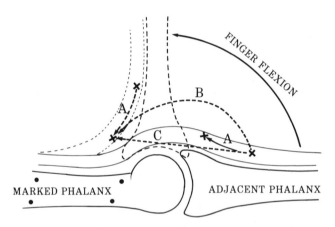

A Distance of tendon gliding = motion in relation to adjacent phalanx
B Motion in relation to marked phalanx
C Tendon motion registered by RSA.

Figure 11. Illustration of the problems to express the motion of a tendon marker positioned over a phalanx other than the marked phalanx. The clinically relevant tendon motion should be the motion in relation to the tendon sheath. Reprinted with permission from Hagberg et Selvik.[10]

REFERENCES

1. R. Barone, C. Pavaux, P.C. Blin, and P. Cuq, "Atlas of Rabbit Anatomy," Masson & Cie, Paris (1973).
2. N.M. Cannon, and J.W. Strickland, Therapy following flexor tendon surgery, *Hand Clinics* 1:147 (1985).
3. K.W. Cullen, P. Tolhurst, D. Lang, and R.E. Page, Flexor tendon repair in zone II followed by controlled active mobilization, *J. Hand Surg.* 14B:392 (1989).
4. H. Creekmore, H. Bellinghausen, V.L. Young, R.C. Wray, P.M. Weeks, and P. Schindeler Grasse, Comparison of early passive motion and immobilization after flexor tendon repairs, *Plast. Reconstr. Surg.* 75:75 (1985).

5. R.J. Duran, and R.G. Houser, Controlled passive motion following flexor tendon repair in zones 2 and 3, in: "American Academy of Orthopaedic Surgeons: Symposium on Tendon Surgery in the Hand," C.V.Mosby Co., St. Louis (1975).

6. R.J. Duran, R.G. Houser, C.R. Coleman, et al., Management of flexor tendon lacerations in zone 2 using controlled passive motion postoperatively, in: "Rehabilitation of the hand, 2nd ed.," Hunter, J.M., Schneider, L.H., Mackin, E.J., et al., eds., C.V.Mosby Co., St. Louis (1984).

7. R.H. Gelberman, D. Amiel, M. Gonsalves, et al., The influence of protected passive mobilization on the healing of flexor tendons: a biomechanical and microangiographic study, *Hand* 13:120 (1981).

8. R.H. Gelberman, S.L. Woo, K. Lothringer, et al., Effects of early intermittent passive mobilization on healing canine flexor tendons, *J. Hand Surg.* 7:170 (1982).

9. R.H. Gelberman, J.S. van de Berg, G.N. Lundborg, and W.H. Akeson, Flexor tendon healing and restoration of the gliding surface; an ultrastructural study in dogs, *J. Bone Joint Surg.* 65A:70 (1983).

10. L. Hagberg, and G. Selvik, Tendon excursion and dehiscence during early controlled mobilization after flexor tendon repair in zone II: an X-ray stereophotogrammetric analysis, Copyright 1991, by Mosby-Year Book, Inc., *J. Hand Surg.*, 16:4, 669 (1991).

11. H.E. Kleinert, J.E. Kutz, and M.J. Cohen, Primary repair of zone II flexor tendon lacerations, in: "American Academy of Orthopaedic Surgeons:Symposium on Tendon Surgery in the Hand," C.V.Mosby Co., St. Louis (1975).

12. J.M. Lane, J. Black, and F.W. Bora, Gliding function following flexor-tendon injury; a biomechanical study of rat tendon function, *J. Bone Joint Surg.* 58:985 (1976).

13. G.D. Lister, H.E. Kleinert, J.E. Kutz, et al., Primary flexor tendon repair followed by immediate controlled mobilization, *J. Hand Surg.* 2:441 (1977).

14. D.A. McGrouther, and M.R. Ahmed, Flexor tendon excursions in "No man's land," *Hand* 13:129 (1981).

15. G. Selvik, P. Alberius, and A.S. Aronson, A roentgen stereophotogrammetric system: construction, calibration and technical accuracy, *Acta Radiol. Diagn.* 24:343 (1983).

16. G. Selvik, Roentgen stereophotogrammetry; a method for the study of the kinematics of the skeletal system, *Acta Orthop. Scand.* Suppl. 60:232 (1989).

17. J.O. Small, M.D. Brennen, and J. Colville, Early active mobilization following flexor tendon repair in zone II, *J. Hand Surg.* 14B:383 (1989).

18. J.W. Strickland, and S.V. Glogovac, Digital function following flexor tendon repair in Zone II: a comparison of immobilization and controlled passive motion techniques, *J. Hand Surg.* 5:537 (1980).

19. M. Tonkin, L. Hagberg, G. Lister, and J. Kutz, Post-operative management of flexor tendon grafting, *J. Hand Surg.* 13B:277 (1988).

EVAMAIN: COMPUTERISED SYSTEM FOR THE EVALUATION

OF THE HAND

Jean-Louis Thonnard,[1,2] Léon Plaghki,[1,2] and Dominique Bragard[2]

[1]Service de Médecine Physique,
Cliniques Universitaires Saint Luc
10 Avenue Hippocrate, 1200 Brussels,Belgium
[2]Unité READ - Université Catholique de Louvain
1200 Brussels, Belgium

INTRODUCTION

The hand is a performant prehensile and sensory tool capable of the strongest grasp and the most delicate touch. The optimal use of our hands in everyday activities depends at the same time on anatomical integrity, joint range of motion, muscle strength, cutaneous sensibility and co-ordination or dexterity. Therefore each of these domains should be explored in a hand test battery. The evaluation of hand function is of critical importance in determining the extent of functional loss in patients with traumatic injuries or destructive diseases and in assessing the outcome of various surgical and rehabilitative procedures.[12,23] Advances in reconstructive surgery and hand therapy depend in part on the quality of the methods used to perform these assessments.[11,20]

Reliability and validity are the first two criteria of good measurement. A reliable measure has the property of yielding consistent results under varying circumstances. Reliability is assessed in several different ways. The two most common types are test-retest reliability, and interrater reliability. The object of test-retest assessments is to determine if repeated applications of the measure on the same subject or group of subjects tends to change the results. Interrater reliability is assessed by comparing the ratings of the same subject or set of subjects by two or more raters. The reliability assessment of a measure depends on the type of measure itself. For example, test-retest reliability is crucial if the measure is to be used as a follow-up tool at two or more time periods. Reliability is usually expressed numerically as some kind of correlation coefficient ranging from 0 to 1, with 0 signifying total unreliability and 1 indicating 100% reliability. Reliability coefficients above 0.85 are generally regarded as high and between 0.60 and 0.85 as moderate. Validity is a more difficult concept. It usually means the degree a measure approximates the actual quantity or quality of the entity being assessed. Validity indicates the truthfulness of an assessment tool. They are many types of validity. For example the most common method of checking the validity of hand strength measurement is to calibrate the device by suspending weights on its handle.

Valid and reliable tests for range of motion, muscle strength, and dexterity are currently available in the field of hand rehabilitation. Most tests of cutaneous sensibility lack validity and reliability coefficients except for the light touch-deep pressure test using the Semmes-Weinstein Pressure Aesthesiometer.[4]

As there are many ways to assess hand function, the aim of our computerised system is to provide an integrated battery of tests based on standardised methods. This goal will be achieved by incorporating reliable and valid testing procedures yet available and secondly by developing new testing devices especially for cutaneous sensibility evaluation.

EVALUATION OF HAND FUNCTIONS

Range of Motion

The range of motion (ROM) is based on the principle that neutral position equals zero degree, as proposed by the American Academy of Orthopaedic Surgeons.[1] In this method, all motion of the joints are measured from defined zero as the starting position to the actual joint position. Active motion is that motion obtained at the joints with maximal voluntary contraction. The possible lack of motivation and the protective reflexes due to pain may introduce limitations in the active ROM scores. Passive ROM measurements require the tester to mobilise the joint by applying certain levels of force necessary to overcome the normal soft tissue resistance. Amis and Miller[2] have observed that this level of force is an extra variable which could alter the measurements considerably.

Apparatus. Manually operated goniometers are used to measure both passive and active ROM.

Procedure. Guidelines have been established that specify both the starting position of the hand during these measurements and the correct placement of the goniometer on the joint being evaluated.[1,26] According to Nicol,[25] the actual steps involved in taking a measurement of joint position can be listed as follows:
- position of subject for the test,
- location of non moving parts of the body,
- alignment of the goniometer in the correct plane of motion,
- the correct identification of bony landmarks,
- the correct application of force for passive range of motion,
- the correct location of landmarks for the second positional location.

Validity and Reliability. Boone et al.[8] have established that a single set of ROM measurements is as reliable as averaging several sets. Yet successive measurements of a joint should be done by the same examiner since intratester reliability is greater than intertester reliability.[8,29] Within these conditions, measurements of the ROM are accepted as valid, accurate and reliable and they can be interpreted with reference to published norms.[1]

Hand Strength

The prehensile activities of the hand are so varied that a simple analysis seems not feasible. On closer examination, however, Napier[24] stated that "this diversity is in fact not so much an expression of a multiplicity of movements but of the vast range of purposive actions involving objects of all shapes and sizes that are handled during everyday activity". His study of the normal hand suggests that there are, in fact, only two distinct patterns of prehensile movements termed power grip and precision grip. Three types of precision grips are the most commonly used in everyday life: tip pinch, palmar pinch and key pinch. Power grip and precision grip are often combined in functional activities. For example, tying a shoe lace involves power grip on the ulnar side of the hand, and precision handling on the radial side of the hand.

Apparatus. The Digital Pinch/Grip Analyser developed by MIE Medical Research (LEEDS, UK.) is used for strength evaluations.[18] The device (Figure 1) consists of two cushioned aluminium bars approximately six inches long. The separation between the two handles is adjustable to suit any hand size or deformity and the design allows measurement of finger pinch and hand grip strength. A supporting handle has been constructed that held the dynamometer on the table top but does not firmly attach it; the dynamometer remains freely movable thanks to a spring system and can be rotated on a pivot for optimal prehension.

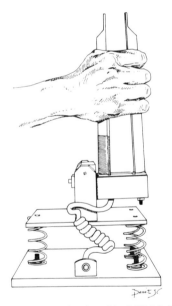

Figure 1. The Digital Pinch/Grip Analyser developed by MIE Medical Research (LEEDS, UK.)

Procedure. One or two trial grips are undertaken to familiarise the subject with the device. Then two kinds of measurements are performed: (1) maximum voluntary strength and (2) muscle endurance. For the measurement of grip or pinch maximum strength, the subject is asked to grip/pinch as fast and as hard as possible after a buzzer is activated and to hold the grip ten seconds until the warning signal for release. The signal of the force transducer is sampled during 15 seconds with a pre trigger of 2 s. The following parameters are computed (Figure 2) : the peak force (F_1) is the maximum force occurring during the measurement period; the rise time is the period necessary to reach 90% of F_1; the fatigue rate is given by the formula $(F_1-F_2).100/F_1.t_i$ where F_1 is the maximum grip strength, F_2 is the grip strength at release and t_i is the time elapsed between F_1 and F_2. All these variables are assessed for power grip, tip pinch, key pinch and palmar pinch. The mean scores of three successive trials is keeped. The mean of three trials is a more accurate ($r > 0.81$; $n=27$) measure of hand strength than one trial or the highest score of three trials.[22] A rest time of two minutes is given to the subject between each trial. For the measurement of hand muscle endurance, the subject is asked to grip or pinch the handle for as long as possible at a level equal to 50% of his maximum strength previously determined. The grip/pinch strength is fed back as a bar graph on a video display in front of the subject. A mark on the display represented the 50% maximum level. The subject has to squeeze the handles, maintaining the bar graph at the 50% level until fatigue caused him to release the handle. The time integral of the strength curve is the impulse (Figure 2) which is considered as a measure of muscle fatigue. The procedure is repeated three times with a resting period between the trials. For each hand strength test, the following position is recommended:[21,22,28] the subject is seated with his shoulder adducted and neutrally rotated, elbow flexed at 90°, forearm in neutral position, and the wrist between 0° and 30° dorsiflexion and between 0° and 15° of ulnar deviation.

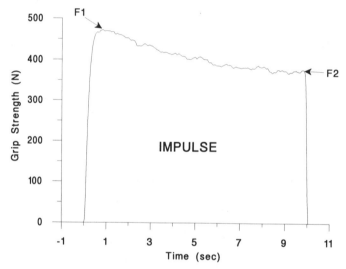

Figure 2. Handgrip strength versus time curve.

Validity and Reliability of Grip/Pinch Maximum Strength Evaluations. Validity of strength evaluations are best assured by periodically checking the calibration accuracy of the grip and pinch strength measurement devices. Helliwell et al.[18] have evaluated the reliability of the maximum grip/pinch hand strength tests described here above. Reproducibility of measurements was investigated by recording three successive attempts at gripping and pinching in a group of 20 subjects with rheumatoid arthritis. Mean values for each parameter at each attempt were calculated and the results analysed. Reproducibility was acceptable for all grip and pinch variables.

Cutaneous Sensibility

Bell[5] reported that in cutaneous sensibility testing there exists at least a twofold objective: the evaluation of sensory acuity and the patient's functioning with this acuity. Tests such as nerve-conduction velocity, pinprick, and Tinel's sign evaluate the peripheral nerve acuity. Yet attempts to directly correlate nerve conduction with diminished functional cutaneous sensation have been largely unsuccessful.[5] Thus it is necessary to evaluate the functional sensibility which enables the hand to engage in full activities of daily living, including those activities in which vision is essentially occluded while the hand manipulates an object. As touch is an active exploratory process of the hand, not merely a passive receptive sense, it can be more accurately assessed if the hand is permitted to actively explore and scan the object presented.[9,16,27] Therefore, except for the light touch/deep pressure testing, we tried to find tests which require active manipulation of objects rather than simply passive recognition of stimuli. In order to provide clinicians with a broad picture of the patient's problem, we have selected the following four tests we will now briefly describe.

Light Touch/Deep Pressure Testing. Light touch/deep pressure testing with monofilaments of increasing forces is one of the most objective test for measuring cutaneous tactility.[4,6,15,32] Light touch and deep pressure sensibility are considered as the two ends of a continuum of cutaneous sensibility, with light touch being perceived by receptors in the superficial skin layers and pressure by receptors in the subcutaneous and deeper tissues.[31] Pressure sensibility is a form of protective sensation, light touch sensibility is a necessary component of fine tactile discrimination. The currently available instrument is the Semmes-Weinstein Aesthesiometer monofilament testing set which contains 20 filaments. The test begins with filaments in the normal threshold level and progresses to filaments of increasing stiffness until touch is identified by the patient. The monofilament is applied perpendicularly and the pressure is increased until the filament begins to bend. The recognition of threshold levels of light touch-deep pressure has been found to be repeatable if the monofilament lengths and diameters are correct.[4]

Active Two-Edges Discrimination. Active Two-Edges Discrimination is an attempt to improve the classical two-point discrimination test which has been reported by Bell & Buford,[7] as a subjective test since, in clinical practice, there is considerable variation in the force and velocity of application among the clinicians. These authors have demonstrated that when a stimulus is applied with a hand-held instrument, the examiner is unable to control for force of application. This is caused, in part by vibration of the hand holding the instrument. In the active two-edges discrimination, the subject is required to actively slide the tip of the index finger over two straight edges which can be separated from 0 to 15 mm (Figure 3). The subject responds on a two-point scale: either "one edge" or "two edges". There are no instructions as to the pressure of the finger tip against the edges although the pressure is monitored by a built-in force transducer.

For the two tests described above we use the method of constant stimuli to determine the detection threshold: five different stimuli (spans or monofilaments) are selected at the beginning of the experiment and applied 20 times in a random order, on the middle of the distal phalanx. The detection threshold is defined as the stimulus intensity corresponding to a 0.5 detection probability. The threshold is computed by interpolating the psycho-physical function adjusted to the experimental data by a Minimum Logit chi[2] method.[3] Both tests mainly explore the peripheral tactile sensibility.

Although there appears to be a relationship between diminishing levels of two-point discrimination and increasing levels of force required by the monofilaments, the two tests cannot be directly equated.[15] The light touch/deep pressure test is classified as a threshold test evaluating a nerve fibre innervating one or several receptors. Two-point discrimination is an innervation-density test in that it measures multiple overlapping peripheral receptive fields.[10]

Figure 3. The Active Two-Edges Discrimination device.

Grip Force Perception. The ability to detect tactile stimuli does not ensure the capacity to perform precise motor tasks. Thus, it is interesting to measure how a patient holds objects and what adjustments are made when manipulating different types of objects. Westling and Johansson[34] found that three factors influence the force control during grip: the weight of the object, the friction between the skin and the object, and a safety margin factor related to the individual. They reported that the applied grip force was critically balanced to optimise the motor behaviour so that slipping was prevented and the grip force did not reach exceedingly high values. In other words, the force in excess to the slipping force defines the safety margin. We have adapted the apparatus and the procedures of Westling & Johansson[34] for quantitative studies of hand function in patients (Figure 4). The task consists of picking up the dynamometer used for strength measures, using a comfortable power grip force, holding for the n seconds, and then gradually reducing the applied grip force until slippage occurs. A vertical force gauge detects slippage. Then the grip force at the moment of slippage is calculated as the minimal force to prevent slippage. This force is called slip force. The difference between sustaining grip force and slippage grip force is reported as the safety margin.

Three loads (P1, P2, P3) are presented six times at random. In figure 5, GF and SF are respectively the mean grip force and the mean slip force for the six trials performed with each load. The force curves are processed by the computer in order to calculate a safety index using the following equation (Figure 5):

$$\text{Safety index} = (S3 + S4/S1 + S2) - 1$$

The areas (S1, S2, S3, S4) are computed by the method of finite differences.

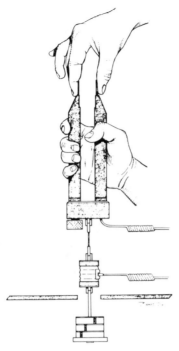

Figure 4. The Grip Force Perception device

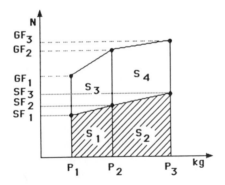

Figure 5. Computation of the Safety Index

Tactile Object Identification. Everyday life perceptions are based on explorations of three-dimensional objects that can be taken into the hand. Phillips[27] reported that active touch discriminates size, shape, indentibility and surface texture by compound perceptions (kinaesthetic, tactual, thermal). The totality must be referred to the brain-models of familiar surfaces and objects, built up during the life of the individual, or fed into the learning processes by which the properties of hitherto-unfamiliar objects are added to the repertory. Tactile object identification appears as being related to high aspects of cognitive processing related to pattern recognition and identification (stereognosis).

The object identification test is performed with sets of objects in four basic shapes (sphere, cylinder, cube, rectangular parallelepiped) over a range of three sizes (small, medium, and large). The objects are arranged on a "carrousel-type device" which is computer driven (Figure 6). Each object is presented ten times at random. The subject is required to identify the object's shape by manipulation without any visual cue and then communicating his choice by a special keyboard. His object identification via the keyboard is fed directly into the computer. After identification, the carrousel is rotated in order to bring the next object in line with his hand. The number of objects correctly identified and the time necessary for identification are the dependent variables.

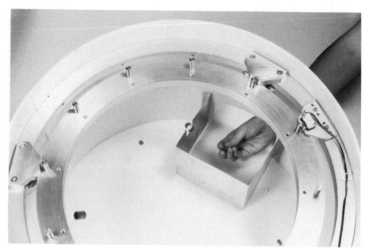

Figure 6. The carrousel-type device used for the Tactile Object Identification.

Dexterity

Dexterity reflects the integration of mobility, strength, tactile perception and it is measured in terms of grip and pinch co-ordination but also in terms of the ability to participate in ADLs tasks. Fleishman & Ellison[13,14] have performed a factor analysis of the intercorrelations among 21 fine manipulative tests which have been administered to 760 Air Force technical school trainees. This study was designed to break down dexterity into more basic functional abilities. According to the capabilities of the patient and to the ability needed to be explored, dexterity is evaluated by choosing one or several of the dexterity tests described below.

Medium Tapping Test. The Medium Tapping Test consists of making three dots within circles that are 3/8 inch in diameter, working as rapidly as possible. The score is the number of circles completed correctly in 30 seconds. This test is principally measuring the wrist-finger speed which is a factor emphasising rapid pendular and/or rotatory wrist movements, where accuracy is not critical.[14]

Purdue Pegboard Tests. The Purdue Pegboard Tests[33] is a well-known test of manipulative dexterity designed to assist in the selection of employees in industrial jobs requiring manipulative dexterity. Four separate test scores may be obtained with the Purdue Pegboard, namely: Right Hand; Left Hand; Both Hands and Assembly. These tests emphasize the Finger Dexterity factor, which is defined as the ability to make rapid, skilful, controlled manipulative movements of small objects, where the fingers are primarily involved.[14]

Minnesota Rate of Manipulation Tests. The two Minnesota subtests are the best definers of the manual dexterity factor which has been defined as the ability to make skillful, controlled arm-hand manipulations of larger objects under speed conditions.[14]

Minnesota Rate of Manipulation-Placing. The subject is required to place 60 cylindrical blocks in the proper holes as rapidly as possible. The score is the number of blocks placed during 45 seconds.

Minnesota Rate of Manipulation-Turning. The subject is required to remove the blocks from the holes with one hand, turn them over with the other hand, and replace them in the same holes, moving from block to block as rapidly as possible. The score is the number of blocks turned in 35 seconds.

Jebsen Test of Hand Function. More comprehensive evaluations of the functional capacities of the hand have also been devised to assess disability and the efficacy of rehabilitation. The Jebsen Test of Hand Function[19] was developed to provide objective measurements of various hand activities. It is regarded as a good test of general hand function,[11,20,30] and is quick to administer. It comprises seven subtests, six of which involve manipulating objects (i.e., turning cards, stacking checkers, lifting objects), and the seventh is a writing test. The time taken to complete each test is recorded for both hands, and these times are then compared with those of the normative sample. Data have been collected on 360 normal subjects and on patient groups. This data suggests that the test can measure a broad spectrum of hand disability (i.e. hemiparesis, rheumatoid arthritis, quadriplegia) and is of value in assessing improvement in hand function gained by therapeutic procedures such as surgery, physical therapy, bracing, and medication.

Validity and Reliability of the dexterity tests. The validity of the dexterity tests described here above has been established by the factor analysis performed by Fleishman & Ellison.[13,14] For each ability we have systematically selected the test with the highest factor loading score. The reliabilities of the dexterity tests described above ranged from 0.68 to 0.94.[13] Estimates of test-retest reliability concerning the Jebsen Test of Hand Function have been obtained from a mixed sample of patients with hand disorders, and these range from 0.60 to 0.99 across the seven subtests, with a mean of 0.84, indicating that this evaluation of hand function does give consistent results.[19]

THE SOFTWARE

We have developed a user friendly software for data acquisition, data treatment, data storage and instrument control (Figure 7). Thus, the tests are conducted in a standard manner so that as many testing variables as possible can be minimised and so that follow-up evaluation can be reliably compared; and knowledgeable interpretation of information gathered.

In addition, the software provides the management of the patient card-index (Figure 7). The investigator records all relevant administrative information: the patient's age, hand dominance, occupation, and avocation are elicited. If the patient has had an injury, the exact mechanism as well as the time and date of the injury and prior treatment can be recorded. All pertinent observation and information provided by the clinical examination is also included.

The diagnoses are encoded according to an international classification (HCIMO code).[17] The names and addresses of the therapist and the evaluator are stored before each evaluation. When all information has been collected, the examiner selects the testing procedures which should be applied to this particular patient. At the end of the evaluation, a final report is systematically printed and stored on magnetic support.

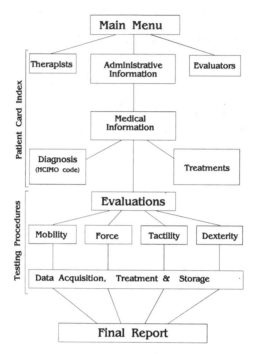

Figure 7. Flowchart of the software

CONCLUSIONS

The combination of careful clinical examination and precise measurement allows the examiner to identify and estimate the patient's rehabilitative potential and the need for therapeutic intervention. Without measurement, perceptions are diffuse and unclear. Evaluation with instruments that measure accurately all of the sensori-motor abilities of the hand permits therapists to correctly identify pathology and dysfunction, assess the effects of treatment, and realistically apprise patients in their progress. Accurate assessment data allows analysis of treatment modalities for effectiveness, provides a foundation for professional communication through research, and eventually influences the scope and the direction of the profession as a whole. The goal of our computerised system is to lessen the different sources of error and variability and to standardise the methodology of evaluation. All patients who are evaluated do not need to be given all of the tests within the assessment battery. Patients' conditions dictate the testing procedures to be used. The frequency of re-evaluation sessions depends entirely on the patient, the progress demonstrated, and the nature of the test itself. The important concept is that change in status be documented with objective measurements at appropriate intervals.

This computerised system concerns many different fields: hand surgery, rehabilitation medicine, occupational medicine, legal medicine and the design of technical aids for the disabled.

REFERENCES

1. American Academy of Orthopedic Surgeons, Joint motion: method of measuring and recording, The Academy, Chicago, (1965).
2. A.A. Amis and J.H. Miller, The elbow - measurement of joint movement, *Clin. Rheum. Dis.* 8:571 (1982).
3. W.D. Ashton, The Logit Transformation with Special Reference to its uses in Bioassay, Griffin, London (1972).
4. J. Bell and E. Tomancik, Repeatability of testing with Semmes-Weinstein mono- filaments, *J. Hand Surg.* 12A:155 (1987).
5. J.A. Bell-Krotoski, Sensibility testing: state of the art, in: Rehabilitation of the Hand: Surgery and Therapy, J.M. Hunter, L.H. Schneider, E.J. Mackin, A.D. Callahan, 3nd ed., The C.V. Mosby Co., St. Louis (1990).
6. J.A. Bell-Krotoski, Light touch-deep pressure testing using Semmes-Weinstein monofilaments, in: Rehabilitation of the Hand: Surgery and Therapy, J.M. Hunter, L.H. Schneider, E.J. Mackin, A.D. Callahan, 3nd ed., The C.V. Mosby Co., St. Louis (1990).
7. J.A. Bell-Krotoski, and W.J. Buford, The force/time relationship of clinically used sensory testing instrument , *J. Hand Ther.* 1:76 (1988).
8. D.C. Boone, S.P. Azen, and C.M. Lim, Reliability of goniometric measurements, *Phys. Ther.* 58:1355 (1978).
9. A.D. Callahan, Sensibility testing: clinical methods, in: Rehabilitation of the Hand: Surgery and Therapy, J.M. Hunter, L.H. Schneider, E.J. Mackin, A.D. Callahan, 3nd ed., The C.V. Mosby Co., St. Louis (1990).
10. A.L. Dellon, Evaluation of Sensibility and Reeducation of Sensation in the Hand, William & Wilkins, Baltimore, (1981).
11. E.E. Fess, The need for reliability and validity in hand assessment instruments, *J. Hand Surg.* 11A:621 (1986).
12. E.E. Fess, Documentation: essential elements of an upper extremity assessment battery, in: Rehabilitation of the Hand: Surgery and Therapy, J.M. Hunter, L.H. Schneider, E.J. Mackin, A.D. Callahan, 3nd ed., The C.V. Mosby Co., St. Louis (1990).
13. E.A. Fleishman, Dimensional analysis of psychomotor abilities, *J. exp. Psychol.* 48: 437 (1954).
14. E.A. Fleishman and G.D. Ellison, A factor analysis of fine manipulative tests, *J. appl. Psychol.* 46:96 (1962).
15. R.H. Gelberman, R.M. Szabo, R.V. Williamson, and M.P. Dimick, Sensibility testing in peripheral-nerve compression syndromes, *J. Bone Joint Surg.* 65A:632 (1983).
16. J. Gibson, Observations on active touch, *Psychol. Rev.* 69:477 (1962).
17. HCIMO, Adaptation hospitalière de la classification internationale des maladies et des opérations (Code ICD-9-CM). Université Catholique de Louvain. Département des Sciences Hospitalières et Médico Sociales. Centre d'informatique Médicale (1990).
18. P. Helliwell, A. Howe, and V. Wright, Functional assessment of the hand: reproducibility, acceptability, and utility of a new system for measuring strength, *Ann. Rheum. Dis.* 46:203 (1987).
19. R.H. Jebsen, N. Taylor, R.B. Trieschmann, M.J. Trotter, and L.A.Howard, An objective and standardized test of hand function, *Arch. Phys. Med. Rehab.* 50:311 (1969).
20. L.A. Jones, The assessment of hand function: A critical review of techniques, *J. Hand Surg.* 14A:221 (1989).
21. G.H. Kraft, Position of function of the wrist, *Arch. Phys. Med. Rehab.* 53:272 (1972).
22. V. Mathiowetz, K. Weber, G. Volland, and N. Kashman, Reliability and validity of grip and pinch strength evaluations, *J. Hand Surg.* 9A:222 (1984).
23. E. Moberg, Objective methods of determining the functional value of sensibility in the hand, *J. Bone Joint Surg.* 40B:454 (1958).
24. J.R. Napier, The prehensile movements of the human hand, J Bone Joint Surg 38B:902 (1956).

25. A.C. Nicol, Measurement of joint motion, *Clin. Rehabilitation* 3:1 (1989).
26. C.C. Norkin and D.J. White, Measurement of Joint Motion: A Guide to Goniometry, 6nd ed., The F.A. Davis Co., Philadelphia, (1988).
27. C.G. Phillips, Movements of the Hand, Liverpool University Press, Liverpool, (1986).
28. J.C. Pryce, The wrist position between neutral and ulnar deviation that facilitates the maximum power grip strength, *J. Biomechanics* 13:505 (1980).
29. J.M. Rothstein, P. Miller, and R. Roettger, Goniometric reliability in a clinical setting: elbow and knee measurements, *Phys. Ther.* 63:1611 (1983).
30. D. Scott and S. Marcus, Hand impairment assessment: some suggestions, *Applied Ergonomics* 22:263 (1991).
31. S. Sunderland, Nerves and Nerve Injuries, 2nd ed., Churchill Livingstone, New York (1978).
32. R.M. Szabo, R.H. Gelberman, and M.P. Dimick, Sensibility testing in patients with carpal tunnel syndrome, *J. Bone Joint Surg.* 66A:60 (1984).
33. J. Tiffin and E.J. Asher, The Purdue Pegboard: Norms and Studies of Reliability and Validity, *J. appl. Psychol.* 32:234 (1948).
34. G. Westling, and R.S. Johansson, Factors influencing the force control during precision grip, *Exp. Brain Res.* 53:277 (1984).

CONTRIBUTORS

Brian D. Adams - Department of Orthopaedics and Rehabilitation, University of Vermont College of Medicine, Given Building, Burlington, VT 05405, USA

Takeshi Akagi - Department of Orthopaedic Surgery, Okayama University Medical School, Shikata-cho 2-5-1, Okayama City, Okayama 700, Japan

Peter C. Amadio - Section of Surgery of the Hand and Orthopedic Biomechanics Laboratory, Mayo Clinic and Mayo Foundation, Rochester, MN 55905, USA

Andrew A. Amis - Biomechanics Section, Mechanical Engineering Department, Imperial College, Exhibition Road, London SW7 2BX, England

Kai-Nan An - Orthopedic Biomechanics Laboratory, Mayo Clinic and Mayo Foundation, Rochester, MN 55905, USA

Yves Andrianne - Service d'Orthopédie-Traumatologie, Cliniques Universitaires de Bruxelles, Hôpital Erasme, 808 route de Lennik, B-1070 Brussels, Belgium

G.A. Ateshian - Orthopaedic Research Laboratory, Departments of Mechanical Engineering and Orthopaedic Surgery, Columbia University, 630 West 168th Street, New York, NY 10032, USA

Thierry Authom - Service d'Orthopédie, Hôpital Universitaire Saint-Pierre, 322 rue Haute, B-1000 Brussels, Belgium

Richard A. Berger - Section of Surgery of the Hand and Orthopedic Biomechanics Laboratory, Mayo Clinic and Mayo Foundation, Mayo Medical School, 200 First Street SW, Rochester, MN 55905, USA

Lawrence Berglund - Orthopedic Biomechanics Laboratory, Mayo Clinic and Mayo Foundation, Mayo Medical School, 200 First Street SW, Rochester, MN 55905, USA

Ranko Bilic - Orthopaedic Hospital, School of Medicine, University of Zagreb, Zagreb 41000, Croatia

Elena V. Biryukova - Institute of Higher Nervous Activity and Neurophysiology of Russian Academy of Sciences, 5-a Buterov Street, Moscow, Russia

Michael J. Botte - Hand and Microvascular Surgery Service, Department of Orthopaedics and Rehabilitation, 8-894 UCSD Medical Center, 225 Dickinson Street, San Diego, CA 92101, USA

Dominique Bragard - Unité READ, Université Catholique de Louvain, 53 avenue Mounier, B-1200 Brussels, Belgium

Franz Burny - Service d'Orthopédie-Traumatologie, Cliniques Universitaires de Bruxelles, Hôpital Erasme, 808 route de Lennik, B-1070 Brussels, Belgium

Federico Casolo - Politecnico di Milano, Instituto degli Azionamenti Meccanici, 32, Piazza Leonardo da Vinci, 20133 Milano, Italy

Edmund Y.S. Chao - Orthopedic Biomechanics Laboratory, Mayo Clinic/Mayo Foundation, Rochester, MN 55905, USA

William P. Cooney III - Section of Surgery of the Hand and Orthopedic Biomechanics Laboratory, Mayo Clinic and Mayo Foundation, Mayo Medical School, 200 First Street SW, Rochester, MN 55905, USA

Steven N. Copp - Hand and Microvascular Surgery Service, Department of Orthopaedics and Rehabilitation, 8-894 UCSD Medical Center, 225 Dickinson Street, San Diego, CA 92101, USA

P.E. Crago - Departments of Orthopaedics and of Biomedical Engineering, Case Western Reserve University, Cleveland, OH 44106, USA

Joseph J. Crisco III - Department of Orthopaedics and Rehabilitation, Yale University School of Medicine, 333 Cedar Street, New Haven, CT 06510, USA

José-Henri David - Laboratory for Functional Anatomy, Université libre de Bruxelles, 808 route de Lennik, C.P. 619, B-1070 Brussels, Belgium

Philippe Delincé - Service d'Orthopédie, Hôpital Universitaire Saint-Pierre, 322 rue Haute, B-1000 Brussels, Belgium

Monique Donkerwolcke - Service d'Orthopédie-Traumatologie, Cliniques Universitaires de Bruxelles, Hôpital Erasme, 808 route de Lennik, B-1070 Brussels, Belgium

Karin Elder - Office of Academic Computing, University of Texas Medical Branch, Galveston, TX 77555, USA

Véronique Feipel - Laboratory for Functional Anatomy, Université libre de Bruxelles, 808 route de Lennik, C.P. 619, B-1070 Brussels, Belgium

Sandro Fioretti - Dipartimento di Elettronica ed Automatica, Facolta' di Ingegneria, Universita' di Ancona, Via Brecce Bianche, I-60131 Ancona, Italy

Ulrich Frank - Department of Anatomy, University of Bonn, Nussallee 10, D-5300 Bonn 1, Germany

Sylvain Gagnon - Hand and Microvascular Surgery Service, Department of Orthopaedics and Rehabilitation, 8-894 UCSD Medical Center, 225 Dickinson Street, San Diego, CA 92101, USA

Marc Garcia-Elias - Hospital General de Catalunya, 08190 Sant Cugat, Barcelona, Spain

Krystyna Gielo-Perczak - Institute of Mechanics and Design, Warsaw University of Technology, 02-524 Warsaw, Poland

David J. Giurintano - Paul W. Brand Biomechanics Laboratory, Gillis W. Long Hansen's Disease Center, 5445 Point Clair Road, Carville, LA 70721, USA

Paul G. Groszewski - Program in Occupational Therapy, Washington University, St-Louis, MO 63110, USA

Lars Hagberg - Department of Hand Surgery, University of Lund, Malmö Allmänna Sjukhus, S-21401 Malmö, Sweden

Dominique C.R. Hardy - Service d'Orthopédie, Hôpital Universitaire Saint-Pierre, 322 Rue Haute, B-1000 Brussels, Belgium

Hiroyuki Hashizume - Department of Orthopaedic Surgery, Okayama University Medical School, Shikata-cho 2-5-1, Okayama City, Okayama 700, Japan

L. Hendrix - Department of Mechanical and Aerospace Engineering, Case Western Reserve University, Cleveland, OH 44106, USA

Maurice Hinsenkamp - Service d'Orthopédie-Traumatologie, Cliniques Universitaires de Bruxelles, Hôpital Erasme, 808 route de Lennik, B-1070 Brussels, Belgium

Kathy A. Holley - Department of Orthopaedics and Rehabilitation, University of Vermont College of Medicine, Given Building, Burlington, VT 05405, USA

Anne M. Hollister - Paul W. Brand Biomechanics Laboratory, Gillis W. Long Hansen's Disease Center, 5445 Point Clair Road, Carville, LA 70721, USA

Emiko Horii - Nagoya University School of Medicine, 1-1-20 Daikominami, Higashiku, Nagoya 461, Japan

Rik Huiskes - Institute of Orthopaedics, Katholieke Universiteit Nijmegen, P.O. Box 9101, 6500 HB Nijmegen, The Netherlands

Toshihiko Imaeda - Orthopedic Biomechanics Laboratory, Mayo Clinic and Mayo Foundation, Mayo Medical School, 200 First Street SW, Rochester, MN 55905, USA

Hajime Inoue - Department of Orthopaedic Surgery, Okayama University Medical School, Shikata-cho 2-5-1, Okayama City, Okayama 700, Japan

Hilaire A.C. Jacob - Biomechanics Unit, Department of Orthopaedic Surgery, Klinic Balgrist, University of Zürich, 340 Forchstrasse, CH-8008 Zürich, Switzerland

Chritian L. Jantea - Department of Orthopaedics, Heinrich-Heine-University, 5 Moorenstrasse, D-4000 Düsseldorf 1, Germany

Tohru Kadowaki - Department of Orthopaedic Surgery - Osaka Kosei-nenkin Hospital - 4-2-78 Fukushima, Fukushima-ku, Osaka 553, Japan

Adalbert I. Kapandji - Orthopédie et Chirurgie de la Main, Clinique de l'Yvette, 43 route de Corbal, F-91160 Longjumeau, France

Lee D. Katz - Department of Diagnostic Imaging, Yale University School of Medicine, 333 Cedar Street, New Haven, CT 06510, USA

John M.G. Kauer - Department of Anatomy and Embryology, Katholieke Universiteit Nijmegen, P.O. Box 9101, 6500 HB Nijmegen, The Netherlands

Hideo Kawai - Department of Orthopaedic Surgery, Hoshigaoka Koseinenkin Hospital, 4-8-1 Hoshigaoka, Hirakata-shi, Osaka 573, Japan

P. Keir - Department of Kinesiology, University of Waterloo, University Avenue, Waterloo, Ontario N2L 3G1, Canada

Rolf Kenn - Department of Radiology, Ludwig-Maximilians-Universität München, 11 Pettenkoferstrasse, 8000 München 2, Germany

Hans P. Kern - Biomechanics Unit, Department of Orthopaedic Surgery, Klinic Balgrist, University of Zürich, 340 Forchstrasse, CH-8008 Zürich, Switzerland

Paul Klein - Laboratory for Functional Anatomy, Université libre de Bruxelles, 808 route de Lennik, C.P. 619, B-1070 Brussels, Belgium

Jan G.M. Kooloos - Department of Anatomy and Embryology, Katholieke Universiteit Nijmegen, P.O. Box 9101, 6500 HB Nijmegen, The Netherlands

Michel Lafontaine - Service d'Orthopédie, Hôpital Universitaire Saint-Pierre, 322 rue Haute, B-1000 Brussels, Belgium

Pascal Ledoux - Centre S.O.S. Main de Bruxelles, Clinique du Parc Léopold, 38 rue Froissart, B-1040 Brussels, Belgium

Marc Lemort - C.R.E.A.R.I.M., 1, rue Héger-Bordet, B-1000 Brussels, Belgium

Tommaso Leo - Dipartimento di Elettronica Ed Automatica, Facolta' di Ingegneria, Universita' di Ancona, Via Brecce Bianche, I-60131 Ancona, Italy

Gan-Tyan Lin - Department of Orthopaedic Surgery, Kaohsiung Medical College, Taiwan, Republic of China

Ronald L. Linscheid - Section of Surgery of the Hand and Orthopedic Biomechanics Laboratory, Mayo Clinic and Mayo Foundation, Mayo Medical School, 200 First Street SW, Rochester, MN 55905, USA

Vittorio Lorenzi - Politecnico di Milano, Instituto degli Azionamenti Meccanici, 32, Piazza Leonardo da Vinci, 20133 Milano, Italy

Stéphane Louryan - C.R.E.A.R.I.M., 1, rue Héger-Bordet, B-1000 Brussels, Belgium

Nicholas Löwer - Anatomische Anstalt, Ludwig-Maximilians-Universität München, 11 Pettenkoferstrasse, 8000 München 2, Germany

Paul R. Manske - Division of Orthopedic Surgery, Washington University, St-Louis, MO 63110, USA

Joseph M. Mansour - Departments of Orthopaedics and of Mechanical and Aerospace Engineering, Case Western Reserve University, Cleveland, OH 44106, USA

Takashi Masatomi - Department of Orthopaedic Surgery, Osaka University Medical School, 1-1-50 Fukushima, Fukushima-ku, Osaka 553, Japan

Antonio Merolli - Clinica Ortopedica dell'Universita' Cattolica, largo Gemelli 8, Roma, I-00168 Italy

Margaret A. Mitchell - Program in Occupational Therapy, Washington University, St-Louis, MO 63110, USA

V.C. Mow - Orthopaedic Research Laboratory, Department of Mechanical Engineering and Orthopaedic Surgery, Columbia University, 630 West 168th Street, New York, NY 10032, USA

Magdalena Müller-Gerbl - Anatomische Anstalt, Ludwig-Maximilians-Universität München, 11 Pettenkoferstrasse, 8000 München 2, Germany

Masakazu Murai - Department of Orthopaedic Surgery, Osaka University Medical School, 1-1-50 Fukushima, Fukushima-ku, Osaka 553, Japan

Tsuyoshi Murase - Department of Orthopaedic Surgery, Hoshigaoka Koseinenkin Hospital, 4-8-1 Hoshigaoka, Hirakata-shi, Osaka 573, Japan

David L. Nelson - Department of Orthopedic Surgery, Room U471, University of California, San Francisco, CA 94143-0728, USA

C.L. Nicodemus - Orthopaedic Surgery Department, University of Texas Medical Branch, Galveston, TX 77555, USA

Vasilije Nikolic - Anatomy Department, School of Medicine, University of Zagreb, Zagreb 41000, Croatia

Minoru Nishino - Department of Orthopaedic Surgery - Osaka Kosei-nenkin Hospital - 4-2-78 Fukushima, Fukushima-ku, Osaka 553, Japan

Takashi Ogura - Department of Orthopaedic Surgery, Okayama University Medical School, Shikata-cho 2-5-1, Okayama City, Okayama 700, Japan

Keiro Ono - Department of Orthopaedic Surgery, Osaka University Medical School, 1-1-50 Fukushima, Fukushima-ku, Osaka 553, Japan

Rita M. Patterson - Division of Orthopaedic Surgery, University of Texas Medical Branch, McCullough Building Room 6.136 (G-92), Galveston, TX 77550, USA

Stephen L. Pennick - Program in Occupational Therapy, Washington University, St-Louis, MO 63110, USA

Bo Peterson - Centre for Biomechanics, Chalmers University of Technology, Horsalv Street, S-41296 Göteborg, Sweden

Léon Plaghki - Service de Medecine Physique et Unité READ, Cliniques Universitaires Saint Luc, 10 avenue Hippocrate, B-1200 Brussels, Belgium

Reinhard Putz - Anatomische Anstalt, Ludwig-Maximilians-Universität München, 11 Pettenkoferstrasse, 8000 München 2, Germany

Jean Quintin - Service d'Orthopédie-Traumatologie, Cliniques Universitaires de Bruxelles, Hôpital Erasme, 808 route de Lennik, B-1070 Brussels, Belgium

D. Ranney - Department of Kinesiology, University of Waterloo, University Avenue, Waterloo, Ontario N2L 3G1, Canada

Marcel Rooze - Laboratory for Functional Anatomy, Université libre de Bruxelles, 808 Route de Lennik, C.P. 619, B-1070 Brussels, Belgium

M.P. Rosenwasser - Orthopaedic Research Laboratory, Department of Mechanical Engineering and Orthopaedic Surgery, Columbia University, 630 West 168th Street, New York, NY 10032, USA

C. Rouvas - Department of Mechanical and Aerospace Engineering, Case Western Reserve University, Cleveland, OH 44106, USA

Leonard K. Ruby - Tufts University School of Medicine, Boston, MA 02111-1854, USA

Philippe Saffar - Institut Français de la Main, Centre Chirurgical Franklin, 15 rue Benjamin Franklin, F-75116 Paris, France

Patrick Salvia - Laboratory for Functional Anatomy, Université libre de Bruxelles, 808 Route de Lennik, C.P. 619, B-1070 Brussels, Belgium

J. Sarangapani - Department of Mechanical and Aerospace Engineering, Case Western Reserve University, Cleveland, OH 44106, USA

Hans H.C.M. Savelberg - Department of Anatomy and Embryology, Katholieke Universiteit Nijmegen, P.O. Box 9101, 6500 HB Nijmegen, The Netherlands

Hans-Martin Schmidt - Department of Anatomy, University of Bonn, Nussallee 10, D-5300 Bonn 1, Germany

Frédéric A. Schuind - Service d'Orthopédie-Traumatologie, Cliniques Universitaires de Bruxelles, Hôpital Erasme, 808 route de Lennik, B-1070 Brussels, Belgium

I. Semaan - Institut Français de la Main, Centre Chirurgical Franklin, 15 rue Benjamin Franklin, F-75116 Paris, France

Gontran R. Sennwald - Chirurgie St. Leonhard, Pestalozzistr. 2, CH-9000 St-Gallen, Switzerland

Ryoichi Shibuya - Department of Orthopaedic Surgery, Hoshigaoka Koseinenkin Hospital, 4-8-1 Hoshigaoka, Hirakata-shi, Osaka 573, Japan

Kozo Shimada - Department of Orthopaedic Surgery - Osaka Koseinenkin Hospital - 4-2-78 Fukushima, Fukushima-ku, Osaka 553, Japan

Douglas K. Smith - Mallinckrodt Institute of Radiology, Washington University School of Medicine, 510 S. Kingshighway Blvd, St-Louis, MO 63110, USA

Yukio Terada - Department of Orthopaedic Surgery, Osaka University Medical School, 1-1-50 Fukushima, Fukushima-ku, Osaka 553, Japan

Jean-Louis Thonnard - Service de Médecine Physique, Cliniques Universitaires Saint Luc, 10 avenue Hippocrate, B-1200 Brussels, Belgium

Paolo Tranquilli Leali - Clinica Ortopedica dell'Universita' Cattolica, Largo Gemelli 8, Roma, I-00168 Italy

Steven F. Viegas - Division of Orthopaedic Surgery, University of Texas Medical Branch, McCullough Building Room 6.136 (G-92), Galveston, TX 77550, USA

Marc A.T.M. Vorstenbosch - Department of Anatomy and Embryology, Katholieke Universiteit Nijmegen, P.O. Box 9101, 6500 HB Nijmegen, The Netherlands

Hiroyoshi Watanabe - Department of Orthopaedic Surgery, Okayama University Medical School, Shikata-cho 2-5-1, Okayama City, Okayama 700, Japan

R. Wells - Department of Kinesiology, University of Waterloo, University Avenue, Waterloo, Ontario N2L 3G1, Canada

Klaus Wilhelm - Department of Hand Surgery, Ludwig-Maximilians-Universität München, 11 Pettenkoferstrasse, 8000 München 2, Germany

Scott W. Wolfe - Department of Orthopaedics and Rehabilitation, Yale University School of Medicine, 333 Cedar Street, New Haven, CT 06510, USA

Tomio Yamamoto - Department of Orthopaedic Surgery - Osaka Kosei-nenkin Hospital - 4-2-78 Fukushima, Fukushima-ku, Osaka 553, Japan

Vera Z. Yourovskaya - Biological Department of Moscow State University, Moscow, Russia

Vilijam Zdravkovic - Department of Orthopaedic Surgery, School of Medicine, University of Zagreb, Zagreb 41000, Croatia

Ephraim M. Zinberg - Hand and Microvascular Surgery Service, Department of Orthopaedics and Rehabilitation, 8-894 UCSD Medical Center, 225 Dickinson Street, San Diego, CA 92101, USA

INDEX

Ulnar nerve paralysis, 225-235
Ulnar styloid process, 431-439
Ulnar variance, 411-412
Upper extremity mathematical model, 95-106

Verbrugge, Jean, 4
Vesalius, 3
Viscoelasticity, carpal ligaments, 457-459

Watson's
  operation, 422
  scaphoid shift test, 457-463, 465-473
Wrist
  circumduction, 313-328
  distal radius fractures, 161-164, 397-399, 417,
      427-429, 431-439
  in-vivo kinematics, 313-328
  kinematic dysfunction, 427-429
  range of motion, 313-334
  ulnar styloid, 431-439